METHODS IN MOLECULAR BIOLOGY™

Series Editor
John M. Walker
School of Life Sciences
University of Hertfordshire
Hatfield, Hertfordshire, AL10 9AB, UK

For other titles published in this series, go to
www.springer.com/series/7651

Protein Chromatography

Methods and Protocols

Edited by

Dermot Walls and Sinéad T. Loughran

School of Biotechnology and National Centre for Sensor Research, Dublin City University, Dublin, Ireland

Humana Press

Editors
Dermot Walls, PhD
School of Biotechnology and National Centre
for Sensor Research
Dublin City University
Dublin, Ireland
dermot.walls@dcu.ie

Sinéad T. Loughran, PhD
School of Biotechnology and National Centre
for Sensor Research
Dublin City University
Dublin, Ireland
sinead.loughran@dcu.ie

ISBN 978-1-4939-6129-0 ISBN 978-1-60761-913-0 (eBook)
DOI 10.1007/978-1-60761-913-0
Springer New York Dordrecht Heidelberg London

Printed on acid-free paper

Humana Press is part of Springer Science+Business Media (www.springer.com)

Preface

Proteins are essential constituents of all organisms, and they participate in virtually every process within cells. These macromolecules are found in roles that are enzymatic, regulatory, structural, and immunological, to name but a few. In order to elucidate the structure and function of any protein, it is first necessary to purify it, and consequently many purification schemes and chromatographic methods for the isolation of native proteins from complex sources have been developed over the years. Every protein has its own particular sequence of amino acids which is determined by the nucleotide sequence of a corresponding gene. The last 30 years or so has witnessed revolutionary changes in experimental biology and specifically in the way that we identify, isolate, and manipulate individual genes and proteins. Thus, the emergence of recombinant DNA technology, genomics, and bioinformatics, in particular, means that now theoretically any protein can be expressed in a "tagged" and rapidly purifiable recombinant form from a heterologous host cell.

As in any *Methods in Molecular Biology*™ volume, the emphasis here is on the provision of clear protocol-style chapters that are suitable for newcomers to the field. We first felt that it was important to include contributions that dealt with generic topics in protein biochemistry, addressing such areas as protein stability and storage, avoiding proteolysis during chromatography, protein quantitation methods including immuno-qPCR, and the contrasting challenges that microfluidics and scale-up production pose to the investigator.

At a glance, it is clear that more than one third of the chapters concern the generation and purification of recombinant proteins, reflecting the major contribution of molecular biology to the field. These largely deal with topics such as recombinant antibody production and the tagging of proteins as a means to enhance their solubility and simplify their purification on an individual scale or in high-throughput systems. Of course, it is also expected that a compilation such as this would include the more "classical" purification methods that are based on exploiting the physico-chemical properties of the target protein. The reader will therefore find protocol-style chapters on many of the more commonly used methodologies, including proteomic/mass spectrometric techniques. We also felt that some topics necessitated treatment in an overview-style format due to the need to encompass a great number of variations that have evolved within these areas (e.g., ion-exchange and immunoaffinity chromatography as well as the tagging of recombinant proteins), often in addition to individual methods-style chapters on those subjects. However, this compendium of methods in protein chromatography does not pretend to be comprehensive, and we plead that an attempt to cover the entire potential menu in one volume would have been futile.

We are indebted to all the authors who have generously given their expertise here. Our aims were to assemble contributions from experienced scientists who have hands-on expertise in the field of protein chromatography and to place particular emphasis on the production of clearly presented step-by-step methodologies, tips, and associated explanatory notes. We hope that those who use these methods will succeed in establishing them in their own laboratories and in troubleshooting any issues that arise. We wish to extend a particular thanks to the series editor, Prof. John Walker, for his patience, advice, and encouragement throughout.

Dublin, Ireland

Dermot Walls
Sinéad T. Loughran

Contents

Preface . . . . *v*
Contributors . . . . *ix*

PART I OVERVIEWS

1 A Digest of Protein Purification 3
Dermot Walls, Robert McGrath, and Sinéad T. Loughran

2 Gel-Filtration Chromatography 25
Ciarán Ó'Fágáin, Philip M. Cummins, and Brendan F. O'Connor

3 Immunoaffinity Chromatography 35
Jenny Fitzgerald, Paul Leonard, Elaine Darcy, and Richard O'Kennedy

4 Avoiding Proteolysis During Protein Chromatography 61
Barry J. Ryan

5 Scale-Up of Protein Purification: Downstream Processing Issues 73
John Joseph Milne

6 Phage Display: A Powerful Technology for the Generation of High Specificity Affinity Reagents from Alternative Immune Sources 87
William J.J. Finlay, Laird Bloom, and Orla Cunningham

7 Engineering Protein Stability 103
Ciarán Ó'Fágáin

8 Microfluidics in Protein Chromatography 137
Frank A. Gomez

9 Tagging Recombinant Proteins to Enhance Solubility and Aid Purification 151
Dermot Walls and Sinéad T. Loughran

PART II PROTOCOLS

10 Storage and Lyophilisation of Pure Proteins 179
Ciarán Ó'Fágáin

11 Differential Precipitation and Solubilization of Proteins 203
Barry J. Ryan

12 Ion-Exchange Chromatography: Basic Principles and Application to the Partial Purification of Soluble Mammalian Prolyl Oligopeptidase. 215
Philip M. Cummins, Oonagh Dowling, and Brendan F. O'Connor

13 Protein Quantitation and Analysis of Purity 229
Eva M. Campion, Sinéad T. Loughran, and Dermot Walls

14 Purification of Proteins Fused to Glutathione S-Transferase 259
Sandra Harper and David W. Speicher

15 Purification of Proteins Fused to Maltose-Binding Protein 281
Mario Lebendiker and Tsafi Danieli

16 Purification of Proteins from Baculovirus-Infected Insect Cells 295
Luke O'Shaughnessy and Sean Doyle

17 Purification of Poly-Histidine-Tagged Proteins 311
Sinéad T. Loughran and Dermot Walls

18 Immobilized Metal Affinity Chromatography/Reversed-Phase Enrichment of Phosphopeptides and Analysis by CID/ETD Tandem Mass Spectrometry. 337
Rosana Navajas, Alberto Paradela, and Juan Pablo Albar

19 Tag Removal by Site-Specific Cleavage of Recombinant Fusion Proteins 349
Adam Charlton and Michael Zachariou

20 Purification of Antibodies Using Affinity Chromatography 369
Elaine Darcy, Paul Leonard, Jenny Fitzgerald, Martin Danaher, and Richard O' Kennedy

21 Optimized Generation of High-Affinity, High-Specificity Single-Chain Fv Antibodies from Multiantigen Immunized Chickens 383
William J.J. Finlay, Laird Bloom, and Orla Cunningham

22 Measuring Protein–Protein Interactions Using Biacore 403
Paul Leonard, Stephen Hearty, and Richard O'Kennedy

23 Ultra-Performance Liquid Chromatography–Mass Spectrometry of Proteins. 419
Timothy R. Croley

24 Hydrophobic Interaction Chromatography 431
Philip M. Cummins and Brendan F. O'Connor

25 Fast Protein Liquid Chromatography 439
Ashkan Madadlou, Siobhan O'Sullivan, and David Sheehan

26 Shotgun Proteomics: A Qualitative Approach Applying Isoelectric Focusing on Immobilized pH Gradient and LC-MS/MS 449
Laurent Geiser, Ali R. Vaezzadeh, Jacques M.P. Deshusses, and Denis F. Hochstrasser

27 Shotgun Proteomics: A Relative Quantitative Approach Using Off-Gel Electrophoresis and LC-MS/MS 459
Laurent Geiser, Loïc Dayon, Ali R. Vaezzadeh, and Denis F. Hochstrasser

28 Clinical Proteomics: Liquid Chromatography–Mass Spectrometry Purification Systems 473
Michael Henry and Paula Meleady

29 Strategies for the Purification of Membrane Proteins 485
Sinead Marian Smith

30 A Multi-Step Chromatographic Strategy to Purify Three Fungal Endo-β-Glucanases 497
Tracey McCarthy and Maria G. Tuohy

Index *525*

Contributors

Juan Pablo Albar • *Laboratorio de Proteómica, Centro Nacional de Biotecnología, Consejo Superior de Investigaciones Científicas (CSIC), Madrid, Spain*
Laird Bloom • *Wyeth Research, Cambridge, MA, USA*
Eva M. Campion • *School of Biotechnology and National Centre for Sensor Research, Dublin City University, Dublin, Ireland*
Adam Charlton • *Commonwealth Serum Laboratories Ltd, Parkville VIC, Australia*
Timothy R. Croley • *US Food and Drug Administration Center for Food Safety and Applied Nutrition 5100 Paint Branch Parkway, College Park, USA*
Philip M. Cummins • *School of Biotechnology, Dublin City University, Dublin, Ireland*
Orla Cunningham • *Wyeth Research Ireland, Dublin, Ireland*
Martin Danaher • *Ashtown Food Research Centre, Dublin, Ireland*
Tsafi Danieli • *Protein Expression and Purification Facility, The Wolfson Centre for Applied Structural Biology, The Hebrew University of Jerusalem, Jerusalem, Israel*
Elaine Darcy • *School of Biotechnology, Dublin City University, Dublin, Ireland*
Loïc Dayon • *Biomedical Proteomics Research Group, Department of Structural Biology and Bioinformatics, Geneva University, Geneva, Switzerland*
Jacques M.P. Deshusses • *Biomedical Proteomics Research Group, Department of Structural Biology and Bioinformatics, Geneva University, Geneva, Switzerland*
Oonagh Dowling • *School of Biotechnology, Dublin City University, Dublin, Ireland*
Sean Doyle • *Department of Biology, National Institute for Cellular Biotechnology, National University of Ireland Maynooth, Co. Kildare, Ireland*
William J.J. Finlay • *Wyeth Research Ireland, Dublin, Ireland*
Jenny Fitzgerald • *School of Biotechnology, Dublin City University, Dublin, Ireland*
Laurent Geiser • *Swiss Centre for Applied Human Toxicology, Geneva University, Geneva, Switzerland*
Frank A. Gomez • *Department of Chemistry and Biochemistry, California State University, Los Angeles, CA, USA*
Sandra Harper • *The Wistar Institute, Philadelphia, PA, USA*
Stephen Hearty • *School of Biotechnology and Biomedical Diagnostics Institute, Dublin City University, Dublin, Ireland*
Michael Henry • *National Institute for Cellular Biotechnology, Dublin City University, Dublin, Ireland*
Denis F. Hochstrasser • *Swiss Centre for Applied Human Toxicology, and Departments of Genetics and Laboratory Medicine, Geneva University Hospitals, School of Pharmaceutical Sciences, Geneva University, Geneva, Switzerland*

MARIO LEBENDIKER • *Protein Expression and Purification Facility, The Wolfson Centre for Applied Structural Biology, The Hebrew University of Jerusalem, Jerusalem, Israel*
PAUL LEONARD • *School of Biotechnology and Biomedical Diagnostics Institute, Dublin City University, Dublin, Ireland*
SINÉAD T. LOUGHRAN • *School of Biotechnology and National Centre for Sensor Research, Dublin City University, Dublin, Ireland*
ASHKAN MADADLOU • *Food Science and Engineering, Tehran University, Karadj, Iran*
TRACEY MCCARTHY • *Department of Biochemistry, School of Natural Sciences, National University of Ireland, Galway, Ireland*
ROBERT MCGRATH • *School of Biotechnology, Dublin City University, Dublin, Ireland*
PAULA MELEADY • *National Institute for Cellular Biotechnology, Dublin City University, Dublin, Ireland*
JOHN JOSEPH MILNE • *BioUetikon Ltd, Dublin City University, Dublin, Ireland*
ROSANA NAVAJAS • *Laboratorio de Proteómica, Centro Nacional de Biotecnología, Consejo Superior de Investigaciones Científicas (CSIC), Madrid, Spain*
BRENDAN F. O'CONNOR • *National Centre for Sensor Research, and Centre for Bioanalytical Sciences (CBAS), School of Biotechnology, Dublin City University, Dublin, Ireland*
CIARÁN Ó'FÁGÁIN • *School of Biotechnology and National Centre for Sensor Research, Dublin City University, Dublin, Ireland*
RICHARD O'KENNEDY • *School of Biotechnology and Biomedical Diagnostics Institute, Dublin City University, Dublin, Ireland*
LUKE O'SHAUGHNESSY • *Department of Biology, National Institute for Cellular Biotechnology, National University of Ireland Maynooth, Co. Kildare, Ireland*
SIOBHAN O'SULLIVAN • *Department of Biological Sciences, Cork Institute of Technology, Cork, Ireland; Department of Biochemistry, University College Cork, Cork, Ireland*
ALBERTO PARADELA • *Laboratorio de Proteómica, Centro Nacional de Biotecnología, Consejo Superior de Investigaciones Científicas (CSIC), Madrid, Spain*
BARRY J. RYAN • *School of Food Science and Environmental Health, Dublin Institute of Technology, Dublin, Ireland*
DAVID SHEEHAN • *Department of Biochemistry, University College Cork, Cork, Ireland*
SINEAD MARIAN SMITH • *Institute of Molecular Medicine and Immunology Research Centre, Trinity College Dublin, Dublin, Ireland*
DAVID W. SPEICHER • *The Wistar Institute, Philadelphia, PA, USA*
MARIA G. TUOHY • *Department of Biochemistry, School of Natural Sciences, National University of Ireland, Galway, Ireland*
ALI R. VAEZZADEH • *Department of Urology, Children's Hospital Boston, Harvard Medical School, Boston, MA, USA*
DERMOT WALLS • *School of Biotechnology and National Centre for Sensor Research, Dublin City University, Dublin, Ireland*
MICHAEL ZACHARIOU • *BioMarin Pharmaceutical Inc, Novato, CA, USA*

Part I

Overviews

Chapter 1

A Digest of Protein Purification

Dermot Walls, Robert McGrath, and Sinéad T. Loughran

Abstract

The isolation of a given protein, free of all other biomolecules, is the primary objective of any protein purification scheme. Classical chromatographic procedures have been designed to exploit particular distinguishing features of individual target proteins, such as size, physico-chemical properties and binding affinity. Advances in molecular biology and bioinformatics have positively contributed at every level to the challenge of purifying individual proteins and more recently have led to the development of high-throughput proteomic platforms. Here, a summation of developments in the field of protein chromatography is given, coupled with a compilation of general resources and tools that are available to assist with protein purification processes.

Key words: Protein, Purification, Chromatography, Proteome

1. Protein Purification

The study of proteins and their associated functions is central to our understanding of virtually all fundamental biological processes. The term "protein purification" refers to a series of procedures that are designed to isolate a single protein type from a complex biological source such as a microbial culture or a biological tissue. Proteins are probably the most commonly purified type of biological molecule as they are integral components of cellular structure and many biological processes, and amongst other roles are to be found as enzymes, scaffold molecules, cell signal transducers and as components of gene regulatory complexes. A successful purification strategy is essential prior to performing structural and functional studies on a protein of interest. The various stages in the purification process may free the protein from a matrix in which it is confined, separate the protein from other non-protein parts of the starting material, and finally separate the desired protein species from all other proteins present.

Dermot Walls and Sinéad T. Loughran (eds.), *Protein Chromatography: Methods and Protocols*, Methods in Molecular Biology, vol. 681, DOI 10.1007/978-1-60761-913-0_1, © Springer Science+Business Media, LLC 2011

The isolation of one protein, free of all other biomolecules, is the primary objective and separation procedures are designed to exploit any distinguishing features of the target protein, such as its size, its physico-chemical properties and its binding affinity.

Proteins consist of long chains of amino acids (see Fig. 1) that are linked to each other by peptide bonds between their carboxyl group (COOH) and the amino group (NH_2) of the adjacent amino acid. At the beginning of the chain, or polypeptide, the first amino acid's amino group is unbound by another amino acid, and at the end of the chain the final amino acid's carboxyl group remains unbound. The start and end of a given polypeptide are therefore referred to as its N terminus (NH_2) and C terminus (COOH), respectively. Each of the 20 amino acids that are to be found in proteins are distinguished from one another by their variable "R group," or side chain, which largely determines the physico-chemical properties of that amino acid (see Table 1). The sequence of amino acids determines the conformation of the protein, and consequently its overall function. In their native states, proteins are highly complex folded molecules whose functions are linked to their three-dimensional structure (1). The folding of polypeptide chains is highly complex and is the end result of multiple intra- and inter-polypeptide chain interactions. Hydrogen bond formation between amino acids yields different repeating patterns along a polypeptide chain, such as α-helices and β-strands, structures that form in order to decrease the energy states of the primary polypeptide sequence. During protein folding, amino acids with hydrophobic side chains can fold inwards away from contact with water. When these hydrophobic amino acids have retreated inward, the rest of the polypeptide chain is free to take up its final conformation, known as its tertiary structure, which is held together by weak non-covalent bonds, including Van Der Waal's interactions, hydrophobic packing, hydrogen bonds and

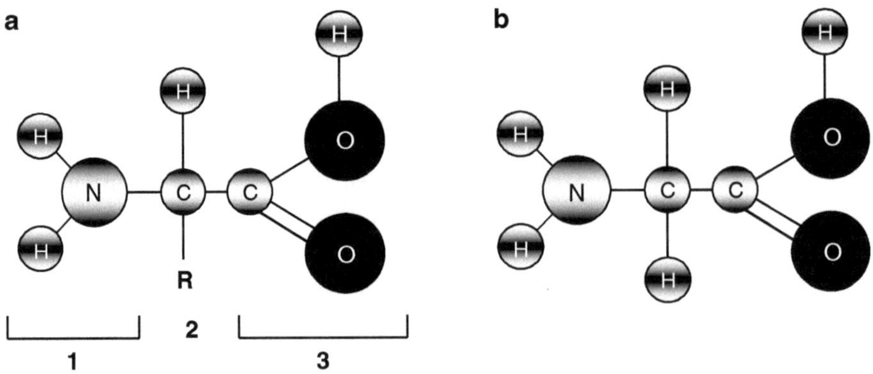

Fig. 1. Amino acid structure (**a**) with the amino group (1) on the left, the amino acid side chain R (2) and the carboxyl group (3) on the right (the acidic group). The amino acid Glycine, shown in (**b**), with only a hydrogen atom as its side chain, is the smallest of the 20 amino acids.

Table 1
Amino acid abbreviations and polarity

Amino acid name	Three-letter code	One-letter abbreviation	Polarity
Alanine	Ala	A	Non-polar, hydrophobic
Arginine	Arg	R	Polar, charged, hydrophilic
Asparagine	Asn	N	Polar, uncharged, hydrophilic
Aspartic Acid	Asp	D	Polar, charged, hydrophilic
Cysteine	Cys	C	Polar, uncharged, hydrophilic
Glutamic Acid	Glu	E	Polar, charged, hydrophilic
Glutamine	Gln	Q	Polar, uncharged, hydrophilic
Glycine	Gly	G	Non-polar, hydrophobic
Histidine	His	H	Polar, charged, hydrophilic
Isoleucine	Ile	I	Non-polar, hydrophobic
Leucine	Leu	L	Non-polar, hydrophobic
Lysine	Lys	K	Polar, charged, hydrophilic
Methionine	Met	M	Non-polar, hydrophobic
Phenylalanine	Phe	F	Non-polar, hydrophobic
Proline	Pro	P	Non-polar, hydrophobic
Serine	Ser	S	Polar, uncharged, hydrophilic
Threonine	Thr	T	Polar, uncharged, hydrophilic
Tryptophan	Trp	W	Non-polar, hydrophobic
Tyrosine	Tyr	Y	Polar, uncharged, hydrophilic
Valine	Val	V	Non-polar, hydrophobic

salt bridges (2, 3). Some proteins are composed of more than one polypeptide, or more that one copy of the same polypeptide, that adapt a stable organized superstructural arrangement known as a quaternary structure. Following translation, the last step in the biosynthesis of many proteins involves post-translational modification. Such modifications can include (1) the addition of functional groups including carbohydrates (glycosylation), acetate (acylation), phosphates (phosphorylation), and lipids; (2) structural changes including proteolytic cleavage of the protein, disulphide bond formation between inter- or intra-chain cysteine residues, or racemization of proline residues; (3) covalent linkage to other polypeptides such as SUMO protein (SUMOylation), ubiquitin (ubiquitination) or ISG15 (ISGylation), and (4) chemical modification of individual amino acids by deamination, deamidation, or eliminylation. Proteins are therefore complex, fragile, and highly interconnected polymers (1).

In order to determine the structure and function of any protein it is first necessary to purify it. The history of protein purification dates back over 200 years to 1789 when Foucroy reported his attempts to isolate substances from plants that had similar properties to egg albumen. One hundred years later Hofmeister obtained the first crystals of a protein, namely ovalbumin, and now 50 years have elapsed since Max Perutz and John Kendrew were the first to use X-ray crystallography to decipher the first protein molecular structures, those of haemoglobin and myoglobin, at Cambridge University (4). Large-scale protein production for commercial purposes first appeared during World War II, driven by an acute need for blood proteins. Thus the Cohen fractionation procedure for the purification of albumin and other serum proteins was first described (5) and was still in recent use. Enzymes in particular became targets for purification and crystallization, with applications being developed in large-scale food, detergent, and tanning industries. In many cases, these enzyme preparations, which were usually obtained from microbial sources and included amylases, lipases, and proteases, were pure enough in terms of activity as was necessary for their intended purpose. The purification processes involved were minimal, and such "process enzymes" were in fact not very pure at all. Most of the methodologies for purifying proteins from native sources were conceived in the 1960s and 1970s enabling small quantities of highly purified proteins to be obtained from large quantities of biological materials such as human/animal tissues or plant material. Such was the quantities of materials needed that in one case, and by extrapolation, it would have necessitated starting with 500,000 sheep brains in order to produce 5 mg of sheep growth hormone. Since that time, the rapid progress made in molecular biology led to the point where theoretically any protein could be produced in recombinant form and in unlimited quantities from heterologous hosts such as bacteria, fungi, cell cultures of insect, plant and animal origin, and even whole animals and plants. In 1988, yeast-derived recombinant bovine chymosin, a new substitute for neo-natal calf rennet in cheese making, was the first recombinant enzyme to gain approval from regulatory authorities for use in food. The revolution in recombinant DNA technology brought additional improvements to almost every stage of recombinant protein production, and was accelerated even further with the introduction of the polymerase chain reaction (PCR) and associated methodologies. Rational approaches to "Protein engineering" by site-directed mutagenesis of the corresponding recombinant gene, gene fusion methods, gene codon content optimization, improvements to expression vectors and tailoring of host-cell genotypes all ensured that many problems with issues such as protein activity, stability, ease of purification, yields, post-translational modifications, folding and downstream processing could be addressed. Recombinant products such as protein hormones and their receptors, haematology-associated proteins including clotting factors and clot-busting molecules, and proteins with immunomodulatory functions including antibodies and vaccine components, are now big business. The modern Biopharma industry uses genetic information to discover, develop, manufacture, and commercialize recombinant biotherapeutics that address significant medical needs.

The physico-chemical properties of proteins, such as size, charge, hydrophobicity, and solubility, have been exploited over the years leading to the development of techniques that enabled individual proteins to be purified from a crude source. ***Protein chromatography*** is a collective term applied to a set of techniques in which individual protein components of a starting mixture of proteins (the "mobile phase") are differentially retained as they pass through a solid medium, or "stationary phase." Differing partition coefficients between protein components in the mixture leads them to move at different rates, thus achieving separation. Thus size differences between proteins can be exploited by methods such as size exclusion chromatography and ultracentrifugation. The overall charge of a protein is the result of all of the individual charges of its amino acids and can vary at different levels of pH. At its isoelectric point (p*I*), the overall charge of a given protein is neutral, as all of the positive charges balance all of the negative charges. At lower pH values, a protein takes on a more positive charge and at higher pH values a protein adopts a more negative overall charge. The charge of a protein can be exploited for the purpose of purification by methods such as ion-exchange chromatography, isoelectric focusing, and chromatofocusing. The hydrophobicity of a protein is determined by the quantity, nature, and distribution of hydrophobic amino acids throughout the polypeptide chain. Although tending to be internalized within the folded protein, external hydrophobic patches or pockets often exist and can aid in the purification of a protein by methods such as hydrophobic interaction chromatography (HIC). Protein solubility in water is the result of interactions between water molecules and hydrophilic amino acids on the polypeptide chain. A protein's solubility, for instance, can be altered by pH, temperature, and ionic strength. This fact can be exploited by methods such as ammonium sulphate precipitation, in which the strength of that salt in the mixture is progressively increased to a point at which the protein of choice starts to precipitate or is said to "salt out" of solution. The biological affinity of a protein can also be utilized for the purposes of purification. Thus enzyme-substrate, ligand-receptor, and antibody–antigen interactions can be exploited for affinity chromatography by first binding the interacting molecule to a solid support and using it to trap the target partner protein molecule. Small-scale protein isolations, with polypeptide sequence determination as the end point objective, demand the recovery of a highly purified product, and acceptable sacrifices can be made in terms of losses during purification steps. High recovery and economy of processing would be primary considerations, however, during the commercial production of large amounts of protein on a continual basis. Protocols for the purification of proteins using many of the methods mentioned above are detailed throughout this book, as are methods for protein quantitation and analysis of protein purity.

Proteins differ to a rather small degree between most of the higher mammals. Thus a protocol designed for the purification of a given native protein from a pig tissue may not need much adjustment for application to the isolation of its human counterpart from the corresponding human tissue. Whether the objective is to generate purified native protein or its equivalent recombinant form, the starting source material should ideally express significant quantities of that protein [and therefore its corresponding messenger RNA (mRNA) in the case of a cloning project]. The investigator should do their homework and be mindful of issues including tissue-specific expression, possible temporal expression and the existence of mRNA splice variants leading to multiple protein isoforms. The choice between yield and purity of final product will obviously depend on the intended purpose. If small amounts of highly pure protein are required for peptide sequencing analysis, ultimately leading on to gene coding sequence identification, then sacrificing yield for the sake of purity may be logical. The advent of major genome sequencing programs such as the Human Genome Project and the development of associated bioinformatics software packages now means that many genes for both known proteins and proteins of unknown function are becoming available and are being directly cloned and expressed in recombinant form from heterologous host cells. The cloning procedure usually ensures that when expressed, these molecules are "tagged" with some sort of additional "affinity handle," a short peptide or small protein domain that serves to facilitate purification of its fusion partner. These tags are the equivalent of molecular Velcro™ and this approach obviates the need to develop purification strategies that exploit the physicochemical properties of the target protein during purification from its native or heterologous host species (see Chapter 9 for a detailed discussion on such tags and their uses). The production of recombinant proteins has greatly facilitated structural studies using Nuclear Magnetic Resonance and X-ray crystallography which require milligram quantities of pure protein. Researchers can now avail of off-the-shelf products for every step of the protein purification process – from cell disruption, to purification resins, and reagents for concentrating and desalting the final product (see Table 3).

The term "Proteome" was first introduced by Marc Wilkins and Keith Williams in 1994 to describe the entirety of all proteins encoded in a single genome expressed under distinct conditions. The development of Matrix-Assisted Laser Desorption/Ionisation (MALDI) and Electron-Spray Ionization (ESI) techniques, in conjunction with mass spectrometric analysis of large and fragile biomolecules, was a major breakthrough in protein analysis that has ultimately enabled rapid high-throughput protein identification on a global scale. Focussing on protein targets that are relevant to disease processes requires first determining where and to what extent a given protein is expressed. Two-dimensional (2D) gel electrophoresis or liquid chromatography followed by mass spectrometric analysis of individual proteins is the most popular method for comparative proteome-wide screening studies. Prior selective removal of the most abundant proteins – the albumins and immunoglobulins in the case of sera or plasma, for example significantly increases the chances of finding proteins that are more rare and probably more interesting.

The ultimate goal of any proteomics study is to determine the roles of all proteins encoded by a given genome. This entails elucidating the structures, functions, and protein–protein interactions of many proteins in parallel. One widely used approach for the simultaneous production of hundreds of proteins starts with their expression in a heterologous host such as *Escherichia coli*. The purification of recombinant proteins expressed from complementary DNAs in such heterologous systems typically involves a host-cell lysis step followed by the affinity purification of the tagged target protein from the lysate using an affinity resin. Simultaneously dealing with large numbers of samples poses a particular challenge however, one that has fostered fruitful academia-industry collaborations. Automation-friendly lysis and affinity-capture reagents are now available for high-throughput protein purification procedures using multi-well platform formats, and many have been validated with a range of robotic liquid handlers. The principal advantages to using automated systems over manual pipetting include the former's greater reproducibility and consistency. The expression, capture, and assay of several hundred tagged proteins can now be performed in just a few hours. However, there are frequent issues with recombinant protein insolubility, structural integrity, and degradation when using cell-based expression systems, and the expression of cDNAs by transcription and translation in a "cell-free" in vitro format is one way around this problem (6). Thus protein arrays are being developed whereby proteins, either tagged or untagged, are produced from cDNAs that have been immobilized on an inert surface.

High-throughput approaches have benefitted much from the advances made in the development of bioinformatics tools for protein identification, characterization, structure prediction, and analysis (see Table 2). For example, a typical two-dimensional

Table 2
Bioinformatics and web-based tools for protein identification, characterization, structure prediction, and analysis

Bioinformatics tool	Function	Accessible at
Protein identification/characterization		
ProtParam	Determination of physico-chemical parameters of a protein sequence (amino acid and atomic compositions, isoelectric point, extinction coefficient, etc.) (7)	http://www.expasy.ch/tools/protparam.html
Compute p*I*/Mw	Compute the theoretical isoelectric point (p*I*) and molecular weight (*Mw*) from a UniProt Knowledgebase entry or for a user sequence (7)	http://www.expasy.ch/tools/pi_tool.html
PeptideCutter	Predicts potential protease and cleavage sites and sites cleaved by chemicals in a given protein sequence (7)	http://www.expasy.ch/tools/peptidecutter/
Popitam (MS/MS data)	Identification and characterization tool for peptides with unexpected modifications (e.g., post-translational modifications or mutations) by tandem mass spectrometry (7)	http://www.expasy.ch/tools/popitam/
Phenyx (MS/MS data)	Protein and peptide identification/characterization from MS/MS data from GeneBio, Switzerland	http://phenyx.vital-it.ch/pwi/login/login.jsp
X!Tandem	Match tandem mass spectra with peptide sequences (8)	http://www.thegpm.org/TANDEM/
Mascot (MS/MS data)	Sequence query and MS/MS ion search from Matrix Science Ltd., London	http://www.matrixscience.com/search_form_select.html
AACompIdent	Identify a protein by its amino acid composition (7)	http://www.expasy.ch/tools/aacomp/
AACompSim	Compare the amino acid composition of a UniProtKB/Swiss-Prot entry with all other entries (7)	http://www.expasy.ch/tools/aacsim/
TagIdent	Identify proteins with isoelectric point (p*I*), molecular weight (*Mw*) and sequence tag, or generate a list of proteins close to a given p*I* and *Mw* (7)	http://www.expasy.ch/tools/tagident.html
MultiIdent	Identify proteins with isoelectric point (p*I*), molecular weight (*Mw*), amino acid composition, sequence tag and peptide mass fingerprinting data (7)	http://www.expasy.ch/tools/multiident/

(continued)

Table 2 (continued)

Bioinformatics tool	Function	Accessible at
DNA to protein		
Translate	Translates a nucleotide sequence to a protein sequence (7)	http://www.expasy.ch/tools/dna.html
Transeq	Nucleotide to protein translation from the EMBOSS package (9)	http://www.ebi.ac.uk/Tools/emboss/transeq/index.html
Graphical Codon Usage Analyser	Displays the codon bias in a graphical manner (10)	http://gcua.schoedl.de/
Reverse Translate	Translates a protein sequence back to a nucleotide sequence (11)	http://www.bioinformatics.org/sms2/rev_trans.html
Similarity searches		
BLAST	Determine sequence similarity with protein and nucleotide databases (7)	http://www.expasy.ch/tools/blast/
Format conversion		
msConvert	Data format conversion from vendor proprietary formats to mzML and mzXML (12)	http://proteowizard.sourceforge.net/index.html
BLAST2FASTA	Converts NCBI BLAST output into FASTA format	http://imed.med.ucm.es/Tools/blast2fasta.html
Three-one-letter amino acid converter	Tool which converts amino acid codes from three-letter to one-letter and vice versa (11)	http://molbiol.ru/eng/scripts/01_17.html
Three To One and One to Three	Convert a three-letter coded amino acid sequence to single letter code and vice versa (11)	http://www.bioinformatics.org/sms2/three_to_one.html
Post-translational modification prediction		
SignalP	Prediction of signal peptide cleavage sites (13)	http://www.cbs.dtu.dk/services/SignalP/
NetPhosK	Kinase-specific phosphorylation sites in eukaryotic proteins (14)	http://www.cbs.dtu.dk/services/NetPhosK/
SUMOplot	Prediction of SUMO protein attachment sites	http://www.abgent.com/tools/sumoplot_login
Subcellular localization prediction		
PSORT	Prediction of protein subcellular localization (15)	http://www.psort.org/
pTARGET	Prediction of subcellular location (16)	http://bioapps.rit.albany.edu/pTARGET/

(continued)

Table 2 (continued)

Bioinformatics tool	Function	Accessible at
Prediction of transmembrane domains		
DAS	Prediction of transmembrane regions in prokaryotes using the Dense Alignment Surface method (17)	http://www.sbc.su.se/~miklos/DAS/
TMpred	Prediction of transmembrane regions and protein orientation (18)	http://www.ch.embnet.org/software/TMPRED_form.html
Structure prediction and analysis		
Jpred	A consensus method for protein secondary structure prediction (19)	http://www.compbio.dundee.ac.uk/www-jpred/
iMolTalk	An Interactive Protein Structure Analysis Server (20)	http://i.moltalk.org/
TopMatch-web	Protein structure comparison (21)	http://topmatch.services.came.sbg.ac.at/
SWISS-MODEL	An automated knowledge-based protein modelling server (22)	http://swissmodel.expasy.org/

liquid chromatography coupled to tandem mass spectrometry (2D-LC-MS/MS) generates in the order of tens of thousands of tandem MS spectra per day making impossible the efficient manual analysis of such data. Data processing using powerful bioinformatics software is necessary to get the most from such sequence information and to save time. Pivotal to peptide and protein identification is the comparison of observed masses and fragmentation patterns following separation, with predicted masses and patterns which are generated in silico. Bioinformatics tools are available to convert dataset files (see Table 2) and for the identification of spectra generated; for example, a typical shotgun proteomics experiment (see Chapters 26 and 27) is carried out by sequence searching using sequence search engines. Experimentally observed spectra are compared with theoretical spectra generated from a list of all possible protein sequences. Sequence search engines include Mascot and X!Tandem, which are commercially available and open source tools, respectively (see Table 2).

In this compendium, the reader will find detailed overviews of protein purification methodologies coupled with user-friendly, step-by-step protocols for both selected classical techniques and methods for isolating recombinant proteins. Throughout the book, suppliers of certain reagents and equipment are noted where pertinent. A more comprehensive list of suppliers of protein purification-associated products and technologies is given here (see Table 3).

Table 3
Products and technologies for protein purification: a compilation of suppliers. Adapted by permission and updated from Macmillan Publishers Ltd (Protein Purification: Table of Suppliers. Nature, 2006, 439, issue 7079, pp 1022–1023)

Company	Products/activity	Location	URL
Affinity-Tag technology			
Active Motif	Ni-TED™ Protein Purification System, spin columns	Carlsbad, CA	http://www.activemotif.com
Alpha Diagnostic International	Antibodies to a number of common fusion-tags	San Antonio, TX	http://www.4adi.com
BD Biosciences	Affinity resins for the purification of tagged proteins	San Diego, CA	http://www.bdbiosciences.com
Bio-Rad	Profinity™ IMAC resin, protein extraction kits	Hercules, CA	http://www.bio-rad.com
Biotech Support Group	NuGEL™ affinity purification, HLPC columns, serum protein analysis tools	North Brunswick, NJ	http://www.biotechsupportgroup.com
Clontech (Takara Bio)	GST and His-fusion protein purification kits, affinity resins	Mountain View, CA	http://www.clontech.com
Covalys Biosciences (NEB)	SNAP-capture™ kits for SNAP-tag™ fusion proteins	Witterswil, Switzerland	http://www.covalys.com
GE Healthcare	GST fusion vectors, various affinity-tag and other protein purification systems	Piscataway, NJ	http://www.gelifesciences.com
IBA BioTAGnology	Strep-tag technology; Strep-Well™ HT Purification plates for automated purification of streptavidin-tagged proteins	Göttingen, Germany	http://www.iba-go.com
Invitrogen	Ni-NTA and ProBond™ purification systems for the purification of His6-tagged proteins. Dynal magnetic beads (Protein A/G coated) for immunoprecipitations	Carlsbad, CA	http://www.invitrogen.com

(continued)

Table 3
(continued)

Company	Products/activity	Location	URL
Macherey-Nagel	Products for chromatography and sample preparation including Protino™ gravity columns for purifying His-tagged and GST-tagged proteins	Düren, Germany	http://www.macherey-nagel.com
Merck	GST-tag, S-tag and His-tag purification kits including Robopop™ kits	San Diego, CA	http://www.merck-chemicals.com
MoBiTec	Columns and kits for protein purification	Göttingen, Germany	http://www.mobitec.de
NeoClone	Softag™ epitope purification system	Madison, WI	http://www.neoclone.com
New England Biolabs	IMPACT, pMAL, and cell-free expression and purification systems	Ipswich, MA	http://www.neb.com
Nunc	Pro-Pur™ protein affinity spin column kits (Protein A, Protein G and IMAC)	Rochester, NY	http://www.nuncbrand.com
Pall Life Sciences	Automation-friendly multi-well filter plates for cleanup, purification, and detection of proteins, reagents for sample cleanup and concentration	East Hills, NY	http://www.pall.com
Pierce Biotechnology	Kits for GST- and His-fusion proteins, purification columns, affinity resins, SwellGel™ discs	Rockford, IL	http://www.piercenet.com
Phenomenex	Chromatography columns	Torrance, CA	http://www.phenomenex.com
PhyNexus	PhyTip™ IMAC and glutathione products	San Jose, CA	http://www.phynexus.com
Prochem	Proteus spin columns for the purification of His6-tagged recombinant proteins	Littleton, MA	http://www.protein-chem.com
Profos	ColTrap™ Proteinnative system for purification of His-tagged proteins expressed in *E. coli* under native conditions	Regensburg, Germany	http://www.profos.de

Promega	Pinpoint Xa vectors and SoftLink resin for tagging with Biotin carboxyl carrier protein; S30 T7 cell-free protein expression system	Madison, WI	http://www.promega.com
R&D Systems	His-tagged protein purification resin	Minneapolis, MN	http://www.rndsystems.com
Roche Applied Science	pIVEX (GST and MBP) vectors	Indianapolis, IN	http://www.roche-applied-science.com
Rockland Immunochemicals	Antibodies with GST and GFP epitope tags, fluorescein and peroxidase antibodies, alkaline phosphatase, and biotin conjugates	Gilbertsville, PA	http://www.rockland-inc.com
Sartorius	Vivapure™ 8-to-96-well cobalt-chelate kit, ion-exchange columns	Göttingen, Germany	http://www.sartorius.com
Sigma-Aldrich	Affinity purification plates and gels, reagents for cell lysis and protein precipitation, His-Select HC nickel magnetic beads	St Louis, MO	http://www.sigmaaldrich.com
Stratagene	InterPlay TAP kits	La Jolla, CA	http://www.stratagene.com
Upstate	Spin column filters	Charlottesville, VA	http://www.upstate.com
QIAGEN	Ni-NTA Superflow™ BioRobot, MagneHis Protein Purification System	Hilden, Germany	http://www.qiagen.com
Tosoh Biosep	Columns and media for chromatography	Montgomeryville, PA	http://www.tosohbiosep.com
Westburg	Expression and purification systems for recombinant proteins	Leusden, The Netherlands	http://www.westburg.eu
ZirChrom Separations	Chromatography columns	Anoka, MN	http://www.zirchrom.com
Zymo Research	His-Spin Protein Miniprep for His-tagged protein purification	Orange, CA	http://www.zymoresearch.com

(continued)

Table 3 (continued)

Company	Products/activity	Location	URL
Magnetic beads and magnetic separators			
Ambion	Magnetic Stand-96 for paramagnetic bead precipitation	Austin, TX	http://www.ambion.com
Bioclone	Recombinant proteins, magnetic beads, magnetic separators	Marrickville, Australia	http://www.bioclone.com.au
Bio-Nobile	QuickPick protein fractionation and purification, PickPen magnetic tool	Turku, Finland	http://www.bio-nobile.com
Bioscience Bead Division	Functionalized Gel Microspheres and magnetic beads for bioaffinity separations	West Warwick, RI	http://www.bioscience-beads.com
Bruker Daltonics	CLINPROT™ magnetic beads for desalting and quick sample preparation	Billerica, MA	http://www.bdal.com
Invitrogen	Dynabeads TALON™ for the isolation of His-tagged proteins	Carlsbad, CA	http://www.invitrogen.com
Europa Bioproducts	Recombinant proteins, expression-ready clones, BcMag magnetic separator and beads for various tags	San Diego, CA	http://www.bioclon.com
Miltenyi Biotec	Paramagnetic MACS Microbeads, ProCatch™ Resins, magnetic separators	Auburn, CA	http://www.miltenyibiotech.com
New England Biolabs	Streptavidin magnetic beads, magnetic separators	Ipswich, MA	http://www.neb.com
Promega	MagneGST Protein Purification System	Madison, WI	http://www.promega.com
Polysciences	BioMag superparamagnetic particles	Warrington, CA	http://www.polysciences.com
StemCell Technologies	EasySep magnet	Vancouver, Canada	http://www.stemcell.com
Thermo Electron Corporation	KingFisher automated system for magnetic particle-based protein purification; Pierce Streptavidin magnetic beads	Waltham, MA	http://www.thermo.com

Chromatography equipment and automated systems			
Artel	Multichannel verification system portable technique to verify the performance of liquid-handling equipment	Westbrook, Maine	http://www.artel-usa.com
Aurora Biomed	VERSA robotic workstation	Vancouver, Canada	http://www.aurorabiomed.com
Agilent Technologies	ZORBAX HPLC columns for protein and peptide separations, combined capillary electrophoresis and mass spectrometry system	Palo Alto, CA	http://www.agilent.com
Applied Biosystems	Protein sequencers, mass spectrometers, chromatography systems	Foster City, CA	http://www.appliedbiosystems.com
Beckman Coulter	Biomek liquid handlers for HT protein purification, ProteomeLab PF2D for protein fractionation; ProteomeLab PA 800 Protein Characterization System	Fullerton, CA	http://www.beckmancoulter.com
Biotage	FLASH chromatography systems for multi-sample purification	Uppsala, Sweden	http://www.biotage.com
BioTek Instruments	Complete line of 96- and 384-well microplate automated pipetting systems, dispensers and washers	Winooski, VT	http://www.biotek.com
Bruker Daltonics	Mass spectrometry instruments and accessories for pharmaceutical, biochemical and chemical research	Billerica, MA	http://www.bdal.com
Caliper Life Sciences	Advanced liquid handling and LabChip technologies Hopkinton	Massachusetts	http://www.caliperLS.com
Cecil Instruments	Products for ion chromatography, HPLC and spectrophotometers	Cambridge, UK	http://www.cecilinstruments.com

(continued)

Table 3 (continued)

Company	Products/activity	Location	URL
Eksigent Technologies	ExpressLC-800 HPLC system, NanoLC-1D and NanoLC-2 proteomics systems for nanoscale chromatography	Livermore, CA	http://www.eksigent.com
GE Healthcare	AKTAExpress automated liquid chromatography platform	Little Chalfont, UK	http://www.gehealthcare.com
Gilson	Solid-phase extraction systems, liquid chromatography	Middleton, WI	http://www.gilson.com
Gyros	Gyrolab Workstation for automated sample preparation for mass spectrometry	Uppsala, Sweden	http://www.gyros.com
Hamilton Company	Automated solid-phase extraction systems	Reno, NV	http://www.hamiltoncompany.com
Hitachi High-Technologies	Process Automation System can automate various laboratory processes ranging from centrifuging to sample partitioning and analysis	Tokyo, Japan	http://www.hitachi-hitec.com/oversea
PerkinElmer Life Sciences	Range of liquid handlers and automated workstations	Boston, MA	http://www.perkinelmer.com
Protedyne (LabCorp)	BioCube automated protein expression and purification	Windsor, CT	http://www.protedyne.com
Sepiatec	Automated HPLC and solid-phase extraction systems for sample separation and isolation	Berlin, Germany	http://www.sepiatec.com
Tecan	Freedom EVO workstation, solid-phase extraction systems	Männedorf, Switzerland	http://www.tecan.com

TTP LabTech	Mosquito liquid-handling robot for setting up protein crystallography screens	Royston, UK	http://www.ttplabtech.com
Varian	Chromatography and mass spectrometry analytical instruments and consumables	Walnut Creek, CA	http://www.varianinc.com
Waters	Products and tools for liquid chromatography and mass spectrometry	Milford, MA	http://www.waters.com
Protein concentration, desalting, and extraction kits			
Ambion	PARIS kit for the isolation of both RNA and native protein from the same sample	Austin, TX	http://www.ambion.com
BioChain	Sepharose and magnetic-bead-based kits for protein extraction	Hayward, CA	http://www.biochain.com
BioVision	Reagents for preparing highly enriched cellular fractions from mammalian cells	Mountain View, CA	http://www.biovision.com
Calbiochem (EMD)	Protein extraction and concentration kits	San Diego, CA	http://www.emdbiosciences.com
Chemicon (Millipore)	Protein extraction and concentration kits	Temecula, CA	http://www.chemicon.com
Eppendorf	Perfect Pure tips for sample preparation for mass spectrometry, centrifuges	Hamburg, Germany	http://www.eppendorf.com
EPICENTRE Biotechnologies	ReadyPreps preparation kit for lysing bacterial cells and reducing viscosity for protein extraction	Madison, WI	http://www.epibio.com
Gerard Biotech	Gerard ProteoCon protein concentration kits for denatured and native proteins	Oxford, OH	http://www.gerardbiotech.com
Norgen Biotek	ProteoSpin kit for concentration, buffer exchange and desalting of protein samples	St Catharines, Canada	http://www.norgenbiotek.com
Novagen (EMD)	iFOLD protein refolding system	Madison, WI	http://www.novagen.com

(continued)

Table 3 (continued)

Company	Products/activity	Location	URL
Profos	ColTrap Protein Unfold and Protein Express, EndoTrap: endotoxin removal, ColTrap: bacterial harvesting and protein preparation	Resensburg, Germany	http://www.profos.de
Proxeon	Prepacked StageTip microcolumns for protein desalting, pre-concentration, and separation of samples for MALDI, LC-MS, and nano ES-MS	Odense, Denmark	http://www.proxeon.com
Sigma-Aldrich	ProteoPrep protein extraction kits	St Louis, MO	http://www.sigmaaldrich.com
Takara Bio USA (Clontech)	Kits for optimizing the refolding conditions of inclusion body proteins	Madison, WI	http://www.takaramirusbio.com
Services			
Abgent	Bacterial and baculovirus protein expression systems	San Diego, CA	http://www.abgent.com
Arvys Proteins	Complete protein expression and purification service	Stamford, CT	http://www.arvysproteins.com
BioGenes	Preparation and purification of antigens and development and production of antibodies	Berlin, Germany	http://www.biogenes.de
Deciphergen Biotechnology	Custom protein and antibody production	Cheshire, CT	http://www.deciphergen.com
Eurogentec	High resolution separation and identification of proteins	Seraing, Belgium	http://www.eurogentec.be
Hypromatrix	Protein–protein interaction assays, protein purification	Worcester, MA	http://www.hypromatrix.com
Invitrogen	Customized protein expression, purification, and characterization	Carlsbad, CA	http://www.invitrogen.com

Nature Technology Corporation	Contract manufacturing of vectors (DNA, viruses, vaccines, plasmids) and recombinant proteins	Lincoln, NE	http://www.natx.com
NextGen Sciences	Protein analytical services	Huntingdon, UK	http://www.nextgensciences.com
Paragon Bioservices	Recombinant protein expression and purification services	Baltimore, MD	http://www.paragonbioservices.com
Protagen	Protein production and analysis services, including proteome analyses	Dortmund, Germany	http://www.protagen.de
Protein Sciences	Cloning, expression, and purification of full-length, biologically active proteins	Meriden, CT	http://www.proteinsciences.com
ProteinX Lab	Contract protein purification services	San Diego, CA	http://www.proteinx.com
Roche Applied Science	Protein production by cell-free and cell-based expression systems	Indianapolis, IN	http://www.roche-applied-science.com
Trenzyme Biotechnology	Protein expression and purification services	Konstanz, Germany	http://www.trenzyme.com
Other proteomics tools			
Alpha Innotech	Chemiluminescence and gel imagers	San Leandro, CA	http://www.alphainnotech.com
AnaSpec	Protein detection kits	San Jose, CA	http://www.anaspec.com
Applied Biosystems	BIOiTRAQ systems for protein biomarker identification and quantitation	Foster City, CA	http://www.appliedbiosystems.com
Bio-Rad	Preparative 2D electrophoresis, isoelectric focusing systems, proteomics workstations	Hercules, CA	http://www.bio-rad.com
Digilab Genomic Solutions	Instrumentation, software, and consumables to isolate and characterize proteins	Holliston, MA	http://www.digilabglobal.com
Eksigent Technologies	Microfluidics-based systems for proteomics analysis	Dublin, CA	http://www.eksigent.com

(continued)

Table 3 (continued)

Company	Products/activity	Location	URL
Eurogentec	Production of custom protein arrays	Seraing, Belgium	http://www.eurogentec.be
JEOL	Magnetic sector mass spectrometers	Tokyo, Japan	http://www.jeol.com
NextGen Sciences	a2DEoptimizer 2D gel electrophoresis system	Huntingdon, UK	http://www.nextgensciences.com
Promega	HaloLink protein arrays	Madison WI	http://www.promega.com
Protein One	Active Protein Array platforms designed for multiplex detection and interaction assessment with DNA, RNA, and/or other ligands	Bethesda, MD	http://www.proteinone.com
Proxeon	Mass spectrometry tools and products	Odense, Denmark	http://www.proxeon.com
Thermo Electron Corporation	Finnigan ProteomeX Integrated Proteomics Workstation for protein and peptide analysis and identification	San Jose, CA	http://www.thermo.com/chromatography
Tyrian Diagnostics	ElectrophoretIQ for first-dimension isoelectric focusing and second dimension SDS–PAGE and blotting in a single bench top unit	North Ryde, Australia	http://www.tyriandiagnostics.com
Whatman	FAST PAK™ protein array kit line	Brentford, UK	http://www.whatman.com

References

1. Sun, P.D., Foster, C.E., and Boyington, J.C. (2004) Overview of protein structural and functional folds. *Curr. Protoc. Protein Sci.* Chapter 17: Unit 17.1.
2. Kang, T.S. and Kini, R.M. (2009) Structural determinants of protein folding. *Cell. Mol. Life Sci.* **66**, 2341–2361.
3. Bartlett, A.I. and Radford, S.E. (2009) An expanding arsenal of experimental methods yields an explosion of insights into protein folding mechanisms. *Nat. Struct. Mol. Biol.* **16**, 582–588.
4. Strandberg, B. (2009) Building the ground for the first two protein structures: myoglobin and haemoglobin. *J. Mol. Biol.* **392**, 2–10.
5. Cohen, E.J., Strong, L.E., Hughes, W.L., Mulford, D.J., Ashworth, J.N., Melin, M., and Taylor, H.L. (1946) Preparation and properties of serum and plasma proteins. IV. A system for the separation into fractions of the proteins and lipoprotein components of biological tissues and fluids. *J. Am. Chem. Soc.* **68**, 459–475.
6. Katzen, F., Peterson, T.C., and Kudlicki, W. (2009) Membrane protein expression: no cells required. *Trends Biotechnol.* **27**, 455–460.
7. Gasteiger, E., Gattiker, A., Hoogland, C., Ivanyi, I., Appel, R.D., and Bairoch A. (2003) ExPASy: the proteomics server for in-depth protein knowledge and analysis. *Nucleic Acids Res.* **31**, 3784–3788.
8. Craig, R., Cortens, J.P., and Beavis, R.C. (2004) Open source system for analyzing, validating, and storing protein identification data. *J. Proteome Res.* **3**(6), 1234–1242.
9. Rice, P., Longden, I., and Bleasby, A. (2000) EMBOSS: the European Molecular Biology Open Software Suite. *Trends Genet.* **16**(6), 276–277.
10. Fuhrmann, M., Hausherr, A., Ferbitz, L., Schödl, T., Heitzer, M., and Hegemann, P. (2004) Monitoring dynamic expression of nuclear genes in *Chlamydomonas reinhardtii* by using a synthetic luciferase reporter gene. *Plant Mol. Biol.* **55**(6), 869–881.
11. Stothard, P. (2000) The Sequence Manipulation Suite: JavaScript programs for analyzing and formatting protein and DNA sequences. *Biotechniques* **28**, 1102–1104.
12. Kessner, D., Chambers, M., Burke, R., Agus, D., and Mallick, P. (2008) ProteoWizard: Open source software for rapid proteomics tools development. Bioinformatics **24**(21), 2534–2536.
13. Bendtsen, J.D., Nielsen, H., von Heijne, G., and Brunak, S. (2004) Improved prediction of signal peptides: SignalP 3.0. *Mol. Biol.* **340**, 783–795.
14. Blom, N., Sicheritz-Ponten, T., Gupta, R., Gammeltoft, S., and Brunak, S. (2004) Prediction of post-translational glycosylation and phosphorylation of proteins from the amino acid sequence. *Proteomics* **4**(6), 1633–1649.
15. Gardy, J.L., Laird, M.R., Chen, F., Rey, S., Walsh, C.J., Ester, M., and Brinkman, F.S.L. (2005) PSORTb v.2.0: Expanded prediction of bacterial protein subcellular localization and insights gained from comparative proteome analysis. *Bioinformatics* **21**(5), 617–623.
16. Guda, C. (2006) pTARGET: a web server for predicting protein subcellular localization. *Nucleic Acids Res.* **34** (web server issue), W210–W213.
17. Cserzo, M., Wallin, E., Simon, I., von Heijne, G., and Elofsson, A. (1997) Prediction of transmembrane alpha-helices in procariotic membrane proteins: the Dense Alignment Surface method. *Prot. Eng.* **10**(6), 673–676.
18. Hofmann, K. and Stoffel, W. (1993) TMbase – A database of membrane spanning proteins segments. *Biol. Chem. Hoppe Seyler* **374**, 166.
19. Cole, C., Barber, J.D., and Barton, G.J. (2008) The Jpred 3 secondary structure prediction server. *Nucleic Acids Res.* **36** (web server issue), W197–W201.
20. Diemand, A.V. and Scheib, H. (2004) MolTalk – a programming library for protein structures and structure analysis. *BMC Bioinformatics*, **5**, 39.
21. Sippl, M.J. and Wiederstein, M. (2008) A note on difficult structure alignment problems. *Bioinformatics* **24**, 426–427.
22. Kiefer, F., Arnold, K., Künzli, M., Bordoli, L., and Schwede, T. (2009). The SWISS-MODEL Repository and associated resources. *Nucleic Acids Res.* **37**, D387–D392.

Chapter 2

Gel-Filtration Chromatography

Ciarán Ó'Fágáin, Philip M. Cummins, and Brendan F. O'Connor

Abstract

Gel-filtration chromatography is a popular and versatile technique that permits the effective separation of proteins and other biological molecules in high yield. Here, the basis of the method is described and typical matrix types are contrasted. The selection of suitable operating conditions and applications of the method are also discussed.

Key words: Gel-filtration chromatography, Gel-permeation, Gel-exclusion, Size-exclusion, Molecular-sieve, Operating conditions, Separations, Molecular mass estimation, Size-exclusion reaction chromatography

1. Introduction

Gel-filtration chromatography is a form of partition chromatography used to separate molecules of different molecular sizes. This technique has also frequently been referred to by various other names, including gel-permeation, gel-exclusion, size-exclusion, and molecular-sieve chromatography. The basic principle of gel-filtration is relatively simple. Molecules are partitioned between a mobile phase and a stationary phase comprising a porous matrix (of defined porosity) as a function of their relative sizes. A column constructed of such a matrix, typically in bead form, will have two measurable liquid volumes: the external volume, consisting of the liquid between the beads, and the internal volume, consisting of the liquid within the beads. The external volume is usually referred to as the void volume (V_0), while the sum of the external and internal volumes is the total volume (V_t). Following sample application, molecules larger than the pores of the stationary phase matrix will be excluded from the internal volume within the beads. They will, therefore, migrate quite rapidly through the

Dermot Walls and Sinéad T. Loughran (eds.), *Protein Chromatography: Methods and Protocols*, Methods in Molecular Biology, vol. 681, DOI:10.1007/978-1-60761-913-0_2, © Springer Science+Business Media, LLC 2011

column, emerging at V_0, while molecules smaller than the matrix pores (as well as those intermediate in size) will equilibrate with both the external and internal liquid volumes, causing them to migrate much more slowly and emerge at a volume (V_e) greater than V_0. Molecules are, therefore, eluted in order of decreasing molecular size. The elution volume, V_e, of a particular molecule depends on the fraction of the stationary phase available to it for diffusion. This can be represented by the constant K_d or K_{av} (also referred to as the partition coefficient). Therefore:

$$V_e = V_0 + K_{av}(V_t - V_0)$$

Rearranging this equation gives:

$$K_{av} = \frac{(V_e - V_0)}{(V_t - V_0)}$$

In addition to molecular size or mass, the flow behaviour of molecules through a gel-filtration column is also a function of their molecular shape, or, to be more precise, hydrodynamic diameter. This is defined as the diameter of the spherical volume (hydrodynamic volume) created by a molecule as it rapidly tumbles in solution. When performing gel-filtration chromatography, one generally assumes that all the molecules within a mixture have the same symmetrical shape, so that the order of elution will be one of decreasing molecular weight. Although this is an acceptable assumption in most cases, one must bear in mind that the operative molecule dimension during gel-filtration is the hydrodynamic volume and, as such, an asymmetrical molecule will appear to elute with an abnormally high molecular weight compared with a symmetrical molecule of similar molecular weight. When separating out proteins, for example, the usual assumption is that all the proteins in the mixture are globular proteins. Asymmetrical proteins (fibrous proteins and certain glycoproteins), however, will appear to elute with an abnormally high molecular weight compared with globular proteins of similar molecular weight.

1.1. Selection of Operating Conditions

Various factors should be considered when designing a gel-filtration system. These include: (1) matrix choice; (2) sample size and concentration; (3) column parameters; (4) choice of eluent; (5) effect of flow rate; and (6) column cleaning and storage.

1.1.1. Matrix Choice

Commonly used gel-filtration matrices consist of porous beads composed of cross-linked polyacrylamide, agarose, dextran (see Table 1) or combinations of these, and are supplied either in suspended form or as dried powders. The matrix should be compatible with the properties of the molecules being separated and its stability to organic solvents, pH, and temperature

Table 1
Some media for gel-filtration

Material	Media[a] and fractionation range (globular proteins)	Advantages	Disadvantages
Dextran	Sephadex G-10 0–700 Da Sephadex G-25 1–5 kDa Sephadex G-50 1.5–30 kDa Sephadex G-100 4–150 kDa Sephadex G-200 5–600 kDa	Smaller fractionation ranges (G-10, G-25) are good for desalting	Expanded forms require low pressures/ hydrostatic heads
Agarose	Sepharose 6B 10–4,000 kDa Sepharose 4B 60–20,000 kDa Sepharose CL-4B 60–20,000 kDa Sepharose CL-2B 70–40,000 kDa	Good for larger molecules. Cross-linked (CL) forms are more robust	Must be kept wet and not allowed to dry out
Allyl dextran-bis-acrylamide	Sephacryl S-200 HR 5–250 kDa Sephacryl S-300 HR 10–1,500 kDa Sephacryl S-400 HR 20–8,000 kDa	Not biodegradable, mechanically robust	Must be kept wet and not allowed to dry out

This table lists some typical media used for gel-filtration. It is not exhaustive, and there are many others, including media for industrial use in special configurations. Users should consult manufacturers' technical information regarding suitability for particular requirements

Da daltons, ***kDa*** kilodaltons

[a]Sephacryl®, Sephadex®, Sepharose®, and Superdex® are registered trademarks of GE Healthcare. Toyopearl resins from Tosoh Bioscience may also be used for size-exclusion chromatography; these are made from polymethacrylate and are available with different fractionation ranges (Toyopearl® is a registered trademark of Tosoh Bioscience)

is also an important consideration. Under separation conditions, matrices should be inert with respect to the molecules being separated in order to avoid partial adsorption of the molecules to the matrix, not only retarding their migration through the column, but also resulting in "tailed" peaks (e.g. see (1)).

When choosing a suitable matrix, one with a molecular mass fractionation range, which will allow the molecule of interest to elute after V_0 and before V_t, should be selected. The most suitable fractionation range, however, will be dictated not only by the molecular mass of the target molecule, but also by the composition of the sample being applied to the column. Therefore, the best separation of molecules within a sample having similar molecular masses is achieved using a matrix with a narrow fractionation range.

1.1.2. Sample Size and Concentration

Maximum resolution in gel-filtration chromatography depends on the application of sample in a small volume, typically 1–5% of the total bed volume. For this reason, gel-filtration chromatography has an inherent low sample-handling capacity and, accordingly, should be performed quite late in a purification procedure when the numbers of different molecules in a sample are relatively low. The concentration of a sample which can be applied to the column will be limited by the viscosity of the sample (which increases with the sample concentration) relative to that of the eluent. A high viscosity will result in irregular sample migration through the column with subsequent loss of resolution and, in some instances, will reduce the column flow rate. When separating proteins by gel-filtration, the sample should not have a protein concentration in excess of 20 mg/mL.

1.1.3. Column Parameters

Maximum resolution in gel-filtration chromatography is obtained with long columns. The ratio of column diameter to length can range from 1:20 up to 1:100.

1.1.4. Choice of Eluent

As gel-filtration chromatography separates molecules only on the basis of their relative sizes, the technique is effectively independent of the type of eluent used. Elution conditions (pH, essential ions, cofactors, protease inhibitors, etc.) which will complement the requirements of the molecule of interest should, therefore, be selected. However, the ionic strength of the eluent should be high enough to minimise protein–matrix and protein–protein associations by electrostatic or van der Waals interactions (e.g. see (1)). The addition of 0.1 M NaCl or KCl to the eluent to avoid these interactions is quite common.

1.1.5. Effect of Flow Rate

Low flow rates offer maximum resolution during gel-filtration chromatography, since flow rate and resolution are inversely related. The optimum flow rate for resolution of proteins is approximately 2 mL/cm^2/h, although much higher flow rates can be used, particularly with rigid matrices such as the Sephacryl HR range from GE Healthcare (30 mL/cm^2/h). Unfortunately, low flow rates mean longer separation times. Therefore, a compromise between desired resolution and speed must be decided upon.

1.1.6. Column Cleaning and Storage

Most gel-filtration matrices can be cleaned with 0.2 M sodium hydroxide or non-ionic detergents. When left unused for long periods of time, matrices should be stored at 4°C in the dark in the presence of an antimicrobial agent (0.02–0.05% [w/v] sodium azide or 20% [v/v] ethanol).

2. Applications of Gel-Filtration Chromatography

One of the principal advantages of gel-filtration chromatography is that separation can be performed under conditions specifically designed to maintain the stability and activity of the molecule of interest without compromising resolution. Absence of a molecule–matrix binding step also prevents unnecessary damage to fragile molecules, ensuring that gel-filtration separations generally give high recoveries of activity.

This separation technique, however, is not without its disadvantages. When separating proteins by gel-filtration chromatography, proteolysis, for example, becomes an increasing problem: the target protein frequently becomes an abundant substrate for proteases that may also be present in the mixture, leading to reduced recovery of activity. Because of the large size of gel-filtration columns, large volumes of eluent are usually required for their operation, often creating excessive running costs. Gel-filtration also has an inherent low resolution compared with other chromatographic techniques, because none of the molecules are retained by the column and non-ideal flow occurs around the beads. In addition, this technique has a low sample-handling capacity dictated by the need to optimise resolution. Despite these disadvantages, gel-filtration chromatography still occupies a key position in the field of biomolecule separation because of its simplicity, reliability, versatility, and ease of scale-up.

2.1. Separation of Proteins and Peptides

Because of its unique mode of separation, gel-filtration chromatography has been used successfully in the purification of literally thousands of proteins and peptides from various sources. These range from therapeutic proteins and peptides, which together constitute a multibillion euro worldwide market, to enzymes and proteins for the brewing, food-processing, and diagnostics industries; some examples of each type are cited below.

Recombinant human granulocyte colony stimulating factor (rhG-CSF) was refolded from inclusion bodies in high yield, with great suppression of aggregates formation, by urea-gradient size-exclusion chromatography on a Superdex 75 column (2). A similar technique was used to purify human interferon-γ, solubilised from inclusion bodies by 8 M urea, to a specific activity of 12,000,000 IU/mg with protein recovery of 67% (3). Luteinising hormone (LH) was purified 46-fold from a crude pituitary extract by gel-filtration on two Sephacryl S-200 columns. The method exploited differential binding of LH (in the crude extract) to blue dextran for the first chromatography step. Before the second step, addition of high salt released LH from the blue dextran, enabling effective purification (4). Fusion ferritin (heavy-chain ferritin plus light-chain ferritin) has also been purified by urea-gradient gel-filtration. In this case, fusion ferritin solubilised from inclusion bodies with 4 M urea was applied to the column. Refolding enhancers were included in the urea-diluent buffer that was subsequently applied to the column to produce properly folded fusion ferritin multimers (5).

A continuous rotating annular size-exclusion chromatography system permitted the purification of crude porcine lipase with productivity of approximately 3 mg lipase per mg gel per hour and an activity recovery of almost 99% (6). Among food-use proteins, hen egg lysozyme has been successfully refolded using both acrylamide- and dextran-based gel columns (Sephacryl S-100 and Superdex 75, respectively) (7). Gel-filtration has also proven useful for the purification of the whey proteins alpha-lactalbumin and beta-lactoglobulin from aqueous two-phase systems (8).

2.2. Separation of Other Biomolecules

Carbohydrates represent a plentiful, but so far only scarcely exploited, reservoir of unique, multifunctional biopolymers which can be readily fractionated by gel-filtration chromatography on the basis of their relative sizes (e.g. see (9, 10)). Various problems, however, have limited the development of gel-filtration methods for oligosaccharides. First, many of the commercially available gel-filtration matrices are themselves carbohydrates (e.g. Sephadex, Sepharose, etc., manufactured by GE Healthcare), shedding milligram quantities of heterodisperse carbohydrate polymers into the mobile phase. Second, non-specific interactions with matrix materials are common, since sugars are essentially amphipathic with a hydrophobic ring structure and hydrophilic functional groups. Despite these problems, however, gel-filtration chromatography still remains an important option for the purification of complex oligosaccharides.

Gel-filtration chromatography has for many years been used to separate various nucleic acid species such as DNA, RNA, and tRNA as well as their constituent bases, adenine, guanine, thymine, cytosine, and uracil. Linear phage lambda DNA and circular double-stranded phage M13 DNA, for example, can be completely separated from chromosomal DNA and RNA by gel-filtration on Sephacryl S-1000 Superfine (11). Plasmid DNA can also be purified by gel-filtration (12), although modern commercial kits often use a centrifugal spin column format for greater convenience. One recent study describes the novel use of two gel-filtration steps, one before and one after a reverse-phase operation, to purify plasmid DNA from a clarified alkaline *Escherichia coli* cell lysate (13).

2.3. Separation of Cells and Virus Particles

Cells of different sizes can be efficiently separated from one another using gel-filtration chromatography. Methods have been developed, for example, to separate both erythrocytes (14) and platelets (15) from blood (most workers now prefer density-gradient media such as Percoll™ or Ficoll™, trademarks of GE Healthcare, for tasks of this nature). Size-exclusion chromatography was used downstream of expanded bed adsorption chromatography to recover active recombinant hepatitis B core antigen (HbcAg) in 45% yield with a purification factor of 4.5 (16). A Sephacryl S-1000 SF proved to be effective and economical in the purification of recombinant *Bombyx mori* nucleopolyhedrosis virus displaying human pro-renin receptor (17). Sephacryl S-1000 gel-filtration chromatography gave more effective purification of turkey coronavirus from infected turkey embryos than did use of a sucrose gradient (18).

2.4. Group Separations

By selecting a matrix pore-size which completely excludes all the larger molecules in a sample from the internal bead volume, but which allows very small molecules to enter this volume easily, one can effect a group separation in a single, rapid gel-filtration step which would traditionally require dialysis for up to 24 h to achieve. Group separation can be used, for example, to effect buffer exchanges within samples, for desalting of labile samples prior to concentration and lyophilisation, to remove phenol from nucleic acid preparations, and to remove inhibitors from enzymes (e.g. see (19)).

2.5. Molecular Mass Estimation

Gel-filtration chromatography is an excellent alternative to SDS-PAGE for the determination of relative molecular masses of proteins, since the elution volume of a globular protein is linearly related to the logarithm of its molecular weight (20). One can prepare a calibration curve for a given column by individually applying and eluting at least five suitable standard proteins (in the correct fractionation range for the matrix) over the column, determining the elution volume for each protein standard, and plotting the logarithm of molecular weight versus V_e/V_0. When a protein of unknown molecular weight is applied to the same column and eluted under the same conditions, one can use the elution volume of the protein to determine its molecular weight from the calibration curve.

2.6. Size-Exclusion Reaction Chromatography: Protein PEGylation

Covalent attachment of PEG (polyethylene glycol; "PEGylation") to a protein can attenuate its antigenicity and/or extend its biological half-life or shelf life. Size-exclusion reaction chromatography (SERC) permits one to control the extent of a reaction (such as PEGylation) that alters molecular size and to separate reactants and products. In SERC, injection of reactants onto a size-exclusion chromatography column forms a moving reaction zone. Reactants and products partition differently within the mobile phase, leading to different flow rates through the column. Thus, products are removed selectively from the reaction zone, shortening their residence time in the reaction zone and separating them into the downstream section of the column. In PEGylation, addition of PEG groups to the protein significantly increases molecular size, allowing the use of SERC to obtain a dominant final PEGylated protein size in high yield. The principle was successfully demonstrated using two model proteins, alpha-lactalbumin and beta-lactoglobulin (21).

3. Conclusion

Despite its disadvantages of sample dilution and the need for a low ratio of sample volume to column volume, gel-filtration remains a popular separation method due to its versatility, the wide range of matrices commercially available, and the mild conditions of operation. It is hoped that this article has given the reader some grasp of the technique's wide range of applications and how to choose appropriate conditions for its gainful use. A useful handbook on gel-filtration is available from GE Healthcare's Life Sciences division through the following URL: http://www6.gelifesciences.com/aptrix/upp00919.nsf/content/LD_153206006-R350.

Other useful chromatography handbooks may be accessed through links from URL: http://www6.gelifesciences.com/aptrix/upp01077.nsf/Content/service_and_support~documents_and_downloads~handbooks.

References

1. Belewa, M., Poratha, J., Fohlmana, J. and Janson, J.-C. (1978) Adsorption phenomena on sephacryl® S-200 superfine. *J. Chromatogr. A.* **147**, 205–212.
2. Wang, C., Wang, L. and Geng, X. (2008) High recovery refolding of rhG-CSF from *Escherichia coli*, using urea gradient size exclusion chromatography. *Biotechnol. Prog.* **24**, 209–213.
3. Guan, Y.X., Pan, H.X., Gao, Y.G., Yao, S.J. and Cho, M.G. (2005) Refolding and purification of recombinant human interferon-gamma expressed as inclusion bodies in *Escherichia coli* using size exclusion chromatography. *Biotechnol. Bioprocess Eng.* **10**, 122–127.
4. Chaudhary, R., Jain, S., Muralidhar, K. and Gupta, M.N. (2006) Purification of bubaline luteinizing hormone by gel filtration chromatography in the presence of blue dextran. *Process Biochem.* **41**, 562–566.
5. Kim, H. and Kim, I.H. (2005) Refolding of fusion ferritin by gel filtration chromatography (GFC). *Biotechnol. Bioprocess Eng.* **10**, 500–504.
6. Vasudevan, P.T., Palekar, A.A. and Yan, S. (2002) Optimization of lipase purification by rotating annular size-exclusion chromatography. *Biocatal. Biotransformation.* **20**, 189–199.
7. Batas, B. and Chaudhuri, J.B. (2001) The influence of operational parameters on lysozyme refolding using size-exclusion chromatography. *Bioprocess Biosyst. Eng.* **24**, 255–259.
8. Rojas, E.E.G., Coimbra, J.M.D., Minim, L.A., Zuniga, A.D.G., Saraiva, S.H. and Minim, V.P.R. (2004) Size-exclusion chromatography applied to the purification of whey proteins from the polymeric and saline phases of aqueous two-phase systems. *Process Biochem.* **39**, 1751–1759.
9. Hodgson, J. (1991) Chromatogrpahy of complex carbohydrates. *Biotechnology (N.Y.).* **9**, 149–150.
10. Ma, Y., Mizuno, T. and Ito, H. (1991) Studies on the host-mediated antitumor polysaccharides. 15. Anti-tumor activity of some polysaccharides isolated from a Chinese mushroom, haungmo, the fruiting body of *Hohenbuehelia serotina*. *Agric. Biol. Chem.* **55**, 2701–2710.
11. Gonzalez, A. and Gomez-Marquez, J., (1990) Purification of bacteriophage DNA by gel-filtration chromatography. *Genet. Anal. Tech. Appl.* 7, 2–4.
12. Kondo, T., Mukai, M. and Kondo, Y. (1991) Rapid isolation of plasmid DNA by LiCl-ethidium bromide treatment and gel-filtration. *Anal. Biochem.* **198**, 30–35.
13. Limonta, M., Marquez, G., Rey, I., Pupo, M., Ruiz, O., Amador-Canizares, Y. and Duenas-Carrera, S. (2008) Plasmid DNA recovery using size-exclusion and perfusion chromatography. *Biopharm Intl.* **21**, 38.
14. Kankura, T., Kurashina, S. and Nakao, M. (1974) Gel-filtration technique for separation of erythrocytes from human blood. *J. Lab. Clin. Med.* **83**, 840–844.

15. Tangen, O., Berman, H.J. and Marfey, P. (1971) Gel filtration – new technique for separation of blood platelets from plasma. *Thromb. Diath. Haemorrh.* **25**, 268–269.
16. Ho, C.W., Tan, W.S. Chong, F.C., Ling, T.C. and Tey, B.T. (2009) A preparative purification process for recombinant hepatitis B core antigen using online capture by expanded bed adsorption followed by size-exclusion chromatography. *J. Microbiol. Biotechnol.* **19**, 416–423.
17. Kato, T., Manoha, S.L., Tanaka, S. and Park, E.Y. (2009) High-titer preparation of *Bombyx mori* nucleopolyhedrovirus (BmNPV) displaying recombinant protein in silkworm larvae by size exclusion chromatography and its characterization. *BMC Biotechnol.* **9**, 55.
18. Loa, C.C., Lin, T.L., Wu, C.C., Bryan, T.A., Thacker, H.L., Hooper, T. and Schrader, D. (2002) Purification of turkey coronavirus by sephacryl size-exclusion chromatography. *J. Virol. Methods.* **104**, 187–194.
19. Helmerhorst, E. and Stokes, G.B. (1980) Microcentrifuge desalting: a rapid, quantitative method for desalting small amounts of protein. *Anal. Biochem.* **104**, 130–135.
20. Andrews, P. (1965) The gel-filtration behaviour of proteins related to their molecular weights over a wide range. *Biochem. J.* **96**, 595–606.
21. Fee, C.J. (2003) Size-exclusion reaction chromatography (SERC): a new technique for protein PEGylation. *Biotechnol. Bioeng.* **82**, 200–206.

Chapter 3

Immunoaffinity Chromatography

Jenny Fitzgerald, Paul Leonard, Elaine Darcy, and Richard O'Kennedy

Abstract

Antibody-based separation methods, such as immunoaffinity chromatography (IAC), are powerful purification and isolation techniques. Antibodies isolated using these techniques have proven highly efficient in applications ranging from clinical diagnostics to environmental monitoring. IAC is an efficient antibody separation method which exploits the binding efficiency of a ligand to an antibody. Essential to the successful design of any IAC platform is the optimisation of critical experimental parameters such as: (a) the biological affinity pair, (b) the matrix support, (c) the immobilisation coupling chemistry, and (d) the effective elution conditions. These elements and the practicalities of their use are discussed in detail in this review. At the core of all IAC platforms is the high-affinity interactions between antibodies and their related ligands; hence, this review entails a brief introduction to the generation of antibodies for use in IAC and also provides specific examples of their potential applications.

Key words: Immunoaffinity chromatography, Antibody, Purification

1. Introduction

1.1. Emergence and Evolution of Immunoaffinity Chromatography

Immunoaffinity chromatography (IAC) is a powerful purification technique relying on the specific recognition between the antibody molecule and a complementary ligand. One of the first IAC procedures performed was reported in 1951 (1) when ovalbumin was covalently immobilised to *p*-aminobenzyl cellulose for the isolation of anti-albumin antibodies. Later, improvements occurred, including precipitating the cellulose matrix into a fine suspension before use (2). Since these early reports, the use of IAC expanded greatly with the development of beaded agarose (3), dextran (4), and polyacrylamide (5) supports, all of which proved to be more efficient than cellulose, resulting in increased efficiency and capacity of IAC matrices. The use of immunoaffinity purification has major advantages when compared to conventional

Dermot Walls and Sinéad T. Loughran (eds.), *Protein Chromatography: Methods and Protocols*, Methods in Molecular Biology, vol. 681, DOI 10.1007/978-1-60761-913-0_3, © Springer Science+Business Media, LLC 2011

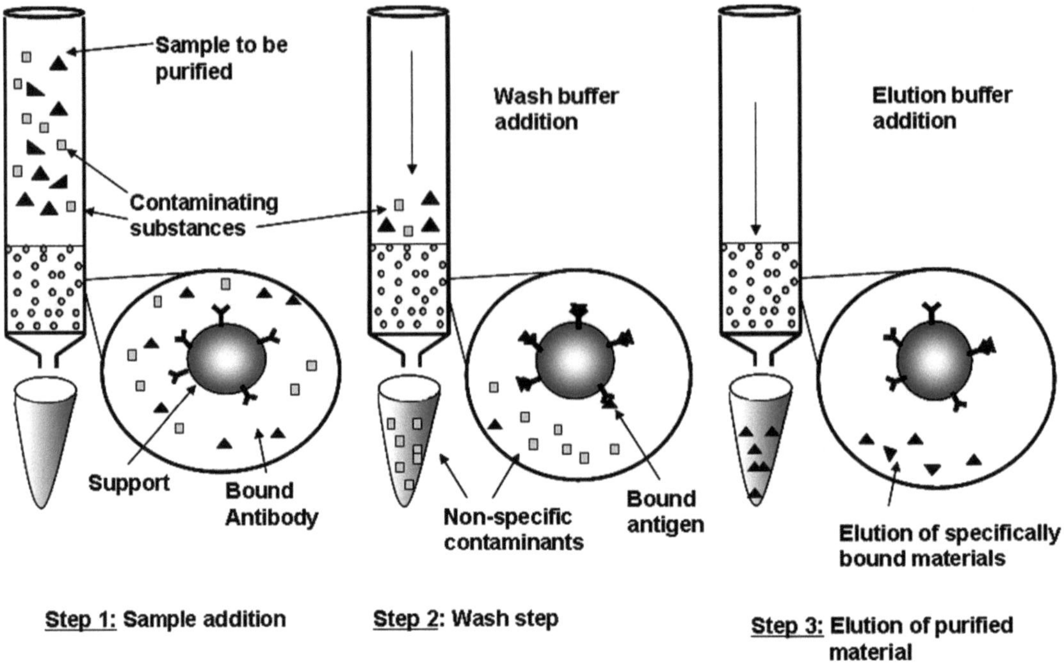

Fig. 1. Basis of Immunoaffinity chromatography. *Step 1*: The sample to be purified is passed through a column containing immobilised antigen or antibody. *Step 2*: Proteins specific to the immobilised ligand will bind to the column, whilst non-specific materials will flow through. *Step 3*: Weakly retained fractions will be removed by washing, and strongly retained fractions will be eluted via an optimised elution strategy.

purification systems due to the high affinity of antigen–antibody interactions, which allow a significant degree of molecular selectivity (6). This binding of analyte to antibody is the result of good spatial complementarity and is a function of the sum of intra-molecular interactions (6). By applying an analyte mixture to a suitably immobilised antigen/antibody column, washing off unbound material and eluting the biological material of interest, both purification (often greater than 1000-fold) and simultaneous concentration can be achieved, as shown in Fig. 1 (7). Hence, IAC has proven to be an extremely useful tool in pharmaceutical and biological laboratories for separation, isolation, or purification of analytes or antibodies, particularly where large-scale production is necessary.

1.2. Antibody: Structure and Production

Antibodies are a class of predominantly soluble globular proteins produced by the immune system in response to a foreign agent, or antigen. The immune repertoire is remarkably diverse, expressing as many as 10^{11} antibodies of varying specificity (8). Antibodies may be found on the surface of B cells (membrane-bound form), functioning as unique receptor proteins for specific antigens or in the blood or lymph (soluble form), directing humoral clearance of invading entities such as bacterial cells (8). The basic structure of all antibody molecules (see Fig. 2) is comprised of two identical light chains (approximately 25 kDa) and two identical heavy

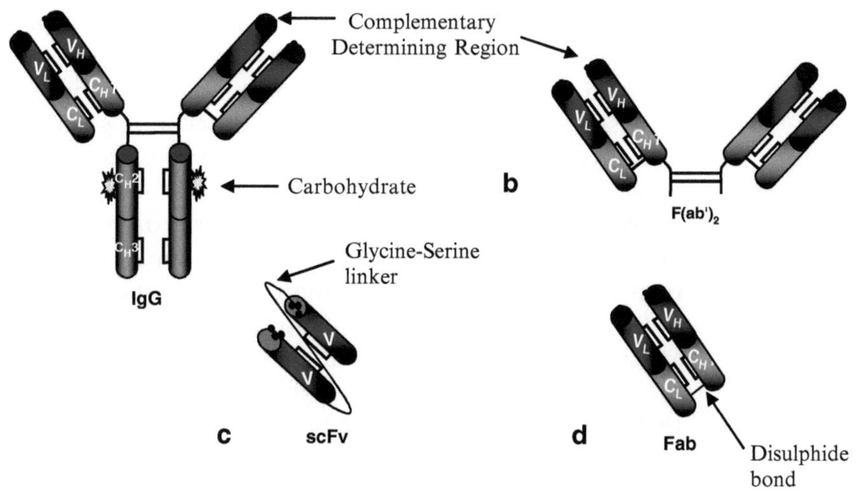

Fig. 2. Structures of an antibody and antibody fragments. (**a**) Immunoglobulin G molecule, consisting of two identical heavy chains and two identical light chains, each of which consists of both variable and constant regions. (**b**) $F(ab')_2$- consisting of two disulphide-linked antigen-binding light chain fragments, with both variable and constant regions. (**c**) ScFv- single chain variable heavy and light chains linked by a glycine–serine sequence. (**d**) Fab- single antibody binding fragment, i.e. one constant and one variable domain from both the heavy and light chains.

chains (approximately 50 kDa). Light chain fragments are connected to the heavy chains by several non-covalent bonds and disulphide bridges. Sequence variation and hence antigen recognition is predominantly focused around three "hypervariable" regions containing residues and structures complementary to their respective antigens. These residues are referred to as "complementarity determining regions" (CDRs). To create the antigen combining site, the CDRs from the variable heavy and variable light chain domains are juxtaposed at one end of the antibody forming a composite surface (9). There are five classes of antibodies, IgA, IgG, IgD, IgE, and IgM, with IgG being the most predominant with serum concentrations of up to 13.5 mg/mL (10). IAC generally incorporates immunoglobulins of the G class and their associated fragments. However, a plethora of IAC methods have utilised antibodies of the various classes for specific purification purposes.

Antibodies may be developed against a broad range of antigenic targets. Several strategies exist for the generation of antibodies, including polyclonal antibody generation naïve/synthetic libraries, immune libraries, hybridoma technology, or the employment of protein engineering to generate a multitude of antibody fragments with enhanced properties. Animals such as rabbits, guinea pigs, goats, sheep, donkeys, and fowl are frequently used for the generation of polyclonal serum. Polyclonal antibodies are extracted from the serum of an animal sensitised with a specific antigen. This method generally results in a heterogeneous mixture of antibodies produced by different cells within the body, with varying affinities and specificities (11).

In 1975, Köhler and Milstein described the production of monoclonal antibodies using their method of somatic cell fusion between a myeloma (plasmacytoma) and an immune lymphoblast expressing a specific gene (12). This technology demonstrated how each individual resultant hybridoma combines the immortality of the myeloma and the antibody production virtues of the lymphocyte, thus serving as a stable source of *mono*-specific antibody (12). Monoclonal antibodies have applications in environmental and food analysis, in immunodiagnostics, and as effective therapeutic agents for the treatment of inflammation, cancer, autoimmune and infectious diseases (13).

The emergence of recombinant antibody phage display technology has provided the opportunity to produce more cost-efficient affinity ligands for use in IAC. Phage display is a molecular diversity technology that allows the presentation of large peptide and protein libraries on the surface of filamentous phage. Phage display allows modifications to be easily introduced in primary antibody sequences. This leads to affinity maturations, the generation of fusion proteins, and the addition of detection and purification tags. The power of phage selection to choose those ligands having the desired biological properties permits us to mimic the immune system for the synthesis of "tailor-made" antibodies for use in diagnostics, immunotherapy, immunoassay development (14), and moreover IAC. Examples of the purification of different antibody formats are reported in Table 1. Several factors have to be considered when performing an IAC procedure. There are three main factors limiting the use of antibodies for affinity chromatography (15). First, antibodies are species- and tissue-specific, and hence, the primary isolation of specific antibodies is necessary for the subsequent isolation or purification of specific antigens. Second, antibodies immobilised on supports may often lose their antigen-binding capabilities (16). Finally, the highly specific binding of antibodies and antigens often complicates the elution of the antigen (15). There are numerous strategies available to circumvent these issues, several of which are discussed in the rest of the chapter.

2. Affinity Ligands

One of the most applied affinity systems for the purification of antibodies is the use of immobilised bacterial surface proteins that interact with the Fc portion of IgG or immunoglobulin light chains, i.e. Protein A/G and Protein L, respectively. Protein A (SpA), a *Staphylococcus aureus* cell wall protein, binds selectively to the Fc region of IgG (17). Protein L (PpL), a *Peptostreptococcus magnus* protein, has selective binding towards antibodies

Table 1
Affinity methodologies for antibody purification

Source	Antibody format	Chromatography method	Ligand	Recovery	Yield (%)	Purity (%)	References
Transgenic tobacco leaves	Mab	IMAC Ni-NTA	Chelating metal ion	–	63	97.2	(130)
Human serum	IgG	Nanoparticles	PHEMA nanoparticles with IMEO	843 mg/L	–	95.3–96.7	(58)
Porcine serum	IgE	Protein G affinity column	Protein G	100 mg/L	–	–	(131)
Escherichia coli	Ig2	IMAC Ni-NTA	Chelating metal ion	21 mg/L	–	–	(132)
Hybridoma C.C.S	Mab	Mab coupled chromatography column	Mab	–	90	97	(133)
Hybridoma C.C.S	IgG	Peptide affinity column	Synthetic peptide of protein-A D-PAM	<60 g/L	>90	>90	(134)
Quadromas (hybrid hybridomas)	IgG2a and IgG1	Protein G and *m*-APBA agarose column	Protein G and HRPO	3.8 mg/L	–	–	(70)
Escherichia coli	ScFv	Ligand capture and magnetic affinity separation	ScFv	–	–	–	(135)
Ovine serum	IgG	Affinity capture resin	Prototype agarose[a]	–	85	90	(136)
Egg yolk	IgY	Antigen capture	Antigen-venom of *E. carinatus*	50–60 mg/egg yolk	5–8/egg yolk	–	(137)
Serum, ascitic fluid, and C.C.S	IgM	Peptide affinity column	Synthetic peptide (TG19318)	7 mg/L	95	85–95	(138)
Human serum	Fab	Reversed flow affinity process	Artificial protein L	–	60	–	(13)

PHEMA poly(hydroxyethyl methacrylate), *IMEO* 3-(2-imidazoline-1-yl) propyl (triethoxysilane), *m-APBA m*-aminophenyleboronic acid, *C.C.S* cell culture supernatant, *IMAC* immobilised metal affinity chromatography, *Ni-NTA* nickel-nitrilotriacetic acid, *HRPO* horseradish peroxidase, *GFP* green fluorescent protein, *PAM* protein A-mimetic synthetic peptide

[a]Novel prototype agarose-based synthetic affinity adsorbent utilising a bicyclic heteroaromatic ligand of less than 500 Da (Millipore)

containing kappa (κ1, κ3, and κ4 excluding κ2) light chains (18). Protein G (SpG) from *Streptococci* groups C and G is a surface IgG binding protein with specific affinity for the Fc region of IgG. However, it also associates with the constant heavy chain (C_H1) domain of the antibody binding fragment (Fab) portion through a β-zipper interaction (19). Protein A/G are often covalently attached directly to the solid support or indirectly attached by biotin/avidin as an intermediate, to provide greater stability (20). Protein G is a popular choice for the purification of serum and human IgG subclass III, due to the poor affinity of Protein A towards that subclass. However, due to the superior stability of Protein A over Protein G, the use of Protein G in industrial purifications is very restricted (17). IAC ligands can be generally classed as biospecific, pseudo-biospecific, or synthetic.

Protein A, L, and G are prime examples of biospecific ligands, which are binding molecules derived from natural sources. These include antigens, antibodies, lectins, or bacterial immunoglobulin binding proteins (13). They generally possess high affinity and selectivity for immunoglobulin molecules (16). However, to generate these ligands, prior isolation and purification is required due to contamination with host DNA or viruses which can prove to be quite expensive and laborious for use in large-scale applications. There is also the added complication of eluting the antigen from these ligands due to their high selectivity and affinity (18). Fassina et al. (21) stated that optimal synthetic ligands should display reduced large-scale production costs, an increased resistance to chemical and biological actions, a reduction in the amount of biological contaminants, and subsequently a higher capacity for large-scale purification and extended selectivity for other classes of immunoglobulins not recognised by Protein A and G (21).

Synthetic ligands are robust low-molecular-weight compounds that can circumvent the drawbacks of natural IgG binding ligands. Their selectivity is comparable to that of natural ligands not to mention their high binding capacity, durability, and cost-effectiveness. Furthermore, they are also highly resistant to chemical and biological degradation (22). One type of synthetic peptide that has come under particular focus more recently are mimetic peptides. The capacity of sorbents derivatized with some mimetic peptides has also been shown to be superior to that of classical ligands (21). A prime example of this includes the novel purification approach employed using Protein A-mimetics. A tetrameric peptide library was designed (21), where four identical peptide chains were assembled starting from a tetradentate lysine core. Screening of this library was performed in a competitive format with Protein A and biotinylated immunoglobulins. After three rounds of biopanning, the most active multimer (Arg-Thr-Tyr)$_4$-K$_2$-K-G (Protein A-mimetic TG19318) was isolated.

Its affinity constant for IgG, as determined by optical biosensor determinations, was close to 0.3 μM (21). Another example of a novel synthetic ligand also reported by Fassina et al. (21) involved the use of cyclic peptides as affinity ligands. They described how cyclic peptides showed an increased resistance to enzymatic degradation and constrained flexibility compared with linear peptides. Libraries of dimeric tripeptides were screened to identify a ligand for mouse IgG purification. The resultant ligand, named "Peptide H," is a cyclic dimeric peptide with a binding capacity for mouse IgG at approximately 1 mg IgG/mL of derivatized support (21). The identification of this "Peptide H" proved highly useful for mouse and rat IgG purifications.

Triazine ligands have also been developed for the purification of antibodies. The presence of a triazine ring system acts as a scaffold that allows the easy introduction of natural or unnatural amino-acid side-chain functional residues, which makes them ideal for use in purifications. A further study showed how a Protein L mimic "ligand 8/7" showed a unique binding ability for the Fab portion of immunoglobulins, resulting in inhibition of Protein L (PpL) with IgG and Fab in competitive ELISA assays (13). Moreover, ligand 8/7 was shown to mimic its template PpL in terms of antibody separation performance while binding to IgG_1 with κ and λ isotypes. Ligand 8/7 adsorbents were also able to isolate immunoglobulins from crude samples, under non-optimised conditions, achieving seven-fold factors and purities of up to 95% (13). Molecularly imprinted polymers (MIPs) have recently emerged as an attractive alternative to antibodies as a recognition matrix for use in applications such as IAC. MIPs can be stored for years at room temperature without any loss of recognition power (23). The main advantage of molecular imprinting is the possibility to synthesise sorbents with selectivity predetermined to a specific analyte (24). Chapuis et al. developed an extraction procedure for triazines from complex matrices using both MIPs and an immunosorbent. They reported comparable selectivity using MIPs to that observed using immunosorbents immobilised with anti-triazine polyclonal antibody (25). Some of the major problems to be overcome to increase the range of applications of MIPs include bleeding of analyte and low sample volumes that can be extracted (26).

The use of synthetic ligands over conventional ligands can provide less laborious and uncomplicated purifications of different classes of immunoglobulins other than IgG (21). Recently, Dong et al. (27) have identified a ligand isolated from a synthetic ligand library which has a high affinity for IgY and is also highly stable to chemical and biological degradation. This ligand has the potential to circumvent the low recoveries obtained from conventional IgY purifications (27). Commercial antibody purification methods are fast becoming an efficient alternative to conventional approaches. Some examples of these include Melon™ Gel and the Eggcellent® chicken IgY purification kit, both of which are manufactured by Pierce. Melon™ Gel is used as an alternative to Protein A and G purifications. Antibodies are purified on the basis of negative selection, where most proteins bind to the support but IgG remains in solution and is eluted in the flow-through. This is useful for purification of IgG from serum samples (28).

Pseudo-bispecific ligands are viewed as cost-effective alternatives to biological ligands with some added advantages over natural antibody binding ligands, such as ease of immobilisation and high adsorption capacities (29). Some examples of this type of ligand include: hydrophobic, thiophilic, hydroxyapatite, chelating metal ions, hydrophobic charge-induced, and mixed mode affinity ligands (13). Hydrophobic charge-induced chromatography (HCIC) may overcome certain restrictions which arise when using HIC, as this process allows molecules to bind at relatively low salt conditions. HCIC is based on the hydrophobic binding of the target protein to the ligand. The desorption is based on ionic repulsion achieved by altering the pH (30). Thiophilic interaction chromatography (TIC) is based on salt-promoted interactions of proteins with 2-mercaptoethanol coupled to divinylsulfone-activated agarose (30). The relatively simple performance, high level of efficiency, and reduced production cost of TIC makes it an attractive option for large-scale purifications (31). Commercial thiophilic affinity matrices are also available such as T-Gel™ adsorbent from Pierce. This is a beaded agarose modified to contain simple sulfone/thioether groups; the main advantage of this matrix is the ability to bind a majority of immunoglobulins that Protein A cannot bind (32). Similarly, hydroxyapatite crystals (Ca_{10} $(PO_4)_6(OH)_2$) are regularly employed for antibody purification (33). Hydroxyapatite comprises functional groups of positively charged pairs of calcium ions and clusters of negatively charged phosphate groups (34). Horenstein (35) reported high yields (80%) and purity (97%) when hydroxyapatite was used in conjunction with Protein A affinity chromatography for the purification of therapeutic antibodies (36). Combining chromatography steps has further improved the purification efficiency of immunoglobulins. This was demonstrated when a 100% pure IgG solution was reported when combining both HIC and size exclusion chromatography for removal of IgG aggregates (37). Several non-affinity purification approaches have been compared with Protein A chromatography for the purification of antibodies (38). Three-step non-affinity purification processes were identified that reached comparable purification levels to that of Protein A. However, due to its efficiency, Protein A chromatography proves less laborious and less time-consuming when compared to these three-step approaches (38).

3. Matrices

The employment of an efficient affinity ligand is of paramount importance in IAC; however, without the incorporation of a suitable surrounding matrix, the functional capacity of the ligand can be severely hampered. Conventional matrices for use in IAC systems have been based on low-pressure resistance, allowing their operation under gravity flow with a slight vacuum or peristaltic flow applied (39). These matrices include natural polymers such as agarose, dextrose, or cellulose (discussed in more detail in Subheading 3). Synthetic support polymers include acrylamide (40), polymethacrylate derivatives (39), polyethersulfone matrices (41), or inorganic materials such as silica (42) and zirconia (43), all of which have been employed for the immobilisation of antibodies, enzymes (44), lectins (45), nucleic acids (46), and hormones (47). The main disadvantage of these low-performance supports is their slow mass transfer, limited hydrodynamic and chemical properties, and limited stability at high pressures and flow rates. In contrast, high-performance supports include derivatized silica (48), glass (49), and certain organic matrices such as azlactone beads (50). The efficiency and enhanced stability of these materials makes them an attractive accompaniment for use in high-performance liquid chromatography (HPLC) methods. This online coupling helps to provide improved speed and precision for analytical applications of IAC (51). Solid supports offering better chemical and hydrodynamic properties than conventional supports have been developed and optimised in recent years; examples of these include immunomagnetic beads (52, 53), porous glass beads with gigapores (34), and monoliths (54, 55). The "sol–gel method" involves immobilising the antibodies in the pores of a silicate glass matrix. This generally involves the transformation of sol from liquid state to the solid-phase gel (39). Advantages of this method include the lack of a fixed antibody orientation which leads to increased access to the CDR regions. Also, no steric hindrance effects are observed because the pores of the matrix are filled with an aqueous solute. This hydrophilic matrix can lead to a reduction in both non-specific adsorption (56) and antibody leaching (57).

Ozturk and co-workers (58) described an alternative method for antibody immobilisation involving reusable poly(hydroxyethyl) methacrylate (PHEMA) nanoparticles. These nanoparticles are a novel type of adsorbent for antibody immobilisation and are produced by a "surfactant-free" emulsion polymerisation technique. They are directly silanizated with IMEO (3-(2-imidazoline-1-yl) propyl (triethoxysilane) – used as a pseudo-specific ligand) which in turn produces up to 160-fold higher binding values than the PHEMA particles alone (58). In addition to both fast binding and desorption kinetics, that was made possible by the specific surface area and absence of internal diffusion resistance (58), this approach exhibited a relatively high binding capacity.

Affi-gel-10 and affi-gel 15 (Bio-Rad) are commercial support matrices, consisting of cross-linked agarose, derivatized with N-hydroxy-succinimide (NHS) ester on either a neutral 10-atom (Affi-gel 10) long spacer arm or a positive 15-atom long spacer arm (Affi-gel 15). The NHS ester allows spontaneous coupling of ligands via free amine groups. Both formats spontaneously couple to ligands in aqueous or non-aqueous solution (59). Digiovanna and co-workers (58) described a protocol for the coupling of peptides to Affi-gel 10 affinity matrix, stating that anhydrous coupling was the most efficient coupling method for peptides (60). However, when the efficiency of Affi-Gel Hz (Cross-linked affi-gel agarose with hydrazide groups) was compared to hydrazide-activated cellulose, for the immobilisation of IgG for cytokine purification, cellulose proved to possess higher binding efficiency (61).

AminoLink gel is another commercial coupling matrix that consists of cross-linked beaded agarose derivatized to contain aldehydes present for attachment. Reducing amination of the resulting Schiff base forms a stable covalent linkage. Up to 85% coupling efficiency was reported when using AminoLink resin for immobilisation of thiabendazole-specific monoclonal antibodies to a solid support (62). When comparing AminoLink gel, SulphoLink gel (A matrix containing iodoacetyl groups which couple through sulphydryls at the hinge region of the antibody molecule), and CarboxyLink gel (A matrix that couples to aldehydes generated by oxidation of carbohydrate moieties) as potential matrices for antibody immobilisation, Domen et al. (63) found that CarboxyLink gel is the most effective at orientating antibody for better antigen purification capability (63). In addition, Cao et al. (64) have recently reported the novel use of magnetic cellulose microspheres for antibody immobilisation. A cellulose binding domain fusion protein with an integrated Protein A binding site was constructed for the linkage of antibodies to the cellulose matrix. They reported over 90% sample binding to the matrix, with purities exceeding 95%. This method shows potential for use in large-scale purifications due to its simplicity and cost-efficient nature.

3.1. Immobilisation Chemistries

Critical to the optimum IAC performance is the functional immobilisation of antibodies to a solid support. The primary factors affecting the specific activity of immobilised antibodies have been studied (65). It was concluded that the most significant factors influencing the specific activity of stochastic coupling of antibodies were multisite attachment, multiple orientations, and steric hindrance caused by crowding of antibodies and the size of the antigen. Antibodies which are bound onto an appropriate sorbent, called "immunosorbents," can be packed into cartridges or columns and used for the highly selective extraction of analytes from the sample mixture (66). During oriented immobilisation the specific activity is affected only by steric hindrance. It was also reported how the specific activity of immunosorbents prepared by immobilisation of F(ab') fragments could be increased to almost 100% by limiting the amount of protein immobilisation and the size of the antigen. Furthermore, they suggested that a higher specific activity may be achieved through the use of an efficient coupling design, and as a result, less antibody is required to prepare a column of desired capacity (65). Many different methods exist for the immobilisation of antibodies and antigens to both high and low performance affinity support matrices, which ideally should be chemically and biologically inert, easily activated, mechanically stable, uniform in particle size and should have large pore sizes to facilitate binding of large antibody molecules (6). Non-site-directed attachment such as physical adsorption methods of antibody immobilisation have been routinely used for IAC purposes. This depends on the non-specific interaction of the antibody with the solid phase, e.g. polyethyleneamine (PEI) (67). The forces attaching the antibody to the matrix can involve hydrophobic interactions, non-covalent interactions such as Van der Waals forces and ionic/hydrogen bonds. However, this type of immobilisation cannot control the antibody orientation, which is highly influenced by the surface characteristics of the antibody, the linking chemistries used, and the chemical properties of the support. Therefore, this type of immobilisation is more prevalent where elution of the antibody is desired or where no elution is performed (68). Streptavidin can also be non-covalently immobilised to a support. Streptavidin is a tetrameric binding protein specific for biotin that may be chemically attached to immunoglobulin for immobilisation. The streptavidin–biotin complex has the highest binding affinity among biological reactions known to date (69).

Recently, Cho and co-workers (69) have described a method to overcome the undefined orientation of IgG binding. They investigated methods for site-directed biotinylation of antibody at the hinge region. Their reports showed that following cleavage of the Fc region, the hinge structure containing the functional groups of the fragment was exposed, allowing easy access to the biotin derivative from the mobile phase. They also compared biotinylation at different functional groups such as amine, sulphydryl and carboxyl moieties, where the use of sulphydryl groups proved to be advantageous due to their localisation at or near the hinge region. Covalent immobilisation of antibodies is often performed by reacting free amine groups on the antibodies with supports that are activated with agents such as *N*,*N′*-carbonyl diimidazole, NHS, and cyanogen bromide (68). These supports provide direct chemical attachment between the functional groups, resulting in coupling via the amino groups of approximately 70 lysine residues in the antibody. This produces both multiple-site attachments and multiple orientations. However, the antibody orientation may result in less than optimum activity due to a loss in binding capacity (68).

Antibodies can also be immobilised via the carbohydrate fragment which is linked to Asn 297 on the Fc and thus at moderate distance from the active sites. As antibodies are mainly glycosylated at a single site in the CH2 domain of each heavy chain, this approach leaves the two Fab antigen binding regions completely free to interact with antigen in the mobile phase (68). Following oxidation with periodate or enzymatic systems, the carbohydrate residues possess aldehyde side groups. These side groups allow attachment of antibodies to amino-derivatized supports such as agarose (70), silica (71), and cellulose (72). Sulphydryl coupling is another site-selective method, whereby the free sulphydryl groups generated during the Fab fragment production are coupled to supports (73). Soluble bi-functional cross-linkers can act as a bridge between functional groups and can be very important in the development of surface plasmon resonance (SPR) biosensor-based immunoassays (67). The advantage of these approaches is the greater accessibility of the antibody binding regions resulting from the defined immobilised points of attachment. This can result in significant improvements in relative binding activity over amine-coupling methods (50).

4. Elution Conditions

The integral elution stage of an IAC procedure involves disruption of the interactions between receptor and ligand. This leads to a reduction of binding affinity, forcing the ligand back into the mobile phase (74). Elution conditions should be tailored to suit each purification process individually. Firer (74) described how an elution buffer should aim to break the various bonds that make up a protein–protein interaction. Hence, using simpler elution buffers is generally a compromise between convenience, target yield, and functionality (74). Several feasible approaches exist for induction of analyte elution including the use of displacer agents, chaotropic agents, temperature increase, pH variations, or water–organic modifier mixtures (6) (see Table 2). Displacer agents act as competitors to the ligand, and hence, addition of excess ligand in the mobile phase will provoke biospecific desorption. The advantages of this method include the mild elution conditions and the specificity of elution. Conversely, the necessity for large molar excesses of ligand, in addition to the dependency of the

Table 2
Frequently used elution conditions for immunoaffinity chromatography

Elution condition	Example	Method	Comment	References
Low pH	Glycine-HCl Citric acid Acetic acid	Disrupt ionic bonds, has hydrophobicity effects, may denature the compound	Proteins have to be pH stable	(80, 138, 139)
High pH	Triethylamine Glycine-NaOH	Disrupt ionic bonds, has hydrophobicity effects, may denature the compound	Proteins have to be pH stable	(79, 140)
Chaotropic	Magnesium chloride Sodium thiocyanate	Interfere with stabilising intra-molecular interactions	High salt concentrations, disrupts the water molecules around the affinity interaction	(86, 131)
Organic	Methanol Ethanol	Hydrophobic effects	Significant elution of small molecules, efficient column re-generation	(62, 86)
Denaturants	Urea Guanidine-HCl	Promotes unfolding	Denaturing effects limit the suitability to highly stable proteins	(88, 89)
Displacer	Antigens Ligands	Competition	Ability to simultaneously isolate several substances, exploits high affinity interactions	(141, 142, 143)

dissociation on the affinity of the receptor–ligand interaction, results in longer elution times when compared to most elution buffers (74).

Several studies have investigated the optimisation of the elution process through the exploitation of the numerous intrinsic properties of both the antibody and antigen. When IMAC is employed for purification, chelating agents such as EDTA and DPTA have proved as efficient eluents when the use of imidazole is undesirable (75).

A comparative study was performed between imidazole-a monomeric displacer and vinyl caprolactam-a polymeric displacer as elution agents for IMAC (76). It was shown that the efficiency of low-molecular-mass displacers could be improved significantly by conversion to a polymeric displacer. Displacement chromatography has been noted for its high loading capacities, ability to simultaneously isolate several substances, resolving power, and sharp elution profiles (77, 78). Other elution methods which include the use of extreme pH buffers that disrupt ionic bonds and high pH buffers such as triethylamine have been proved popular (79). Low-pH buffers, such as glycine-HCl, are also frequently used for elution (80, 81). McMahon and O'Kennedy previously reported how, prior to purification, their M11 monoclonal antibody displayed monoreactive properties, yet after purification, polyreactivity was observed. It was concluded that this characteristic was a direct consequence of the limited denaturation of the M11 antibody upon exposure to low pH conditions during the affinity purification (82).

Singh et al. (83) compared 20 different elution methods in an ELISA-based elution assay, including ionic substances, detergents, solvents, and extreme pH buffers to determine the optimum eluent for dissociating their small-molecule immune complex. The eluent with the highest recorded elution efficiency was found to be 25% methanol at pH 11.2. Organic solvents such as methanol are frequently used as successful eluents in IAC (84, 85). Similarly, chaotropic reagents are often employed as elution agents. They affect elution by changing the structure of water in and around the site of the affinity interaction (83). Durkee et al. (86) reported that when the elution of a monoclonal antibody was compared in both acid alkaline and chaotropic conditions (magnesium chloride), magnesium chloride was the most effective eluent. Narhi (87) previously reported how a higher elution efficiency was observed when denaturants urea or guanidine-HCl were used as eluents, when compared to glycine-HCl, for the recovery of several monoclonal antibodies (88). Shukla and co-workers (89) have recently presented an efficient method to circumvent the aggregation issues often presented when eluting antibodies from Protein A columns (89). The addition of urea at moderate concentrations to the purifications proved successful in overcoming problems with aggregation. Furthermore, operating the purification at low temperatures helped to stabilise the rigidity of the structure, leading to minimal aggregation occurrences (89). While several amino acids for eluting antibodies from Protein A columns have been compared (90), it has been shown that recovery of antibody was significantly increased when arginine was used as an eluent (91). Proline, lysine, glycine, and histidine were also compared for their elution properties. All except histidine showed zero percentage recoveries, with histidine giving a 50% recovery. This 50% is modest when compared to the 80% recovery achieved when arginine was employed. The elution properties of arginine were also directly compared with $MgCl_2$ (an ingredient in conventional commercial antibody eluents). This study proved interesting as a higher recovery and higher protein concentration in purified fractions were achieved when arginine was used in comparison to the commonly used $MgCl_2$ (90).

5. Characterisation

Antibodies can be characterised by a wide variety of physiochemical and biochemical features. Characterisation procedures provide information on the integrity, affinity, sensitivity, purity, and functionality of the antibody. Fractions collected from the purification of the antibody can be characterised in several different ways to determine the location of the protein-rich fractions. Qualitative approaches include the use of SDS-PAGE, western blotting, and isoelectric focusing.

Quantitative approaches include spectrophometric approaches, determining the amount of protein present via measuring the absorbance of the fractions using assays such as the Bradford, Lowry, and bicinchoninic acid assays (92). The concentration of a specific antibody can also be quantified using an immunoassay system such as an enzyme-linked immunosorbent assay (ELISA) (93). Friguet et al. (94) reported in 1985 that it is possible to theoretically measure antibody affinity by ELISA. Bobrovnik suggested a revised method with applicability to more complex cases such as the affinity evaluation of bivalent antibodies, or at least two monovalent antibodies in a mixture (95).

Since the advent of SPR technology, SPR systems have been used widely for the characterisation and identification of antibodies. Monitoring antigen–antibody interactions in "real-time" in a "label free" environment has many advantages over conventional detection systems. The affinity of an antibody for an analyte can also be readily measured (96) (see Chapter 22). Isoelectric focusing is a technique which separates proteins on the basis of their isoelectric point. This is also useful for the characterisation of different isoforms of a protein (97) or where there are different levels of glycosylation. Isoelectric focusing used in conjunction with SDS-PAGE or immunoblotting is commonly employed as an antibody characterisation technique (97–99). It is generally carried out in a polyacrylamide gel into which low-molecular-weight amphoteric compounds have been incorporated to produce a pH gradient (93). Immunodot blot is a relatively economical fast method to determine the presence of target protein from numerous samples. This is performed by spotting/immobilising antigen onto membranes like nitrocellulose, probing this with analyte/sample fractions (e.g. eluted fractions purified from an IAC column, to determine which fractions contain the purified material) and detecting the antigen–antibody complex using a secondary labelled antibody (99, 100).

ELISAs are the foundation of immunoassays and can be used to determine the concentration, sensitivity, affinity, specificity, or cross-reactivity of an antibody.

Mass spectrophotometry (MS) involves ionising the samples to be analysed which confers a charge on the compound and then measures the charge-to-mass ratio. MS has emerged as an essential tool for the characterisation of proteins; this is due to its superior resolution when compared to other analytical techniques (101, 102). The methods chosen for characterisation of an antibody are generally dependent on the intended use of the antibody.

5.1. Applications

IAC encompasses a vast range of applications from clinical therapeutics to environmental monitoring. IAC can function as a purification, extraction, detection, or separation tool, depending on the intended application. The technique of immunoextraction refers to the use of IAC for the removal of a specific solute or group of solutes from a sample prior to qualitative/quantitative determination by a second analytical method (51). Immunoextraction essentially involves the typical IAC procedure, combined either online or offline with a method for quantification of the analytes. Online methods involve direct coupling of immunoextraction with quantitative detection methods such as HPLC (99, 103) gas chromatography (104), capillary electrophoresis (79) or, more recently, with methods like laser-induced fluorescence detection (105). Immunoextraction efficiently combines key distinct steps including conditioning, percolation of the sample, washing, elution of the target analytes, and regeneration of the matrix employed (106).

One of the most widely used applications of IAC is for the extraction/detection of contaminants in foodstuffs, particularly for mycotoxin contamination. These mainly include the aflatoxins (107, 108), ochratoxins (109, 110), zearalenone (111), and fumonisins (112). Other contaminants also extracted from foodstuffs include quinolone and sulphonamide antibiotics extracted from animal tissues (113) and the herbicide atrazine extracted from soil and foodstuffs (114). Further, environmental applications include the detection of microcystins (115) and alkylbenzene sulfonates (116) in fresh water.

Diagnostic applications include the detection of certain drugs like testosterone in athletes (117), determination of estrogen in urine of pregnant women (118), or detection of tetrodotoxin in poisoned patients (119). The clinical applications include the detection of specific biomarkers for a variety of different targets in biological fluids. These include the evaluation of nicotine exposure by the isolation of anti-nicotine metabolite antibodies (120) and the measurement/isolation of several different biomarkers including: nitrotyrosine (121), C-reactive protein (122), type II collagen peptides (123), and polycyclic aromatic hydrocarbon (PAH) metabolites (124). IAC is frequently employed for the determination of allergens in food, for example, the detection of β-lactoglobulin in whey proteins (125) or *Dactylis glomerata* pollen allergens (126). Therapeutic applications of IAC range from purification of monoclonal antibodies to isolation of recombinant proteins for therapeutic use (127).

IAC is generally used as a routine procedure for sample clean-up prior to detection, e.g. the quantitative detection of aflatoxins in different food matrices is performed after extraction and clean-up via IAC methods (105, 106).

Immunoaffinity columns can be used for sample clean-up, extraction, and pre-concentration prior to quantification (117). The determination of contaminants or isolation of desired components in foodstuffs, biological fluids, and environmental matrices requires standardised procedures which include an initial extraction procedure. These extraction methods often involve immunoaffinity column clean-up procedures, which offer the extraction of contaminants/components using simple methods such as using aqueous solvent mixtures (128). Schuhmacher et al. compared the efficiencies of both an immunoaffinity-column clean-up method and a conventional clean-up method for the determination of zealerone in corn. It was concluded that whilst the conventional clean-up procedure was more cost-effective than the IAC method, the IAC method gave 14% higher recoveries than the conventional approach and also circumvented the use of chlorinated solvents such as chloroform (129). Several additional applications are listed in Table 3.

Table 3
Immunoaffinity chromatography applications

Field	Applications	Example of use	References
Clinical	Determination of biological markers	Development of a procollagen biomarker for measuring collagen type II synthesis	(144)
Diagnostic	Determination of presence of substance in biological fluids	Measurement of urinary 8-epi prostaglandin $f^{2\alpha}$ an index of lipid peroxidation	(145–147)
Environmental	Management and detection of environmental pollutants	Detection of aflatoxins in peanut/milk samples	(105, 106, 148–150),
Therapeutic Production and sample clean-up	Isolation/purification of therapeutic agents Purification and clean-up of biological materials/analytes	Isolation of anti-venom antibodies Purification of biopharmaceuticals/analysis of mycotoxin contamination	(137, 151–153)(1, 22)

6. Conclusion

IAC is a fundamental isolation and purification tool which is incorporated in a vast range of applications. Exploiting affinity interactions results in a highly efficient method for separating desired components from a heterogeneous mixture based on their specific affinities. The application of IAC methodologies and their use in conjunction with online detection systems represents a powerful and sensitive technology for the rapid isolation/detection of bio-recognition elements. The number of FDA-approved therapeutic monoclonal antibodies has continued to increase rapidly every year (139) and has resulted in the need for optimised and efficient IAC strategies for the purification of these antibodies. The ever-expanding niche incorporating antibodies in diagnostic assays and moreover in therapeutics necessitates the requirement of these sophisticated affinity procedures for the production of purified antibodies.

Acknowledgments

This material is based upon works supported by the Food Institutional Research Measure (FIRM) of the Department of Agriculture and Food (Ireland), the Science Foundation Ireland under Grant No. 05/CE3/B754 (Ireland), the Biomedical Diagnostics Institute (BDI), and Enterprise Ireland, all of which are gratefully acknowledged.

References

1. Campbell, D. H., Luescher, E., and Lerman, L. S. (1951) Immunologic Adsorbents. I. Isolation of Antibody by Means of a Cellulose-Protein Antigen. *Proc. Natl. Acad. Sci.* **37**, 575–578.
2. Gurvich, A. E., Kapner, R. B., and Nezlin, R. S. (1959) Isolation of Pure Antibodies with the Aid of Fixed on Cellulose Antigens and Studies on their Properties. *Biokhimiia.* **24**, 144–156.
3. Hjerten, S. (1964) The Preparation of Agarose Spheres for Chromatography of Molecules and Particles. *Biochim. Biophys. Acta.* 79, 393–398.
4. Van Tunen, A. J. and Mol, J. N. M. (1987) A Novel Purification Procedure for Chalcone Flavanone Isomerase from Petunia Hybrida and the use of its Antibodies to Characterize the Po Mutation. *Archiv. Biochem. Biophys.* **257**, 85–91.
5. Mizejewski, G. J., Simon, R., and Vonnegut, M. (1979) Purification of α-Fetoprotein from Mouse Amniotic Fluid by Gel-Entrapped Antibody Filtration. *J. Immunol. Methods.* **31**, 333–339.
6. Delaunay, N., Pichon, V., and Hennion, M. C. (2000) Immunoaffinity Solid-Phase Extraction for the Trace-Analysis of low-Molecular-Mass Analytes in Complex Sample Matrices. *J. Chromatogr. B.* **745**, 15–37.
7. Burgess, R. R. and Thompson, N. E. (2002) Advances in Gentle Immunoaffinity Chromatography. *Curr. Opin. Biotech.* **13**, 304–308.
8. Murphy, K. P., Travers, P., Walport, M., Janeway, C. A., Ehrenstein, M., Mauri, C. (2007) *Immunobiology*, 7th edition. Current Biology/Garland, London.
9. Barbas, C. F. III, Barbas, C. F., Burton, D. R., Scott, J. K., and Silverman, G. J. (2004) *Phage Display: A Laboratory Manual.* Cold

Spring Harbor Laboratory Press, New York, USA.

10. Shantha, R. T. (2003) Glycosylation Variations with Expression Systems and their Impact on Biological Activity of Therapeutic Immunoglobulins. *Bioprocess Int.* **1**, 44–53.
11. Fitzpatrick, J., Fanning, L., Hearty, S., Leonard, P., Manning, B. M., Quinn, J. G., and O'Kennedy, R. (2000) Applications and Recent Developments in the use of Antibodies for Analysis. *Anal. Lett.* **33**, 2563–2609.
12. Köhler, G. and Milstein, C. (1975) Continuous Cultures of Fused Cells Secreting Antibody of Predefined Specificity. *Nature (Lond.).* **256**, 495–497.
13. Roque, A. C. A., Silva, C. S. O., and Taipa, M. Â. (2007) Affinity-Based Methodologies and Ligands for Antibody Purification: Advances and Perspectives. *J. Chromatogr. A.* **1160**, 44–55.
14. Hoogenboom, H. R., de Bruïne, A. P., Hufton, S. E., Hoet, R. M., Arends, J., and Roovers, R. C. (1998) Antibody Phage Display Technology and its Applications. *Immunotechnology.* **4**, 1–20.
15. Muronetz, V. I. and Korpela, T. (2003) Isolation of Antigens and Antibodies by Affinity Chromatography. *J. Chromatogr. B.* **790**, 53–66.
16. Turková, J. (1999) Oriented Immobilization of Biologically Active Proteins as a Tool for Revealing Protein Interactions and Function. *J. Chromatogr. B.* **722**, 11–31.
17. Hober, S., Nord, K., and Linhult, M. (2007) Protein A Chromatography for Antibody Purification. *J. Chromatogr. B.* **848**, 40–47.
18. Nilson, H. K., Lögdberg, L., Kastern, W., Björck, L., and Åkerström, B. (1993) Purification of Antibodies using Protein L-Binding Framework Structures in the Light Chain Variable Domain. *J. Immunol. Methods.* **164**, 33–40.
19. Roque, A. C., Lowe, C. R., and Taipa, M. A. (2004) Antibodies and Genetically Engineered Related Molecules: Production and Purification. *Biotechnol. Prog.* **20**, 639–654.
20. Polzius, R., Schneider, T., Biert, F. F., Bilitewski, U., and Koschinski, W. (1996) Optimization of Biosensing using Grating Couplers: Immobilization on Tantalum Oxide Waveguides. *Biosens Bioelectron.* **11**, 503–514.
21. Fassina, G., Ruvo, M., Palombo, G., Verdoliva, A., and Marino, M. (2001) Novel Ligands for the Affinity-Chromatographic Purification of Antibodies. *J. Biochem. Biophys. Methods.* **49**, 481–490.
22. Lowe, C. R., Lowe, A. R., and Gupta, G. (2001) New Developments in Affinity Chromatography with Potential Application in the Production of Biopharmaceuticals. *J. Biochem. Biophys. Methods.* **49**, 561–574.
23. Weiss, R., Freudenschuss, M., Krska, R., and Mizaikoff, B. (2003) Improving Methods of Analysis for Mycotoxins: Molecularly Imprinted Polymers for Deoxynivalenol and Zearalenone. *Food. Addit. Contam.* **20**, 386–395.
24. Andersson, L. I. (2000) Molecular Imprinting for Drug Bioanalysis: A Review on the Application of Imprinted Polymers to Solid-Phase Extraction and Binding Assay. *J. Chromatogr. B.* **739**, 163–173.
25. Chapuis, F., Pichon, V., Lanza, F., Sellergren, B., and Hennion, M. (2004) Retention Mechanism of Analytes in the Solid-Phase Extraction Process using Molecularly Imprinted Polymers: Application to the Extraction of Triazines from Complex Matrices. *J. Chromatogr. B.* **804**, 93–101.
26. Caro, E., Marcé, R. M., Borrull, F., Cormack, P. A. G., and Sherrington, D. C. (2006) Application of Molecularly Imprinted Polymers to Solid-Phase Extraction of Compounds from Environmental and Biological Samples. *TrAC Trends Anal. Chem.* **25**, 143–154.
27. Dong, D., Liu, H., Xiao, Q., and Li, R. (2008) Affinity Purification of Egg Yolk Immunoglobulins (IgY) with a Stable Synthetic Ligand. *J. Chromatogr. B.* **870**, 51–54.
28. Tsagkogeorgas, F., Ochsenkühn-Petropoulou, M., Niessner, R., and Knopp, D. (2006) Encapsulation of Biomolecules for Bioanalytical Purposes: Preparation of Diclofenac Antibody-Doped Nanometer-Sized Silica Particles by Reverse Micelle and sol-gel Processing. *Anal. Chim. Acta.* **573–574**, 133–137.
29. Pişkin, E. (2004) Molecularly Designed Water Soluble, Intelligent, Nanosize Polymeric Carriers. *Int. J. Pharm.* **277**, 105–118.
30. Coffinier, Y. and Vijayalakshmi, M. A. (2004) Mercaptoheterocyclic Ligands Grafted on a Poly (Ethylene Vinyl Alcohol) Membrane for the Purification of Immunoglobulin G in a Salt Independent Thiophilic Chromatography. *J. Chromatogr. B.* **808**, 51–56.
31. Qian, H., Chen, M. Q., and Lin, Z. Y. (2008) The Effective and Specific Isolation of Antibodies from Human Serum by using Thiophilic Paramagnetic Polymer Beads. *Colloids Surf. B Biointerfaces.* **67**, 224–229.
32. Palmer, D. A. and Miller, J. N. (1995) Thiophilic Gels: Applications in Flow-Injection Immunoassays for Macromolecules and Haptens. *Anal. Chim. Acta.* **303**, 223–230.
33. Luellau, E., von Stockar, U., Vogt, S., and Freitag, R. (1998) Development of a

Downstream Process for the Isolation and Separation of Monoclonal Immunoglobulin A Monomers, Dimers and Polymers from Cell Culture Supernatant. *J. Chromatogr. A.* **796**, 165–175.

34. Yu, S., Geng, J., Zhou, P., Wang, J., Chen, X., and Hu, J. (2008) New Hydroxyapatite Monolithic Column for DNA Extraction and its Application in the Purification of *Bacillus subtilis* Crude Lysate. *J. Chromatogr. A.* **1183**, 29–37.
35. Denizli, A., Alkan, M., Garipcan, B., Özkara, S., and Pişkin, E. (2003) Novel Metal-Chelate Affinity Adsorbent for Purification of Immunoglobulin-G from Human Plasma. *J. Chromatogr. B.* **795**, 93–103.
36. Horenstein, A. L., Crivellin, F., Funaro, A., Said, M., and Malavasi, F. (2003) Design and Scaleup of Downstream Processing of Monoclonal Antibodies for Cancer Therapy: From Research to Clinical Proof of Principle. *J. Immunol. Methods.* **275**, 99–112.
37. Azevedo, A. M., Rosa, P. A. J., Ferreira, I. F., and Aires-Barros, M. R. (2008) Integrated Process for the Purification of Antibodies Combining Aqueous Two-Phase Extraction, Hydrophobic Interaction Chromatography and Size-Exclusion Chromatography. *J. Chromatogr. A.* **1213**, 154–161.
38. Follman, D. K. and Fahrner, R. L. (2004) Factorial Screening of Antibody Purification Processes using Three Chromatography Steps without Protein A. *J. Chromatogr. A.* **1024**, 79–85.
39. Schuster, M., Wasserbauer, E., Neubauer, A., and Jungbauer, A. (2000) High Speed Immuno-Affinity Chromatography on Supports with Gigapores and Porous Glass. *Bioseparation.* **9**, 259–268.
40. Verdoliva, A., Pannone, F., Rossi, M., Catello, S., and Manfredi, V. (2002) Affinity Purification of Polyclonal Antibodies using a New all-D Synthetic Peptide Ligand: Comparison with Protein A and Protein G. *J. Immunol. Methods.* **271**, 77–88.
41. Bellofiore, P., Petronzelli, F., De Martino, T., Minenkova, O., Bombardi, V., Anastasi, A. M., Lindstedt, R., Felici, F., De Santis, R., and Verdoliva, A. (2006) Identification and Refinement of a Peptide Affinity Ligand with Unique Specificity for a Monoclonal Anti-Tenascin-C Antibody by Screening of a Phage Display Library. *J. Chromatogr. A.* **1107**, 182–191.
42. Katoh, S., Imada, M., Takeda, N., Katsuda, T., Miyahara, H., Inoue, M., and Nakamura, S. (2007) Optimization of Silica-Based Media for Antibody Purification by Protein A Affinity Chromatography. *J. Chromatogr. A.* **1161**, 36–40.
43. Mowry, M. C., Meagher, M., Smith, L., Marks, J., and Subramanian, A. (2004) Production and Purification of a Chimeric Monoclonal Antibody Against Botulinum Neurotoxin Serotype A. *Protein Expr. Purif.* **37**, 399–408.
44. Ehle, H. and Horn, A. (1990) Immunoaffinity Chromatography of Enzymes. *Bioseparation.* **1**, 97–110.
45. Hortin, G. L. and Trimpe, B. L. (1990) Lectin Affinity Chromatography of Proteins Bearing O-Linked Oligosaccharides: Application of Jacalin-Agarose. *Anal. Biochem.* **188**, 271–277.
46. Gadgil, H., Oak, S. A., and Jarrett, H. W. (2001) Affinity Purification of DNA-Binding Proteins. *J. Biochem. Biophys. Methods.* **49**, 607–624.
47. Uhlik, O., Kamlar, M., Kohout, L., Jezek, R., Harmatha, J., and Macek, T. (2008) Affinity Chromatography Reveals RuBisCO as an Ecdysteroid-Binding Protein. *Steroids.* **73**, 1433–1440.
48. Bouzerna, N., Hammadi, H., Richard, C., Formstecher, P., and Dautrevaux, M. (1988) Improvements in the Purification of an Antisteroid Antiserum Resistant to Conventional Immunoadsorption Chromatography. *J. Immunol. Methods.* **112**, 251–259.
49. Prince, A. E., Fan, T. S., Skoczenski, B. A., and Bushway, R. J. (2001) Development of an Immunoaffinity-Based Solid-Phase Extraction for Diazinon. *Anal. Chim. Acta.* **444**, 37–49.
50. Hermanson, G. T., Mattson, G. R., and Krohn, R. I. (1995) Preparation and use of Immunoglobulin-Binding Affinity Supports on Emphaze Beads. *J. Chromatogr. A.* **691**, 113–122.
51. Hage, D. S. (1998) Survey of Recent Advances in Analytical Applications of Immunoaffinity Chromatography. *J. Chromatogr. B.* **715**, 3–28.
52. Paoli, G. C., Kleina, L. G., and Brewster, J. D. (2007) Development of *Listeria monocytogenes*-Specific Immunomagnetic Beads using A Single-Chain Antibody Fragment. *Foodborne Pathog. Dis.* **4**, 74–83.
53. Widjojoatmodjo, M. N., Fluit, A. C., Torensma, R., and Verhoef, J. (1993) Comparison of Immunomagnetic Beads Coated with Protein A, Protein G, or Goat Anti-Mouse Immunoglobulins. Applications in Enzyme Immunoassays and Immunomagnetic Separations. *J. Immunol. Methods.* **165**, 11–19.

54. Rucevic, M., Clifton, J. G., Huang, F., Li, X., Callanan, H., Hixson, D. C., and Josic, D. (2006) Use of Short Monolithic Columns for Isolation of Low Abundance Membrane Proteins. *J. Chromatogr. A.* **1123**, 199–204.
55. Uzun, L., Say, R., and Denizli, A. (2005) Porous Poly (Hydroxyethyl Methacrylate) Based Monolith as a New Adsorbent for Affinity Chromatography. *React. Funct. Polym.* **64**, 93–102.
56. Cichna, M., Knopp, D., and Niessner, R. (1997) Immunoaffinity Chromatography of Polycyclic Aromatic Hydrocarbons in Columns Prepared by the Sol-Gel Method. *Anal. Chim. Acta.* **339**, 241–250.
57. Yang, H., Zhu, Q., Qu, H., Chen, X., Ding, M., and Xu, J. (2002) Flow Injection Fluorescence Immunoassay for Gentamicin using Sol-Gel-Derived Mesoporous Biomaterial. *Anal. Biochem.* **308**, 71–76.
58. Öztürk, N., Bereli, N., Akgöl, S., and Denizli, A. (2008) High Capacity Binding of Antibodies by Poly (Hydroxyethyl Methacrylate) Nanoparticles. *Colloids Surf. B Biointerfaces.* **67**, 14–19.
59. Simpson, R. J. (2004) *Purifying Proteins for Proteomics: A Laboratory Manual*, Chapter 10. Cold Spring Harbor Laboratory Press, New York, USA.
60. Digiovanna, M. P., Roussel, R. R., and Stern, D. F. (2001) Production of Antibodies that Recognize Specific Tyrosine-Phosphorylated Peptides. *Curr. Protoc. Mol. Biol.* Chapter **18**, Unit 18.6.
61. Vaková, R., Gaudinová, A., Süssenbeková, H., Dobrev, P., Strnad, M., Holík, J., and Lenfeld, J. (1998) Comparison of Oriented and Random Antibody Immobilization in Immunoaffinity Chromatography of Cytokinins. *J. Chromatogr. A.* **811**, 77–84.
62. Horne, E., Coyle, T., O'Keeffe, M., and Brandon, D. L. (1999) Immobilisation of Thiabendazole-Specific Antibodies on an Agarose Matrix for Application in Immunoaffinity Chromatography. *Analyst.* **124**, 87–90.
63. Domen, P. L., Nevens, J. R., Mallia, A. K., Hermanson, G. T., and Klenk, D. C. (1990) Site-Directed Immobilization of Proteins. *J. Chromatogr.* **510**, 293–302.
64. Cao, Y., Zhang, Q., Wang, C., Zhu, Y., and Bai, G. (2007) Preparation of Novel Immunomagnetic Cellulose Microspheres Via Cellulose Binding Domain-Protein A Linkage and its use for the Isolation of Interferon α-2b. *J. Chromatogr. A.* **1149**, 228–235.
65. Wimalasena, R. L. and Wilson, G. S. (1991) Factors Affecting the Specific Activity of Immobilized Antibodies and their Biologically Active Fragments. *J. Chromatogr. A.* **572**, 85–102.
66. Pichon, V., Bouzige, M., Mičge, C., and Hennion, M. C. (1999) Immunosorbents: Natural Molecular Recognition Materials for Sample Preparation of Complex Environmental Matrices. *Trends Anal. Chem.* **18**, 219–235.
67. Chatrathi, M. P., Wang, J., and Collins, G. E. (2007) Sandwich Electrochemical Immunoassay for the Detection of *Staphylococcal enterotoxin* B Based on Immobilized Thiolated Antibodies. *Biosens. Bioelectron.* **22**, 2932–2938.
68. Nisnevitch, M. and Firer, M. A. (2001) The Solid Phase in Affinity Chromatography: Strategies for Antibody Attachment. *J. Biochem. Biophys. Methods.* **49**, 467–480.
69. Cho, I. H., Paek, E. H., Lee, H., Kang, J. Y., Kim, T. S., and Paek, S. H. (2007) Site-Directed Biotinylation of Antibodies for Controlled Immobilization on Solid Surfaces. *Anal. Biochem.* **365**, 14–23.
70. Bhatnagar, P. K., Das, D., and Suresh, M. R. (2008) Sequential Affinity Purification of Peroxidase-tagged Bispecific Anti-SARS-CoV Antibodies on Phenylboronic Acid Agarose. *J. Chromatogr. B.* **863**, 235–241.
71. Hu, X., O'Hara, L., White, S., Magner, E., Kane, M., and Gerard Wall, J. (2007) Optimisation of Production of a Domoic Acid-Binding scFv Antibody Fragment in *Escherichia coli* using Molecular Chaperones and Functional Immobilisation on a Mesoporous Silicate Support. *Protein Expr. Purif.* **52**, 194–201.
72. Peng, L., Calton, G. J., and Burnett, J. W. (1987) Evaluation of Activation Methods with Cellulose Beads for Immunosorbent Purification of Immunoglobulins. *J. Biotechnol.* **5**, 255–265.
73. Spitznagel, T. M., Jacobs, J. W., and Clark, D. S. (1993) Random and Site-Specific Immobilization of Catalytic Antibodies. *Enzyme Microb. Technol.* **15**, 916–921.
74. Firer, M. A. (2001) Efficient Elution of Functional Proteins in Affinity Chromatography. *J. Biochem. Biophys. Methods.* **49**, 433–442.
75. Müller, K. M., Arndt, K. M., Bauer, K., and Plückthun, A. (1998) Tandem Immobilized Metal-Ion Affinity Chromatography/ Immunoaffinity Purification of His-Tagged Proteins – Evaluation of Two Anti-His-Tag Monoclonal Antibodies. *Anal. Biochem.* **259**, 54–61.

76. Arvidsson, P., Ivanov, A. E., Galaev, I. Y., and Mattiasson, B. (2001) Polymer Versus Monomer as Displacer in Immobilized Metal Affinity Chromatography. *J. Chromatogr. B.* **753**, 279–285.
77. Freitag, R. and Breier, J. (1995) Displacement Chromatography in Biotechnological Downstream Processing. *J. Chromatogr. A.* **691**, 101–112.
78. Kumar, A., Galaev, I. Y., and Mattiasson, B. (2000) Polymer Displacement/Shielding in Protein Chromatography. *J. Chromatogr. B.* **741**, 103–113.
79. Mazuet, C., Lerouge, D., Poul, M., and Blin, N. (2006) Breast Carcinoma Specific Antibody Selection Combining Phage Display and Immunomagnetic Cell Sorting. *Biochem. Biophys. Res. Commun.* **348**, 550–559.
80. Gomes, P. C., Machado de Ávila, R. A., Selena Maria, W., Richardson, M., Fortes-Dias, C. L., and Chávez-Olórtegui, C. (2007) The Co-Purification of a Lectin (BJcuL) with Phospholipases A2 from *Bothrops jararacussu* Snake Venom by Immunoaffinity Chromatography with Antibodies to Crotoxin. *Toxicon.* **49**, 1099–1108.
81. Hoos, J. S., Damsten, M. C., de Vlieger, J. S. B., Commandeur, J. N. M., Vermeulen, N. P. E., Niessen, W. M. A., Lingeman, H., and Irth, H. (2007) Automated Detection of Covalent Adducts to Human Serum Albumin by Immunoaffinity Chromatography, on-Line Solution Phase Digestion and Liquid Chromatography–Mass Spectrometry. *J. Chromatogr. B.* **859**, 147–156.
82. McMahon, M. J. and O'Kennedy, R. (2000) Polyreactivity as an Acquired Artefact, rather than a Physiologic Property, of Antibodies: Evidence that Monoreactive Antibodies may Gain the Ability to Bind to Multiple Antigens After Exposure to Low pH. *J. Immunol. Methods.* **241**, 1–10.
83. Singh, K. V., Kaur, J., Raje, M., Varshney, G. C., and Suri, C. R. (2003) An ELISA-Based Approach to Optimize Elution Conditions for Obtaining Hapten-Specific Antibodies. *Anal. Bioanal. Chem.* **377**, 220–224.
84. Li, L., Wang, J., Zhou, S., and Zhao, M. (2008) Development and Characterization of an Immunoaffinity Monolith for Selective On-Line Extraction of Bisphenol A from Environmental Water Samples. *Anal. Chim. Acta.* **620**, 1–7.
85. Rejeb, S. B., Cléroux, C., Lawrence, J. F., Geay, P., Wu, S., and Stavinski, S. (2001) Development and Characterization of Immunoaffinity Columns for the Selective Extraction of a New Developmental Pesticide: Thifluzamide, from Peanuts. *Anal. Chim. Acta.* **432**, 193–200.
86. Durkee, K. H., Roh, B. H., and Doellgast, G. J. (1993) Immunoaffinity Chromatographic Purification of Russell's Viper Venom Factor X Activator using Elution in High Concentrations of Magnesium Chloride. *Protein Expr. Purif.* **4**, 405–411.
87. Zhou, G., Gui, L., Li, Z., Yuan, X., and Zhang, Q. (2008) Generation and Characterization of Monoclonal Antibodies Against the Flounder *Paralichthys olivaceus rhabdovirus*. *J. Virol. Methods.* **148**, 205–210.
88. Narhi, L. O., Caughey, D. J., Horan, T., Kita, Y., Chang, D., and Arakawa, T. (1997) Effect of Three Elution Buffers on the Recovery and Structure of Monoclonal Antibodies. *Anal. Biochem.* **253**, 236–245.
89. Shukla, A. A., Gupta, P., and Han, X. (2007) Protein Aggregation Kinetics during Protein A Chromatography: Case Study for an Fc Fusion Protein. *J. Chromatogr. A.* **1171**, 22–28.
90. Ejima, D., Yumioka, R., Tsumoto, K., and Arakawa, T. (2005) Effective Elution of Antibodies by Arginine and Arginine Derivatives in Affinity Column Chromatography. *Anal. Biochem.* **345**, 250–257.
91. Arakawa, T., Philo, J. S., Tsumoto, K., Yumioka, R., and Ejima, D. (2004) Elution of Antibodies from a Protein-A Column by Aqueous Arginine Solutions. *Protein Expr. Purif.* **36**, 244–248.
92. Berges, J. A., Fisher, A. E., and Harrison, P. J. (1993) A Comparison of Lowry, Bradford and Smith Protein Assays using Different Protein Standards and Protein Isolated from the Marine Diatom Thalassiosira Pseudonana. *Mar. Biol.* **115**, 187–193.
93. Delves, P. J. (1996) Antibody Applications: Essential Techniques Edited by PJ Delves, J Wiley & Sons Ltd, 1995, *Biochimie.* **78**, 290.
94. Friguet, B., Chaffotte, A. F., Djavadi-Ohaniance, L., and Goldberg, M. E. (1985) Measurements of the True Affinity Constant in Solution of Antigen–Antibody Complexes by Enzyme-Linked Immunosorbent Assay. *J. Immunol. Methods.* 77, 305–319.
95. Bobrovnik, S. A. (2003) Determination of Antibody Affinity by ELISA. Theory. *J. Biochem. Biophys. Methods.* **57**, 213–236.
96. Leonard, P., Säfsten, P., Hearty, S., McDonnell, B., Finlay, W., and O'Kennedy, R. (2007) High Throughput Ranking of

Recombinant Avian scFv Antibody Fragments from Crude Lysates using the Biacore A100. *J. Immunol. Methods.* **323**, 172–179.

97. Amadeo, I., Oggero, M., Zenclussen, M. L., Robles, L., Pereira, D., Kratje, R., and Etcheverrigaray, M. (2004) A Single Monoclonal Antibody as Probe to Detect the Entire Set of Native and Partially Unfolded rhEPO Glycoforms. *J. Immunol. Methods.* **293**, 191–205.
98. Coleman, L. and Mahler, S. M. (2003) Purification of Fab Fragments from a Monoclonal Antibody Papain Digest by Gradiflow Electrophoresis. *Protein Expr. Purif.* **32**, 246–251.
99. Rubin, G. L., Sharp, S., Jones, A. L., Glatt, H., Mills, J. A., and Coughtrie, M. W. H. (1996) Design, Production and Characterization of Antibodies Discriminating between the Phenol- and Monoamine-Sulphating Forms of Human Phenol Sulphotransferase. *Xenobiotica.* **26**, 1113–1119.
100. Zhao, M., Liu, Y., Li, Y., Zhang, X., and Chang, W. (2003) Development and Characterization of an Immunoaffinity Column for the Selective Extraction of Bisphenol A from Serum Samples. *J. Chromatogr. B.* **783**, 401–410.
101. Zhang, Z., Pan, H., and Chen, X. (2009) Mass Spectrometry for Structural Characterization of Therapeutic Antibodies. *Mass Spectrom. Rev.* **28**, 147–176.
102. Collins, M. O. and Choudhary, J. S. (2008) Mapping Multiprotein Complexes by Affinity Purification and Mass Spectrometry. *Curr. Opin. Biotechnol.* **19**, 324–330.
103. Maurer, H. H., Schmitt, C. J., Weber, A. A., and Kraemer, T. (2000) Validated Electrospray Liquid Chromatographic–Mass Spectrometric Assay for the Determination of the Mushroom Toxins α- and β-Amanitin in Urine After Immunoaffinity Extraction. *J. Chromatogr. B.* **748**, 125–135.
104. Aguilera, R., Becchi, M., Grenot, C., Casabianca, H., and Hatton, C. K. (1996) Detection of Testosterone Misuse: Comparison of Two Chromatographic Sample Preparation Methods for Gas Chromatographic-Combustion/Isotope Ratio Mass Spectrometric Analysis. *J. Chromatogr. B.* **687**, 43–53.
105. Delaunay-Bertoncini, N. and Hennion, M. C. (2004) Immunoaffinity Solid-Phase Extraction for Pharmaceutical and Biomedical Trace-Analysis – Coupling with HPLC and CE – Perspectives. *J. Pharm. Biomed. Anal.* **34**, 717–736.
106. Lee, J. E., Kwak, B., Ahn, J., and Jeon, T. (2009) Occurrence of Aflatoxin M1 in Raw Milk in South Korea using an Immunoaffinity Column and Liquid Chromatography. *Food Control.* **20**, 136–138.
107. Hussain, I., Anwar, J., Asi, M. R., Munawar, M. A., and Kashif, M. (2010) Aflatoxin M1 Contamination in Milk from Five Dairy Species in Pakistan. *Food Control.* **21**, 122–124.
108. Ariño, A., Herrera, M., Estopañan, G., Rota, M. C., Carramiñana, J. J., Juan, T., and Herrera, A. (2009) Aflatoxins in Bulk and Pre-Packed Pistachios Sold in Spain and Effect of Roasting. *Food Control.* **20**, 811–814.
109. Kabak, B. (2009) Ochratoxin A in Cereal-Derived Products in Turkey: Occurrence and Exposure Assessment. *Food Chem Toxicol.* **47**, 348–352.
110. Bento, J. M. V., Pena, A., Lino, C. M., and Pereira, J. A. (2009) Determination of Ochratoxin A Content in Wheat Bread Samples Collected from the Algarve and Bragança Regions, Portugal: Winter 2007. *Microchem. J.* **91**, 165–169.
111. Manova, R. and Mladenova, R. (2009) Incidence of Zearalenone and Fumonisins in Bulgarian Cereal Production. *Food Control.* **20**, 362–365.
112. Silva, L., Fernández-Franzón, M., Font, G., Pena, A., Silveira, I., Lino, C., and Mañes, J. (2009) Analysis of Fumonisins in Corn-Based Food by Liquid Chromatography with Fluorescence and Mass Spectrometry Detectors. *Food Chem.* **112**, 1031–1037.
113. Li, C., Wang, Z., Cao, X., Beier, R. C., Zhang, S., Ding, S., Li, X., and Shen, J. (2008) Development of an Immunoaffinity Column Method using Broad-Specificity Monoclonal Antibodies for Simultaneous Extraction and Cleanup of Quinolone and Sulfonamide Antibiotics in Animal Muscle Tissues. *J. Chromatogr. A.* **1209**, 1–9.
114. Chuang, J. C., Van Emon, J. M., Jones, R., Durnford, J., and Lordo, R. A. (2007) Development and Application of Immunoaffinity Column Chromatography for Atrazine in Complex Sample Media. *Anal. Chim. Acta.* **583**, 32–39.
115. Mhadhbi, H., Ben-Rejeb, S., Cléroux, C., Martel, A., and Delahaut, P. (2006) Generation and Characterization of Polyclonal Antibodies Against Microcystins – Application to Immunoassays and Immunoaffinity Sample Preparation Prior to Analysis by Liquid Chromatography and UV Detection. *Talanta.* **70**, 225–235.

116. Sánchez-Martínez, M. L., Aguilar-Caballos, M. P., Eremin, S. A., and Gómez-Hens, A. (2005) Determination of Linear Alkylbenzenesulfonates in Water Samples by Immunoaffinity Chromatography with Fluorescence Detection. *Anal. Chim. Acta.* **553**, 93–98.
117. Chen, H., Huang, T., and Zhang, X. (2009) Immunoaffinity Extraction of Testosterone by Antibody Immobilized Monolithic Capillary with On-Line Laser-Induced Fluorescence Detection. *Talanta.* **78**, 259–264.
118. Su, P., Zhang, X., and Chang, W. (2005) Development and Application of a Multi-Target Immunoaffinity Column for the Selective Extraction of Natural Estrogens from Pregnant Women's Urine Samples by Capillary Electrophoresis. *J. Chromatogr. B.* **816**, 7–14.
119. Kawatsu, K., Shibata, T., and Hamano, Y. (1999) Application of Immunoaffinity Chromatography for Detection of Tetrodotoxin from Urine Samples of Poisoned Patients. *Toxicon.* **37**, 325–333.
120. Wielkoszyński, T., Tyrpieć, K., and Szumska, M.(2009) The Enzyme-Linked Immunosorbent Assay (ELISA) Method for Nicotine Metabolites Determination in Biological Fluids. *J. Pharm. Biomed. Anal.* **49**, 1256–1260.
121. Radabaugh, M. R., Nemirovskiy, O. V., Misko, T. P., Aggarwal, P., and Mathews, W. R. (2008) Immunoaffinity Liquid Chromatography–Tandem Mass Spectrometry Detection of Nitrotyrosine in Biological Fluids: Development of a Clinically Translatable Biomarker. *Anal. Biochem.* **380**, 68–76.
122. Peoples, M. C., Phillips, T. M., and Karnes, H. T. (2008) Demonstration of a Direct Capture Immunoaffinity Separation for C-Reactive Protein using a Capillary-Based Microfluidic Device. *J. Pharm. Biomed.* **48**, 376–382.
123. Berna, M., Schmalz, C., Duffin, K., Mitchell, P., Chambers, M., and Ackermann, B. (2006) Online Immunoaffinity Liquid Chromatography/TandemMassSpectrometry Determination of a Type II Collagen Peptide Biomarker in Rat Urine: Investigation of the Impact of Collision-Induced Dissociation Fluctuation on Peptide Quantitation. *Anal. Biochem.* **356**, 235–243.
124. Lee, C., Cho, S., Kang, J., Lee, S., Ju, Y., Sung, J., Strickland, P. T., and Kang, D. (1999) Comparison of Three Analytical Methods for 1-Hydroxypyrene Glucuronide in Urine After Non-Occupational Exposure to Polycyclic Aromatic Hydrocarbons. *Toxicol. Lett.* **108**, 209–215.
125. Puerta, A., Diez-Masa, J. C., and de Frutos, M. (2005) Development of an Immunochromatographic Method to Determine β-Lactoglobulin at Trace Levels. *Anal. Chim. Acta.* **537**, 69–80.
126. Leduc-Brodard, V., Inacio, F., Jaquinod, M., Forest, E., David, B., and Peltre, G. (1996) Characterization of Dac g 4, a Major Basic Allergen from *Dactylis glomerata* Pollen. *J. Allergy Clin. Immunol.* **98**, 1065–1072.
127. Sharma, S. K. (1989) Key Issues in the Purification and Characterization of Recombinant Proteins for Therapeutic use. *Adv. Drug. Deliv. Rev.* **4**, 87–111.
128. Scott, P. M. and Trucksess, M. W. (1997) Application of Immunoaffinity Columns to Mycotoxin Analysis. *J. AOAC Int.* **80**, 941–949.
129. Schuhmacher, R., Krska, R., Grasserbauer, M., Edinger, W., and Lew, H. (1998) Immuno-Affinity Columns Versus Conventional Clean-Up: A Method-Comparison Study for the Determination of Zearalenone in Corn. *Fresenius. J. Anal. Chem.* **360**, 241–245.
130. Platis, D., Drossard, J., Fischer, R., Ma, J. K. C., and Labrou, N. E. (2008) New Downstream Processing Strategy for the Purification of Monoclonal Antibodies from Transgenic Tobacco Plants. *J. Chromatogr. A.* **1211**, 80–89.
131. Rupa, P., Hamilton, K., Cirinna, M., and Wilkie, B. N. (2008) Porcine IgE in the Context of Experimental Food Allergy: Purification and Isotype-Specific Antibodies. *Vet. Immunol. Immunopathol.* **125**, 303–314.
132. Drees, C., Stürmer, C. A. O., Möller, H. M., and Fritz, G. (2008) Expression and Purification of Neurolin Immunoglobulin Domain 2 from *Carrassius auratus* (Goldfish) in *Escherichia coli. Protein Expr. Purif.* **59**, 47–54.
133. Zhuang, R., Zhang, Y., Zhang, R., Song, C., Yang, K., Yang, A., and Jin, B. (2008) Purification of GFP Fusion Proteins with High Purity and Yield by Monoclonal Antibody-Coupled Affinity Column Chromatography. *Protein Expr. Purif.* **59**, 138–143.
134. D'Agostino, B., Bellofiore, P., De Martino, T., Punzo, C., Rivieccio, V., and Verdoliva, A. (2008) Affinity Purification of IgG Monoclonal Antibodies using the D-PAM Synthetic Ligand: Chromatographic Comparison with Protein A and Thermodynamic Investigation

of the D-PAM/IgG Interaction. *J. Immunol. Methods.* **333**, 126–138.

135. Reiersen, H., Berntsen, G., Stassar, M., and Cochlovius, B. (2007) Screening Human Antibody Libraries Against Carcinoma Cells by Affinity Purification and Polymerase Chain Reaction. *J. Immunol. Methods.* **330**, 44–56.
136. Chhatre, S., Francis, R., Titchener-Hooker, N. J., Newcombe, A. R., and Keshavarz-Moore, E. (2007) Evaluation of a Novel Agarose-Based Synthetic Ligand Adsorbent for the Recovery of Antibodies from Ovine Serum. *J. Chromatogr. B.* **860**, 209–217.
137. Paul, K., Manjula, J., Deepa, E. P., Selvanayagam, Z. E., Ganesh, K. A., and Subba Rao, P. V. (2007) Anti-Echis Carinatus Venom Antibodies from Chicken Egg Yolk: Isolation, Purification and Neutralization Efficacy. *Toxicon.* **50**, 893–900.
138. Pal, S., Chatterjee, M., Bhattacharya, D. K., Bandhyopadhyay, S., and Mandal, C. (2000) Identification and Purification of Cytolytic Antibodies Directed Against O-Acetylated Sialic Acid in Childhood Acute Lymphoblastic Leukemia. *Glycobiology.* **10**, 539–549.
139. Guerrier, L., Flayeux, I., and Boschetti, E. (2001) A Dual-Mode Approach to the Selective Separation of Antibodies and their Fragments. *J. Chromatogr. B.* **755**, 37–46.
140. Palenzuela, O., Sitja-Bobadilla, A., and Álvarez-Pellitero, P. (1996) Isolation and Partial Characterization of Serum Immunoglobulins from Sea Bass (*Dicentrarchus labraxL.*) and Gilthead Sea Bream (*Sparus aurataL.*). *Fish Shellfish Immunol.* **6**, 81–94.
141. Greene, G. L., Nolan, C., Engler, J. P., and Jensen, E. V. (1980) Monoclonal Antibodies to Human Estrogen Receptor. *Proc. Natl. Acad. Sci. USA.* 77, 5115–5119.
142. Lee, J. R., Hahn, H. S., Yu, J. R., Kim, S. T., Yang, J. M., and Hahn, M. J. (2005) Immunoaffinity Purification of SRT-Tagged Human Creatine Kinase by Peptide Elution. *J. Biotechnol.* **117**, 287–291.
143. Yu, H., Dong, X. Y., and Sun, Y. (2004) An Alternating Elution Strategy for Screening High Affinity Peptides from a Phage Display Peptide Library. *Biochem. Eng. J.* **18**, 169–175.
144. Nemirovskiy, O. V., Sunyer, T., Aggarwal, P., Abrams, M., Hellio Le Graverand, M. P., and Mathews, W. R. (2008) Discovery and Development of the N-Terminal Procollagen Type II (NPII) Biomarker: A Tool for Measuring Collagen Type II Synthesis. *Osteoarthritis Cartilage.* **16**, 1494–1500.
145. Reig, M., Mora, L., Navarro, J. L., and Toldrá, F. (2006) A Chromatography Method for the Screening and Confirmatory Detection of Dexamethasone. *Meat Sci.* **74**, 676–680.
146. Hoos, J. S., Sudergat, H., Hoelck, J., Stahl, M., de Vlieger, J. S. B., Niessen, W. M. A., Lingeman, H., and Irth, H. (2006) Selective Quantitative Bioanalysis of Proteins in Biological Fluids by On-Line Immunoaffinity Chromatography–Protein Digestion–Liquid Chromatography–Mass Spectrometry. *J. Chromatogr. B.* **830**, 262–269.
147. Bachi, A., Zuccato, E., Baraldi, M., Fanelli, R., and Chiabrando, C. (1996) Measurement of Urinary 8-Epi-Prostaglandin f2α, a Novel Index of Lipid Peroxidation in Vivo, by Immunoaffinity extraction/gas Chromatography-Mass Spectrometry. Basal Levels in Smokers and Nonsmokers. *Free Radic. Biol. Med.* **20**, 619–624.
148. Harris, B. (1999) Exploiting Antibody-Based Technologies to Manage Environmental Pollution. *Trends Biotechnol.* **17**, 290–296.
149. Farré, M., Kantiani, L., and Barceló, D. (2007) Advances in Immunochemical Technologies for Analysis of Organic Pollutants in the Environment. *Trends Anal. Chem.* **26**, 1100–1112.
150. Holtzapple, C. K., Buckley, S. A., and Stanker, L. H. (2001) Determination of Fluoroquinolones in Serum using an on-Line Clean-Up Column Coupled to High-Performance Immunoaffinity-Reversed-Phase Liquid Chromatography. *J. Chromatogr B.* **754**, 1–9.
151. Nuntaprasert, A., Mori, Y., Fujita, K., Yoneda, M., Miura, R., Tsukiyama-Kohara, K., and Kai, C. (2005) Expression and Purification of Recombinant Swine Interleukin-4. *Comp. Immunol. Microbiol. Infect. Dis.* **28**, 17–35.
152. Wannemacher, K. M., Terskiy, A., Bian, S., Yadav, P. N., Li, H., and Howells, R. D. (2008) Purification and Mass Spectrometric Analysis of the κ Opioid Receptor. *Brain Res.* **1230**, 13–26.
153. Camponova, P., Baud, S., Mattras, H., Duroux-Richard, I., Bonnafous, J., and Marie, J. (2007) High-Level Expression and Purification of the Human Bradykinin B2 Receptor in a Tetracycline-Inducible Stable HEK293S Cell Line. *Protein Expr. Purif.* **55**, 300–311.

Chapter 4

Avoiding Proteolysis During Protein Chromatography

Barry J. Ryan

Abstract

All cells contain proteases, which effect catalytic hydrolysis of the peptide bond between amino acids in the protein backbone. Typically, proteinases are prevented from nonspecific proteolysis by regulation and physical separation into different subcellular compartments; however, this segregation is not retained during cell lysis to release a protein of interest. Prevention of proteolysis during protein purification often takes the form of a two-pronged approach; first the inhibition of proteolysis in situ, followed by the separation of the protease from the protein of interest via chromatographical purification. Proteinase inhibitors are routinely used to limit the effect of the proteinases before they are physically separated from the protein of interest via column chromatography. Here, commonly used approaches to reducing proteolysis during chromatography are reviewed.

Key words: Protease, Proteolysis, Proteinase inhibitor buffer, Protein purification

1. Introduction

Protein stability can be defined as "the persistence of molecular integrity or biological function despite adverse influences or conditions, such as heat or other deleterious conditions" (1). One of the key deleterious conditions during protein chromatography is the presence of proteolytic substances, often referred to as proteases. Proteolysis is the directed degradation of proteins by specific proteases. Proteases have been referred to as "*Nature's Swiss Army knife*" due to their diverse applications in protein cleavage (2). Proteases belong to the hydrolase class of enzyme (Enzyme Classification 3.4), which catalyse the hydrolysis of various bonds with the participation of a water molecule. The proteolytic process involves the hydrolysis of the peptide bonds that link amino acids together in the polypeptide chain. Proteases are defined as either exopeptidases (detach the terminal amino

Dermot Walls and Sinéad T. Loughran (eds.), *Protein Chromatography: Methods and Protocols*, Methods in Molecular Biology, vol. 681, DOI 10.1007/978-1-60761-913-0_4, © Springer Science+Business Media, LLC 2011

acids from the protein chain, examples include aminopeptidases and carboxypeptidase) or endopeptidases (which target internal peptide bonds of a protein, common examples here include trypsin, chymotrypsin, pepsin, and papain) (3). Proteases are also divided into four major groups according to their catalytic active site and mode of action: serine proteinases, cysteine (thiol) proteinases, aspartic proteinases, and metalloproteinases (4).

Serine proteases, as the name suggests, have a serine residue as part of its catalytic site. Subtilisin (EC 3.4.21.62, an endopeptidase sourced from *Bacillus subtilis*) is one of the most common serine proteases examples cited (5). Cysteine proteases have a nucleophilic cysteine thiol as part of their active site. Papain (EC 3.4.22.2, an endopeptidase sourced from *Carica papaya*) is a frequently cited example of a cysteine protease (6). Aspartic proteinases use an aspartate residue for catalysis and, in general, they have two highly conserved aspartate residues in the active site and are optimally active at acidic pH. Plasmepsin (EC 3.4.23.39, an endopeptidase produced by the *Plasmodium* parasite) is an example of an aspartic proteinase (7). Metalloproteinases contain a catalytic mechanism involving a metal; most contain zinc, however, cobalt centers are also observed. Adamalysin (EC 3.4.24.46, an endopeptidase from the rattlesnake *Crotalus adamanteus*) is an example of a metalloproteinase (8).

Proteases are employed by all living cells to maintain a particular rate of protein turnover by continuous degradation and synthesis of proteins. Catabolism of proteins provides a ready pool of amino acids that can be reused as precursors for protein synthesis. Intracellular proteases participate in executing correct protein turnover for the cell: in *Escherichia coli*, the ATP-dependent protease La, the *lon* gene product, is responsible for hydrolysis of abnormal proteins (9). The turnover of intracellular proteins in eukaryotes is also affected by a pathway involving ATP-dependent proteases (10).

As such, proteases are essential components in all life forms, and in normal circumstances proteases are typically packaged into specialized organelles to minimize the chance of nonspecific proteolytic activity. Within these organelles, there are specific regulators associated with each protease, controlling the action of the protease. However, when cells are disrupted for chromatography purification, proteases that are normally located in a different subcellular compartment are separated from their regulator molecules and exposed to the protein of interest, thus increasing the probability of undesired proteolysis (11). Realistically, it is impossible to remove all proteinases that present in a chromatography sample preparation; however, careful selection of host cell (if protein of choice is recombinantly expressed) or cell type (if the protein of choice is native) in conjunction with specific sample preparation protocols can reduce unwanted proteolysis during purification (3).

Proteases are ubiquitous and play a crucial role in normal and abnormal physiological conditions in all living things by effecting catalysis throughout many metabolic pathways. However, there is an uneven distribution of proteinases depending on which cell type (bacterial or eukaryotic) or tissue is disrupted. During heterologous protein expression, the recombinant protein of interest may be exposed to a host proteinase to which it is particularly susceptible. Simply altering the host may reduce recombinant proteolysis. Many commercial companies offer protease-deficient strains for heterologous protein expression; for example *E. coli* BL21 is deficient in two proteases encoded by the *lon* (cytoplasmic protease) and *ompT* (periplasmic protease) genes (see Table 1). Additionally in mammalian tissues, liver and kidney samples contain a much higher concentration of proteolytic enzymes compared with skeletal or cardiac muscle (12). Once the source of the protein of choice has been optimized, a commonly used approach toward the prevention of further unwanted proteolysis

Table 1
Some commercially available protease-deficient *E. coli* strains that are used to express recombinant proteins

Strain name	Protease deficiency	Supplier
UT5600	Deficient in *OmpT* (an outer membrane protease that cleaves between sequential basic amino acids)	New England Biolabs Inc.
CAG626	Deficient in Lon (a protease that degrades abnormal/misfolded proteins)	New England Biolabs Inc.
CAG597	Stress-induced proteases at high temperature	New England Biolabs Inc.
CAG629	Stress-induced proteases at high temperature and Lon protease	New England Biolabs Inc.
PR1031	Deficient in DnaJ – a chaperone that can promote protein degradation	New England Biolabs Inc.
KS1000	Deficient in Prc (Tsp), a periplasmic protease	New England Biolabs Inc.
Rosetta	Deficient in Lon and OmpT	Novagen
Rosetta-gami B	Deficient in Lon and OmpT	Novagen
Origami B	Deficient in Lon and OmpT	Novagen
BL21 Star (DE3) pLysS	Deficient in Lon and OmpT	Invitrogen
BL21 Star (DE3)	Deficient in Lon and OmpT	Invitrogen
BL21-AI	Deficient in Lon and OmpT	Invitrogen

during protein isolation is to include proteinase inhibitors during sample preparation, purification, and characterization.

1.1. Proteinase Inhibitor Selection and Preparation

Judicious inhibitor choice will depend on the correct empirical identification of the proteinase involved. Classification of the proteinase(s) can be carried out in several ways, however, the simplest method is to incubate the sample of choice with a single inhibitor from the group of inhibitors (Serine, Cysteine, Thiol, etc.) listed in Table 2. The degree of proteolysis can be identified from sodium dodecyl sulfate–polyacrylamide gel electrophoresis (SDS–PAGE) analysis of the protein sample postinhibitor incubation; increased protein band smearing on the gel or a change in expected protein size will indicate potential proteolysis. Proteolysis inhibition, indicated by a maintenance of correct protein size with no protein band smearing after a given incubation period with inhibitor, will permit the identification of a suitable inhibitor group for the sample preparation. Once the protease has been identified, individual inhibitors can be chosen from Table 2 or a typical *general-use* proteinase inhibitor mix can be prepared immediately before use from the stock concentrations outlined in Table 3. Proteinase inhibitor solutions must be correctly stored after they have been prepared. Aliquot the stock of inhibitor and store at the correct temperature (see Table 2) to maintain the properties of the inhibitor. Make small, single use aliquots to reduce the risk

Table 2
Protease inhibitors: stock solutions and storage conditions

Inhibitor activity	Inhibitor	Solvent	Molarity	Storage
Serine	PMSF[a]	Dry methanol or propanol	200 mM	-20°C
Serine	3,4-DCL	Dimethylsulfoxide	10 mM	-20°C
Serine	Benzamidine	Water	100 mM	-20°C
Cysteine	Iodoacetic acid	Water	200 mM	Prepare fresh
Cysteine	E64-c	Water	5 mM	-20°C
Thiol (serine and cysteine)	Leupeptin	Water	10 mM	-20°C
Metallo	1,10-Phenanthroline	Methanol	100 mM	RT or 4°C
Metallo	EDTA[b]	Water	0.5 M	RT or 4°C
Acid proteases	Pepstatin	DMSO	10 mM	-20°C
Aminopeptidase	Bestatin	Water	5 mM	-20°C

RT room temperature
[a]PMSF is toxic. Weigh this compound in a fume hood, and wear appropriate personal protective equipment
[b]Does not inhibit pancreatic elastase

Table 3
General proteinase inhibitor mix

Stock inhibitor	Volume (μL)
PMSF (100 mM) or 3,4-DCI (10 mM) or benzamidine (5 mM)	200
Iodoacetate (200 mM) or E64-c (5 mM)	200
1,10-Phenanthroline (100 mM) or EDTA (500 mM) or leupeptin (10 mM)	100
Pepstatin (10 mM)	100
Double distilled water	400
Final volume	1,000

of stock contamination. Ensure that the proteinase inhibitor/inhibitor mix is combined with the cell sample prior to cell disruption. If the individual proteinase inhibitor/inhibitor mix is to be prepared fresh, then it must be used within 1 h of preparation.

It should be noted that the generic proteinase inhibitor cocktail outlined here is not guaranteed to work in all circumstances. The success of any mix will depend on the correct empirical identification of the proteinase involved.

1.2. Commercially Available Universal Protease Inhibitor Mixes

There are several types of commercially available "Universal Proteinase Inhibitors" that may also be used (e.g. *Complete Protease Inhibitor Cocktail Tablets*, Roche Applied Science). Additionally, many companies offer inhibitor panels, such as the *Protease Inhibitor Panel* (Sigma-Aldrich), which is a cost-effective method for personalized proteinase cocktail inhibitor generation (13).

1.3. Supplementary Protease Inhibitor Components

1.3.1. Additional Inhibitors

If a particular protease is dominant within a sample preparation, the cocktail mix may be supplemented with additional specific proteinase inhibitors (12, 14–21). Commonly used specific individual protease inhibitor components are outlined in Table 4.

Phosphatase inhibitors may be required also as many enzymes are activated by phosphorylation, hence dephosphorylation must be inhibited if enzyme activity is to be maintained. Again, an empirical approach is required to identify if a phosphatase inhibitor is required (see Subheading 1.1 and Table 5). Protein phosphatases can be divided into two main groups: protein tyrosine phosphatases and protein serine/threonine phosphatases, which remove phosphate from proteins (or peptides) containing phosphotyrosine or phosphoserine/phosphothreonine, respectively (22). Inhibitors commonly used here include: *p*-bromotetramisole,

Table 4
Additional inhibitors that can be used to supplement protease inhibitor mixes

Inhibitor	Solvent	Molarity	Storage
Serine protease inhibitors			
Aprotinin (does not inhibit thrombin or factor Xa)	Water	300 mM	-20°C (at pH 7)
Chymostatin (inhibits chymotrypsin-like serine proteases such as chymase cathepsins A, B, D, and G. Also inhibits some cysteine proteases such as papain)	DMSO	10 mM	-20°C
Antithrombin III (inhibits thrombin, kallikreins, plasmin, trypsin, and factors Ixa, Xa, and Xia)	Water	10 U/mL	-20°C (at pH 7)
TLCK (inhibits chymotrypsin-like serine proteases)	1 mM HCl	100 μM	Prepare fresh
TPCK (inhibits chymotrypsin-like serine proteases)	Ethanol	10 mM	4°C
DIFP (highly toxic cholinesterase inhibitor. Broad spectrum serine protease inhibitor. Hydrolyzes rapidly in aqueous solutions)	Anhydrous isopropanol	200 mM	-20°C
Antipain (inhibits serine proteases such as plasmin, thrombin and trypsin. Also inhibits some cysteine proteases such as calpain and papain)	Water	10 mM	-20°C
α2-Macroglobulin (broad spectrum protease inhibitor)	Water	100 mM	-20°C
Cysteine protease inhibitors			
N-ethylmaleimide	Water	100 mM	Prepare fresh
Metalloproteinase inhibitors			
Phosphoramidon (strong inhibitor of metall oendoproteinases, thermolysin, and elastases, but a week inhibitor of collagenase)	Water	1 mM	-20°C

cantharidin, microcystin LR (Ser/Thr protein phosphatases and alkaline phosphatase L-isozymes) and imidazole, sodium molybdate, sodium orthovanadate, sodium tartrate (Tyr protein phosphatases and acid and alkaline phosphatases, see Table 5). There are also a number of commercially available Phosphatase Inhibitor Mastermixes (e.g. PhosphataseArrest™ Phosphatase Inhibitor Cocktail, Geno Technologies Ltd). These are often supplied in convenient, ready-to-use 100× solutions that are simply added to the protein extraction buffer or individual samples.

Table 5
Commonly used phosphatase inhibitors

Name	Typical working molarity range	Stock molarity	Typical inhibitory targets
p-Bromotetramisole	0.1–1.5 mM	100 mM	Alkaline phosphatases (25, 26)
Cantharidin	20–250 μM	2.5 mM	Protein phosphatase 2-A (25, 27)
Microcystin LR	20–250 nM	2.5 μM	Protein phosphatase 1 and 2-A (25, 28)
Imidazole	50–200 mM	1 M	Alkaline phosphatases (29, 30)
Sodium molybdate	50–125 mM	1 M	Acid phosphatases and phosphoprotein phosphatases (27, 30)
Sodium orthovanadate	50–100 mM	1 M	ATPase inhibition, protein tyrosine phosphatases, phosphate-transferring enzymes (30, 31)
Sodium tartrate	50–100 mM	1 M	Acid phosphatases (28, 30).

These mixes can be sourced as either broad spectrum phosphatase inhibitor cocktails or as phosphatase inhibitors for targeting particular set of phosphatases.

1.4. Supplementary Chemical Compounds Including Enzymes

The addition of supplementary chemical components to disrupt proteinase activity first should be carefully assessed on a small scale as often these components will alter the function/stability of the target protein (see Table 6). Furthermore, additional protease inhibitors should be introduced to the sample with caution as protein modifications, such as alteration of protein charge, may occur. These alterations may interfere with further protein characterization studies. For example, 2-mercaptoethanol will reduce the activity of cysteine proteinases, but will also unfold target proteins containing disulphide bridges. EDTA is included in many proteinase inhibitor buffers as metal ions are frequently involved in proteolysis, thus their removal will impede proteolysis. However, if one is purifying poly-Histidine-tagged proteins or metalloproteins, then the chelating effect of EDTA will dramatically alter purification yields, and the EDTA should be removed from the buffer by dialysis or a buffer exchange resin. Inclusion of 2 M thiourea may also prevent proteolyis. Castellanos-Serra and Paz-Lago (23) noted the proteolysis inhibitory effects of its addition in conjunction with its efficiency in solubilizing proteins. DNase (100 U/mL), although not itself a protease inhibitor, can be included in the cell lysis buffer as this will reduce the overall viscosity of the crude lysate. The reaction is allowed to proceed for 10 min at 4°C in the presence of 10 mM $MgCl_2$.

Table 6
Supplemental chemical/enzyme additions to protease inhibitor buffer (32)

Item and typical working concentration	Advantages	Disadvantages	Uses/typical protease targets
2-Mercaptoethanol (1 mM)	Reduction cysteine proteinase activity	Unfolding of target proteins containing disulphide bridges	Cysteine proteases
EDTA (5 mM)	Removal of metal ions involved in proteolysis impeding proteolysis	The chelating effect of EDTA will affect the structure of metalloproteins and dramatically reduce the purification of poly-Histidine-tagged proteins	Non-His tagged protein targets or nonmetallo-protein targets
Thiourea (2 M)	Proteolysis inhibitory effects, in conjunction with improved protein solubilization	Thiourea is considered a possible human carcinogen and mutagen	General purpose protease inhibitor
DNase (100 U/mL)	Reduction in the crude lysate viscosity	Requires further incubation step of 10 min at 4°C in the presence of 10 mM $MgCl_2$	Can be included in the cell lysis buffer for optimal efficiency

1.5. Protease Inhibition During Chromatography

The introduction of contaminating proteases from your own skin, nonsterile water etc. can be avoided by sterilizing all plasticware and by wearing appropriate personal protective equipment. All buffers should be sterile filtered (0.2 μm) into autoclaved bottles (sterile filtering will not remove contaminating proteases, but will remove any protease secreting microorganisms). Additionally, sterile filters the protein eluate, once purification is complete.

Cell disruption, as with all other parts of the purification procedure, should take place at 2–8°C. This temperature will not only reduce the activity of proteinases, but also aid in stabilizing the target protein (reduction in thermal denaturation). Kulakowska-Bodzon et al. (24) provide an excellent review on protein preparation from various cell types for proteomic work. In general, all buffers and materials should be prechilled to 2–8°C. Rapid purification at this lower temperature will reduce the risk of unwanted proteolysis. Do not store such samples at 2–8°C for more than 1 day between purification steps, instead store at −20°C.

Gel filtration (size exclusion chromatography, see Chapter 2) is often used as the final step in protein purification and it can be used to desalt and buffer exchange the protein (therefore no need for dialysis). Contaminating proteinases can also be separated from the protein of choice if there is significant separation between elution peaks for the protease and the protein of choice. This is based on the presumption that there is a considerable difference between the size of the protease and the size of the protein of interest.

1.6. Postchromatography Analysis

Protease inhibition can be either reversible or irreversible. The majority of serine and cysteine proteinase inhibitors are irreversible, whereas the aspartic and metalloproteinase inhibitors are reversible. Even when the inhibitors are added at an early stage, they may be lost during purification and subsequent handling steps, resulting in proteolysis postpurification. The further readdition of proteinase inhibitors may therefore be required after purification.

Even with increased numbers of purification steps, very few protocols will remove all contaminants from a sample preparation; however, one can achieve an adequate reduction in the level of these contaminants. Each purification protocol will have a unique definition of "*adequate protease reduction*" based on a number of variables including the activity of the remaining proteases, further downstream applications of the protein of choice and the cost of further protease removal. Increased purification steps often result in a reduced final yield, as such the trade-off between contaminant reduction and yield must be optimized. A pure protein that gives a single band on a Coomassie-stained SDS–PAGE gel should be reanalyzed over time to ensure minimal proteinase activity exists in the sample. This may be carried out by simply storing an aliquot of the purified protein solution at room temperature and analyzing samples of this by SDS–PAGE at regular intervals. If the protein is being degraded (indicated by a smear or a reduced size of the protein of choice), proteinase contamination is present and an additional purification step (or supplemental inhibitor addition) is required.

Some purification protocols require the addition of specific proteases such as *enterokinase* (recognition site D-D-D-K) or *TEV protease* (recognition site E-N-L-Y-F-Q-G) to remove polypeptide tags from recombinant proteins. Ensure that any proteinase inhibitor containing buffer is exchanged, by dialysis or a suitable buffer exchange resin, prior to the addition of the desired proteinase (see also Chapter 19).

Care must be taken to rule out the possible loss of enzyme activity due to other destabilizing factors during protein purification. These other factors include, but are not limited to, thermal denaturation, oxidative damage, and column surface adherence. Thermal denaturation of proteins is the decreased stability of a protein caused by extremes of temperature experienced by the protein of interest. Thermal denaturation can be reduced if the purification procedure is carried out at 2–8°C, all buffers and chromatography columns/resins are prechilled to 2–8°C and if the purified protein is stored at the correct temperature postpurification. Oxidative damage to proteins can be divided into a number of sections; however, improper disulfide formation is the most pertinent here. Thiol oxidation is crucial for correct protein folding, resulting in a stabilized 3-D protein structure in proteins containing disulphide bridges. The formation of incorrect intra- or intermolecular disulfides is a detrimental process that can often result in loss of activity and/or aggregation. Thiol oxidative damage can be avoided by not exposing the protein of choice to thiol reducing compounds (e.g. β-mercaptoethanol) during purification, thus maintaining the correct folded state of the protein. Column surface adherence is caused by the attraction of the protein of choice to the surface of the purification column by the protein's intrinsic physico-chemical properties (e.g. surface charge or hydrophobicity). Nonspecific protein adherence can cause sheer stress damage to the protein during purification; however, this can be circumvented by careful selection of the purification column (type/grade of glass or plastic) and purification resin.

2. Conclusion

The presence of proteolytic enzymes can result in target protein degradation during protein chromatography. Careful selection of source organism/tissue, along with judicious use of protease inhibitors, can reduce these degrading effects. Commonly used inhibitors are listed here in tabular format (see Tables 3 and 4), along with supplemental compounds (see Tables 5 and 6) for easy selection. Protease inhibitors can be added individually or as part of a mix, however, optimal inhibitor selection is an empirical process.

References

1. O'Fágáin, C. (1997) Protein stability and its measurement, in *Stabilising Protein Function* (O'Fágáin, C., ed.), Springer Press, Berlin, pp. 1–14.
2. Seife, C. (1997) Blunting nature's Swiss army knife. *Science* **277**, 1602–1603.
3. Sandhya, C., Sumantha, A., and Pandey, A. (2004) Proteases, in *Enzyme Technology* (Pandey, A., Webb, C., Soccol, C.R., and Larroche, C., eds.), Asiatech Publishers Inc., New Delhi, India, pp. 312–325.
4. Rawlings, N.D., Morton, F.R., Kok, C.Y., Kong, J., and Barrett, A.J. (2008) *MEROPS*: the peptidase database. *Nucleic Acids Res.* **36**, D320–D325.
5. Rawlings, N.D. and Barrett, A.J. (1994) Families of serine peptidases. *Methods Enzymol.* **244**, 19–61.
6. Bühling, F., Fengler, A., Brandt, W., Welte, T., Ansorge, S., and Nägler, D.K. (2000) Review: novel cysteine proteases of the papain family. *Adv. Exp. Med. Biol.* **477**, 241–254.
7. Dame, J.B., Reddy, G.R., Yowell, C.A., Dunn, B.M., Kay, J., and Berry, C. (1994) Sequence, expression and modeled structure of an aspartic proteinase from the human malaria parasite

Plasmodium falciparum. *Mol. Biochem. Parasitol.* **64**, 177–190.

8. Edwards, D.R., Handsley, M.M., and Pennington, C.J. (2008) The ADAM metalloproteinases. *Mol. Aspects Med.* **29**, 258–289.
9. Chung, C.H. and Goldberg, A.L. (1981) The product of the lon (capR) gene in *Escherichia coli* is the ATP-dependent protease, protease La. *Proc. Natl. Acad. Sci. USA* **78**, 4931–4935.
10. Hershko, A., Leshinsky, E., Ganoth, D., and Heller, H. (1984) ATP-dependent degradation of ubiquitin-protein conjugates. *Proc. Natl. Acad. Sci. USA* **81**, 1619–1623.
11. Vanaman, T.C. and Bradshaw, R.A. (1999) Proteases in cellular regulation. *J. Biol. Chem.* **274**, 20047.
12. Beynon, R.J. and Oliver, S. (2004) Avoidance of proteolysis in extracts, in *Protein Purification Protocols, Methods in Molecular Biology* (Cutler, P., ed.), Humana, Totowa, NJ, **244**, pp. 75–85.
13. http://www.sigmaaldrich.com/life-science/metabolomics/enzyme-explorer/learning-center/protease-inhibitors.html.
14. Beynon, R.J. (1998) Prevention of unwanted proteolysis, in *Methods in Molecular Biology: New Protein Techniques* (Walker, J.M., ed.), Humana, Totowa, NJ, **3**, pp. 1–23.
15. Frank, M.B. (1997) "Notes on Protease Inhibitors" from a Bionet Newsgroup described in *Molecular Biology Protocols* (http://omrf.ouhsc.edu/~frank/protease.html).
16. Harper, J.W., Hemmi, K., and Powers, J.C. (1985) Reaction of serine proteases with substituted isocoumarins: discovery of 3,4-dichloroisocoumarin, a new general mechanism based serine protease inhibitor. *Biochemistry* **24**, 1831–1841.
17. Hassel, M., Klenk, G., and Frohme, M. (1996) Prevention of unwanted proteolysis during extraction of proteins from protease-rich tissue. *Anal. Biochem.* **242**, 274–275.
18. North, M.J. and Benyon, R.J. (1994) Prevention of unwanted proteolysis, in *Proteolytic Enzymes: A Practical Approach* (Beynon, R.J. and Bond, J.S., eds.), Oxford University Press, New York, USA, pp. 241–249.
19. Sreedharan, S.K., Verma, C., Caves, L.S.D., Brocklehurst, S.M., Gharbia, S.E., Shah, H.N., and Brocklehurst, K.M. (1996) Demonstration that 1-trans-epoxysuccinyl-L-leucylamido-(4-guanidino)butane (E-64) is one of the most effective low Mr inhibitors of trypsin-catalysed hydrolysis. Characterization by kinetic analysis and by energy minimization and molecular dynamics simulation of the E-64–b-trypsin complex. *Biochem. J.* **316**, 777–786.
20. Salvensen, G. and Nagase, H. (1989) Inhibition of proteolytic enzymes, in *Proteolytic Enzymes: A Practical Approach* (Beynon, R.J. and Bond, J.S., eds.), Oxford University Press, New York, USA, pp. 83–104.
21. North, M.J. (1989) Prevention of unwanted proteolysis, in *Proteolytic Enzymes: A Practical Approach* (Beynon, R.J. and Bond, J.S., eds.), IRL Press, Oxford, pp. 105–124.
22. Barford, D. (1996) Molecular mechanisms of the protein serine/threonine phosphatases. *Trends Biochem. Sci.* **21**, 407.
23. Castellanos-Serra, L. and Paz-Lago, D. (2002) Inhibition of unwanted proteolysis during sample preparation: evaluation of its efficiency in challenge experiments. *Electrophoresis* **23**, 1745–1753.
24. Kulakowska-Bodzon, A., Bierczynska-Krzysik, A., Dylag, T., Drabik, A., Suder, P., Noga, M., Jarzebinska, J., and Silberring, J. (2007) Methods for sample preparation in proteomic research. *J. Chromatogr. B* **849**, 1–31.
25. Pendyala, P.R., Ayong, L., Eatrides, J., Schreiber, M., Pham, C., Chakrabarti, R., Fidock, D., Allen, C.M., and Chakrabarti, D. (2008) Characterization of a PRL protein tyrosine phosphatase from *Plasmodium falciparum*. *Mol. Biochem. Parasitol.* **158**, 1–10.
26. Kuwana, T. and Rosalki, S.B. (1991) Measurement of alkaline phosphatase of intestinal origin in plasma by *p*-bromotetramisole inhibition. *J. Clin. Pathol.* **44**, 236–237.
27. Jain, M.K. (1982) *Handbook of Enzyme Inhibitors*, John Wiley and Sons, New York, p. 222.
28. Jain, M.K. (1982) *Handbook of Enzyme Inhibitors*, John Wiley and Sons, New York, p. 334.
29. Jain, M.K. (1982) *Handbook of Enzyme Inhibitors*, John Wiley and Sons, New York, pp. 189–190.
30. http://www.emdbiosciences.com/html/cbc/Phosphatase_Inhibitor_Cocktail_Sets.htm.
31. Gordon, J.A. (1991) Use of vanadate as protein-phosphotyrosine phosphatase inhibitor. *Methods Enzymol.* **201**, 477–482.
32. Bodzon-Kulakowska, A., Bierczynska-Krzysik, A., Dylag, T., Drabik, A., Suder, P., Noga, M., Jarzebinska, J., and Silberring, J. (2007) Methods for samples preparation in proteomic research. *J. Chromatogr. B* **849**, 1–31.

Chapter 5

Scale-Up of Protein Purification: Downstream Processing Issues

John Joseph Milne

Abstract

Large-scale chromatography operations continue to occupy a key position in the overall strategy for the downstream processing and purification of protein products for therapeutic use. Increasing product titres from mammalian cell culture has resulted in a trend to identify ways of improving the economics of product recovery and purification processes. In commercial manufacturing, a requirement exists to increase the scale of the chromatography operations, which are typically developed and optimised in small-scale experiments. This short review discusses the key factors in the chromatography process that need to be considered as the scale of the purification step is increased in order to maintain the purity and integrity of the product purified at smaller scale.

Key words: Protein purification, Downstream processing, Chromatography, Scale-up

1. Introduction

The market for biopharmaceutical products represents an increasingly significant proportion of total pharmaceutical annual sales resulting mainly from the considerable interest in monoclonal antibodies as therapeutic agents (1). The system of choice for the production of complex recombinant proteins remains the mammalian cell with expression in Chinese Hamster Ovary (CHO) cells grown in suspension culture being the most popular choice to date for manufacturers (2, 3). Expression of recombinant proteins in mammalian cells offers advantages in terms of protein folding and post-translational modifications when compared with other systems (4). Tremendous developments have taken place in mammalian cell culture in the areas of cell-line development, new media formulations and bioreactor design (5) which has led to a

Dermot Walls and Sinéad T. Loughran (eds.), *Protein Chromatography: Methods and Protocols*, Methods in Molecular Biology, vol. 681, DOI 10.1007/978-1-60761-913-0_5, © Springer Science+Business Media, LLC 2011

concomitant increase in cell titres with industrial claims of product titres in the range 5–10 g/L. Due to these advances in industrial cell-culture processes, the focus for biopharmaceutical manufacturers has shifted to those unit operations downstream of the bioreactor where new innovations in process design will be required in order to improve process capacity, speed, and economics (6). Downstream operations are typically carried out at a scale that is dependent on the amount of product required and in order to avoid increased capital investment and optimise facility usage, process improvements, and new innovations will be required in the future (7).

Production of a licenced biopharmaceutical therapeutic product involves several disciplines with downstream processing occupying a central position in the overall process of ensuring the ultimate safety of the target product. Downstream processing can be defined as a series of separation and purification activities that together can produce a protein product fit for its intended use. Protein purification using chromatography is currently the central enabling technology in all biopharmaceutical downstream processing (8). Large-scale process chromatography has played a key role in the purification of therapeutic proteins and although alternative techniques such as the use of membrane adsorption are currently receiving attention preparative chromatography will most likely remain the primary purification platform within the biopharmaceutical industry (7, 9). However, due to the increased volumes of cell-culture harvests (>10,000 L) and associated high levels of protein expression (>10 g/L) that are predicted to occur in the near future, increased productivity will be required to alleviate the real bottleneck in downstream processing that will exist over the next few years (10, 11).

1.1. Protein Purification Using Chromatography

Following cell-culture expression, the purification of proteins proceeds through a series of sequential unit operations downstream of the bioreactor to achieve solid–liquid fractionation after harvesting. One such operation is chromatography in which a sample dissolved in a mobile phase is passed through a stationary phase held in a column shell. In preparative chromatography, the stationary phase is selected on the basis of its ability to purify the target molecule from a panel of contaminating proteins. In the case of non-affinity-based capture chromatography, product concentration using ultrafiltration and diafiltration is used to condition the feed material produced in the bioreactor prior to separation. The purification phase of protein manufacture is designed to specifically take the target molecule from the upstream stage to the final formulation stage, achieving the required level of purity and recovery and to do so safely, reliably, and economically. In order to achieve a product licence, regulatory agencies require that process characterisation, risk mitigation, and validation are carried out prior to market approval. Each of these disciplines presents distinct challenges and requires a thorough knowledge of manufacturing capabilities (12).

Within industry, the primary method of protein purification currently used is process chromatography which is typically performed on a large scale. Such chromatography must be robust and reproducible to be capable of dealing with the variations inherent in the use of cell-based protein expression systems. When chromatography is selected as a method of purification, process development will typically involve distinct phases of bench-scale development, optimisation of chromatographic parameters, and scale-up to the required final scale. Scale-up of chromatography presents unique challenges in terms of both the media selected and equipment used and is a critical issue for the successful commercialisation of biopharmaceuticals (13).

When practitioners set out to design a purification process, the basis of each chromatography step in the overall scheme should be assessed in small-scale laboratory experiments. Purification schemes for currently licenced products typically involve at least three orthogonal chromatography steps, each offering a different selectivity and exploiting alternative attributes of the target product and/or contaminants in order to maximise the overall purity of the product (14). It is important that as a purification process is being developed at small scale, reference is made to the predicted final scale of the process, such that specific issues relating to the scale-up can be fully appraised. This is particularly important, if the protein is to be purified according to the guidelines of current good manufacturing practice (GMP) as defined by regulatory authorities (15).

Chromatography is generally regarded as being composed of three distinct phases namely capture, intermediate purification, and final polishing. Each of these phases requires a different focus and different challenges will be presented as the process is scaled-up. The initial capture step can involve protein isolation from crude feedstock and thus requires a medium with a high capacity and high operating flow rate. The binding capacity of the medium for the target protein in the presence of many impurities, rather than resolution, is the most critical factor. As process streams can be large, fast flow resins possessing a large bead size range are used to improve overall processing times and thus stabilise the protein product quickly. Economic constraints generally require the resin to be recycled for a validated number of uses when large-scale capture applications are involved in which the volume of medium is large. Due to the crude nature of the starting material, the ability to clean and sanitise media effectively is an important issue to be considered. Resolution of the target protein from host-cell contaminants becomes more important in intermediate purification applications. To ensure productivity and process economics, the binding capacity of the medium for the target protein is also important. Since bead size correlates with resolution, smaller bead sizes are often more appropriate at this stage of purification. Speed is usually less critical in intermediate purification applications since impurities affecting stability of the target will have been removed in a successfully optimised capture step. In polishing chromatography, the main issue is resolution, to ensure removal of trace contaminants and also structural variants of the target that can lead to immunogenicity concerns following therapeutic administration (16). High efficiency media with small bead size are used in final polishing chromatography. As the use of a smaller bead size will result in increased pressure, lower flow rates are used and process times can be longer. In overall terms, the performance criteria to be considered in developing an optimised process are speed, recovery, capacity, and resolution. The priority given to these attributes will be dependent on the nature of the chromatographic step and in particular whether the purification step is operating in capture, intermediate, or polishing mode.

1.2. Scale-Up of Protein Chromatography

Following small-scale scouting trials, a variety of media will have been screened in order to select a suitable medium that can provide a basis for separation of the target protein from impurities in the sample feed material. Once a medium has been defined, the next stage in the process is to design specific chromatographic procedures that can optimise the dual requirements of product recovery and purity, by defining the buffer solutions used during sample conditioning, media equilibration, and product elution. Typically, at this stage, the purification process will be capable of producing milligram quantities of protein. The main result of process optimisation is that product information and process parameters will have been considered in detail and where possible defined to ensure smooth technical transfer when the process is ultimately scaled-up (see Table 1). When the running conditions and chromatography parameters have been determined for any column purification step, the final issue to resolve is how the chromatographic process can be scaled appropriately to a level required for required large-scale production (17). Scaling from laboratory to pilot plant can involve scale-up factors of 50- to 100-fold, while further increasing from pilot plant to commercial production scale will involve another 10- to 50-fold scale-up (18). Scale-up of chromatography should involve a detailed consideration of chromatographic, non-chromatographic, and equipment-related issues, and the role of these factors can play in successful scale-up.

An increase in the scale of chromatographic purification is most typically achieved by increasing the column diameter and volumetric flow rate, while at the same time ensuring that the media bed height and linear flow rate remain constant (19)

Table 1
Parameters to be defined during the optimisation phase of development

Process parameters	Sample pre-conditioning required Sample concentration (load/ml resin) Sample volume (load/ml resin) Product concentration (load/ml resin) Resin bed height (cm) Linear flow rates (cm/h) Process volumes Pressure drop Packed column qualification Buffer conditions (pH, conductivity) Fractionation scheme if used Stability of resin to cleaning and sanitisation agents
Product parameters	Stability Solubility in buffers Storage conditions Storage time

Table 2
Status of various process parameters during chromatography scale-up

Remain constant	Column bed height Linear flow rates (cm/h) at all stages Sample attributes (concentration and conditioning) Ratio of sample volume to media volume Ratio of gradient volume to media volume Buffer specifications
Increased	Column diameter Volumetric flow rate (ml/min) Sample volume proportionally Gradient volume proportionally Buffer volumes proportionally
Issues to be addressed	Decreased wall support as column diameter increases Quality of packing/bed instability Increased pressure drop Uneven flow distribution Zone broadening

(see Table 2). This ensures that the overall residence time of the target molecule on the small-scale and large-scale columns remains constant. In theory, the respective separations should be very similar and therefore provide the basis for successful scale-up.

However, as the column scale increases other factors such as choice of buffers and their preparation, media packing, column engineering, and process hygiene can present limitations and often problems particularly at the scale of manufacturing required in commercial biopharmaceutical manufacturing (20). Thus proper scale-up is a crucial consideration in process design involving the optimisation of many factors in order to ensure robust and reproducible purification of a protein therapeutic over the lifetime of the product licence. The remainder of this review will consider some of the main issues in chromatography scale-up and the approaches taken by industry to provide acceptable solutions.

2. Downstream Processing Issues

Preserving the quality of resolution and the level of purification determined at laboratory scale is the key to successful chromatography scale-up (21). Whenever a requirement exists to increase the scale of a chromatography process from small-scale scouting experiments through pilot scale and on to full commercial manufacturing adhering to the conditions of GMP, a series of key issues need to be considered to determine their effect on the overall scale-up process. Such issues can be categorised into those relating to the choice of medium itself and to the chromatographic separation process. In commercial manufacture, equipment and hardware issues can often present challenges due to the need to use large columns and automated systems.

2.1. Chromatography Media

In protein purification applications where chromatography is utilised, it is the physical and chemical properties of the medium used in the process that are the main route to achieving the required purity. The rational selection of chromatography media prior to screening experiments is critical for successful downstream processing for industrial chromatography applications (21). Modern processes require media that are not only consistent and selective but also offer high-dynamic capacities and low-cycle times to be compatible with the increasing large volumes and high expression levels of current cell-culture processes. The main considerations when choosing a medium for inclusion in a production process for therapeutic products are a route to scalable and robust processing, while at the same time ensuring that procedures can be validated. As the volume of medium required increases the available commercial supply of media with demonstrated lot-to-lot consistency and full regulatory support documentation is mandatory. Such issues should be noted prior to early small-scale trials and if not resolved might dictate that a different media should be considered for use. There are many commercial suppliers who offer a large portfolio of media with differing functionalities that are compatible with the requirements of industry and regulators alike.

Media and capital costs at large scale should not be underestimated, and cost analysis is frequently an important component when considering an approach to scale-up. Affinity-based separations while offering clear advantages for overall effectiveness can often be very expensive particularly as column volumes are increased. Protein A, a protein naturally found in the cell wall of *Staphylococcus aureus* has the ability to selectively interact with the fraction-crystallizable (Fc) domain of immunoglobulins. A regularly cited limitation with the use of Protein A media for the purification of monoclonal antibodies is the cost of the media itself which is an order of magnitude more expensive when compared with media using non-proteinaceous ligands (22). The particle size distribution of chromatography media impacts directly on scalability due to its effect on column pressures. The particle size chosen for any purification step will depend on the nature of the feed sample and on the degree of resolution required. The desired resolution will depend on whether the chromatography is operating in capture, intermediate, or final polishing mode.

The inherent stability of the medium chosen is an important attribute in large-scale manufacture, which can be overlooked during small-scale development studies. An ideal medium for large-scale applications should exhibit good chemical and physical stability profiles. The chemical instability of packed column beds can be caused by leakage of the ligand following repeated purification cycles and from the deterioration in the quality of the media following necessary treatments with harsh cleaning agents. In large-scale manufacturing operations, resins are re-cycled due to economic necessity for a validated number of uses, often >50 times. A medium with low-ligand leakage that can withstand harsh cleaning-in-place/regeneration protocols will lower production costs. At the same time, assurance can be given that contamination arising from the medium itself or from carryover from the feed material will be minimised in the product stream.

In order to satisfy regulatory scrutiny, concerning the important issue of process validation full documentation and technical support is required for media used in a process. Column media should be considered as a raw material and new batches must be tested before being introduced into a production cycle. Currently for therapeutic products, the use of media prepared using animal-free components and validated manufacturing technologies are mandated by regulatory agencies. A guarantee of the supply of the chosen media along with full and proper vendor certification should also be ensured prior to specifying a medium for use.

2.2. Column Packing at Large Scale

Ensuring reproducibility of media packing as the diameter of the column increases is often regarded as the most problematic aspect in scale-up. It is not surprising that small-scale columns are easier to pack reproducibly. One limitation in the approach to scale-up by increasing the column diameter, while maintaining packed bed height, is that commercial manufacturers typically supply column diameters of a set defined size. From a practical viewpoint, the packing of large-diameter columns becomes more difficult and the stability of the resultant packed bed can also be reduced. Instability in the media bed at large scale is due in the main to the decreased physical support offered by the wall of the column as the diameter increases. This can result in high pressures caused by compression of the medium via drag forces exerted by fluid flow through the bed. In general terms, decreased stability may result in unpredictable flow distribution during processing leading to hysteresis, edge effects, and media compression (11) all of which will affect process reproducibility.

The packing procedure itself can become quite challenging when large-diameter columns are used. To satisfy the need for larger-production capacities, biopharmaceutical facilities will typically use columns with diameters in the range 1–2 m for the initial capture step. The size of column required will depend on the volume of the feed material and the concentration of binding protein contained therein. Insufficient quality in column packing can cause channelling in the bed, which will lead to broadening or splitting of peaks and an obvious decrease in the resultant resolution. Manual packing of chromatography columns is used within industry and is often appropriate if the column diameter is relatively small (<0.5 m). It is performed by pouring the required amount of prepared medium slurry into the column shell and pumping an appropriate buffer through the column adaptor thus inducing the medium bed to pack. This operation is obviously quite manual and at stages is open to the surrounding environment. Thus media packed using this method is usually sanitised with a caustic solution prior to equilibration and processing to remove any potential bioburden that could have been introduced. Automated packing-in-place systems can be utilised which allow for more consistent and efficient packing operations when large-diameter columns are used. Such systems also allow for packing operations to be carried out in a more closed and controlled manner and are composed of diaphragm pumps and associated valving that serves to direct the flow of buffers or resin slurries to appropriate valves on the column or slurry vessels, respectively (23).

All chromatography media will require different conditions to achieve optimum packing and trials should take place in order to develop robust and optimal packing procedures prior to scale-up (20). When a column is packed, the integrity of that packing is usually assessed by estimating parameters such as the asymmetry factor (As) and the height equivalent to the theoretical plate (HETP). Both of these factors are determined from an analysis of the chromatogram that is generated when a small volume (<0.025 CV) of a low molecular weight tracer compound (NaCl or acetone) is applied and eluted from the packed column (24). Continued monitoring of asymmetry and HETP is useful and may be used as a first indication of the deterioration of column integrity as the number of processing cycles is increased during the lifetime of a packed column.

2.3. Cleaning of Chromatography Media

The maintenance of proper sanitary conditions is crucial to current biopharmaceutical processing operations. Whatever media and column systems are chosen for a particular application at large scale, the issue of cleaning and sanitisation is therefore of high importance. If a chromatography medium is to be re-used, assessment of the cleaning regime employed between cycles will form the major part of the required resin lifetime study (25). The potential for carryover of impurities from one column run to the next must be minimised and any cleaning regime must be capable or removing residual contamination. Indeed a validated cleaning strategy for packed chromatography columns and associated equipment is a major component in the overall process validation requirements requested by regulatory agencies.

Cleaning and sanitisation protocols are most usually performed after every each batch of material is processed on the column. These activities are designed to mitigate the risk by minimising exposure of the medium to contaminating bioburden thus prolonging the lifetime of the medium and reducing downtime due to the need to re-pack with fresh medium. Notwithstanding the time involved, repacking of columns can necessitate the use of large quantities of water and buffers and a re-qualification of the packed bed quality should also be performed.

Cleaning-in-place (CIP) strategies that are often performed after every cycle must be designed having regard to the stability of the medium used, the properties of the feed material applied and the position of the step in the purification sequence. The important decision to re-use a medium in a process can often depend on the ability to clean the medium satisfactorily. Media used in the first capture step can be more difficult to clean due to the nature of the feedstock applied. However, at the capture stage, media volumes are generally large and re-use presents the only option, particularly given that the medium may be very expensive. During polishing operations feed samples should be much cleaner and hence column volumes will be generally be smaller. However, at this late stage in manufacturing, carryover and any potential ligand leaching could significantly compromise product quality and a decision should be made as to the merits of a re-use strategy in this case (26). Sanitisation cycles typically use alkali (0.5–1 M NaOH) to reduce the build up of endotoxins and bioburden. If the CIP protocol is carefully designed, it is possible to re-use a medium many times without negative effects. In one study, a Protein A medium was used for 150 cycles with cleaning-in-place being performed after every cycle. It was demonstrated in this study that the yield and purity were consistent after each cycle. In addition, there was no detectable carryover and the binding capacity for the target IgG_1 antibody remained at a high level (>85%) when compared with the initial cycle (10).

2.4. Production Equipment

While in theory at least chromatographic processes can be scaled efficiently by taking note of key chromatography parameters and scaling accordingly, one aspect of large-scale downstream processing that will clearly be an issue is the nature of the equipment used. Glass and acrylic columns are used as scale is increased but for very large-diameter columns stainless steel is usually the material of construction. Column adaptors at large scale (0.5–2.0 M) are very heavy and contingency must exist within a facility for packing and maintenance operations. This will invariably involve the use of hydraulic lifting equipment to protect column hardware and to ensure operator safety. When a column is specified for inclusion in a GMP process, the manufacture of all components of the column and the materials of construction therein must be fully traceable to minimise the risk associated with leaching, particularly given the harsh cleaning solutions that will be used during processing.

Large-scale chromatography systems are generally automated systems that use a series of pipings, inlet and fraction valves, flow meters, air detectors, air-traps, pressure sensors, and buffer mixing capability for application requiring gradient formation. Such equipment may give rise to increased dead volumes, leading to dilution, higher pressure drops, and peak broadening, which will cause extra dilution of the product fraction or even loss of resolution if the application is sensitive to variations in plate number in the system used. Chromatography systems are typically constructed from inert piping and are configured to minimise hold-up volumes and to ensure that the fluid path can be properly cleaned and sanitised to minimise contamination. Proper qualification of chromatography equipment can identify problems prior to undertaking production activities (13). It is generally necessary to further refine previously optimised parameters (e.g. flow rates) upon scale-up as a direct result of differing equipment design.

2.5. Non-chromatographic Factors Applicable at Large Scale

Several factors not directly related to the chromatography process per se become an issue when the downstream process increases to large scale. These factors can affect the purity and yield achieved from a purification cycle when the resultant product fraction is characterised and compared with the corresponding purification step performed at small scale. As the size of the cell-culture bioreactor increases, both the concentration of the target molecule and the distribution of the associated panel of contaminants present will change due to the natural variation that will take place during cell culture. In general, during scale-up design, culture feedstock should be supplied from reactor runs performed close to the predicted final scale so that the effect of bioreactor conditions on product quality can be determined. When a chromatography column step itself is up-scaled, there will be an obvious increase in the amount of buffer solutions required to achieve purification. As volumes increase, the storage space taken up by such solutions in a facility can be problematic. Reproducibility in buffer preparation at large scale becomes very important and realistic specifications for pH and conductivity must be set in order to ensure that buffer manufacturers can effectively prepare solutions that will perform robustly during processing. The use of expensive organic solvents is often required to achieve the desired level of purification in reverse-phase chromatography applications. In addition, processing suites where such activities take place within a facility will have to be designed carefully to ensure that the room is properly vented and risk of explosion is removed. This will ensure that operators and personnel entering the area can be protected from the effects of large volumes of concentrated solvents. Such requirements can add considerable expense to a downstream process and will require that facilities must be dedicated to such activities, which will reduce overall flexibility.

If a purification step has been carefully optimised, the volumes of buffers required at each stage of the chromatography should have been minimised. If scale-up is planned, it is more appropriate to have the target product molecule elute in 1–2 column volumes rather than in 5–10 column volumes. This feature will decrease the volume of buffers required, reduce the resultant storage required, and lower accompanying material costs. A model study has shown the effect of increasing elution volume per cycle as column volume is increased for the affinity purification of a monoclonal antibody (20). Furthermore, if the volume of a bioreactor is increased and the associated capture chromatography step has to be scaled up by increasing the number of individual cycles rather than the diameter of the column, the additional volumes of buffer required will necessitate higher costs. Membrane adsorption now offers the potential of a cost-effective alternative to conventional chromatography, for flow through and some polishing applications due to the lower consumption of buffers that results and the obvious cost benefit that accrues when compared with the chromatography process (11).

3. Summary

Due to the recent developments in cell-culture technology, there is an increasing demand on biopharmaceutical manufacturers to develop robust and reproducible downstream processes to cater for the increased product titres that are now routinely achieved. Scaling-up of chromatography operations represents an increasingly important issue in the successful commercial manufacture of therapeutic proteins. Only when the rational design of a chromatography process has taken place can effective scale-up of that purification process be realised. Such design must have regard to all factors that can affect scale-up from media properties, packing operations, column hardware, ancillary equipment, and facility issues. This review has described some of the key issues that can present themselves whenever a decision is made to increase the scale of a purification step involving chromatography. Due to the central role chromatography plays in the manufacturing of the current pipeline of high dose protein products in development, the impetus to find new innovations in large-scale bioprocessing will remain a real and necessary challenge.

References

1. Reichert, J. and Pavlou, A. (2004) Monoclonal antibodies market. *Nature Reviews. Drug Discovery* **3**, 383–384.
2. Kretzmer, G. (2002) Industrial processes with animal cells. *Applied Microbiology and Biotechnology* **59**, 135–142.
3. Wurm, F. (2004) Production of recombinant protein therapeutics in cultivated mammalian cells. *Nature Biotechnology* **22**, 1393–1398.
4. Moreira, A. R. (2007) The evolution of protein expression and cell culture. *Biopharm International* **20**(10), 56–68.
5. Birch, J. R. and Racher, A. J. (2006) Antibody production. *Advanced Drug Delivery Reviews* **58**, 671–685.
6. Langer, E. (2007) Dowstream production challenges in 2007. *Bioprocess International* **5**(6), 22–28.
7. Low, D., O'Leary, R., and Pujar, N. S. (2007) Future of antibody purification. *Journal of Chromatography B* **848**, 48–63.
8. Curling, J. and Gottschalk, U. (2007) Process chromatography: five decades of innovation. *Biopharm International* **20**(10), 70–93.
9. Thommes, J. and Etzel, M. (2007) Alternatives to chromatographic separations. *Biotechnology Progress* **23**, 42–45.
10. Sofer, G. and Chirica, L. C. (2007) Improving productivity in downstream processing. *Pharmaceutical Technology Europe* **19**(4), 35–41.
11. Gottschalk, U. (2008) Bioseparation in antibody manufacturing: the good, the bad and the ugly. *Biotechnology Progress* **24**, 496–503.
12. Sofer, G. (2005) Guidelines to process validation, in *Process Validation in Manufacturing of Biopharmaceuticals* (Rathore, A. S. and Sofer, G., ed.), CRC Press, Taylor and Francis Group, Florida.
13. Edwards, J. (1997) Large-scale column chromatography – a GMP manufacturing perspective, in *Handbook of Downstream Processing* (Goldberg, E., ed.), Blackie Academic & Professional, London, pp. 167–184.
14. O'Neil, P. and Sellick, I. (2004) Innovations that improve process column chromatography. *Filtration and Separation* **41**, 24–27.
15. International Conference on Harmonisation (ICH) Topic Q7. (2000) Good manufacturing practice guide for active pharmaceutical ingredients, CPMP/ICH/4106/00.
16. Schellekens, H. (2002) Bioequivalence and the immunogenicity of biopharmaceuticals. *Nature Reviews. Drug Discovery* **1**, 457–462.

17. Sofer, G. and Hagel, L. (1997) Purification design, optimisation and scale-up, in *Handbook of Process Chromatography: A Guide to Optimisation, Scale-Up and Validation* (Sofer, G. and Hagel, L., ed.), Academic Press, New York, pp. 27–113.
18. Rathore, A. and Velayudhan, A. (2003) Guidelines for optimisation and scale-up in preparative chromatography. *Biopharm International* **16**(1), 34–42.
19. Janson, J. C. and Hedman, P. (1982) Large-scale chromatography of proteins, in *Advances in Biochemical Engineering* (Fiechter, A., ed.), Springer Verlag, Berlin, **25**, pp. 43–99.
20. Aldington, S. and Bonnerjea, J. (2007) Scale-up of monoclonal antibody purification processes. *Journal of Chromatography B* **848**, 64–78.
21. Graumann, K. and Ebenbichler, A.-A. (2005) Development and scale-up of preparative HIC for the purification of a recombinant therapeutic protein. *Chemical Engineering and Technology* **28**, 1398–1406.
22. Ghose, S., Hubbard, B., and Cramer, S. M. (2006) Evaluation and comparison of alternatives to protein A chromatography, mimetic and hydrophobic charge induction chromatographic stationary phases. *Journal of Chromatography A* **1122**, 144–152.
23. Bloomingburg, G. and Gandhi, P. (2005) Engineering design considerations for column packing in large-scale biotechnology facilities. *Bioprocess International* **3**(6), 44–51.
24. Rathore, A. S., Kennedy, R. M., O'Donnell, K., Bemberis, I., and Kaltenbrunner, O. (2003) Qualification of a chromatographic column: why and how to do it? *Biopharm International* **16**(3), 30–40.
25. Rathore, A. S. and Sofer, G. (2005) Life span studies for chromatography and filtration media, in *Process Validation in Manufacturing of Biopharmaceuticals* (Rathore, A. S. and Sofer, G., ed.), CRC Press, Taylor and Francis Group, Florida, pp. 169–203.
26. Sofer, G. and Yourkin, J. (2007) Cleaning and cleaning validation in process chromatography. *Bioprocess International* **5**(10), 72–82.

Chapter 6

Phage Display: A Powerful Technology for the Generation of High Specificity Affinity Reagents from Alternative Immune Sources

William J.J. Finlay, Laird Bloom, and Orla Cunningham

Abstract

Antibodies are critical reagents in many fundamental biochemical methods such as affinity chromatography. As our understanding of the proteome becomes more complex, demand is rising for rapidly generated antibodies of higher specificity than ever before. It is therefore surprising that few investigators have moved beyond the classical methods of antibody production in their search for new reagents. Despite their long-standing efficacy, recombinant antibody generation technologies such as phage display are still largely the tools of biotechnology companies or research groups with a direct interest in protein engineering. In this chapter, we discuss the inherent limitations of classical polyclonal and monoclonal antibody generation and highlight an attractive alternative: generating high specificity, high affinity recombinant antibodies from alternative immune sources such as chickens, via phage display.

Key words: Chicken, scFv, Phage display, Chromatography

1. Introduction

The rapid expansion of the genomics, proteomics, and biotechnology fields has led to a growing demand for affinity reagents that can specifically recognize proteins, peptides, carbohydrates, and haptens. Affinity reagents of high specificity are routinely required for diverse protein drug targets, members of newly discovered biochemical pathways, posttranslationally modified proteins, protein cleavage products, and even small molecules such as drugs of abuse and toxins. Individual biomedical researchers will often need to monitor, quantify and purify proteins of interest via affinity chromatography, but there may not be any commercially available antibody reagents to allow them to do so (1). Indeed, even in

Dermot Walls and Sinéad T. Loughran (eds.), *Protein Chromatography: Methods and Protocols*, Methods in Molecular Biology, vol. 681, DOI 10.1007/978-1-60761-913-0_6, © Springer Science+Business Media, LLC 2011

situations where there are commercially available antibodies, these reagents are often expensive, poorly characterized, and/or simply not appropriate for demanding applications. Compounding this problem, the technical difficulty of monoclonal antibody generation by the untrained researcher and the high cost (~$15,000) of a commercial monoclonal antibody generation program leads many researchers to the default solution of producing polyclonal hyperimmune sera in hosts such as rabbits. The net result of this is that researchers often settle for reagents that lack the necessary specificity to perform the applications for which they were intended.

In this review, we will outline the limitations of classical antibody generation technologies and illustrate an attractive alternative: the use of phage display libraries of recombinant antibodies built on immunoglobulin repertoires from nonmammalian animals. In particular, we will highlight the advantages of libraries derived from the domestic chicken *Gallus gallus*, which offers a relatively inexpensive and technically accessible route to high-quality monoclonal reagents. If, like many people, you have purchased (or paid to generate) a costly and "specific" antibody, but subsequently found that it is actually polyreactive and of dubious quality, phage display from immunized chickens may offer an attractive alternative.

1.1. Historical Difficulties in Antibody Generation Technology

Hyperimmune sera from rabbits, sheep, or other mammals may be produced in large quantities, but they do not offer the consistency of monoclonal antibodies and need to be regularly replenished and recharacterized. Serum antibodies are also polyclonal and frequently polyspecific, even when purified over an antigen column, rendering them suboptimal for the specific recognition of a single component in a complex matrix. One illuminating study (2) has demonstrated that when used to probe a comprehensive yeast proteome chip, unpurified polyclonal antibody preparations could recognize up to 1,770 different proteins, with some monoclonal antibodies and antigen-column purified polyclonal antibodies also recognizing multiple proteins (related and unrelated).

The arrival of monoclonal antibody technology (3) was a major step forward in generating high-specificity reagents, but the reliance on the murine immunoglobulin system frequently leads to a number of practical difficulties: (1) Monoclonal antibodies are raised on the basis of an inefficient fusion of splenic B-cells to an immortalized mouse myeloma line, followed by limiting dilution of the cell population. Target-specific antibodies are randomly identified, often by a simple direct ELISA, where few preconditions can be set to determine which antibodies are identified and one must "take what one can get" during the screening process. (2) It is often desirable to have multiple monoclonal antibodies with specificity for different epitopes on the same target molecule, but the difficulty in sequencing monoclonals does not allow the rapid identification of unique clones early in the screening process. (3) Humans and rodents are relatively closely related phylogenetically. Many proteins of interest are highly conserved among mammals and this can frequently lead to thymic tolerance restricting the antibody response after immunization. (4) When an immune response to a human protein is raised in mice, the large regions of sequence similarity between murine and human proteins may lead to a restricted number of immunogenic epitopes. (5) To generate antibodies that cross-react with homologs from multiple species of mammal is particularly tricky, as the common epitopes among mammals are the very ones that are unlikely to provoke a strong immunoglobulin response in the mouse. (6) Tolerance issues can become even harder to circumvent when the protein of interest is from a mouse or rat. Creating "knockout" mice, in which the endogenous copy of the gene for the target protein has been disabled, can often break tolerance, but this is a highly laborious and time-consuming process that few laboratories have the resources to undertake. These factors all hinder the generation of high-quality antibody reagents and thereby limit one's experimental options when developing antibodies for purifying or tracking novel proteins.

2. Display Technologies as an Alternative Source of Specific Antibodies

To bypass the limitations in polyclonal and monoclonal antibody generation, several groups have turned to in vitro display technologies such as phage display, ribosome display, or yeast display libraries to generate recombinant antibodies. The more recent technologies of yeast and ribosome display are becoming highly established, but phage display is currently the most robust, well characterized, and reliable of these methods. Antibodies derived from these technologies are cloned in microbial hosts and are therefore monoclonal from the start, with their production being easily scaled up. Critically, selection and screening efforts can be directed toward specific epitopes or species cross-reactivity and away from polyspecific binding. Phage display allows a researcher to do in a single tube what would be unfathomable in traditional hybridoma work: interrogate libraries of millions to tens of billions of antibodies on the basis of their binding specifically to a target of interest.

All of this is possible due to the ingenious concept of "genotype-to-phenotype linkage," which is exemplified by phage display. Phage display was originally described as a rapid method for cloning gene fragments that encode a specific protein (4). By cloning gene fragments into the genome of a filamentous *E. coli* bacteriophage, Smith et al. were able to generate libraries of gene fusions with the key phage coat protein p3. After transformation into *E. coli*, the viral replication system packaged the genome (and therefore the cloned gene) into a highly stable complex carrying the gene product on the tip of the phage particle, as a fusion protein with p3. By subsequently selecting expressed virions on an immobilized antibody with specificity for a known protein that had been cloned in the phage genome, they were able to show 1,000-fold enrichment of the gene product. This set of experiments showed that "genotype-to-phenotype linkage" could be achieved and thereby defined the basis of all display technologies developed since.

The subsequent development of a method for the effective cloning of antibody V-gene sequences via PCR allowed the capture of antibody sequences in a recombinant form (5), removing the need to immortalize B-cells via hybridoma fusion as required in traditional monoclonal antibody generation. This discovery was combined with the phage display process to make an efficient method of isolating antibodies and their corresponding gene sequences simultaneously (6). In the antibody phage display process, libraries of diverse V-gene sequences are cloned into an appropriate expression vector in *E. coli*, creating an in-frame fusion with the p3 protein or, as favored in most recently described libraries, a truncated form of p3.

The phage display of protein libraries was originally performed using "phage" vectors (i.e., built upon the phage genome itself), but due to practical difficulties in handling these libraries, more recent phage methods have mostly used so-called phagemid expression vectors. In this case, the DNA backbone is a stable, small plasmid such as pUC, and the p3 gene plus f1 phage packaging origin are the only phage-derived DNA sequences (7). These phagemid libraries are more easily handled and more stable than phage libraries, as the plasmid is incapable of causing phage production by itself. The libraries can therefore be more simply cloned, expanded, and controlled than phage libraries.

By infecting "helper" phage (based on M13) into a growing culture of *E. coli* harboring a phagemid library, the phage propagation machinery is provided, but due to a mutation introduced in the origin of replication on the helper phage genome, preferential packaging of the phagemid DNA and p3 fusion occurs during phage replication (7). Phage production is thereby induced, genotype-to-phenotype linkage is created, and the expressed phage particles are interrogated for the presence of useful protein sequences via target binding. This selective step may be performed by simply immobilizing the target protein on, for example, a protein-binding plastic surface such as an ELISA plate, adding phage, and allowing binding to occur via the antibody-p3 fusion proteins. Nonbinding phage are removed by washing the immobilized surface, and the remaining bound phage are eluted. The eluted phage is then reinfected back into a fresh culture of *E. coli* to retrieve the selected gene sequences. This process is not perfect, however, and two to four rounds of selection/reexpression/selection are normally performed iteratively to remove all unwanted clones and enrich the binding population from the background of the library. Nonetheless, this process can be spectacularly powerful, massively enriching specific antibodies in a single selection round (8). Furthermore, the use of multiple forms of the target antigen in sequential selection rounds and the inclusion of competitor proteins can drive the selected pool toward a highly specific set of epitopes.

Much of the evolution and progression of phage display technology has been driven by the remarkable discovery that this method can be used to mine large libraries of combinatorial human antibody diversity, theoretically removing the need for animals in antibody production (9). By random recombination of V_H and V_L sequences from human lymphocyte cDNA (9), or, by using degenerate oligonucleotides to create diversity of DNA sequence in the loops associated with target binding (10), several groups have created very large libraries of antibody gene sequences for phage display. These libraries are analogous to the naïve antibody repertoire in an animal, and selecting from them can result in the identification of antibody fragments that exhibit high specificity and occasionally high affinity for the target protein. Several major studies have proven that with appropriate application of this technology, specific antibodies can be raised to proteins, peptides, haptens, and even carbohydrates. In the most exemplary studies, antibodies have been raised with equivalent affinities to those associated with a strong humoral immune response (11–14).

2.1. Recombinant Antibody Formats Used in Phage Display

Recombinant antibodies are typically displayed and expressed in "fragment" forms. The simplest and most commonly used fragment is the single-chain fragment variable (scFv) where a flexible peptide sequence links the V-regions of antibodies between the C terminus of one domain and the N terminus of the other, thereby combining both V-domains into a single polypeptide (15). The scFv may be assembled in V_L–V_H or V_H–V_L orientations, with V_H–V_L being the most heavily used format historically. The flexible linker helps to make the scFv simple to express, but must be sufficiently long and flexible to allow effective association of the V-regions to form a functional antigen-combining site. As long as this is true, the classical hydrophobic pairing of the V-regions will stabilize the structure. By far the most common linkers are based on glycine–serine repeat structures such as GGGGS × 3.

The second most commonly used recombinant antibody format is the Fab (fragment antigen binding) molecule. This structure is a complete binding "arm" of an antibody and comprises the full immunoglobulin light chain, expressed in conjunction with the V_H–C_H1 region from the heavy chain (16). Fabs obligately form predominantly monomeric, monovalent fragments. They are the most "natural" of the recombinant antibody fragments and it has been shown that the presence of the constant regions can often help to stabilize antibody variable regions (17). The Fab format is the less commonly used of the two main recombinant antibody formats, however, as its dual polypeptide structure is generally more difficult to express and display in *E. coli* than the scFv (17).

2.2. Why Is Phage Display not more Heavily Used?

Much of the underuse of phage display may be due to experiences with the early libraries derived from naïve or synthetic human antibody diversity, which were donated to academic laboratories that were not specifically invested in antibody engineering. Unfortunately, these forays into display technologies have often left investigators somewhat disappointed. Many people have accepted the viewpoints of recombinant antibody technology experts that these libraries can yield useful high-affinity antibodies to any form of antigen. In general (mostly to those highly skilled in the field), this is indeed true, but the average antibody generated from these libraries is often of disappointingly low affinity to those who are used to high-sensitivity antibodies from immunized sources. Human recombinant antibodies from naïve library sources can require technically challenging in vitro molecular evolution, if they are to perform the demanding "real world" functions required of many reagent antibodies (8). Molecular evolution is far from trivial to perform and is usually beyond both the scope and interest level of the average researcher. Small wonder then that most people either ignore, or at worst disparage, phage display technology itself.

Nevertheless, phage display can be a relatively simple technology to use and when employed to harness natural repertoires of antibodies from immunized animals, it can offer a rapid path to highly specific, high-affinity antibodies against problematic antigens. While the most successful naïve antibody libraries contain over 10^{10} members and are often the domain of biotechnology companies, typical immune libraries are in the 10^7–10^8 range and are easily assembled by a single investigator (18, 19). When an immunized rabbit or sheep has raised a significant serum immunoglobulin titer, the common endpoint to the experiment is to exsanguinate the animal and harvest the serum. However, harvesting B-cell-rich lymphoid tissues from the animal, such as the spleen and bone marrow, allows the isolation of total RNA and the subsequent generation of cDNA (18). This is a simple method with which many biomedical researchers are familiar, and commercial kits are available to simplify most steps of the process.

The immunoglobulin gene sequences of many animals are now known and the cDNA from immune tissues can subsequently be used for the RT-PCR amplification and cloning of the animal's variable region sequence repertoire (18). These cloned variable region sequences can then be assembled into a display library format such as scFv or chimeric Fab (using human C_H1 and $C_{\kappa/\lambda}$ regions) (20). These targeted immune libraries thereby offer a potentially huge advantage over monoclonal antibodies, as libraries of $>10^8$ variants may be built, allowing the effective sampling of a much broader range of antibodies than the hundreds (occasionally thousands) of clones usually examined in a monoclonal antibody screen.

The resulting library can be interrogated for specific binding proteins via phage display and the retrieved antibody fragments expressed very simply in bacteria (20). This process has been used to successfully harness the antibody repertoires of a large number of immune host species, including mice (18), rabbits (19), sheep (21), camelids (22), and sharks (23). Of greater interest to us, however, is to exploit this approach to harvest the novel immunoglobulin repertoires of the domestic chicken (*G. gallus*), which is as simple to use as mice and rabbits, but also highly phylogenetically distant from mammals.

3. Why Chickens (G. gallus)?

Gallus, Scottish dialect for (1) self-confident, daring, cheeky. (2) Stylish, impressive. Origin derogatory, meaning wild; deserving to be hanged (from the Gallows). *Gallus*, the Latin word for "cockerel."

Avians can circumvent many of the common problems encountered with mammalian immunizations described above. As a fully domesticated small animal, chickens are an attractive host for immunization as they are highly accessible, very affordable and easily housed in a generic animal house. Most importantly, however, the amino acid homology between the mammalian and avian orthologs of a given protein is typically lower than between the mammals commonly used for antibody generation, and indeed, some mammalian proteins may not even exist in avians. The immunoglobulin response of chickens to highly conserved mammalian proteins is reliably robust, generally exhibits high avidity, and potentially targets a broad spectrum of epitopes on protein immunogens (24–26).

Chickens therefore have a potentially major advantage over other common immune hosts: they can produce a high affinity cross-reactive antibody response targeting an epitope that is conserved across multiple orthologs of a mammalian protein.

This can lead to significant savings in time and resources as, if a single, broadly applicable cloned reagent can be identified, it can then be used to generate a single affinity column for the capture of the target protein from multiple species. Chicken immunoglobulins have also shown beneficial biophysical properties: they exhibit high stability to changes in pH and temperatures up to 70°C (27, 28), provide functional coating on latex microspheres (29) and demonstrate functional direct covalent coupling to a dextran layer for the detection of serum proteins by surface plasmon resonance (30). Furthermore, as chickens are small animals, very little protein immunogen is required to raise a strong immunoglobulin response. Approximately 200 μg/bird of purified protein is sufficient to carry out a full-immunization regime (31, 32).

These observations have led to the regular use of chickens as an immune host for production of the polyclonal antibody termed IgY (egg yolk antibody), in both research and commercial settings. Laying hens will export significant quantities of polyclonal IgY into the egg (~100 mg of IgY per yolk), in a process analogous to mammalian placental IgG transfer, which allows direct screening of their antibody response without the need for serum sampling (33). Once a strong immune response has been raised, large quantities of polyclonal antibody are easily prepared from the yolk. These polyclonal antibodies have been successfully applied in research immunochemistry (34), diagnostics (35), and affinity column purification (36). Indeed, immunodepletion resins based on chicken IgY can be used to remove high-abundance proteins from serum and are now commercially available (Genway Seppro®).

Unfortunately, polyclonal IgY does still suffer from the same issues of ill-defined specificity that all polyclonal antibody preparations do. In addition, there have been several studies describing successful chicken hybridoma monoclonal antibody generation to antigens such as human peptides (37), sporozoite proteins (38), and prion protein (39), but the low antibody expression and instability associated with chicken myeloma cell lines (37, 40) led to the underuse of this species as a source of monoclonal antibodies. Today, however, the progress in chicken antibody phage display has circumvented these problems and made recombinant chicken antibody reagents readily accessible, as we describe below.

3.1. Harnessing the Chicken Immune Response via Phage Display

The chicken immunoglobulin repertoire is almost ideally suited to antibody phage display, as chickens generate their immunoglobulin repertoire from a single set of V_H and V_L germ-line sequences (41). Diversity in the V-regions is created by both V-D-J recombination and somatic hypermutation, with the additional influence of "gene conversion," where multiple upstream pseudogenes are recombined into the functional sequence. This germ-line V-gene system means that the entire chicken antibody repertoire can be captured using only four PCR primers (42), making chicken libraries highly representative of the induced immunoglobulin response. This is in direct contrast with immune hosts such as mice, which have diverse germ-line V-gene sequences and therefore require complex mixes of PCR primers (18). Additionally, the two V-gene germ-line sequences found in chickens are highly homologous to the human V_λ and V_H3 germ-line families (43), which are both associated with creating V-domains with high stability and solubility. Indeed, chicken scFvs can be stable in crude bacterial culture supernatants for up to 1 month at room temperature (44).

The initial work of Davies et al. (42) showed that a simple recombinant chicken antibody library could be displayed on phage. While this small library was nonimmune and derived from the bursa cells of a single young chicken, the group was able to select target-specific scFv sequences recognizing lysozyme, serum albumin, and thyroglobulin. The potential of chicken recombinant antibodies was further highlighted by a study (45), which used an scFv library derived from the spleens of immunized chickens and successfully generated highly specific scFv antibodies that targeted both mouse and rat serum albumins, where tolerance issues limit the ability to generate murine monoclonal antibodies.

A major study (20) subsequently demonstrated that chickens could be a useful source of scFv and chimeric Fab antibodies with specificity for hapten molecules. However, none of these early studies characterized the antibodies for their affinity or their function as practical reagents. More recent studies have shown that scFv antibodies derived from immunized chickens are highly effective reagents in diverse settings such as diagnostic ELISA for Infectious Bural Disease Virus (46), the diagnosis of prion disease (47), immunodetection of haptenic shellfish toxins (48), immunostaining of SARS-infected cells (49), biosensing of cardiac biomarkers (50), and the measurement of ApoB protein in mouse and human sera (51). Raats et al. (52) have also illustrated that anti-idiotype scFvs from an immune chicken scFv library were of considerably higher sensitivity than those derived from a human antibody library in the same study. The generation of highly selective scFvs towards the PrP protein, which is highly conserved in mammals, further demonstrates the advantage of the chicken as an immune model, as isolated scFvs were shown to react with murine, ovine, and bovine orthologs of the protein (47).

The cloning of chicken antibodies via phage display has also allowed the precise dissection of the specificity and affinity of the chicken immunoglobulin response. High-throughput affinity measurements for panels of chicken scFvs to the inflammatory biomarker C-reactive protein have identified clones that preferentially recognize the multimeric and monomeric forms of the protein (50). In the same study, clones with affinities as high as 350 pM were generated from an immune phage display library of only 3×10^7 total clones. In addition, chicken anti-PrP scFv have been reported to have affinities up to 15 pM, making them among the highest affinity scFvs reported to date (53).

What may be of particular practical interest to many researchers is that chickens can serve as a host for simultaneous immunization with multiple proteins of interest, with as many as eight proteins being used successfully in a single immunization scheme (31, 44, 54). The target proteins of interest are mixed in a single adjuvant preparation and each immunized animal receives all proteins simultaneously. Spleen and bone marrow tissues from the immunized animals are then used to generate relatively small phage display libraries and specific antibodies are derived via selection of the library separately on the individual proteins originally used for immunization (31). The immunized chickens appear to react to the proteins fully independently, as the phage display libraries generate individual scFv antibody clones that are fully specific by western blot and ELISA, showing no reactivity to their co-immunogens (31, 54). This approach has major benefits practically and ethically, as it allows the use of a single library to derive high affinity antibodies to a group of proteins of interest. Multi-immunization methods also simultaneously minimize animal use and raise the likelihood of success in generating an immediately useful reagent (31, 54).

Multitarget immunization regimes should be designed with one of two objectives in mind. First, the simplest scenario combines multiple unrelated proteins, which leads to unrelated B-cell responses after immunization. To derive antibodies of greatest specificity during a multitarget immunization of this kind, it is important to ensure that each of the protein immunogens is highly purified and that no closely related proteins are co-immunized into a single animal. Secondly, to derive antibodies that are cross-reactive to orthologs of a conserved protein from multiple species, it is likely to be beneficial, but not necessarily essential, to include each ortholog in the mix of immunogens given to each animal. Iterative selection rounds that change ortholog each time can then be used to bias toward the isolation of cross-reactive antibodies.

The generation and selection of chicken recombinant antibodies is extremely reliable using the methods described in detail in the accompanying chapter. The subsequent identification and sequencing of antibodies displaying the characteristics desired can also be performed simply. In general, scFvs isolated from immunized chicken libraries exhibit high affinity and can be assayed via a direct ELISA, using crude periplasmic extracts from the protein expressing *E. coli* clones. The level of further downstream analysis carried out on positive hits identified during the binding ELISA depends on what the end user requires. Specific binding function may suffice for scFvs that are to be used simply as reagent antibodies for in vitro analysis of samples via ELISA or western blotting. For antibodies to be used in affinity chromatography, the antibody fragments must be purified and tested for their function after being coupled to a solid matrix and for their specificity during purification.

4. Toward Affinity Chromatography with Antibody Fragments

Few studies have been performed using antibody fragments in affinity chromatography, but several potential approaches have been described. For the antibody binding site(s) to be fully solvent-exposed and active, the antibody fragment must be directionally captured onto the solid matrix. Even for full-length IgG, nonspecific adsorption or covalent coupling onto solid supports can lead to denaturation, reducing or negating antigen-binding function (55). Antibody fragments may be slightly more prone to chemical or physical denaturation than full-length immunoglobulins, but scFv and Fab have been used successfully in affinity chromatography and in the creation of SPR sensing surfaces, which can go through serial rounds of binding and regeneration (56).

McElhiney et al. (57) created a simple scFv-based affinity column for the concentration and cleanup of microcystin toxins from environmental samples, by transiently coupling the His-tagged scFv to a disposable nickel chelate column. The scFv was thereby coupled directionally, maximizing the functional antibody content on the column, and the analyte for purification was co-eluted with the scFv before quantification by reverse-phase HPLC. However, this method can only be used under a limited number of conditions, as the interaction of the His-tag with nickel is noncovalent and pH dependent. Other possibly useful low-affinity expression tags include *E. coli* maltose binding protein and glutathione S-transferase, which have both been successful in protein purification (58, 59) (see also Chapter 9 for a discussion on protein tagging).

High affinity, highly stable linkage via affinity tagging may also be achieved by site-specific biotinylation of antibody fragments and their immobilization onto a matrix that has been passively or covalently coated with avidin. Bacterial expression vectors are now available that introduce biotin into specific peptide tags (AviTag), which can be produced on the termini of recombinant proteins. A similar method was proven to be efficient in the production of Fabs that are specifically biotinylated in vivo during bacterial expression, via C-terminal fusion of the Fabs to the *E. coli* acetyl-CoA carboxylase (60). Importantly, these biotinylated Fabs were successfully used to purify recombinant TNF-alpha from bacterial lysates, via a streptavidinated column. The peptide tagging method has also been used successfully to label both Fab and scFv antibodies for their oriented immobilization and use as capture antibodies in clinical diagnostic ELISAs (61, 62). These studies suggest that biotin–streptavidin coupling is a simple and rapid method for the stable, directional capture of recombinant antibody fragments.

While the covalent coupling of recombinant antibody fragments via their reactive lysine side chains is likely to be disruptive to their function, some alternative covalent coupling methods have been identified. In the simplest example, the disulfide bonds linking the two constant regions of a Fab can be reduced using a mild agent to expose cysteine thiols. These thiol groups can then be used to covalently couple the fragment to a thiol-activated surface (63). More elegant versions of this approach have expressed antibody fragments with a C-terminal cysteine group, then gently applied the same chemistry, to preferentially reduce the exposed disulphide groups (56). The exposed terminal thiols are again an efficient reactive group for covalent attachment. It is also possible to express scFv fused to the constant regions of human IgG light chains as another source of usable cysteine residues external to the V-regions (64).

Whether any of the above attachment methods are appropriate for a given affinity purification application may be decided upon by the individual investigator. In cases where stable linkage has been achieved and the column is to be reused, it is prudent for the investigator to examine multiple clones for their stability under repeated cycles of elution and regeneration. While chicken scFvs are built upon naturally stable frameworks, the stability of different clones cannot be taken for granted. In cases where stability remains an issue, the appropriate chicken V-regions can be cloned into an Fc-fusion (65) or IgG (66) expression vector to produce full-length antibody in mammalian, yeast, and even plant culture systems (67).

References

1. Kusnezow, W. & Hoheisel, J. D. (2002) Antibody microarrays: promises and problems. *Biotechniques* **Suppl**, 14–23.
2. Michaud, G. A., Salcius, M., Zhou, F., Bangham, R., Bonin, J., Guo, H., Snyder, M., Predki, P. F. & Schweitzer, B. I. (2003) Analyzing antibody specificity with whole proteome microarrays. *Nat Biotechnol* **21**, 1509–12.
3. Kohler, G. & Milstein, C. (1975) Continuous cultures of fused cells secreting antibody of predefined specificity. *Nature* **256**, 495–7.
4. Smith, G. P. (1985) Filamentous fusion phage: novel expression vectors that display cloned antigens on the virion surface. *Science* **228**, 1315–7.
5. Orlandi, R., Gussow, D. H., Jones, P. T. & Winter, G. (1989) Cloning immunoglobulin variable domains for expression by the polymerase chain reaction. *Proc Natl Acad Sci U S A* **86**, 3833–7.
6. McCafferty, J., Griffiths, A. D., Winter, G. & Chiswell, D. J. (1990) Phage antibodies: filamentous phage displaying antibody variable domains. *Nature* **348**, 552–4.
7. Hoogenboom, H. R., Griffiths, A. D., Johnson, K. S., Chiswell, D. J., Hudson, P. &

Winter, G. (1991) Multi-subunit proteins on the surface of filamentous phage: methodologies for displaying antibody (Fab) heavy and light chains. *Nucleic Acids Res* **19**, 4133–7.

8. Hoogenboom, H. R. (2005) Selecting and screening recombinant antibody libraries. *Nat Biotechnol* **23**, 1105–16.
9. Marks, J. D., Hoogenboom, H. R., Bonnert, T. P., McCafferty, J., Griffiths, A. D. & Winter, G. (1991) By-passing immunization. Human antibodies from V-gene libraries displayed on phage. *J Mol Biol* **222**, 581–97.
10. Barbas, C. F., 3rd, Bain, J. D., Hoekstra, D. M. & Lerner, R. A. (1992) Semisynthetic combinatorial antibody libraries: a chemical solution to the diversity problem. *Proc Natl Acad Sci U S A* **89**, 4457–61.
11. Fellouse, F. A., Esaki, K., Birtalan, S., Raptis, D., Cancasci, V. J., Koide, A., Jhurani, P., Vasser, M., Wiesmann, C., Kossiakoff, A. A., Koide, S. & Sidhu, S. S. (2007) High-throughput generation of synthetic antibodies from highly functional minimalist phage-displayed libraries. *J Mol Biol* **373**, 924–40.
12. Hoet, R. M., Cohen, E. H., Kent, R. B., Rookey, K., Schoonbroodt, S., Hogan, S., Rem, L., Frans, N., Daukandt, M., Pieters, H., van Hegelsom, R., Neer, N. C., Nastri, H. G., Rondon, I. J., Leeds, J. A., Hufton, S. E., Huang, L., Kashin, I., Devlin, M., Kuang, G., Steukers, M., Viswanathan, M., Nixon, A. E., Sexton, D. J., Hoogenboom, H. R. & Ladner, R. C. (2005) Generation of high-affinity human antibodies by combining donor-derived and synthetic complementarity-determining-region diversity. *Nat Biotechnol* **23**, 344–8.
13. Knappik, A., Ge, L., Honegger, A., Pack, P., Fischer, M., Wellnhofer, G., Hoess, A., Wolle, J., Pluckthun, A. & Virnekas, B. (2000) Fully synthetic human combinatorial antibody libraries (HuCAL) based on modular consensus frameworks and CDRs randomized with trinucleotides. *J Mol Biol* **296**, 57–86.
14. Vaughan, T. J., Williams, A. J., Pritchard, K., Osbourn, J. K., Pope, A. R., Earnshaw, J. C., McCafferty, J., Hodits, R. A., Wilton, J. & Johnson, K. S. (1996) Human antibodies with sub-nanomolar affinities isolated from a large non-immunized phage display library. *Nat Biotechnol* **14**, 309–14.
15. Huston, J. S., Levinson, D., Mudgett-Hunter, M., Tai, M. S., Novotny, J., Margolies, M. N., Ridge, R. J., Bruccoleri, R. E., Haber, E., Crea, R. & et al. (1988) Protein engineering of antibody binding sites: recovery of specific activity in an anti-digoxin single-chain Fv analogue produced in *Escherichia coli*. *Proc Natl Acad Sci U S A* **85**, 5879–83.
16. Barbas, C. F., 3rd, Crowe, J. E., Jr., Cababa, D., Jones, T. M., Zebedee, S. L., Murphy, B. R., Chanock, R. M. & Burton, D. R. (1992) Human monoclonal Fab fragments derived from a combinatorial library bind to respiratory syncytial virus F glycoprotein and neutralize infectivity. *Proc Natl Acad Sci U S A* **89**, 10164–8.
17. Rothlisberger, D., Honegger, A. & Pluckthun, A. (2005) Domain interactions in the Fab fragment: a comparative evaluation of the single-chain Fv and Fab format engineered with variable domains of different stability. *J Mol Biol* **347**, 773–89.
18. Krebber, A., Bornhauser, S., Burmester, J., Honegger, A., Willuda, J., Bosshard, H. R. & Pluckthun, A. (1997) Reliable cloning of functional antibody variable domains from hybridomas and spleen cell repertoires employing a reengineered phage display system. *J Immunol Methods* **201**, 35–55.
19. Li, Y., Cockburn, W., Kilpatrick, J. B. & Whitelam, G. C. (2000) High affinity ScFvs from a single rabbit immunized with multiple haptens. *Biochem Biophys Res Commun* **268**, 398–404.
20. Andris-Widhopf, J., Rader, C., Steinberger, P., Fuller, R. & Barbas, C. F., 3rd. (2000) Methods for the generation of chicken monoclonal antibody fragments by phage display. *J Immunol Methods* **242**, 159–81.
21. Charlton, K., Harris, W. J. & Porter, A. J. (2001) The isolation of super-sensitive anti-hapten antibodies from combinatorial antibody libraries derived from sheep. *Biosens Bioelectron* **16**, 639–46.
22. Arbabi Ghahroudi, M., Desmyter, A., Wyns, L., Hamers, R. & Muyldermans, S. (1997) Selection and identification of single domain antibody fragments from camel heavy-chain antibodies. *FEBS Lett* **414**, 521–6.
23. Dooley, H., Flajnik, M. F. & Porter, A. J. (2003) Selection and characterization of naturally occurring single-domain (IgNAR) antibody fragments from immunized sharks by phage display. *Mol Immunol* **40**, 25–33.
24. Ikemori, Y., Peralta, R. C., Kuroki, M., Yokoyama, H. & Kodama, Y. (1993) Research note: avidity of chicken yolk antibodies to enterotoxigenic *Escherichia coli* fimbriae. *Poult Sci* **72**, 2361–5.
25. Lemamy, G. J., Roger, P., Mani, J. C., Robert, M., Rochefort, H. & Brouillet, J. P. (1999) High-affinity antibodies from hen's-egg yolks against human mannose-6-phosphate/insulin-like growth-factor-II receptor (M6P/IGFII-R): characterization and potential use in clinical cancer studies. *Int J Cancer* **80**, 896–902.

26. Stuart, C. A., Pietrzyk, R. A., Furlanetto, R. W. & Green, A. (1988) High affinity antibody from hen's eggs directed against the human insulin receptor and the human IGF-I receptor. *Anal Biochem* **173**, 142–50.

27. Hatta, H., Tsuda, K., Akachi, S., Kim, M. & Yamamoto, T. (1993) Productivity and some properties of egg yolk antibody (IgY) against human rotavirus compared with rabbit IgG. *Biosci Biotechnol Biochem* **57**, 450–4.

28. Shimizu, M., Nagashima, H., Sano, K., Hashimoto, K., Ozeki, M., Tsuda, K. & Hatta, H. (1992) Molecular stability of chicken and rabbit immunoglobulin G. *Biosci Biotechnol Biochem* **56**, 270–4.

29. Davalos-Pantoja, L., Ortega-Vinuesa, J. L., Bastos-Gonzalez, D. & Hidalgo-Alvarez, R. (2001) Colloidal stability of IgG- and IgY-coated latex microspheres. *Colloids Surf B Biointerfaces* **20**, 165–175.

30. Vikinge, T. P., Askendal, A., Liedberg, B., Lindahl, T. & Tengvall, P. (1998) Immobilized chicken antibodies improve the detection of serum antigens with surface plasmon resonance (SPR). *Biosens Bioelectron* **13**, 1257–62.

31. Finlay, W. J., deVore, N. C., Dobrovolskaia, E. N., Gam, A., Goodyear, C. S. & Slater, J. E. (2005) Exploiting the avian immunoglobulin system to simplify the generation of recombinant antibodies to allergenic proteins. *Clin Exp Allergy* **35**, 1040–8.

32. Gassmann, M., Thommes, P., Weiser, T. & Hubscher, U. (1990) Efficient production of chicken egg yolk antibodies against a conserved mammalian protein. *FASEB J* **4**, 2528–32.

33. Polson, A., Coetzer, T., Kruger, J., von Maltzahn, E. & van der Merwe, K. J. (1985) Improvements in the isolation of IgY from the yolks of eggs laid by immunized hens. *Immunol Invest* **14**, 323–7.

34. Larsson, A., Karlsson-Parra, A. & Sjoquist, J. (1991) Use of chicken antibodies in enzyme immunoassays to avoid interference by rheumatoid factors. *Clin Chem* **37**, 411–4.

35. Tsen, Y. C., Kao, G. Y., Chang, C. L., Lai, F. Y., Huang, C. H., Ouyang, S., Yu, M. H., Wang, C. P. & Chiou, Y. N. (2003) Evaluation and validation of a duck IgY antibody-based immunoassay for high-sensitivity C-reactive protein: avian antibody application in clinical diagnostics. *Clin Chem* **49**, 810–3.

36. Qian, W. J., Kaleta, D. T., Petritis, B. O., Jiang, H., Liu, T., Zhang, X., Mottaz, H. M., Varnum, S. M., Camp, D. G., 2nd, Huang, L., Fang, X., Zhang, W. W. & Smith, R. D. (2008) Enhanced detection of low abundance human plasma proteins using a tandem IgY12-SuperMix immunoaffinity separation strategy. *Mol Cell Proteomics* 7, 1963–73.

37. Michael, N., Accavitti, M. A., Masteller, E. & Thompson, C. B. (1998) The antigen-binding characteristics of mAbs derived from in vivo priming of avian B cells. *Proc Natl Acad Sci U S A* **95**, 1166–71.

38. Sasai, K., Lillehoj, H. S., Matsuda, H. & Wergin, W. P. (1996) Characterization of a chicken monoclonal antibody that recognizes the apical complex of *Eimeria acervulina* sporozoites and partially inhibits sporozoite invasion of CD8+ T lymphocytes in vitro. *J Parasitol* **82**, 82–7.

39. Matsushita, K., Horiuchi, H., Furusawa, S., Horiuchi, M., Shinagawa, M. & Matsuda, H. (1998) Chicken monoclonal antibodies against synthetic bovine prion protein peptide. *J Vet Med Sci* **60**, 777–9.

40. Nakamura, N. A. Y., Horiuchi, H., Furusawa, S., Yamanaka, H. I., Kitamoto, T. & Matsuda, H. (2000) Construction of recombinant monoclonal antibodies from a chicken hybridoma line secreting specific antibody. *Cytotechnology* **32**, 191–8.

41. McCormack, W. T., Tjoelker, L. W. & Thompson, C. B. (1993) Immunoglobulin gene diversification by gene conversion. *Prog Nucleic Acid Res Mol Biol* **45**, 27–45.

42. Davies, E. L., Smith, J. S., Birkett, C. R., Manser, J. M., Anderson-Dear, D. V. & Young, J. R. (1995) Selection of specific phage-display antibodies using libraries derived from chicken immunoglobulin genes. *J Immunol Methods* **186**, 125–35.

43. Tsurushita, N., Park, M., Pakabunto, K., Ong, K., Avdalovic, A., Fu, H., Jia, A., Vasquez, M. & Kumar, S. (2004) Humanization of a chicken anti-IL-12 monoclonal antibody. *J Immunol Methods* **295**, 9–19.

44. Chiliza, T. E., Van Wyngaardt, W. & Du Plessis, D. H. (2008) Single-chain antibody fragments from a display library derived from chickens immunized with a mixture of parasite and viral antigens. *Hybridoma (Larchmt)* **27**, 413–21.

45. Yamanaka, H. I., Inoue, T. & Ikeda-Tanaka, O. (1996) Chicken monoclonal antibody isolated by a phage display system. *J Immunol* **157**, 1156–62.

46. Sapats, S., Gould, G., Trinidad, L., Parede, L. H., David, C. & Ignjatovic, J. (2005) An ELISA for detection of infectious bursal disease virus and differentiation of very virulent strains based on single chain recombinant chicken antibodies. *Avian Pathol* **34**, 449–55.

47. Nakamura, N., Shuyama, A., Hojyo, S., Shimokawa, M., Miyamoto, K., Kawashima, T., Aosasa, M., Horiuchi, H., Furusawa, S. &

Matsuda, H. (2004) Establishment of a chicken monoclonal antibody panel against mammalian prion protein. *J Vet Med Sci* **66**, 807–14.

48. Finlay, W. J., Shaw, I., Reilly, J. P. & Kane, M. (2006) Generation of high-affinity chicken single-chain Fv antibody fragments for measurement of the *Pseudonitzschia pungens* toxin domoic acid. *Appl Environ Microbiol* **72**, 3343–9.
49. Lee, Y. C., Leu, S. J., Hung, H. C., Wu, H. H., Huang, I. J., Hsieh, W. S., Chiu, W. T., Hsieh, M. S., Cheng, T. F. & Yang, Y. Y. (2007) A dominant antigenic epitope on SARS-CoV spike protein identified by an avian single-chain variable fragment (scFv)-expressing phage. *Vet Immunol Immunopathol* **117**, 75–85.
50. Leonard, P., Safsten, P., Hearty, S., McDonnell, B., Finlay, W. & O'Kennedy, R. (2007) High throughput ranking of recombinant avian scFv antibody fragments from crude lysates using the Biacore A100. *J Immunol Methods* **323**, 172–9.
51. Sato, Y., Nishimichi, N., Nakano, A., Takikawa, K., Inoue, N., Matsuda, H. & Sawamura, T. (2008) Determination of LOX-1-ligand activity in mouse plasma with a chicken monoclonal antibody for ApoB. *Atherosclerosis* **200**, 303–9.
52. Raats, J., van Bree, N., van Woezik, J. & Pruijn, G. (2003) Generating recombinant anti-idiotypic antibodies for the detection of haptens in solution. *J Immunoassay Immunochem* **24**, 115–46.
53. Nishibori, N., Horiuchi, H., Furusawa, S. & Matsuda, H. (2006) Humanization of chicken monoclonal antibody using phage-display system. *Mol Immunol* **43**, 634–42.
54. Hof, D., Hoeke, M. O. & Raats, J. M. (2008) Multiple-antigen immunization of chickens facilitates the generation of recombinant antibodies to autoantigens. *Clin Exp Immunol* **151**, 367–77.
55. Nielsen, U. B., Cardone, M. H., Sinskey, A. J., MacBeath, G. & Sorger, P. K. (2003) Profiling receptor tyrosine kinase activation by using Ab microarrays. *Proc Natl Acad Sci U S A* **100**, 9330–5.
56. Torrance, L., Ziegler, A., Pittman, H., Paterson, M., Toth, R. & Eggleston, I. (2006) Oriented immobilisation of engineered single-chain antibodies to develop biosensors for virus detection. *J Virol Methods* **134**, 164–70.
57. McElhiney, J., Drever, M., Lawton, L. A. & Porter, A. J. (2002) Rapid isolation of a single-chain antibody against the cyanobacterial toxin microcystin-LR by phage display and its use in the immunoaffinity concentration of microcystins from water. *Appl Environ Microbiol* **68**, 5288–95.
58. Pavlickova, P., Schneider, E. M. & Hug, H. (2004) Advances in recombinant antibody microarrays. *Clin Chim Acta* **343**, 17–35.
59. Lichty, J. J., Malecki, J. L., Agnew, H. D., Michelson-Horowitz, D. J. & Tan, S. (2005) Comparison of affinity tags for protein purification. *Protein Expr Purif* **41**, 98–105.
60. Weiss, E., Chatellier, J. & Orfanoudakis, G. (1994) In vivo biotinylated recombinant antibodies: construction, characterization, and application of a bifunctional Fab-BCCP fusion protein produced in *Escherichia coli*. *Protein Expr Purif* **5**, 509–17.
61. Saviranta, P., Haavisto, T., Rappu, P., Karp, M. & Lovgren, T. (1998) In vitro enzymatic biotinylation of recombinant fab fragments through a peptide acceptor tail. *Bioconjug Chem* **9**, 725–35.
62. Warren, D. J., Bjerner, J., Paus, E., Bormer, O. P. & Nustad, K. (2005) Use of an in vivo biotinylated single-chain antibody as capture reagent in an immunometric assay to decrease the incidence of interference from heterophilic antibodies. *Clin Chem* **51**, 830–8.
63. Vikholm-Lundin, I. & Albers, W. M. (2006) Site-directed immobilisation of antibody fragments for detection of C-reactive protein. *Biosens Bioelectron* **21**, 1141–8.
64. Raats, J. M. & Hof, D. (2005) Recombinant antibody expression vectors enabling double and triple immunostaining of tissue culture cells using monoclonal antibodies. *Eur J Cell Biol* **84**, 517–21.
65. Ono, K., Kamihira, M., Kuga, Y., Matsumoto, H., Hotta, A., Itoh, T., Nishijima, K., Nakamura, N., Matsuda, H. & Iijima, S. (2003) Production of anti-prion scFv–Fc fusion proteins by recombinant animal cells. *J Biosci Bioeng* **95**, 231–8.
66. Akamatsu, Y., Pakabunto, K., Xu, Z., Zhang, Y. & Tsurushita, N. (2007) Whole IgG surface display on mammalian cells: application to isolation of neutralizing chicken monoclonal anti-IL-12 antibodies. *J Immunol Methods* **327**, 40–52.
67. Wieland, W. H., Lammers, A., Schots, A. & Orzaez, D. V. (2006) Plant expression of chicken secretory antibodies derived from combinatorial libraries. *J Biotechnol* **122**, 382–91.

Chapter 7

Engineering Protein Stability

Ciarán Ó'Fágáin

Abstract

This article defines protein stability, emphasizes its importance and surveys some notable recent publications (2004–2008) in the field of protein stability/stabilization. Knowledge of the factors stabilizing proteins has emerged from denaturation studies and from study of thermophilic (and other extremophilic) proteins. One can enhance stability by protein engineering strategies, the judicious use of solutes and additives, immobilization, and chemical modification in solution. General protocols are set out on how to measure the kinetic thermal stability of a given protein and how to undertake chemical modification of a protein in solution.

Key words: Protein stability, Thermal stability measurement, Thermophile, Protein engineering, Stabilizing additives, Immobilization, Chemical modification

1. Recent Progress in Protein Stability/Stabilization

1.1. The Vital Importance of Protein Stability

The usefulness of enzymes and proteins as analytical tools and industrial catalysts is often limited by their requirements for "mild" reaction conditions. Deterioration of protein preparations over extended storage periods (i.e., a limiting "shelf life") is another serious drawback. Protein stability has great practical consequence throughout the biotechnology industry and is an important topic of interest for protein scientists. Even more importantly, however, biological fitness and evolution may depend on protein stability (despite the general acceptance that evolution selects for a protein's biological activity rather than its stability). In order to function, a protein must fold into a stable native structure and remain in that folded state under physiological conditions. Thus, a protein must attain a minimal stability threshold in order to function and evolve. Higher stability makes a protein more robust to the effects of mutations, since a given mutation is less

Dermot Walls and Sinéad T. Loughran (eds.), *Protein Chromatography: Methods and Protocols*, Methods in Molecular Biology, vol. 681, DOI 10.1007/978-1-60761-913-0_7, © Springer Science+Business Media, LLC 2011

likely to cause the protein to fall below the minimum stability threshold. Such a protein can tolerate a wider range of mutations (not only potentially beneficial but also perhaps destabilizing) while still maintaining its native structure. In this way, extra stability promotes evolvability (1). Other work has shown that decreased stability of any protein coded by an essential gene could lead to a lethal phenotype. Zeldovich et al. demonstrate a relationship between mutation rate, genome size, and thermodynamic changes in proteins due to point mutations. Molecular evolution rates are subject to a universal "speed limit", above which populations become extinct due to lethal mutagenesis. For mesophiles, this critical rate is approximately six mutations per essential part of genome per replication; for thermophiles, it is one to two mutations per genome per replication. This relationship has implications for populations as well as individuals: RNA viruses are close to the speed limit, while DNA viruses and many bacteria are well below it (2).

1.2. Definition and Measurement of Protein Stability

The term *stability* refers to a protein's resistance to adverse influences such as heat or denaturants, that is, to the persistence of its molecular integrity or biological function in the face of high temperatures or other deleterious influences. A folded, functional monomeric protein can lose its biological activity in vitro by *unfolding* of its tertiary structure to a disordered polypeptide, in which key residues are no longer aligned closely enough for continued participation in functional or structure-stabilizing interactions. Such unfolding is termed *denaturation*. It is usually cooperative and may be reversible if the denaturing influence is removed, since the polypeptide chain has not undergone any chemical changes (3).

In addition to unfolding, an isolated protein in an aqueous system can suffer various adverse reactions over time (due to physical, chemical, and biological factors) that lead to an irreversible loss of activity or *inactivation* (3). Typical physical phenomena are aggregation (4, 5) and precipitation, while an individual chain "attempting" to refold may enter an incorrect, kinetically trapped conformation from which it cannot emerge. There are many deleterious chemical reactions involving the side chains of amino-acid residues, notably asparagine, aspartic acid, and cysteine/cystine (4–6) or the glycation of lysine residues with reducing sugars (the Maillard reaction) (7, 8). Biological deterioration can result from loss of an essential cofactor or from the action of proteolytic enzymes, either endogenous or arising from microbial contamination. An unfolded, extended polypeptide will be much more prone to proteolysis than a tightly packed, globular protein (9).

These different molecular phenomena give rise to two distinct definitions of in vitro protein stability. These are *thermodynamic* (or *conformational*) stability and *long-term* (or *kinetic*) stability. Thermodynamic stability concerns the resistance of the folded protein conformation to denaturation (i.e., its Gibbs free energy of unfolding), while long-term stability measures the resistance to irreversible inactivation (i.e., persistence of biological activity under adverse conditions of temperature, pH, solvents, salt concentration, and so on). Both types can be represented in a single scheme (10, 11):

$$\mathrm{N} \overset{K}{\leftrightarrow} \mathrm{U} \overset{k}{\rightarrow} \mathrm{I}$$

where N stands for the native, U for the unfolded (denatured), and I for the irreversibly inactivated forms of the protein. K is the equilibrium constant for the reversible N $\leftrightarrow$ U transition, while k is the rate constant for the irreversible U $\rightarrow$ I reaction. The N $\leftrightarrow$ U transition corresponds to thermodynamic stability, while the U $\rightarrow$ I reaction represents long-term (kinetic) stability. Table 1 sets out the main features of various indices of protein stability. Measurement of the Gibbs energy value is beyond the scope of this chapter, though see Table 1 (and its references); see also Note 1.

Table 1
Indices of protein stability

Parameter	Measures	Constraints	How estimated	Ref.
T_{50}	Temperature of half-inactivation (over a fixed time)	None	$\text{Activity}_{\text{Elevated temperatures}} / \text{Activity}_{\text{“Normal” temperature}} \times 100\%$	–
% Activity	Residual activity at time t (at a fixed temperature)	None	$\text{Activity}_{\text{time } t} / \text{Activity}_{\text{time } 0} \times 100\%$	(10)
Half life	Time to reach 50% activity	First order decay	$0.693/k$ (k = first order rate constant)	(10)
Accelerated degradation	Predicts lifetime at a given temperature	First order decay at all temperatures used	Extrapolation of plot ln k vs. $1/T$ (Kelvin)	(12–14)
T_m	“Melting” temperature	None	Temperature where unfolding = 50%	(15)
$\Delta G_{(25°C)}$	Conformational stability	Two-state unfolding	Thermal denaturation curve	(16)
$\Delta G_{(H2O)}$	Conformational stability	Two-state unfolding	Denaturant unfolding curve	(16, 17)
T_S	Temperature of maximum stability	Two-state unfolding	Stability curve	(18)

Kinetic stability is distinct from (and need not correspond with) *thermodynamic* stability. It involves measuring the persistence of catalytic (or other biological) activity with time under adverse conditions of temperature, pH, solvents, salt concentration, and so on (or, to put it another way, the progressive loss of function). It can be represented by the scheme

$$N \xrightarrow{k_{in}} I$$

where N is the native, functional protein, I is an irreversibly inactivated form, and k_{in} is the rate constant for the inactivation process. The equation $V_{in} = -d[N]/dt = k_{in}[N]$ describes the process mathematically, where V_{in} is the experimentally observed rate of disappearance of the native form (10). Experimentally, determination of kinetic stability (usually equated with the molecule's long-term stability (10)) is fairly simple: see Subheading 3.2. If the loss of activity with time fits a single exponential function (i.e., if a plot of ln (activity) versus time is linear), then one may use the first-order rate constant k to calculate a true half-life (see Note 2).

1.3. Scope of Review

The following sections attempt to give a very selective overview of developments during the 5 years (2004–2008) in the field of protein stability/stabilization (see Note 3). Emphasis is mainly on discoveries regarding extremophiles/thermophiles and protein engineering (by genetic manipulation); however, stability enhancements achieved by other means (use of additives, immobilization, chemical modification of proteins in solution) are also described. In addition, some new or refined methods for protein stability measurement are outlined (Table 2). The reader may notice some overlap between these themes, as the boundaries between them are not watertight. Previous reviews on this field include references (27–29).

Manning and Colón (30) suggest that the physical basis for kinetic stability (very slow unfolding due to a high barrier to unfolding) is a rigid protein structure. While the detergent sodium dodecyl sulphate (SDS) unfolds most proteins (hence its widespread use in gel electrophoresis), a few are resistant. SDS resistance seems to be an attribute of kinetic, rather than folding, stability. These workers prepared SDS-containing solutions of different proteins and boiled a portion of each. Boiled and unboiled samples were electrophoresed on polyacrylamide gel and their migration distances compared. Eight proteins were notably resistant to unfolding by SDS. These same proteins also unfolded very slowly in guanidine hydrochloride (confirming their kinetic stability) and were very resistant to digestion by proteinase K. It is thought that a rigid protein structure may account

Table 2
Additional techniques for protein stability measurement.

Technique	Protein studied	Remarks	Ref.
Differential scanning fluorimetry	Hen egg lysozyme, pig heart citrate synthase	A dye compound shows increased fluorescence on binding to newly exposed hydrophobic regions of an unfolded protein. Thermal unfolding is conducted and monitored in a real-time PCR machine. A preferred dye is SYPRO Orange. Requires single-step transition but only small amounts of protein (75 μg/mL). Allows screening of potentially stabilizing ligands. Detailed procedural instructions are given	(19)
Differential scanning fluorimetry	Y220C oncogenic mutant of p53 tumour suppressor protein	p53 is only marginally stable and many oncogenic mutants, including Y220C, are very unstable. This study tested the stabilizing effects of small ligands on the Y220C protein. SYPRO Orange dye was used to bind to 10 μM protein samples at scan rate 270 K/h due to the irreversibility of the unfolding. Thermal unfolding was also followed by differential scanning calorimetry (scan rate 250 K/h) and unfolding kinetics by time-dependent Trp fluorescence	(20)
SPROX (stability of proteins from rates of oxidation)	Four model proteins (RNase, ubiquitin, cyclophilin A, carbonic anhydrase)	H_2O_2 (at constant concentration) oxidizes proteins, in presence of increasing concentrations of a chemical denaturant, for a given time before quenching. Extent of oxidation (irreversible, unlike H/D exchange) is determined by mass spectrometry. Underlying assumptions are set out. Oxidation rates are consistent with Met oxidation to sulfoxide. Allows analysis of proteins in complex mixtures. Also applicable to protein–ligand complexes	(21)
High-throughput stability analysis using yeast surface two-hybrid system	Fibronectin type III domain	Combines yeast surface two-hybrid and fragment reconstruction methods to quantify protein stability. Can estimate tolerance of amino acid substitutions at a single position by analysis of a randomized library. Uses Trp fluorescence to probe denaturation by guanidine thiocyanate	(22)
Nuclear magnetic resonance	Nine test proteins. Software validated against five sample proteins	Defines and calculates an autocorrelation function C(ω) of the one-dimensional proton NMR spectrum. The value of C(ω) at 0.5 ppm differs between folded, partly folded and random coil proteins: value is >0.5 for folded, and <0.4 for partly folded or unfolded, proteins. No isotope labelling is required	(23)

(continued)

Table 2 (continued)

Technique	Protein studied	Remarks	Ref.
Pulse proteolysis	*E. coli* Rnase H (and variants I53A, I53D); ligand binding to Maltose binding protein	Unfolded proteins are much more prone to proteolysis than folded proteins. Pulse proteolysis (using excess thermolysin over timescale of minutes) follows overnight incubation of target protein in various urea concentrations. Only the unfolded fraction is digested. Proteolysis is quenched with EDTA and normal SDS–PAGE is used for separations. ΔG values for unfolding agree with those obtained by other methods. May be a useful high-throughput method and can be used on crude lysates, although suitable pulse times and conditions need to be ascertained. 30–500 μg sample may suffice	(24)
Capillary Iso-Electric Focusing with Whole-Column Imaging Detection (CIEF-WCID)	β-Lactoglobulin B (p*I* 5.2), trypsin inhibitor (p*I* 4.5), phosphorylase b (p*I* 6.3), trypsinogen (p*I* 9.3); also with phosphatidyl choline (zwitterionic) and phosphatidyl serine (acidic)	Denaturation at 60°C yields altered CIEF profile, with fast separations, for all four proteins. The two phospholipids interact with the proteins in different ways. Image of entire column is detected by a camera. The phospholipids form vesicles which can influence stability	(25)
96-Well microtitre plates	Bovine and equine cytochrome c, bovine serum albumin	Uses microplate-reader fluorimeter to monitor Trp fluorescence at 340 nm. High-throughput method uses serial additions of denaturant to a single well; more accurate medium-throughput method uses one well per denaturant concentration. Both methods give correct protein stability rank orders	(26)

for these diverse resistances and that SDS resistance may serve as a simple means of identifying proteins with kinetically stable native conformations. Interestingly, most of the kinetically stable SDS-resistant proteins have oligomeric β-sheet structures.

1.4. Extremophilic and Thermophilic Proteins

1.4.1. Comparisons with Mesophilic Proteins

Comparisons between mesophilic and thermophilic homologues are often undertaken to probe the special stabilizing factors pertaining to these remarkably robust proteins. Lee et al. (31) studied denaturant-induced unfolding of homotetrameric L-arabinose isomerase (AI) from *Bacillus halodurans* (mesophile), *Geobacillus stearothermophilus* (thermophile), and *Thermotoga maritima* (hyperthermophile) by circular dichroism spectroscopy. Each AI underwent a two-state transition between native tetramer and unfolded monomers, so the usual linear extrapolation method could be used to estimate intrinsic stabilities. Stability curves of both thermophilic holo-AIs were broader than their apo-counterparts, indicating higher melting temperatures (T_m) due to lower heat capacity (ΔC_p). Also, the difference in ΔC_p value between the holo- and apo-proteins was greater for the thermophiles than for the mesophile. It seems that the increased structural stability of the thermophilic AIs is metal-dependent and that the extent of this dependence correlates with oligomerization.

Schafer et al. (32) examined the X-ray crystal structure of the heat- and acid-stable maltose-binding protein (MBP) of *Alicyclobacillus acidocaldarius* and compared it with the corresponding protein in a mesophile (*E. coli*) and in two hyperthermophiles (*Pyrococcus furiosus* and *Thermococcus litoralis*). The *A. acidocaldarius* MBP has fewer charged residues, but more polar residues, than the other three. It has a higher number of basic residues exposed on its surface (giving it a very positive surface charge), while most acidic residues are buried in the core. Nevertheless, the content of acidic and basic residues is nearly equal. There are fewer buried salt bridges than in the other MBPs, but the number of surface-exposed salt bridges is not unusual. These properties seem to relate to *A. acidocaldarius* MBP's acid tolerance rather than to its thermal stability. Examination of cavities within the protein structures indicates that the extremophile proteins are more closely packed than is the mesophile protein. The three extremophile proteins have a slightly higher Pro content, and Pro is more common at the second position within a β-turn. These factors probably contribute to thermostability.

Yokota et al. (33) undertook a comparative analysis of thermophilic and mesophilic proteins using a large dataset of 47 homologous polypeptide pairs. The thermophilic proteins showed higher frequencies of Arg, Glu, and Tyr on their surfaces, while Ala often occurred in the cores. Surface Gln and Met are unsuitable for thermal stability, as are Cys and Ser in the core.

Further insight into protein thermophilicity has been gained from a biophysical study of protein structure graphs (PSGs), based on the strengths of noncovalent interactions between amino-acid side chains, for 232 proteins (34). Aromatic residues, together with Arg, His, and Met, form strong hubs (i.e., they are residues with a strong preference to be highly connected), while Ile and Leu form weak hubs. These hubs help bring together different secondary structural elements in the tertiary structures of the proteins and contribute to the enhanced stability that is a feature of thermophilic proteins.

Pechkova et al. (35) performed a bioinformatic analysis of protein homologous pairs from mesophilic and thermophilic organisms to investigate the role of the aqueous environment in protein structure. Thermophilic and mesophilic bacterial thioredoxins were then compared in self-assembly and Langmuir–Blodgett studies at increasing temperatures using nanogravimetry. Both approaches pointed to a role for inner bound water in determining protein thermal stability.

1.4.2. Features of Particular Thermophilic Proteins

Genomic evidence shows that disulfide bonds play a very important role in the structural stabilization of the intracellular proteins of a distinct subset of thermophilic prokaryotes. A specific protein, protein disulfide oxidoreductase, is thought to play a key role in intracellular disulfide formation in these organisms (36). Boutz et al. (37) found a highly unusual arrangement in the homodimeric citrate synthase of *Pyrobaculum aerophilum*, a hyperthermophilic archaeon. Each subunit has a single intramolecular disulfide, the formation of which cyclizes the protein chain. Remarkably, this cyclization links the polypeptide topologically to its partner subunit (which undergoes a similar cyclization) such that they cannot separate.

Another unusual arrangement exists in the antioxidant Dps protein of *Thermosynechococcus elongatus*, a thermophilic cyanobacterium (38). These DNA binding proteins comprise 12 identical subunits that assemble with 2,3 symmetry in a compact cage-like arrangement. The dimeric and trimeric interfaces in this thermophilic Dps (featuring salt bridges and hydrogen bonds) are quite different from the interfacial hydrophobic interactions of their mesophilic counterparts. A noteworthy aspect of the *T. elongatus* Dps is the chloride ion that coordinates with Arg residues that line the opening of the Dps-like pore towards the internal cavity.

The structural stability of trehalose–maltose binding protein MalE1 of *Thermus thermophilus* was studied by Fourier infrared spectroscopy and *in silico* methods. This 48 kDa monomer has mainly α-helical structure with a small amount of P-structure. Stability was investigated in the presence and absence of (1) maltose and (2) SDS at neutral pH and pH 0.8. Both MalE1 α-helices were very thermostable in the absence of SDS, even at basic pH values. Reversible changes to β-sheet structure occurred at high temperature; these reverted to the previous structure when temperature was decreased. Hydrophobic interactions are thought to be important contributors to MalE1's high thermal stability (39).

Cattoni et al. (40) studied the kinetic thermal stability of recombinant CopA, a P-1B-type Cu^{+}-ATPase from the hyperthermophile *Archaeoglobus fulgidus*, expressed in *E. coli*, using Trp fluorescence, Phe UV spectroscopy, far-UV circular dichroism, and binding of the hydrophobic probe 1-aniline-8-naphthalenesulfonate. The enzyme was maximally active at 75°C, but in the absence of substrates, it underwent an irreversible exponential loss of activity between 66 and 85°C. This behavior indicated a two-state unfolding process involving fully active and inactive molecules. Loss of activity was associated with irreversible partial unfolding of the polypeptide, but the inactive protein retained large hydrophobic regions and much secondary structure. Although CopA inactivated much more slowly than similar mesophilic P-type ATPases, its activation energy value was close to that of mesophiles.

An avidin-related gene expressed as a recombinant protein AVR4 was found to be hyperthermostable and to bind biotin with high affinity. The enhanced thermal stability likely arises from the conformation of the L3,4 loop, within which there are a tandem Pro-Gly sequence and an Asp-Arg ion pair. These act to rigidify the protein in both its apo and biotin-complexed forms. In addition, unlike avidin and streptavidin, AVR4 has Tyr115 at its 1-3 monomer–monomer interface. This permits additional interactions between monomers (specifically, a Tyr-Tyr pi-pi interaction and a hydrogen bond with Lys92) that correlate closely with AVR4's enhanced thermal stability (41).

Some extremophilic proteins, such as the α-amylase from *Halothermothrix orenii*, need to combine halophilicity with thermophilicity. Usually, halophilic proteins have an acidic surface, but the *H. orenii* amylase lacks this feature. Instead, it is unique in forming reversible polydispersed oligomers that possess unusually high thermal stability (42).

1.4.3. A Stability-Activity Trade-Off?

There has long been a debate as to whether proteins are forced to compromise between stability and optimal activity/function. This continues despite literature examples of improved function *and* stability obtained by DNA shuffling (e.g., (43)). Mukaiyama et al. (44) describe an example of a hyperthermophilic protein, ribonuclease HII of *Thermococcus kodakarensis*, that can be said to undergo a stability–function trade-off. They examined the enzymatic activity and thermostability of active site mutants D7N, E8A, E8Q, D105A, and D135A in comparison with wild type. Differential scanning calorimetry was used to monitor thermal unfolding, while guanidine hydrochloride-induced denaturation was observed by circular dichroism (220 nm). Both heat- and denaturant-induced unfolding were highly reversible. Except for E8Q (Glu → Gln), all mutations notably increased stability (by 7.0–11.1 kJ/mol at 50°C) but at the cost of greatly decreased catalytic activity. E8Q had little catalytic effect and enhanced stability by only 2.5 kJ/mol.

1.5. Protein Engineering for Enhanced Stability

1.5.1. Rational Mutations

A protein may need to tolerate adverse factors other than high temperatures. Chemical or oxidative stability can be of great interest and importance also. Valderrama et al. achieved notable improvements in the resistance of iso-1-cytochrome *c* to oxidation by following a redox-based design strategy. This heme-containing protein can react with hydrogen peroxide but is vulnerable to oxidative inactivation by excess peroxide substrate. Introduction of five mutations (N52I, W59F, Y67F, K79A, F82G) endowed the protein with full stability against catalytic peroxide concentrations but with a total turnover number 15 times greater than the native protein. Their study clearly shows that rational changes to the intramolecular electron transfer network can prevent suicide inactivation of a heme peroxidase. Note that most of the changes involve the substitution of the chemically reactive side chains of Asn, Trp, Tyr, and Lys (45).

Stability at extremes of pH can be important in industrial situations, especially where cleaning-in-place of chromatography media must take place between purification runs. Protein G, a widely used affinity chromatography ligand for antibody purification, needs to withstand passage of caustic alkali (0.5 M NaOH) solutions. Palmer et al. (46) achieved an eightfold gain in alkaline stability of protein G by replacing its three Asn residues. Electrostatic calculations indicated that deprotonation at high pH of Tyr, Lys, and Arg residues would be destabilizing. Incorporation of the further triple mutation Y3F/T16I/T18I gave a further stability gain (6.8 cal/mol) and the resulting protein G unfolded at around pH 13, 1.5 units higher than wild type.

Kinetic stability is particularly important for industrial enzymes such as lipase, a broadly specific (and hence very versatile) enzyme. Rodriguez-Larrea et al. studied heat denaturation of wild-type *Thermomyces lanuginosa* lipase, of four single-site mutants and of two very stable multiple-site mutants. Denaturation was two-state, irreversible, and kinetically controlled in all cases. Mutations greatly affected activation enthalpy and entropy, but not the kinetic urea *m*-value. Authors concluded that the mutations affected some structural feature of the transition state for irreversible denaturation that is not related to solvent accessibility changes. They went on to propose that a solvation barrier (due to a time separation between the breaking of internal contacts and penetration by water) may contribute to the *T. lanuginosa* lipase's stability (47).

Membrane proteins are often challenging research targets due to their immersion in a hydrophobic lipid bilayer instead of aqueous surroundings. Minetti et al. have reviewed recent work on membrane protein folding and stability. Representative examples of α-helical and β-barrel structures, viral receptors, and pore-forming toxins are discussed, as are techniques used in the field (48).

1.5.2. Directed Evolution

Wunderlich et al. (49) applied their Proside directed-evolution method to the *Bacillus subtilis* cold shock protein CspB. They achieved an increase over wild type of 31.2°C in the midpoint of thermal unfolding by combining the mutations M1R, E3K, K65L, and E66L. The effects of mutations were strongly interdependent. The stabilizing effects of charge mutations did not correspond with net changes in the protein's charge, nor with ion-pair formation. The Proside method exploits the enhanced resistance of stabilized protein variants to protease digestion and links this with the infectivity of a filamentous phage. Protease concentrations were increased in later selection rounds to increase the selection pressure.

Shortly afterwards, Proside was used to identify stabilizing changes to domain 1 of streptococcal protein G (Gβ1), in parallel with in vitro evolution of gene libraries via codon randomization at positions 16, 18, 25, and 29. Notable increases in thermal stability were found among 11 selections from five independent libraries (50).

Barakat et al. describe an in vivo screen for protein stability where a test protein's thermal stability correlates directly to the transcriptional regulation of a reporter gene. The technique is based on an engineered chimera comprising a constant N-terminal DNA-binding domain, a variable test protein in the middle, and a constant C-terminal transcription activation domain. The system was proven with a panel of nine variants of the β1 domain of streptococcal protein G (Gβ1), spanning a thermal stability range of 38°C to >100°C, as the test proteins. With a low-stability variant, transcription of the reporter gene is upregulated compared with more stable, less flexible variants. Transformation efficiency and thermal stability are likewise inversely related. (This may be due to peculiarities of the phage display system used and/or to alterations in the physiologically necessary interactions of Gβ1 with other RNA polymerase components.) Results of the in vivo screen correlated with T_m values and susceptibility to proteolysis. Trp43 and Tyr45 play an important role in stabilizing the Gβ1 wild-type domain: mutation of either residue to Ala leads to a dramatic decrease in T_m. Thermal and structural properties of each variant were characterized by spectroscopic means. The screen was combined with *in silico* methods to select mutants with greater structural integrity in a library of randomized variants. A drawback of the screen is that less-stable variants are more successfully transcribed and hence have a selective advantage over more-stable variants. Possibly, replacement of the antibiotic-resistance reporter gene by one yielding a toxic phenotype (that would kill cells hosting less-stable mutants) would give a more effective screen (51).

G protein-coupled receptors mediate many important signal transduction events. Sarkar et al. subjected a mammalian G protein-coupled receptor to directed evolution and achieved notably increased heterologous expression levels in both prokaryotic and eukaryotic host cells. The evolved variant was also more stable to solubilization and purification but functioned like the wild type biochemically. The multiple small changes that led to these improvements would not have been easily achieved by rational design. Later, a single amino-acid change was identified that abolished binding of an antagonist while retaining agonist-binding affinity (52).

1.5.3. Inference and Prediction

Many computer-based bioinformatic approaches have been brought to bear on protein stability problems, especially to the prediction of the effects on stability of amino-acid changes in engineered proteins. A detailed discussion is beyond the scope of this chapter, though some recent developments are summarized in Table 3.

Combinatorial protein engineering tends to generate very large libraries. Chaparro-Riggers et al. review how, in what they term "data-driven" protein engineering, existing knowledge (e.g., the consensus approach) can be used to limit library size without preventing the identification of very influential unpredictable substitutions (61). The review by Bommarius et al. (62) is also of interest.

Introduction of "consensus" mutations into a protein has led to notably improved stability in numerous cases. This strategy, reviewed by its originator Steipe, requires only the scrutiny of a number of related, aligned, homologous sequences; no 3-D structural information is necessary. Simply put, the method postulates that the amino acid occurring most frequently at any position in the protein chain must contribute to the stable folding of the global protein sequence and that its frequency arises from some sort of natural selection. Hence, the most frequent amino acid at that position is expected to stabilize the protein. Incorporation of further consensus mutations should lead to an additive effect (63).

Consensus-designed ankyrin repeat proteins (ARPs) have notable thermodynamic stability. Analysis of the structure of designed ARP E3_5 (64) shows that its stability arises from a regular fold with highly conserved structural motifs and hydrogen bonding networks. In contrast, ARP E3_19 is much less stable than E3_5 despite the two ARPs sharing 88% sequence identity and many structural features. E3_19's surface charge distribution, however, is quite different from that of E3_5; the former has clusters of charged residues and more exposed hydrophobic residues (and an unstable C-terminal cap; (65)). Merz et al. continued this work with a further refined ARP library, solving the structure of a full-consensus designed ARP that contains an N-capping repeat, three identical internal repeats, and a C-capping repeat. Structural comparison with E3_5 and E3_19 indicates that surface salt bridges, arranged in a regular network (and complemented by three sulphate ions in the crystal structure), increase thermostability of the full-consensus designed ARP NI3C such that it can withstand boiling. These results underpin the notion that surface-exposed electrostatic interactions and regular charge

Table 3
Predictive/bioinformatic methods relating to protein stability

Method and comments	Ref.
Combines machine learning with structure-based computational mutagenesis. Measures environmental perturbation at the altered position and at its six nearest neighbours in the 3-D protein structure. Extensive training sets of experimentally derived mutant data from a range of proteins are used to instruct machine learning. This combined approach yields better-performing predictive models	(53)
Poisson–Boltzmann theory is used to predict stability changes via a free energy describing both electrostatic (charged) and hydrophobic (uncharged) interactions. Model is fast and computationally simple and worked well with the four proteins tested	(54)
Uses an interpretable prediction tree method (iPTREE-2) to predict stability changes from sequence-based information. Tree gave r^2 values of 0.70 between experimental and predicted values for a dataset of 1,859 single point mutations from ProTherm database	(55)
Computational mutations to estimate protein stability (CMEPS) is a first-principles method that estimates the importance of amino acid R-groups to the stability of a protein fold. Experimental and CMEPS values of $\Delta\Delta G$ gave r^2 values of 0.73 with two validating proteins. CMEPS was then extended to replace each amino acid in the insulin monomer by the small-volume Ala residue, permitting the correlation of stability changes with increased cavity volume. CMEPS correctly identified five residues that are known to be important for stability of the monomeric insulin fold	(56)
Support vector machines predict stability changes from single amino acid replacements based on both sequence and structural information. Predictions were 84% accurate with regard to the sign of the stability changes. Predictive accuracy based on sequence alone is close to that obtained with structural information (although identical/redundant mutations at particular sites should be removed first), so method can be used where 3-D structure is not known. A window size of 7, centred on the target residue, worked best with the sequence-only information. Performance is compared with other published methods. URL for this Mupro tool is http://mupro.proteomics.ics.uci.edu/	(57)
Applies linear algebra to biomacromolecular design. A training dataset of Ala mutants of the Arc repressor was used to derive a quantitative model that correctly distinguished less-stable Ala mutants from those with near-wildtype stability in 40/41 cases. In a test set, 11/12 cases were correctly predicted	(58)
Method analyses voids in proteins (empty cavities not accessed by solvent), calculating total void volume and maximum void size. Despite its low thermal stability, the p53 tumor suppressor protein has average packing. Detrimental p53 mutations occurring in cancer tend to lead to large voids	(59)
Method predicts the geometry and relative stability of point mutants. Geometry optimization was tested against >2,000 pairs of protein structures differing by a single mutation. Training dataset included many small-to-large mutations. Achieved r^2 values of up to 0.82. Gives insights into protein stability improvement in absence of structural information. Large-to-small residues mutations are more accurately predicted than small-to-large ones	(60)

networks are highly stabilizing and imply that the charge interaction network is a design feature of the ARP family (66). A molecular dynamics study focused on N3C, which has a structure that closely resembles that of E3_19 but loses little secondary structure upon unfolding, unlike E3_19. The results indicate that stability of designed ARPs in aqueous solution is increased by a homogeneous charge distribution over the repeat units (65).

Tripp and Barrick (67) added consensus-derived modules to *Drosophila* Notch ARP domains to enhance stability. Insertions between repeats 5 and 6 promoted structure retention at increased denaturant concentrations but had little effect on either far- or near-UV CD spectra, indicating little change to secondary or tertiary structure. The unfolding transition becomes three-state with insertions, however, due to an unfolding event within the C-terminal repeats. Removal of Notch ARP repeats 6 and 7 (i.e., these C-terminal repeats) restores the two-state transition and shows that the high stability of the consensus repeats propagates into the N-terminal, naturally occurring Notch ARPs. Kloss et al. have reviewed the folding and unfolding of repeat proteins (i.e., those with regular repetitive tertiary structures), including ARPs (68).

Another bioinformatic technique, improved configurational entropy (ICE), uses sequence alignments together with optimization of local structural entropy (LSE) to achieve increased protein stability. Sequence segments that likely exist within secondary structural elements will have lower LSE values than unstructured segments. LSE values of all possible four amino acid segments of the overall protein sequence were calculated; these correlate quite well with thermal stability within a given protein family. The technique led to stabilized variants (in its CORE domain) of a mesophilic adenylate kinase, using the sequence information of just a single psychrophilic (*less* stable) homologue. Like conventional consensus methods, ICE requires neither a 3D structure nor a large number of homologous sequences. The method is the subject of a patent application (69).

The contribution of cation-pi interactions to protein stability is being increasingly recognized. Gromiha studied the stabilizing roles of conventional noncovalent and of cation-pi interactions among 62 nonredundant DNA binding proteins. A consensus method based on hydrophobicity, long-range interactions, and amino-acid conservation was used to identify 138 stabilizing residues. A geometric approach, involving distance and energy criteria, was used to pinpoint 196 stabilizing cation-pi interactions. Curiously, the "consensus" stabilizing residues do not contribute to cation-pi interactions, while nearly all the residues forming cation-pi interactions were not identified as stabilizing (70).

Yin et al. proposed a methodology, Eris, to compute the effects on stability of mutations modelled in their protein-modelling suite, Medusa. Using 595 mutants from five structurally unrelated proteins, they evaluated the stability changes resulting from mutations and found significant correlation between experimental and predicted values. The method, which can also model backbone flexibility, is based on physical descriptions of atomic interaction, without reliance on parameter training using experimental data (71). Eris is freely available from a web server (http://eris.dokhlab.org or http://troll.med.unc.edu/eris/login.php).

A key problem in protein engineering is to identify sites or regions that may be randomized to improve stability without compromising either a protein's function or its intrinsic, endogenous stability. Wiederstein and Sippl used knowledge-based potentials to predict such prospectively stabilizing regions. Knowledge-based potentials use information from a database of protein structures to calculate quantitative estimates of forces that contribute to folding. Numerous residues are varied simultaneously and a large number of mutants evaluated. The method is distinct from the use of database-derived potentials to predict effects of changes at a single site. In nine out of ten cases tested, regions of the proteins that yield stable variants upon randomization (usually on the surface) were successfully identified (72).

The ability to predict the stability of a given mutant protein would save time and resources by decreasing the number of experiments to be undertaken. There has been some progress in this direction. Saraboji et al. used the ProTherm database to obtain three different datasets of 1,791, 1,396, and 2,204 mutants and analyzed these for thermal stability (ΔT_m) and for Gibbs energy changes due to thermal ($\Delta\Delta G$) and denaturant-induced ($\Delta\Delta G_{H2O}$) unfolding. Mutants were categorized into 380 possible substitutions; stability was assigned to each based on information obtained with similar types of mutation. This assignment could distinguish stabilizing from destabilizing mutations to an accuracy of 70–80%. Accuracy increased to an average of 82% with a correlation of 0.56 when secondary structure and solvent accessibility were taken into account. Inclusion of information on residues' locations (interior, partially buried, or surface) gave a prediction accuracy of 81% with a correlation of 0.59. Gibbs energy changes ($\Delta\Delta G$) due to mutations could be predicted within a deviation of 0.64 kcal/mol (73).

Most unfolding studies concern classic two-state proteins, since this is a necessary condition for thermodynamic analysis and calculation of ΔG values (see Table 1 and also Note 1). Apoflavodoxin from *Anabaena* PCC7119 is a "three-state" protein with a distinct intermediate during its thermal unfolding process. The structure of its thermal unfolding intermediate is known. (Unfolding by urea is two-state, however.) Campos et al. (74) considered that the overall conformational stability of a three-state protein is due to a "relevant" term (Native-to-Intermediate transition; "relevant" because protein function is likely lost during this step) and a "residual" one (Intermediate-to-Denatured transition). Large increases in apoflavodoxin's *overall* (global) stability resulted from mutations designed from electrostatic calculations or from analysis of sequence conservation. (Electrostatic calculations were more reliable.)

Mapping of these mutations onto the apoflavodoxin thermal unfolding intermediate structure showed that those mutations augmenting the relevant stability (E61K, D126K; two of six mutants prepared) lie in the intermediate's small unstructured region, while the others occur within the native-like region. This illustrates the importance and usefulness of focusing on the Native-to-Intermediate partial unfolding and the structure of protein intermediates.

1.6. Additives and Osmolytes

1.6.1. Peptergents and TMAO

Yeh et al. describe the stabilizing effects of peptergents, self-assembling small peptides with detergent properties, on an integral membrane flavoprotein, glycerol-3-phosphate dehydrogenase. This protein, with six transmembrane spans, is readily solubilized by detergents such as n-octyl-β-D-glucopyranoside but loses activity after a few days when reconstituted in detergent micelles. The enzyme preserves activity up to ten times longer in peptergents. A soluble flavoenzyme, NADH peroxidase, also showed extended activity in presence of peptergents. These positive features may be due to antioxidant properties, although further work is required to clarify exactly how peptergents exert their beneficial effects (75).

TMAO (trimethylamine N-oxide) was used in a conformational stability study of hyperthermophilic ribonuclease HII from *Thermococcus kodakarensis.* At 0.5 M, TMAO conferred higher stability on this 228-residue monomeric protein at all temperatures examined and had similar effects on the kinetics of both unfolding (slower) and refolding (faster). Results show that even thermophilic proteins can benefit from the stabilizing effects of osmolytes. Interestingly, without any TMAO, HII ribonuclease was completely unfolded at 90°C, the optimum growth temperature of *T. kodakarensis.* This indicates that osmolytes may be used in vivo to stabilize this and other proteins (76).

Prion proteins (PrP) have been the subject of intense study due to their role in amyloidogenesis and disease in humans and animals. Granata et al. showed that TMAO destabilized both a shortened human PrP and a full-length sheep PrP. This occurred at pH values both above and below TMAO's pKa of approx. 4.7. TMAO decreased PrP thermal stability at low pH but was also strongly denaturing at room temperature. This is likely due to interaction of TMAO's cationic form with the protein backbone, together with weakened electrostatic interactions that are important for PrP folding (77).

1.6.2. Carbohydrates and Derivatives

In addition to harboring intrinsically stable proteins, extremophile organisms may possess novel stabilizing osmolytes that could be used in vitro to enhance mesophilic protein stability. A negatively charged osmolyte, 2-O-α-mannosylglycerate, that occurs widely among (hyper)thermophiles effectively suppressed the unfolding of recombinant nuclease A from *Staphylococcus aureus*. Faria et al. (78) found a 7°C greater T_m in presence of 0.5 M potassium mannosylglycerate together with a twofold greater unfolding heat capacity. The unfolding pathway was not affected. The results from picosecond time-resolved fluorescence spectroscopy were confirmed by differential scanning calorimetry. It appears that the denatured state is destabilized by preferential exclusion of the solute from the protein hydration shell upon unfolding, while specific interactions stabilize the native state.

Poddar et al. compared the stabilizing effects of oligosaccharides and of equimolar amounts of their constituent monosaccharides. They undertook thermal denaturation of ribonuclease A at pH values between 2.0 and 6.0 in presence of glucose, fructose (monosaccharides), galactose, sucrose (disaccharides), raffinose (trisaccharide), and stachyose (tetrasaccharide). Results indicated that each sugar stabilized the protein in terms of T_m and $\Delta G_{(D)}$ degrees, under enthalpic control, that oligosaccharide-mediated stabilization was much less than that of an equimolar mixture of the constituent monosaccharides and that stabilization by monosaccharides in a mixture is fully additive (79).

Tissues of plants undergoing stress can accumulate polyhydroxy compounds (especially cyclic types known as cyclitols) at concentrations up to 0.4 M. Using six model proteins (bovine serum albumin, *E. coli* glutamine synthetase, pig heart malate dehydrogenase (MDH), phospholipase Cγ1 SH2 domain, SH2_Myc and GST_MycMax fusion proteins), Ortbauer and Popp examined stabilizing effects of numerous polyhydroxy compounds. The compounds D-pinitol (1D-3-O-methyl-chiro-inositol), L-quebrachitol (1L-2-O-methyl-chiro-inositol), myo-inositol, D-chiro-inositol, mannitol, glucose, and trehalose improved the structural and thermal stability of each protein. With 0.4 M myo-inositol, T_m of each protein increased by 3.3–4.7°C (T_m of MDH rose by 5.7°C with 0.8 M myo-inositol). Glycerol and conduritol (1,4/2,3-cyclohexanetetrol) were without effect. The O-methylated cyclitols, D-pinitol and L-quebrachitol, were better preservatives than the less hydrophobic, nonmethylated myo-inositol and D-chiro-inositol. Hydrophobic cyclitols were the most effective stabilizers (80).

1.6.3. Nonsugar Polyols

Mishra et al. studied the effects of polyols, including glycerol, on the folding of phage P22 tailspike protein, a homotrimer. P22 folding is temperature sensitive and depends on the stability of monomeric folding intermediates. Glycerol (1–4 M) greatly increased refolding yields at nonpermissive temperatures and, at low temperature, could offset the refolding-prevention/destabilizing effects of urea. It seems that the beneficial effects of the polyol are due to its stabilizing crucial folding intermediates rather than increased solvent viscosity and that the "chaperone" effect of polyols on the folding of large proteins is due to preferential hydration that favors formation of structure in folding intermediates (81).

The protection afforded by a particular additive may be protein- and even mutant-specific. Diglycerol phosphate, a major osmolyte in the hyperthermophile *Archaeoglobus fulgidus*, has very different effects on rubredoxins from *Desulfovibrio gigas* (has a hairpin loop) and from *D. desulfuricans* (no hairpin loop). A series of loop deletion mutants of *D. gigas* rubredoxin was prepared to explore the role of this loop in the protein's stability. Proton NMR revealed that most structural features were maintained even in the mutant having the greatest deletion (Δ17\29) and that concentrations of the solutes diglycerol phosphate or mannosylglycerate led to subtle conformational changes with temperature. The effect of each solute on kinetic stability was examined by the decrease in absorbance at 494 nm, an index of iron release. The degree of protection given by each solute depended on the particular mutant: 0.1 M diglycerol phosphate destabilized mutant Δ23\29 but endowed wild-type *D. gigas* rubredoxin with a threefold longer half-life. Protein structure seems to be compact in presence of the solutes (82).

1.6.4. Osmolytes in General

Bolen (83) highlights an osmophobic effect that needs to be taken into account together with the better-known excluded volume and preferential interaction parameters to arrive at a full description of osmolyte effects on protein stability and solubility. Osmolytes can "force" proteins to fold or can lead to denaturation. Street et al. use a key observation that the transfer Gibbs energy (ΔG_{tr}) of a protein backbone from water to a water–osmolyte solution is negatively correlated with the osmolyte's fractional polar surface area to attempt to develop a theoretical explanation of the effects of osmolyte–protein interactions on protein stability. This correlation implies that the protein backbone interacts more favorably with polar groups of the osmolyte than with nonpolar groups. Street et al. devised a quantitative solvation model in which interactant polarity and surface influence, respectively, backbone/solvent interaction energy and the number of energetically equivalent ways of achieving a given interaction. Values of ΔG_{tr} calculated using the model correlated strongly (coefficient value 0.99) with measured values, and the model correctly predicted that protecting osmolytes would be preferentially excluded from the protein backbone, while denaturing osmolytes would accumulate at the backbone (84). Auton et al. proposed the use of two experimental metrics to describe the effects of osmolytes on protein properties. Solvophobic/solvophilic effects satisfactorily account for protein stability and the dimensions of the denatured state, while surface tension alone does not (85).

The question arises: do osmolytes influence each other's performance in a mixture? Holthauzen et al. addressed this question (86). They used the *m*-value, an experimental quantity that measures an osmolyte's ability to cause a protein to fold or unfold, as an index of an osmolyte's efficacy. They devised an experimental system, comprising direct and indirect methods, that permitted the evaluation of *m*-values of osmolytes in presence or absence of a second osmolyte. This system is based on a marginally stable protein held at the midpoint of its native-to-denatured transition (34°C in 10 mM buffer, pH 7, 200 mM salt). These conditions allow the evaluation of *m*-values of both protecting (sarcosine) and denaturing (urea) osmolytes in presence or absence of a second osmolyte. In this way, one can assess whether, and to what extent, the two osmolytes affect each other's efficacy. Neither compound changed the other's efficacy in forcing folding or unfolding of the protein; they had independent effects even at multimolar concentrations. This is thought to be due to low occupancy of the large number of osmolyte interaction sites on the protein, such that the two osmolytes neither cooperate nor compete in their interactions with the protein.

Sometimes a given additive can have dual effects. Tween 80, a widely used nonionic surfactant, adversely affected storage stability of interleukin-2 mutein with respect to both oxidation and aggregation in temperature- and formulation-dependent ways. In contrast, Tween 80 notably inhibited shaking-induced aggregation of the same protein (87).

1.6.5. Miscellaneous

It may be stretching matters to regard a partner protein as an additive, but the situation of the important human protein paraoxonase-1 (PON1) greatly influences its stability. PON1 is nowadays thought to protect against vascular disease but was earlier believed to protect against poisoning by toxic organophosphates. The enzyme mainly occurs on HDL (high-density lipoprotein) particles, and both its activity and stability depend on this complex and constantly changing molecular environment. In addition, the recently discovered human phosphate binding protein, HPBP, associates with PON1 and greatly influences the latter's oligomerization, thermal and storage stability. Pure PON1 comprises a mixture of at least two states, with a tendency to form oligomers in the absence of HPBP. Outside the natural ambience, HPBP stabilizes the active conformation(s) of PON1. Although PON1 is naturally very thermostable, association with HPBP greatly slows the denaturation rate. A hybrid recombinant PON1 (from *E. coli*) was more thermostable than the human enzyme, and its stability was not influenced by HPBP. Stabilized PON1 variants have the potential to be used as bioscavengers of toxic organophosphates. Any engineered variant for administration would need to preserve PON1's known vascular protective effect (88, 89).

A solvent system rich in ionic liquid, which does not form ice, provided long-term protection against aggregation and hydrolysis of lysozyme (>200 mg/mL). Reversible thermal unfolding and refolding of the enzyme could also be accomplished (90).

The long-term storage of therapeutic proteins for administration poses major challenges. Any additives or excipients used in the formulation must be nontoxic and nonimmunogenic, and protein aggregation must be avoided. Frokjaer and Otzen have reviewed this topic (91).

1.7. Immobilization

In the case of a therapeutic protein for administration, stability in vivo is just as important as that in vitro. The preparation of biodegradable carriers for the in vivo stabilization and sustained release of therapeutic proteins is a highly desirable goal. Towards this end, Torres et al. (92) described the preparation of microspheres from copolymers of the anhydrides 1,6-bis(p-carboxyphenoxy)hexane (CPH) and 1,8-bis(p-carboxyphenoxy)-3,6-dioxaoctane (CPTEG) formulated in different compositions. Structures of two model proteins, lysozyme and ovalbumin, were maintained upon encapsulation in the amphiphilic polymeric microspheres. Biological activity remained intact following release from the microspheres.

Cross-linked enzyme aggregates, or CLEAs (CLEA is a registered ® trade mark), are stable, recyclable immobilized enzyme preparations. The enzyme of interest is precipitated from aqueous buffer using a nondenaturing precipitant (such as ammonium sulphate) and the enzyme aggregate subjected to cross-linking with a suitable bifunctional reagent. The resulting insoluble preparation provides a high concentration of active catalyst (active sites) and is easily used and recycled. Combi-CLEAs of two or more enzymes can also be prepared for use in one-pot, multistep syntheses (93). The method has been applied to many enzymes (93) including penicillin amidase (94) and subtilisin (95).

Kreiner et al. described the stabilities and catalytic performance of protein-coated microcrystals (PCMC) in organic solvents. These comprise micron-sized water-soluble crystalline biocomposite particles coated with the biocatalyst of interest. The single-step, fast formation procedure both dehydrates the enzyme and immobilizes it onto the microcrystal. One simply adds a concentrated solution of a crystal-forming, water-soluble compound (e.g., a salt; K_2SO_4 is particularly suitable) to the enzyme in buffer solution. This mixture is slowly added dropwise to a vigorously shaken 1-propanol/1% w/w water system. Crystals form and precipitate; these may be stored at room temperature. PCMC of subtilisin Carlsberg and of lipase B from *Candida antarctica* each retained almost 90% of initial activity after one year's storage at room temperature in 1-propanol/1 % w/w water. These impressive stability results seem to be due more to the propanol–water system than to immobilization (but immobilization permits easier handling and use). Temperature effects on catalytic activity (range 25–60°C) and on operational stability depended on the solvent, with better results in the propanol system compared with other solvents tested (96).

Often, the stability of an immobilized enzyme increases with the number of attachment points between the protein and the insoluble phase. Abian et al. (97) mutated three surface residues of penicillin acylase to Lys to provide additional sites for attachment to glyoxyl agarose particles. The Lys substitutions did not alter either the stability or kinetic properties of the enzyme. Wild-type and mutant enzymes were immobilized onto glyoxyl agarose by their Lys residues. The mutant was 4–11 times more stable than wild type under different stress conditions (at low pH, at elevated temperatures, or in 60% dimethylformamide).

Beta-galactosidase is an important enzyme used for lactose hydrolysis in the dairy industry and for other purposes. Unfortunately, the enzyme is not very stable, but some stability gains have resulted from immobilization. Hydrophilization of a previously immobilized beta-galactosidase preparation led to significant stabilization in water-miscible organic cosolvents such as ethanol and acetone. Beta-galactosidase from bacterial, yeast, and fungal sources were immobilized on glutaraldehyde agarose and then treated with a polyaldehyde-dextran polymer followed by a polyamine-dextran polymer. Half-life of the fungal enzyme in 50% v/v acetone increased by 25-fold (98).

A hyperthermophilic single-domain 30 kDa endoglucanase from *Pyrococcus furiosus* (optimum activity at 104°C) became even more stable following adsorption onto polystyrene (hydrophobic) or silica (hydrophilic) surfaces (99). Adsorption was irreversible in both cases, without loss of activity. The soluble protein's denaturation temperature of 108°C increased to 116°C on silica and to 135°C on polystyrene, perhaps the highest protein denaturation temperature yet reported. A heat-tolerant endoglucanase would have great potential for the treatment of cellulosic wastes at high temperature. Specific activity decreased, however, upon immobilization (56% on polystyrene, 51% on silica). This was ascribed to random orientation of the enzyme on the surface, i.e., where the active site faces the solid surface and cannot interact with the polymeric substrate.

Immobilization does not always lead to increased protein stability. While tethering to a surface encourages protein folding by an entropic effect, this may be offset by a destabilizing enthalpic effect of the surface itself. Since both the entropic and enthalpic contributions to stability specifically depend on the topology and placement of tethering, one cannot predict stability a priori. Depending on whether the tethering increases or decreases the protein's ability to form a correct tertiary structure, that protein may be either stabilized or destabilized on a surface, as shown for protein A (100).

1.8. Chemical Modification and Cross-linking

Kim and Stites used cross-linkers varying in length from 10.5 to 21.3 Å to examine the effects of molecular crowding and excluded volume on the stability of single-Cys mutants of staphylococcal nuclease.

Dimers with short linkers showed notable three-state denaturation behavior, in contrast to the two-state denaturation of monomeric controls. Where linker length exceeded 17 Å, few three-state characteristics remained. The three-state profile for closely tethered dimers likely arises from interconversion of native dimer protein (N-N) with a dimer where one protein is denatured (N-D). The remaining native protein of N-D can then undergo denaturation to give the doubly denatured dimer D-D. It seems that excluded volume effects on the denatured state limit its size and flexibility. Curiously, none of the cross-linked dimers showed enhanced stability (101).

Posttranslational glycosylation is a form of chemical modification of a protein. An *in silico* experiment studied the folding of 63 engineered SH3 domain variants that had been glycosylated at different sites of the protein surface with varying numbers of conjugated polysaccharide chains. Thermal stabilization increased with the number of sugar chains attached and depended more on the site of attachment than on the size of the polysaccharide. The enhanced protein stability arises from destabilization of the unfolded state, which is forced by the bulky sugar chains to adopt a more extended conformation instead of a residual structure. The glycosylation-induced thermodynamic stabilization is coupled with kinetic stabilization (102). Remarkably, O-glycosylation destabilized the macrophage activating factor Gc-MAF to the extent of about 1 kcal/mol (103).

1.9. Conclusion

Therapeutic protein substances can be of immense value to patients and are becoming increasingly prominent in the pharmaceutical industry. These proteins must be produced to a very high specification and quality, and the processes involved can affect protein stability. A detailed discussion of this topic is well beyond the scope of this chapter, but a 2008 review summarizes the effects of formulation, fill and finish operations on the stability of biopharmaceuticals and suggests ways of minimizing such problems (104).

Aside from therapeutics, enzymes (and other proteins) find use across a wide spectrum of modern and traditional biotechnology-based industries (105, 106). Enzymes are also ideal tools for green chemistry reactions (107–109). Stability issues are often important in these situations, as with therapeutic proteins. Immense progress has been made in stabilizing proteins (e.g., derivatives that can withstand boiling (110)). Protein stability will continue to attract much scientific interest, especially in light of its evolutionary implications (1, 2). It is hoped that the literature reviewed in this article, and the methods set out below, will assist protein scientists in their work.

2. Materials

Required for Subheadings 3.1 and 3.2:

1. Water bath with adjustable heater.
2. Thermometer.
3. Stop clock.
4. Test tube rack(s).
5. Test tubes (with good heat transfer characteristics and low heat capacity).
6. Adjustable-volume automatic pipette(s) with suitable disposable tips.
7. 96-Well transparent microtiter plates (see Note 7).
8. Ice bath.
9. Assay mix (to measure catalytic or other biological activity of the protein of interest).
10. Personal computer with spreadsheet and graphics application.
    Additional materials for Subheading 3.3:
11. Screw-capped bijou bottles are convenient for the preparation and mixing of small-volume chemical modification reaction mixtures.

3. Methods

Take care to avoid burns or scalds when working with samples, water baths, or other apparatuses at elevated temperatures.

3.1. Thermal Profile

One can determine a thermal profile for an enzyme by placing samples for a fixed, short period at successively increasing temperatures and then assaying the samples' residual activities. This procedure also allows one to select a suitable temperature for a thermal inactivation experiment (Subheading 3.2 below).

1. Incubate a sample of the protein in question plus a sample of native or untreated protein at one of a range of increasing temperatures for a fixed period. Suitable incubation period is 10–15 min, while 10°C increments will do for a first run (see Notes 4–6).
2. When the incubation period at each temperature has elapsed, remove the samples onto ice to cool them rapidly (see Notes 7 and 8).

3. Upon completion of all of the temperature incubations, rewarm the entire complement of samples to room (or assay) temperature and assay the remaining biological activity by your usual method.
4. Defining the activity at the "normal" temperature of assay as 100%, express the remaining activity at each of the progressively rising temperatures as a percentage of this. Plot percent activity remaining against temperature and observe the resulting graph. There will likely be a sharp fall in activity over a narrow temperature range. Inspection of the plot will reveal the temperature where approx. 50% of the original activity remains, i.e., the temperature of half-inactivation, T_{50}, analogous to T_m (Table 1). Be aware that this T_{50} is a purely empirical value and the conditions under which it is obtained must be specified for each case (see Note 9).
5. For a more accurate determination, perform further experiments over narrower increments of 5°C (or even 2°C over the steepest part of the plot) (see Note 10). Possibly, activation effects will occur, where the activity increases with temperature (i.e., to values >100%) until one reaches a threshold temperature where a sharp decline in activity is observed (see also Note 10).

3.2. Thermal Inactivation

Once a suitable fixed inactivation temperature has been ascertained (i.e., one where activity loss is neither too rapid nor inconveniently slow; typically the T_{50} – see Subheading 3.1 above), one can perform a thermal inactivation over time (see Notes 4–7 and 11).

1. When comparing different treatments, include an untreated (control) protein sample in each experiment. This can provide an internal validation and a qualitative result even in cases of experimental variation, of one (or a few) missed time point, or of data that fit poorly to model equations.
2. Place the samples in the water bath or heating block, starting a stop clock at the same time (see Note 6). Immediately remove a "time zero" sample (=100% activity) onto ice. Withdraw further samples onto ice at short intervals; as the experiment progresses, one can extend the intervals between samplings (see Note 12).
3. At the end of the experiment, rewarm all samples and assay under optimal conditions by the customary method.
4. If activity at time zero is defined as 100%, a plot of percent activity remaining against time will allow one to estimate and compare stabilities. Frequently, but not always, the loss of activity will follow a first-order decay to yield an exponential plot of % remaining activity vs. time. Such a graph declines steeply at first before tracking a slower rate of decline.

5. Fit the experimental data to a first-order decay equation using a computer with suitable statistical software, paying special attention to goodness-of-fit, confidence limits, and other appropriate quality indices. If the fit is good, a semilog plot of ln or log (% activity] vs. time will yield a straight line of slope *k*, and one may calculate a true half-life from the first-order rate constant *k* (see Note 13). More complex decay functions will not yield a straight line in semilog form. Nevertheless, even in these cases, empirical comparisons of stability can be made from plots of % activity vs. time (see also Note 11).

3.3. Chemical Modification of a Protein

1. Assemble as much information as possible concerning the protein to be modified: functionally/biologically essential residues (to be avoided or protected), number of Lys, Cys, (Asp + Glu) or other modifiable residues per polypeptide molecule or subunit, molecular mass (to calculate molarity), suitable/unsuitable buffers and storage conditions, and assay method(s). Literature searches will be helpful here.
2. Decide which type of residue(s) to modify and which chemical reagent(s) to use: type of chemistry, single-step or multi-step reaction, monofunctional modifier or bifunctional cross-linker (what length? homo- or hetero-bifunctional?), suitable protocol and scale (see Table 4 and refs. 127, 128).
3. Place the protein of interest in a buffered reaction mixture that is suitable both for it and for the chemical reaction envisaged. In parallel, run a protein sample that is not exposed to the chemical modifier but is otherwise treated similarly to the test sample(s).
4. Add the modifying reagent in considerable molar excess over the number of target groups on each protein molecule. (For example, if a 1 mg/mL protein solution represents a 50 μM concentration of the protein to be modified and there are 10 Lys residues per protein molecule, then the molar concentration of Lys (the residue to be modified, and assuming that all are available for reaction) is 500 μM. The modifying reagent should be used in excess: here, a final modifier concentration of 5 mM in the reaction mixture represents a tenfold excess of reagent over target residue (see Note 14).
5. Ensure that any elevated temperature used will not inactivate the protein.
6. Terminate the reaction after a set time by means of a chemical step or a rapid separation technique such as centrifugal gel filtration. (Dialysis is also effective but requires more time.)
7. Test the catalytic or biological activity of the treated and untreated (activity = 100%) protein samples. Estimate any loss of activity arising from exposure to the chemical(s) concerned.

Table 4
Amino-acid side chains and reagents for their modification

Amino acid	Side chain	Reagents	Reaction	Ref.
Cysteine	Thiol	5,5′-Dithiobis(2-nitrobenzoic acid)[a] Maleimido compounds *p*-Mercuribenzoate[a]	Disulfide Formation Alkylation (112, 113)[a] Addition (112, 113)[a]	(111)
Lysine	Amino	Trinitrobenzene sulphonate[a] Imidates Acid anhydrides (Cyano)borohydride + aldehyde	Addition Amidination Acylation Reductive (117) alkylation	(112, 114)[a] (115) (116)
Arginine	Guanidino	Dicarbonyls 9,10-Phenanthrenequinone[a] Camphorquinone-10-sulfonic acid (119)	Not fully known	(112, 118) (112)[a]
Histidine	Imidazole	Diethylpyrocarbonate[a]	Addition	(112)[a], (120)
Aspartic acid, glutamic acid	Carboxyl	Carbodiimides Trialkoxonium salts	Amidation Esterification (122)	(121)
Tyrosine	Phenol	Tetranitromethane Iodine	Nitration Iodination (124)	(123)
Tryptophan	Indole	N-Bromosuccinimide	Oxidation	(125)
Methionine	Thioether	Hydrogen peroxide	Oxidation	(126)

This table presents only some of the more common reagents for protein R-group modification; it is by no means exhaustive

Reaction of Arg with the dicarbonyl 2,3-butanedione should be carried out in the dark (112)

[a]These compounds may be used for the spectrophotometric (or, in the case of Arg, fluorimetric) quantitation of the content of the reactive R-group in a protein: see respective references

8. If possible, estimate the number of modified/unmodified target residues using a suitable diagnostic chemical reaction (e.g., TNBS for Lys residues, DTNB for Cys; see Table 4 and references therein) or other techniques.
9. Compare the stabilities of the treated and untreated samples as described in Subheadings 3.1 and 3.2.

4. Notes

1. T_m (the melting, or unfolding, temperature) is a robust index of folding stability: it is purely empirical and involves no underlying assumptions. Measurement of a protein's T_m is often carried out spectrophotometrically at a single diagnostic

wavelength, using a thermal programmer which gradually raises the cell temperature over a set range. Folding stability can also be measured by a variety of biophysical techniques, but many of these methods (circular dichroism, hydrogen-deuterium exchange, fluorimetry, differential scanning calorimetry) require the use of specialized (and sometimes quite expensive) equipment. RT-PCR machines are becoming commonplace and can be used for differential scanning fluorimetry (19), while pulse proteolysis (24) needs no specialized apparatus; see Table 2. Calculation of Gibbs energies may be performed on thermal data of this sort or on unfolding studies using urea or guanidine hydrochloride but requires care. Pace (16, 17) has described the experimental and data analysis requirements for reliable Gibbs energy estimations.

2. *k*-Values can also be informative when correlated with other data such as the number of attachment sites to a solid phase (immobilized enzyme) or changes in protein hydrophilicity (10). The activity loss will often be first order, although more complex inactivation patterns are well documented (129). It is possible, however, that an apparently unimolecular first-order time course of inactivation may mask a more complex set of inactivating molecular events (129).
3. A search undertaken on 7 Jan 2009 on ISI Web of Science for the terms "protein AND stabili*" in the TITLES of published papers revealed 1,624 publications for the 5 years 2004–2008. This very broad category is not limited to in vitro proteins and includes reports describing for example effects on in vivo protein stability due to altered ubiquitination patterns, or effects of food proteins on foam stability. While further refinements are needed to filter out reports not relevant to this chapter, it is nevertheless clear that protein stability/stabilization is a field of great interest. The following terms were used separately to refine the initial 1,624 hits, with the number of published papers in brackets: additive (5), chemical modif* (14), consensus (17), crosslink (10), directed evolution (25), extremophil* (2), immobili* (21), osmolyte (17), shuffling (0), site-directed (33), surface modif* (19), thermophil* (31).
4. All samples to be tested should be at uniform protein concentration in an identical buffer composition. Certain ions can stabilize or destabilize proteins, e.g., calcium has stabilizing effects on amylase, peroxidase, and some proteolytic enzymes. Variations in protein concentration can also influence stability.
5. Place a thermometer as close as possible to the samples being incubated so that the temperature indicated accurately reflects that of the samples. Conditions may not be uniform throughout the water bath.

6. When testing multiple samples, stagger the insertion of each one into the water bath by 10–15 s; removal of samples in a similar sequence will ensure the exposure of each one to high temperature for the exact period required.
7. Removal of aliquots at various time points from different samples can lead to a considerable number of samples for assay. We find it very convenient to dispense individual aliquots/time points into the wells of a 96-well microtiter plate that is held on ice until the end of the thermoinactivation experiment. The plate's 12 × 8 grid allows one to arrange the samples in a preplanned fashion, to rewarm them uniformly to assay temperature and, in many cases, to assay in situ if a microassay protocol is possible.
8. One may obtain different values for T_{50} depending on the exact protocol followed: subjecting different aliquots of a protein sample to a single high-temperature incubation (followed by their withdrawal onto ice) is not equivalent to subjecting a single aliquot/sample to progressively increasing temperatures and withdrawing portions of this onto ice after the fixed time has elapsed at each temperature of measurement. We prefer the former procedure. Whichever way one decides to conduct the experiment, use that procedure consistently.
9. The T_{50} and T_m values may well be equivalent in the case of a simple monomeric protein, but one should be aware that they need not agree, especially with oligomeric proteins or proteins containing prosthetic groups.
10. In the case of a thermophilic protein, there may be little or no decline in activity even approaching 100°C. In such a case, use of moderate concentrations of a denaturant, of a reducing and/or chelating agent, or a combination of these, can reduce the T_{50} to a suitable, measurable value.
11. We have always held inactivation samples on ice prior to simultaneous assay of all samples, but others sometimes assay each inactivation sample immediately following removal from elevated temperature. Since proteins inactivate according to the model $N \leftrightarrow U \rightarrow I$, where N is the native (and only active) form, U is reversibly unfolded, and I is irreversibly inactivated (29), it is possible that some refolding may occur during the ice storage stage so that the measured residual activities may represent (N + refolded U) and not N alone. If one observes no difference between data obtained from immediate or delayed assay of time samples in a control thermoinactivation experiment, refolding is unlikely and the observed activities are probably due to N alone.
12. To obtain a good statistical fit of experimental data, it is particularly important to sample as frequently as possible during

the initial steep decline to approx. 40–50% of the starting (time zero) activity. After this period, the curve will be flatter; it is this feature that allows one to extend the intervals between samplings. Practice and familiarity will allow one to refine the sampling intervals and the duration of the experiment.

13. Recall that the equation for a first-order [single] exponential decay is $A_t = A_0 \exp[-kt]$, where A_0 and A_t represent the activities at times zero and t respectively and k is the first order rate constant. Half-life is defined as $0.693/k$.

14. Protein cross-links accomplished with bifunctional reagents may be *intra*molecular (within the same protein molecule) or *inter*molecular (between different protein molecules/subunits). The length or span of the cross-linking reagent, the properties of the protein itself, and the reaction conditions will each influence which type of cross-link forms. With respect to the experimental conditions, low concentrations of highly charged protein (influenced by the buffer pH) with excess amounts of cross-linker will favor the formation of an intramolecular cross-link, while greater concentrations of minimally charged protein (buffer pH close to the protein's p*I*) will tend to favor intermolecular links (130). SDS–polyacrylamide gel electrophoresis will allow one to distinguish between these two possibilities in a stabilized (or otherwise altered) protein that has been treated with a cross-linker: an intrapolypeptide cross-link will not alter the electrophoretic mobility (apparent molecular mass), whereas an interpolypeptide one will lead to decreased migration (due to increased molecular mass/radius).

Acknowledgement

The author thanks Dr Barry J. Ryan for helpful and informative discussions.

References

1. Bloom, J.D., Labthavikul, S.T., Otery, C.R., Arnold, F.H. (2006) Protein stability promotes evolvability. *Proc. Natl. Acad. Sci. USA* **103**, 5869–5874.
2. Zeldovich, K.B., Chen, P., Shakhnovich, E.I. (2007) Protein stability imposes limits on organism complexity and speed of molecular evolution. *Proc. Natl. Acad. Sci. USA* **104**, 16152–16157.
3. Mozhaev, V.V., Martinek, K. (1982) Inactivation and reactivation of enzymes. *Enzyme Microb. Technol.* **4**, 299–309.
4. Liu, W.R., Langer, R., Klibanov, A.M. (1991) Moisture-induced aggregation of lyophilized proteins in the solid state. *Biotechnol. Bioeng.* **37**, 177–184.
5. Costantino, H.R., Langer, R. & Klibanov, A.M. (1995) Aggregation of a lyophilized pharmaceutical protein, recombinant human albumin. *Biotechnology* **13**, 493–496.
6. Volkin, D.B., Middaugh, C.R. (1992) The effect of temperature on protein structure, in *Stability of protein pharmaceuticals, part A: chemical and physical pathways of protein*

degradation (Ahern, T.J., Manning, M.C., eds) Plenum, New York, pp. 215–247.
7. Hageman, M.J. (1992) Water sorption and solid-state stability of proteins, in *Stability of protein pharmaceuticals, part A: chemical and physical pathways of protein degradation* (Ahern, T.J., Manning, M.C., eds) Plenum, New York, pp. 273–309.
8. Quax, W.J. (1993) Thermostable glucose isomerases. *Trends Food. Sci. Technol.* **4**, 31–34.
9. Parsell, D.A., Sauer, R.T. (1989) The structural stability of a protein is an important determinant of its proteolytic susceptibility in *Escherichia coli*. *J. Biol. Chem.* **264**, 7590–7595.
10. Mozhaev, V.V. (1993) Mechanism-based strategies for protein thermostabilization. *Trends Biotechnol.* **11**, 88–95.
11. Zale, S.E., Klibanov, A.M. (1983) On the role of reversible denaturation (unfolding) in the irreversible thermal inactivation of enzymes. *Biotechnol. Bioeng.* **25**, 2221–2230.
12. Jerne, N.K., Perry, W.L.M. (1956) The stability of biological standards. *Bull. World Health Organ.* **14**, 167–182.
13. Kirkwood, T.B.L. (1984) Design and analysis of accelerated degradation tests for the stability of biological standards III. Principles of design. *J. Biol. Stand.* **12**, 215–224.
14. Franks, F. (1994) Accelerated stability testing of bioproducts: attractions and pitfalls. *Trends Biotechnol.* **12**, 114–117.
15. Baldwin, R.L., Eisenberg, D.E. (1987) Protein stability, in *Protein engineering* (Oxender, D.L., Fox, C.F., eds) Alan R. Liss, New York. pp. 127–148.
16. Pace, C.N. (1990) Measuring and increasing protein stability. *Trends Biotechnol.* **8**, 93–98.
17. Pace, C.N. (1986) Determination and analysis of urea and guanidine hydrochloride denaturation curves. *Methods Enzymol.* **131**, 266–280.
18. Becktel, W.J., Schellman, J.A. (1987) Protein stability curves. *Biopolymers* **26**, 1859–1877.
19. Niesen, F.H., Berglund, H., Vedadi, M. (2007) The use of differential scanning fluorimetry to detect ligand interactions that promote protein stability. *Nat. Protoc.* **2**, 2212–2221.
20. Boeckler, F.M., Joerger, A.C., Jaggi, G., Rutherford, T.J., Veprintsev, D.B., Fersht, A.R. (2008) Targeted rescue of a destabilized mutant of p53 by an in silico screened drug. *Proc. Natl. Acad. Sci. USA* **105**, 10360–10365.
21. West, G.M., Tang, L., Fitzgerald M.C. (2008) Thermodynamic analysis of protein stability and ligand binding using a chemical modification- and mass spectrometry-based strategy. *Anal. Chem.* **80**, 4175–4185.
22. Dutta, S., Koide, A., Koide, S. (2008) High-throughput analysis of the protein sequence-stability landscape using a quantitative yeast surface two-hybrid system and fragment reconstruction. *J. Mol. Biol.* **382**, 721–733.
23. Hoffmann, B., Eichmüller, C., Steinhauser, O., Konrat, R. (2005) Rapid assessment of protein structural stability and fold validation via NMR. *Methods Enzymol.* **394**, 142–174.
24. Park, C.W. and Marqusee, S. (2005) Pulse proteolysis: a simple method for quantitative determination of protein stability and ligand binding. *Nat. Methods* **2**, 207–212.
25. Bo, T., Pawliszyn, J. (2006) Protein thermal stability and phospoprotein-lipid interaction investigated by capillary isoelectric focusing with whole column imaging detection. *J. Sep. Sci.* **29**, 1018–1025.
26. Aucamp, J.P., Cosme, A.M., Lye, G.J., Dalby, P.A (2005) High-throughput measurement of protein stability in microtiter plates. *Biotechnol. Bioeng.* **89**, 599–607.
27. Ó'Fágáin, C. (2003) Enzyme stabilization: recent experimental progress. *Enzyme Microb. Technol.* **33**, 137–149.
28. Ó'Fágáin, C. (1997) *Stabilizing protein function.* Springer, Berlin.
29. Ó'Fágáin, C. (1995) Understanding and increasing protein stability. *Biochim. Biophys. Acta* **1252**, 1–14.
30. Manning, M., Colón, W. (2004) Structural basis of protein kinetic stability. *Biochemistry* **43**, 11248–11254.
31. Lee, D.W., Hong, Y.H., Choe, E.A., Lee, S.J., Kim, S.B., Lee, H.S., Oh, J.W., Shin, H.H., Pyun, Y.R. (2005) A thermodynamic study of mesophilic, thermophilic and hyperthermophilic L-arabinose isomerases. *FEBS Lett.* **579**, 1261–1266.
32. Schafer, K., et al. (2004) X-ray structures of the maltose-maltodextrin-binding protein of the thermoacidophilic bacterium *Alicyclobacillus acidocaldarius* provide insight into acid stability of proteins. *J. Mol. Biol.* **335**, 261–274.
33. Yokota, K., Satou, K., Ohki, S. (2006) Comparative analysis of protein thermal stability: differences in amino acid content and substitutions at the surfaces and in the core regions of thermophilic and mesophilic proteins. *Sci. Tech. Adv. Mater.* **7**, 255–262.
34. Brinda, K.V., Vishveshwara, S. (2005) A network representation of protein structures: implications for protein stability. *Biophys. J.* **89**, 4159–4170.
35. Pechkova, E., Sivozhelezov, V., Nicolini, C. (2007) Protein thermal stability: the role of

protein structure and aqueous environment. *Arch. Biochem. Biophys.* **466**, 40–48.
36. Beeby, M., O'Connor, B.D., Ryttersgaard, C., Boutz, D.R., Perry, L.J., Yeates, T.O. (2005) The genomics of disulfide bonding and protein stabilization in thermophiles. *PLoS Biol.* **3**, 1549–1558.
37. Boutz, D.R., Cascio, D., Whitelegge, J., Perry, L.J., Yeates, T.O. (2007) Discovery of a thermophilic protein complex stabilized by topologically interlinked chains. *J. Mol. Biol.* **368**, 1332–1344.
38. Franceschini, S., Ceci, P., Alaleona, F., Chiancone, E., Ilari, A. (2006) Antioxidant Dps protein from the thermophilic cyanobacterium *Thermosynechococcus elongatus*: an intrinsically stable cage-like structure endowed with enhanced stability. *FEBS J.* **273**, 4913–4928.
39. Scire, A., Marabotti, A., Aurilia, V., Staiano, M., Ringhieri, P., Iozzino, L., Crescenzo, R., Tanfani, F., D'Auria, S. (2008) Molecular strategies for protein stabilization: the case of a trehalose/maltose-binding protein from *Thermus thermophilus*. *Proteins* **73**, 839–850.
40. Cattoni, D.I., Flecha, F.L.G., Arguello, J.M. (2008) Thermal stability of CopA, a polytopic membrane protein from the hyperthermophile *Archaeoglobus fulgidus*. *Arch. Biochem. Biophys.* **471**, 198–206.
41. Eisenberg-Domovich, Y., Hytonen, V.P., Wilchek, M., Bayer, E.A., Kulomaa, M.S., Livnah, O. (2005) High-resolution crystal structure of an avidin-related protein: insight into high-affinity biotin binding and protein stability. *Acta Crystallogr. D Biol. Crystallogr.* **61**, 528–538.
42. Sivkumar, N., Li, N., Tang, J.W., Patel, B.K.C., Swaminathan, K. (2006) Crystal structure of AmyA lacks acidic surface and provides insights into protein stability at polyextreme condition. *FEBS Lett.* **580**, 2646–2652.
43. Cherry, J.R., Lamsa, M.H., Schneider, P., Vind, J., Svendsen, A., Jones, A., Pedersen, A.H. (1999) Directed evolution of a fungal peroxidase. *Nat. Biotechnol.* **17**, 379–384.
44. Mukaiyama, A., Haruki, M., Ota, M., Koga,Y., Takano, K., Kanaya, S. (2006) A hyperthermophilic protein acquires function at the cost of stability. *Biochemistry* **45**, 12673–12679.
45. Valderrama, B., Garcia-Arellano, H., Giansanti, S., Baratto, M.C., Pogni, R., Vazquez-Duhalt, R. (2006) Oxidative stabilization of iso-1-cytochrome c by redox-inspired protein engineering. *FASEB J.* **20**, 1233–1235.
46. Palmer, B., Angus, K., Taylor, L., Warwicker, J., Derrick, J.P. (2008) Design of stability at extreme alkaline pH in streptococcal protein G. *J. Biotechnol.* **134**, 222–230.
47. Rodriguez-Larrea, D., Minning, S., Borchert, T.V., Sanchez-Ruiz, J.M. (2006) Role of solvation barriers in protein kinetic stability. *J. Mol. Biol.* **360**, 715–724.
48. Minetti, C.A., Remeta, D.P. (2006) Energetics of membrane protein folding and stability. *Arch. Biochem. Biophys.* **453**, 32–53.
49. Wunderlich, M., Martin, A., Schmid, F.X. (2005) Stabilization of the cold shock protein CspB from *Bacillus subtilis* by evolutionary optimization of coulombic interactions. *J. Mol. Biol.* **347**, 1063–1076.
50. Wunderlich, M., Martin, A., Staab, C.A., Schmid, F.X. (2005) Evolutionary protein stabilization in comparison with computational design. *J. Mol. Biol.* **351**, 1160–1168.
51. Barakat, N.H., Barakat, N.H., Carmody, L.J., Love, J.J. (2007) Exploiting elements of transcriptional machinery to enhance protein stability. *J. Mol. Biol.* **366**, 103–116.
52. Sarkar, C.A., Dodevski, I., Kenig, M., Dudli, S., Mohr, A., Hermans, E., Plueckthen, A. (2008) Directed evolution of a G protein-coupled receptor for expression, stability and binding selectivity. *Proc. Natl. Acad. Sci. USA* **105**, 14808–14813.
53. Masso, M., Vaisman, I.I. (2008) Accurate prediction of stability changes in protein mutants by combining machine learning with structure-based computational mutagenesis. *Bioinformatics* **24**, 2002–2009.
54. Tan, Y.H., Luo, R. (2008) Protein stability prediction: a Poisson-Boltzmann approach. *J. Phys. Chem. B* **112**, 1875–1883.
55. Huang, L.T., Gromiha, M.M., Ho, S.Y. (2007) Sequence analysis and rule development of predicting protein stability change upon mutation using decision tree model. *J. Mol. Model.* **13**, 879–890.
56. Zoete, V., Meuwly, M. (2006) Importance of individual side chains for the stability of a protein fold: computational alanine scanning of the insulin monomer. *J. Comput. Chem.* **27**, 1843–1857.
57. Cheng, J.L., Randall, A., Baldi, P. (2006) Prediction of protein stability changes for single-site mutations using support vector machines. *Proteins* **62**, 1125–1132.
58. Marrero-Ponce, Y., Medina-Marrero, R., Castillo-Garit, J.A., Romero-Zaldivar, V., Torrens, F., Castro, E.A. (2005) Protein linear indices of the 'macromolecular pseudograph alpha-carbon atom adjacency matrix'

in bioinformatics. Part 1. *Bioorg. Med. Chem.* **13**, 3003–3015.

59. Cuff, A.L., Martin, A.C.R. (2004) Analysis of void volumes in proteins and application to the stability of the p53 tumour suppressor protein. *J. Mol. Biol.* **344**, 1199–1209.
60. Bordner, A.J., Abagyan, R.A. (2004) Large-scale prediction of protein geometry and stability changes for arbitrary single point mutations. *Proteins* **57**, 400–413.
61. Chaparro-Riggers, J.F., Polizzi, K.M., Bommarius, A.S. (2007) Better library design: data-driven protein engineering. *Biotechnol. J.* **2**, 180–191.
62. Bommarius, A.S., Broering, J.M., Chaparro-Riggers, J.F., Polizzi, K.M. (2006) High-throughput screening for enhanced protein stability. *Curr. Opin. Biotechnol.* **17**, 606–610.
63. Steipe, B. (2004) Consensus-based engineering of protein stability: from intrabodies to thermostable enzymes. *Methods Enzymol.* **388**, 176–186.
64. Binz, H.K., Kohl, A., Plueckthun, A., Gruetter, M.G. (2006) Crystal structure of a consensus-designed ankyrin repeat protein. *Proteins* **65**, 280–284.
65. Winger, M., van Gunsteren, W.F. (2008) Use of molecular dynamics simulation for optimizing protein stability. *Helv. Chim. Acta* **91**, 1605–1613.
66. Merz, T., Wetzel, S.K., Firbank, S., Plueckthun, A., Gruetter, M.G., Mittl, P.R.E. (2008) Stabilizing ionic interactions in a full-consensus ankyrin repeat protein. *J. Mol. Biol.* **376**, 232–240.
67. Tripp, K.W., Barrick, D. (2007) Enhancing the stability and folding rate of a repeat protein through the addition of consensus repeats. *J. Mol. Biol.* **365**, 1187–1200.
68. Kloss, E., Courtemanche, N., Barrick, D. (2008) Repeat protein folding: new insights into origins of cooperativity, stability and topology. *Arch. Biochem. Biophys.* **469**, 83–89.
69. Bae, E., Bannen, R.M., Phillips, G.N. (2008) Bioinformatic method for protein thermal stabilization by by structural entropy optimization. *Proc. Natl. Acad. Sci. USA* **105**, 9594–9597.
70. Gromiha, M.M. (2005) Distinct roles of conventional non-covalent and cation-pi interactions in protein stability. *Polymer* **46**, 983–990.
71. Yin, S., Ding, F., Dokholyan, N.V. (2007) Modelling backbone flexibility improves protein stability estimation. *Structure* **15**, 1567–1576.
72. Wiederstein, M., Sippl, M.J. (2005) Protein sequence randomization: efficient estimation of protein stability using knowledge-based potentials. *J. Mol. Biol.* **345**, 1199–1212.
73. Saraboji, K., Gromiha, M.M., Ponnuswamy, M.N. (2006) Average assignment method for predicting the stability of protein mutants. *Biopolymers* **82**, 80–92.
74. Campos, L.A., Garcia-Mira, M.M, Godoy-Ruiz, R., Sanchez-Ruiz, J.M., Sancho, J. (2004) Do proteins always benefit from a stability increase? *J. Mol. Biol.* **344**, 223–237.
75. Yeh, J.I., Du, S.C., Tortajada, A., Paulo, J., Zhang, S.G. (2005) Peptergents: peptide detergents that improve stability and functionality of a membrane protein, glycerol-3-phosphate dehydrogenase. *Biochemistry* **44**, 16912–16919.
76. Mukaiyama, A., Koga,Y., Takano, K., Kanaya, S. (2008) Osmolyte effect on the stability and folding of a hyperthermophilic protein. *Proteins* **71**, 110–118.
77. Granata, V., Palladino, P., Tizzano, B., Negro, A., Berisio, R., Zagari, A. (2006) The effect of the osmolyte trimethylamine N-oxide on the stability of the prion protein at low pH. *Biopolymers* **82**, 234–240.
78. Faria, T.Q., Lima, J.C., Bastos, M., Macanita, A.L., Santos, H. (2004) Protein stabilization by osmolytes from thermophiles. *J. Biol. Chem.* **279**, 48680–48691.
79. Poddar, N.K., Ansari, Z.A., Singh, R.K.B., Moosavi-Movahedi, A.A., Ahmad, F. (2008) Effect of monomeric and oligomeric sugar osmolytes on delta $G_{(D)}$, the Gibbs energy of stabilization of the protein at different pH values. *Biophys. Chem.* **138**, 120–129.
80. Ortbauer, M., Popp, M. (2008) Functional role of polyhydroxy compounds on protein structure and thermal stability studied by circular dichroism spectroscopy. *Plant Physiol. Biochem.* **46**, 428–434.
81. Mishra, R., Bhat, R., Seckler, R. (2007) Chemical chaperone-mediated protein folding: stabilization of P22 tailspike folding intermediates by glycerol. *Biol. Chem.* **388**, 797–804.
82. Pais, T.M., Lamosa, P., dos Santos, W., LeGall, J., Turner, D.L., Santos, H. (2005) Structural determinants of protein stabilization by solutes. *FEBS J.* **272**, 999–1011.
83. Bolen, D.W. (2004) Effects of naturally occurring osmolytes on protein stability and solubility. *Methods* **34**, 312–322.
84. Street, T.O., Bolen, D.W., Rose, G.D. (2006) A molecular mechanism for osmolyte-

induced protein stability. *Proc. Nat. Acad. Sci. USA* **103**, 13997–14002.

85. Auton, M., Ferreon, A.C.M., Bolen, D.W. (2006) Metrics that differentiate the origins of osmolyte effects on protein stability. *J. Mol. Biol.* **361**, 983–992.
86. Holthauzen, L.M.F., Bolen, D.W. (2007) Mixed osmolytes: the degree to which one osmolyte affects the protein stabilizing ability of another. *Protein Sci.* **16**, 293–298.
87. Wang, W., Wang, Y.J., Wang, D.Q. (2008) Dual effects of Tween 80 on protein stability. *Int. J. Pharm.* **347**, 31–38.
88. Rochu, D., Chabriere, E., Renault, F., Clery-Barraud, C., Chabriere, E., Masson, P. (2007) Stability of highly purified human paraoxonase (PON1): association with human phosphate binding protein (HPBP) is essential for preserving its active conformation(s). *Biochem. Soc. Trans.* **35**, 1616–1620.
89. Rochu, D., Renault, F., Elias, M., Clery-Barraud, C., Masson, P. (2007b) Stabilization of the active form(s) of human paraoxonase by human phosphate-binding protein. *Biochim. Biophys. Acta* **1774**, 874–883.
90. Byrne, N., Wang, L.M., Belieres, J.P., Angell, C.A. (2007) Reversible folding-unfolding, aggregation protection and multi-year stabilization, in high concentration protein solutions, using ionic liquids. *Chem. Commun. (Camb)* (41), 2714–2716.
91. Frokjaer, S., Otzen, D.E. (2005) Protein drug stability: a formulation challenge. *Nature Rev. Drug Discov.* **4**, 298–306.
92. Torres, M.P, Determan, A.S., Anderson, G.L., Mallapragada, S.K., Narasimhan, B. (2007) Amphiphilic polyanhydrides for protein stabilization and release. *Biomaterials* **28**, 108–116.
93. Sheldon, R. (2007) Crosslinked enzyme aggregates (CLEA®s): stable and recyclable biocatalysts. *Biochem. Soc. Trans.* **35**, 1583–1587.
94. Illanes, A., Wilson, L., Altamirano, C., Cabrera, Z., Alvarez, L., Aguirre, C. (2007) Production of cephalexin in organic medium at high substrate concentrations with CLEA of penicillin acylase and PGA-450. *Enzyme Microb. Technol.* **40**, 195–203.
95. Sangeetha, K., Abraham, T.E. (2008) Preparation and characterization of cross-linked enzyme aggregates (CLEA) of subtilisin for controlled release applications. *Int. J. Biol. Macromol.* **43**, 314–319.
96. Kreiner, M., Fernandes, J.F.A., O'Farrell, N., Halling, P.J., Parker, M.C. (2005) Stability of protein-coated microcrystals in organic solvents. *J. Mol. Catal. B* **33**, 65–72.
97. Abian, O., Grazu, V., Hermoso, J., Gonzalez, R., Garcia, J.L., Fernandez-Lafuente, R., Guisan, J.M. (2004) Stabilization of penicillin G acylase from *Escherichia coli*: site-directed mutagenesis of the protein surface to increase multipoint covalent attachment. *Appl. Environ. Microbiol.* **70**, 1249–1251.
98. Irazoqui, G., Giacomini, C., Batista-Viera, F., Brena, B.F. (2007) Hydrophilization of immobilized model enzymes suggests a widely applicable method for enhancing protein stability in polar organic co-solvents. *J. Mol. Catal. B* **46**, 43–51.
99. Koutsopoulos, S., van der Oost, J., Norde, W. (2004) Adsorption of an endoglucanase from the hyperthermophilic *Pyrococcus furiosus* on hydrophobic (polystyrene) and hydrophilic (silica) surfaces increases protein heat stability. *Langmuir* **20**, 6401–6406.
100. Knotts, T.A., Rathore, N., de Pabolz, J.J. (2008) An entropic perspective of protein stability on surfaces. *Biophys. J.* **94**, 4473–4483.
101. Kim, Y.H., Stites, W.E. (2008) Effects of excluded volume upon protein stability in covalently cross-linked proteins with variable linker lengths. *Biochemistry* **47**, 8804–8814.
102. Shental-Bechor, D., Levy, Y. (2008) Effect of glycosylation on protein folding: a close look at thermodynamic stabilization. *Proc. Natl. Acad. Sci. USA* **105**, 8256–8261.
103. Spiriti, J., Bogani, F., van der Vaart, A., Ghirlanda, G. (2008) Modulation of protein stability by O-glycosylation in a designed Gc-MAF analog. *Biophys. Chem.* **134**, 157–167.
104. Rathore, N., Rajan, R.S. (2008) Current perspectives on stability of protein drug products during formulation, fill and finish operations. *Biotechnol. Prog.* **24**, 504–514.
105. Klibanov, A.M. (2001) Improving enzymes by using them in organic solvents. *Nature* **409**, 241–246.
106. Godfrey, T., West, S. (1996) *Industrial enzymology*, 2nd ed. Macmillan, London.
107. Breuer, M., Ditrich, K., Habicher, T., Hauer, B., Kesseler, M., Stuermer, R., Zelinski, T. (2004) Industrial methods for the production of optically active intermediates. *Angew. Chem. Int. Ed. Engl.* **43**, 788–824.
108. Miyazaki, M., Maeda, H. (2006) Microchannel enzyme reactors and their applications for processing. *Trends Biotechnol.* **24**, 463–470.
109. Schmid, A., Hollmann, F., Park, J.B., et al. (2002) The use of enzymes in the chemical industry in Europe. *Curr. Opin. Biotechnol.* **13**, 359–366.
110. Van den Burg, B., Vriend, G., Veltman, O.R., Eijsink, V.G.H. (1998) Engineering an enzyme to resist boiling. *Proc. Natl. Acad. Sci. USA* **95**, 2056–2060.

111. Riddles, P.W., Blakely, R.L., Zerner, B. (1983) Reassessment of Ellman's reagent. *Methods Enzymol.* **91**, 49–60.
112. Roig, M.G., Kennedy, J.F. (1992) Perspectives for chemical modifications of enzymes. CRC *Crit. Rev. Biotechnol.* **12**, 391–412.
113. Riordan, J.F., Vallee, B.L. (1972a) Reactions with *N*-ethylmaleimide and *p*-mercuribenzoate. *Methods Enzymol.* **25**, 449–456.
114. Fields, R. (1972) The rapid determination of amino groups with TNBS. *Methods Enzymol.* **25**, 464–468.
115. Inman, J.K., Perham, R.N., DuBois, G.C., et al. (1983) Amidination. *Methods Enzymol.* **91**, 559–569.
116. Klapper, M.H., Klotz, I.M. (1972) Acylation with dicarboxylic acid anhydrides. *Methods Enzymol.* **25**, 531–536.
117. Jentoft, N., Dearborn, D.G. (1983) Protein labeling by reductive alkylation. *Methods Enzymol.* **91**, 570–579.
118. Yankeelov, J.A. (1972) Modification of arginine by diketones. *Methods Enzymol.* **25**, 566–579.
119. Pande, C.S., Pelzig, M., Glass, J.D. (1980) Camphorquinone-10-sulfonic acid and derivatives: convenient reagents for reversible modification of arginine residues. *Proc. Natl. Acad. Sci. USA* 77, 895–899.
120. Dominici, P., Tancini, B., Voltattorni, C.B. (1985) Chemical modification of pig kidney 3,4-dihydroxyphenylalanine decarboxylase with diethyl pyrocarbonate. *J .Biol. Chem.* **260**, 10583–10589.
121. Carraway, K.L., Koshland, D.E. (1972) Carbodiimide modification of proteins. *Methods Enzymol.* **25**, 616–623.
122. Wilcox, P.E. (1972) Esterification. *Methods Enzymol.* **25**, 596–616.
123. Riordan, J.F., Vallee, B.L. (1972) Nitration with tetranitromethane. *Methods Enzymol.* **25**, 515–521.
124. Morrison, M. (1970) Iodination of tyrosine: isolation of lactoperoxidase (bovine). *Methods Enzymol.* **17**, 653–660.
125. Spande, T.F., Witkop, B. (1967) Determination of the tryptophan content of protein with N-bromosuccinimide. *Methods Enzymol.* **11**, 498–532.
126. Neumann, N.P. (1972) Oxidation with hydrogen peroxide. *Methods Enzymol.* **25**, 393–401.
127. Lundblad, R.L. (2005) *Chemical reagents for protein modification*, 3rd ed. CRC, Boca Raton. ISBN 9780849319839.
128. Means, G.E., Feeney, R.E. (1990) Chemical modification of proteins: history and applications. *Bioconjug. Chem.* **1**, 2–12.
129. Sadana, A. (1988) Enzyme deactivation. *Biotechnol. Adv.* **6**, 349–446.
130. Wold, F. (1972) Bifunctional reagents. *Methods Enzymol.* **25**, 623–651.

Chapter 8

Microfluidics in Protein Chromatography

Frank A. Gomez

Abstract

The development of microfluidics and its utilization in a myriad of applications has grown exponentially over the past 15 years. One area that has benefited from the great strides in fabrication of microelectromechanical systems (MEMS) is separations chemistry. Most studies have focused on small molecule and DNA separations; few on protein chromatographic techniques on microchips. This review details recent developments in protein separations on microfluidic platforms and how MEMS have the potential for revolutionizing protein chromatography.

Key words: Microfluidics, Protein chromatography, Lab-on-a-chip, Separations, Proteins

1. Introduction

Microfluidics involves the study of fluid movement through micron-sized chambers and channels called a microelectromechanical system or MEMS. Microfluidics can be considered a combination of fluid mechanics, surface science, chemistry, and biology (and in many cases optics, microscopy, electronics, control systems, and microfabrication as well). Analogous to the integration of processes on-chip, research in this field involves an integration of many disciplines. Lab-on-a-chip technologies represent a revolution in laboratory experimentation, bringing the benefits of miniaturization, integration, and automation to many research-based industries. The use of microfluidic chips has attracted great attention as microchips offer several advantages over nonmicrofluidic-based techniques including small sample volume requirements, portability, low production costs per device, speed to analysis, versatile format for integration of various detection schemes, low power consumption, parallel processing of samples and ability to multiplex, and compatibility with other techniques (1–35).

Dermot Walls and Sinéad T. Loughran (eds.), *Protein Chromatography: Methods and Protocols*, Methods in Molecular Biology, vol. 681, DOI 10.1007/978-1-60761-913-0_8, © Springer Science+Business Media, LLC 2011

Microfluidics has been the motivation for various biochemical application advancements in point-of-care (POC) diagnostics, bioterrorism detection, and drug discovery. Potential applications include biotechnology, pharmaceuticals, the life sciences, defense, public health, and agriculture. For example, microfluidic technologies have been embraced by the pharmaceutical industry because of the benefits of miniaturization, integration, and automation to research and development. At present, nearly every company and institution doing research and development is aware of this revolution in portable instrumentation and modifying long-term strategic plans with miniaturization in mind. The expense, inefficiency, and high maintenance and space requirements of robotic automation systems present barriers to performing experiments. By contrast, microfluidic chips are inexpensive, require little maintenance and space, and need only minute amounts of samples and chemical inputs to make experiments work, making them more efficient and potentially cheaper to use. A query of PubMed indicates that the citation rate for the word "microfluidics" increased fivefold between 2001 and 2004 and doubled to 500 between 2003 and 2004. In 1998, the number of issued patents detailing the use of a microfluidic platform in a commercial application was less than 25. In 2004 that number had risen to over 350. The microfluidics market for life science applications reached $750 million in 2004 and is expected to reach $2 billion by 2010. Specific applications of life sciences companies include in vitro diagnostics, drug delivery, and drug delivery markets. This review will highlight the most salient works pertaining to protein chromatography on microchips. Commentary will focus on the benefits of newly developed microfluidic techniques involving protein separations and its promise in revolutionizing the life sciences.

2. Highlights in Protein Chromatography on Microchips

Poly(dimethylsiloxane) (PDMS), glass, and silicon are the materials of choice when fabricating microfluidic devices (MDs). In general, they are low cost, able to be machined, and offer a number of other properties amenable to their use in a variety of applications. The use of thermoplastics in microfluidics is still relatively new, yet they have the potential to be an alternative to the traditional materials used in MDs due to their low cost and long history of the plastics industry and associated manufacturing methodologies. A number of plastic materials have been used in MDs, including polymethyacrylate (PMMA) (36, 37), polycarbonate (38, 39), polyester (40, 41), fluorinated ethylene propylene (42), and cyclic olefin copolymers (COC) (43). Das et al. used

COC fabricated with an array of microvalves for two-dimensional protein separations (44). The devices were made by compression molding whereby a plastic resin is introduced into an open mold and formed under heat and pressure. They used in situ gel polymerization to create microvalves at intersections of orthogonal channels. COC has the advantage of being more solvent resistant than PDMS and, like other polymers, has higher optical clarity and reduced absorption of moisture. Using isoelectric focusing (IEF), green fluorescent protein (GFP) and R-phycoerythrin (RPE) were separated and detected using laser-induced fluorescence (LIF). In IEF, a pH gradient is created by carrier ampholytes. Under an electric field, proteins migrate along the pH gradient and focus where the pH value equals its isoelectric point (p*I*). Since proteins have unique values for p*I*, they can be separated along the pH gradient. Although only two proteins were examined, this study does demonstrate the ability of COC-based MDs to efficiently separate biomolecules and ease of fabrication.

Several years ago, Bjorkman et al. demonstrated the use of polycrystalline diamond in microfluidics to separate several proteins (45). In this work, diamond was chemically vapor deposited (CVD) onto silicon and the channels of the microchips were filled with continuous polymer beds and used for anion-exchange chromatography of proteins. Four acidic proteins (horse heart myoglobin, conalbumin, ovalbumin, and trypsin inhibitor) were separated by gradient elution in less than 30 s. Hemoglobin A_o and β-lactalbumin were also separated in less than 20 s. Diamond has a number of properties amenable for use in microfluidic technologies: high-thermal conductivity, electrical insulating capability, and mechanical strength. These properties make diamond a natural choice for capillary electrophoretic separations on chips as high-field strengths should be possible resulting in shorter analysis times without loss of separation efficiency due to little Joule heating. Furthermore, diamond transmits electromagnetic radiation in a broad wavelength span, allowing for UV and fluorescent detection through the chips.

Recently, chips fabricated from synthetic polymers have gained in popularity due to the simplicity of the processes including hot embossing or injection molding (46, 47). Furthermore, they are much less costly than chips made of glass, quartz, or silica. Porous polymer monolithic materials have also shown great usage as stationary phases, in particular, for HPLC and capillary electrochromatography (CEC). Monoliths offer higher porosity than packed materials thereby allowing for high flow velocities and faster separations. In addition, monoliths are more resistant to mass transfer than their packed material counterparts thereby yielding higher efficiencies at higher flow rates. Levkin et al. developed poly(lauryl methacrylate-*co*-ethylene dimethacrylate) (LaMA-EDMA) and poly(styrene-*co*-divinylbenzene) (ST-DVB) stationary phases in a monolithic format to separate a mixture of proteins including ribonuclease A, myoglobin, cytochrome *c*, and ovalbumin (Fig. 1) (48). The monoliths were prepared by thermally initiated free radical polymerization. The effects of flow rate and gradient steepness were examined and it was found that no appreciable decrease in efficiency of separation was observed at higher flow velocities and separation was accelerated by decreasing the gradient time. The permeability of the two types of monoliths was calculated to be 1.4 and 0.63×10^{-10} for the LaMA-EDMA and ST-DVB, respectively, equating to columns packed with approximately 4 and 2–3 μm particles. The monoliths were very stable and runs were highly reproducible over time.

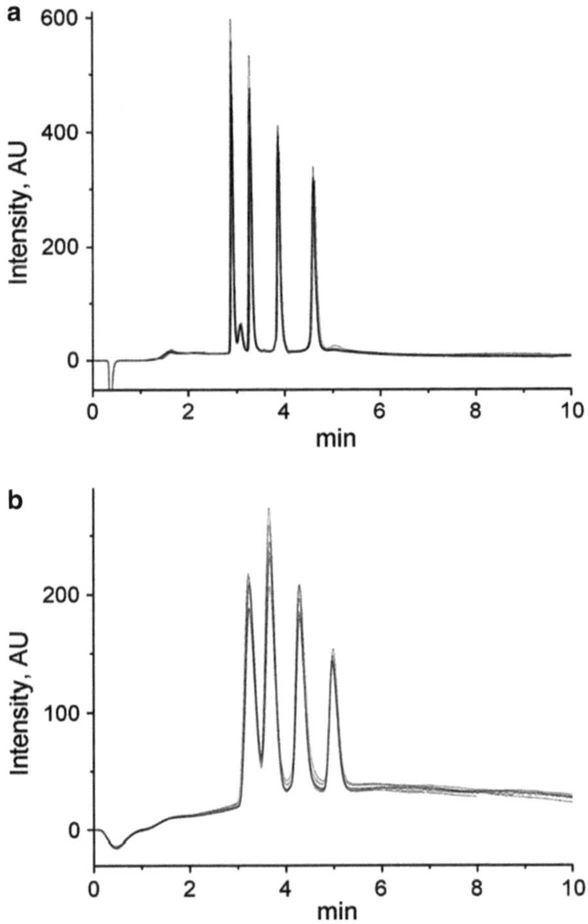

Fig. 1. Overlaid chromatograms obtained by ten successive gradient separations of ribonuclease A, cytochrome *c*, myoglobin, ovalbumin (order of elution) using chips containing both monolithic poly(styrene-*co*-divinylbenzene) (**a**) and poly(lauryl methacrylate-*co*-ethylene dimethacrylate) (**b**) stationary phases. Gradient: 0–60% of acetonitrile in water (0.05% v/v of formic acid) for 10 min. Concentration of proteins in sample solution: 0.2 and 0.1 mg/mL for poly(lauryl methacrylate-*co*-ethylene dimethacrylate) and poly(styrene-*co*-divinylbenzene) monoliths, respectively. Injection volume 40 nL. Flow rate 4 μL/min. Detection: UV at 210 nm. Taken with permission from ref. 48.

While analysis of DNA by miniaturized analysis systems (μ-TAS) has been made possible due to advances in speed and parallelism, similar successes in protein separations have been slow due to solubility issues, range of sizes, native charges, and the large range of concentrations (e.g., in cells and proteomic work) that require detection. Traditionally, multidimensional separation techniques are the technique of choice when probing the proteome. Two-dimensional electrophoresis (2DE) is the benchmark method that initially separates proteins based on p*I* by IEF then secondly by mass by denaturing electrophoresis. Protein microarrays are an alternative technique to multidimensional separations but suffer from construction concerns. Microfabricated systems, on the other hand, offer fast analysis of nucleic acids, coupling of sample processing and analysis steps, and high-throughput capabilities. Emrich et al. developed a platform for performing two-dimensional differential gel electrophoretic (DIGE) separations of cellular protein mixtures produced by induced protein expression in *Escherichia coli* (49). Here, the first- and second-dimension channels are connected with passive valve structures made of shallower and narrower microchannels to form a microfluidic interface (MFI). The reduced channel cross section of the MFI allows it to act as a barrier to fluid flow and diffusion-driven transport. The micro-DIGE analyzer is fabricated from borosilicate glass and consists of a single 3.75 cm long channel for IEF which is sampled in parallel by 20 channels yielding the second dimension separation by electrophoresis. Proteins (e.g., maltose binding protein) were labeled with Cy2 and Cy3 and detected simultaneously with a fluorescence scanner. This technique can be extended by increasing the number of second-dimension separation channels. Furthermore, it should be possible to incorporate the labeling, mixing, and sample loading steps all on-chip, thereby saving time and reagent cost needed for high-throughput applications.

Two-dimensional polyacrylamide gel electrophoresis (2D-PAGE) is the standard technique used for protein separations in proteomics. 2D-PAGE, though, is laborious requiring 2 or more days to complete. A number of groups have detailed the use of zone electrophoresis and IEF on microfluidic platforms. IEF performance in these samples is marginal when compared to ampholyte-based, liquid-phase IEF separations. Cui et al. reported a protocol that improves the resolving power of IEF in a polymeric microfluidic chip (50). In this work, several stages of IEF are coupled in series by first focusing proteins in a straight channel followed by refocusing segments of the first channel into secondary channels that branch from the first one. Using fluorescently labeled proteins (GFP, allophycocyanin, and RPE), it was found that the proteins that had focused within a segment of the straight channel during the first stage refocused at higher resolution due to the shallower pH gradient and higher electrical field. This work has great potential if and when it is coupled to mass spectrometry.

While the merits of MDs have been well documented, their integration into the mainstream vis-à-vis POC diagnostics, chemical warfare agent and bioterrorism detection, and agricultural, just to name a few, has been limited since they are generally connected to bulky detection and fluid-handling systems. For the benefits of MDs to become a true reality in the solving of societal problems, the development and integration of world-to-chip interfaces and control systems must be a major focus of research. Renzi et al. described the design and fabrication of a hand-held microchip-based instrument for the detection of fluorescently labeled proteins and other biomolecules (51). LIF detection allowed for picomolar sensitivity and nanomolar sensitivity for fluorescamine-labeled proteins. Specifically, γ-lactalbumin (LACT), carbonic anhydrase B (CAB), ovalbumin (OVA), bovine serum albumin (BSA), and mouse IgG were separated and analyzed (Fig. 2). Separations performed over a day showed high reproducibility with a maximum migration time standard error of 1.4%. Chip-to-chip migration times varied ~2–3%. A major benefit of the instrument's architecture was that it could easily be modified for integration with IEF, HPLC, CEC, immunoassays, and PCR analysis.

Most microfluidic-based separation systems utilize gels or polymer solutions to separate nucleic acids and proteins. In recent years, micro-/nanofabricated molecule-sorting structures

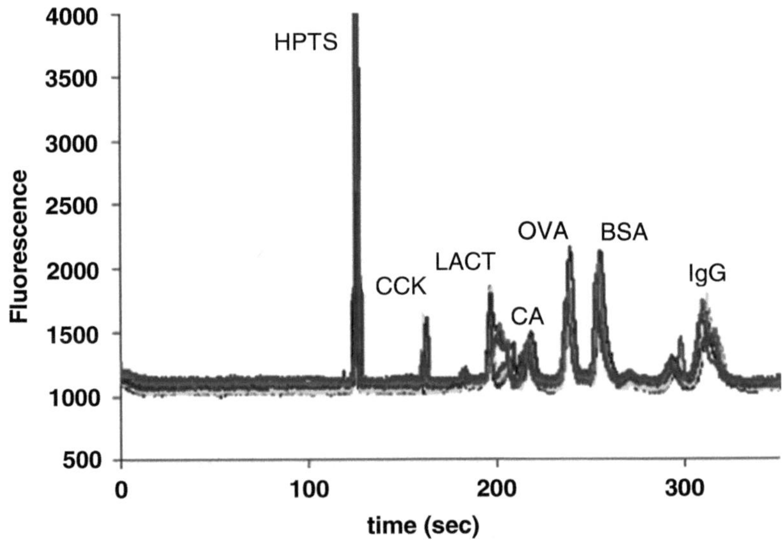

Fig. 2. Overlay of nine CGE separations performed using constant current mode and taken over a 6-h time period. Analytes are HPTS dye, CCK peptide, γ-lactalbumin (LACT), carbonic anhydrase (CA), ovalbumin (OVA), bovine serum albumin (BSA), and mouse IgG. Taken with permission from ref. 51.

have been shown to improve the efficiency and speed of biomolecular separations thereby making them a viable alternative to traditional matrix materials. Unlike, gel and polymer matrices, micro-/nanomachined sieves feature regular and precisely engineered geometry, mechanical robustness, and easy integration into microfluidic platforms. Zeng and Harrison detailed gel-free size fractionation of DNA and proteins within ordered colloidal arrays (52). The colloids are embedded into microfluidic channels as a three-dimensional nanofluidic sieving structure. The pore size of the colloidal arrays can be tuned to maximize separation performance for a given molecule. Proteins smaller than 10 nm (20–200 kDa) to micrometer-sized DNA (0.5–50 kbp) coils can be separated. Specifically, five proteins (trypsin inhibitor, ovalbumin, BSA, phosphorylase B, and β-galactosidase) were separated using a 160-nm silica particle array. Typical separations are complete in minutes and separation efficiencies are comparable to gel/polymer-filled systems. The use of self-assembled colloidal arrays, like those detailed here, may be ideal for understanding mechanisms underlying biomolecular separation in porous media.

Schulze et al. used native fluorescence detection in microfluidic CE (MCE) by coupling a deep UV Nd:YAG laser to a epifluorescence microscope (53). The technique allowed for increased detection limits and broadened the application range of fluorescence detection in microchips. A mixture of basic proteins (lysozyme, trypsinogen, and chymotrypsinogen) were easily separated (within 3 min) on application of a voltage (1.5 kV). Detection limits were 0.9, 0.5, and 0.5 μM for lysozyme, trypsinogen, and chymotrypsinogen, respectively. The major advantage of this technique is that the proteins (or other biomolecules for that matter) do not require labeling.

While LIF is a common detection technique in lab-on-a-chip devices, systems are generally bulky and require delicate light coupling components thereby lessening the inherent advantages in sensitivity afforded to these systems. Portability of such systems can be problematic thereby limiting broad usage of microfluidic devices. Vieillard et al. developed a hybrid device fabricated from glass and PDMS incorporating optical waveguides to separate proteins (carbonic anhydrase B and β-lactoglobulin) as well as small molecules (streptavidin and Cy3) using both capillary zone electrophoresis (CZE) and capillary gel electrophoresis (CGE) (54). The optical waveguide was fabricated in soda-lime glass by the Na^+/K^+ ion exchange method which resulted in a local change to the refractive index of the glass. The efficiency was marginal (30,000 theoretical plates) and was less when compared to classical CE and a commercial instrument. On the other hand, the resolution of the proteins was better on-chip than using classical CE.

High-throughput analysis is now an integral component of the chemical industry and especially in the drug discovery sector. The huge multitude of compounds synthesized on a daily basis requires instrumentation that can both purify and identify synthesized substances. There are now a number of commercial instruments based on CE that contain many capillary columns required for high-throughput analysis. Additional problems lie in how to integrate sample handling, injection, and detection schemes. Naturally, evolution to CE on-a-chip array-type platforms was to be expected. At present, though, viable detection modes are limited making full incorporation of chip-based systems for high-throughput screening still years ahead. Shadpour et al. developed a 16-channel microfluidic chip incorporating a contact conductivity sensor array (55). The chip contained 16 separation channels hot embossed into polycarbonate (PC). A gold sensor array was lithographically patterned onto a PC cover plate set in the fluidic chip via thermal bonding. Electrodes were incorporated into each of the channels serving as contact conductivity detectors. Platinum wires were printed onto a circuit board to provide high voltage to all the reservoirs. Chymotrypsinogen A, cytochrome *c*, and BSA were systematically separated on the chip as were mixtures of peptides and amino acids. Further work is focused on expanding the number of channels fabricated into the microchip.

The ease of which multiple layers of a polymeric material can be fabricated to form microfluidic devices lends the technology to multifaceted applications. For example, new techniques that can illicit separation via molecular differentiation in two dimensions (2D) are at a premium. There have been a number of papers that detail multidimensional peptide and protein separations. Li et al. integrated a protein concentration/separation system combining non-native IEF with sodium dococyl sulfate (SDS) gel electrophoresis on a microchip (56). In this work, proteins (actin, BSA, trypsin inhibitor, and parvalbumin) are electrokinetically transferred into an array of orthogonal microchannels then resolved by SDS gel electrophoresis (Figs. 3 and 4). Separations were complete within 10 min with peak capacities of ~1,700. This work proves the efficacy of microfluidics as a substitute for conventional 2D-PAGE.

In many situations, sufficient quantities of sample are not always available to be examined in microfluidic platforms. This problem is not unique to MDs for, in many cases, traditional spectroscopic techniques also face similar limitations. The advantage of MDs is that, even at dilute concentrations, if preconcentration can occur outside the device, analysis by microfluidic means is still a possibility. On the other hand, if preconcentration could occur on the device it would both hasten the analysis and lessen the propensity for sample contamination. Kim et al.

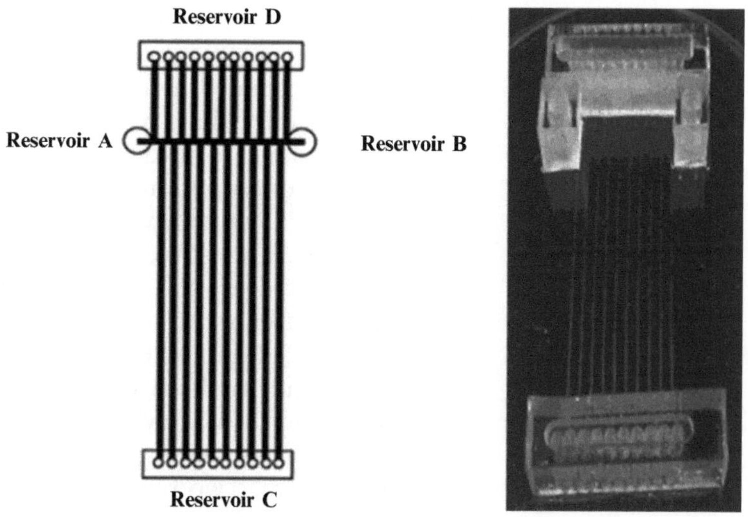

Fig. 3. Schematic and image of 2-D protein separation platform using plastic microfluidics. Taken with permission from ref. 56.

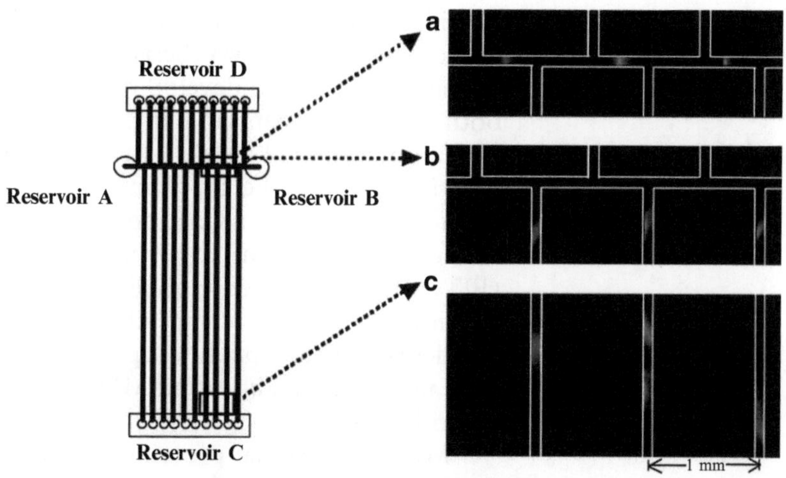

Fig. 4. Fluorescent images of on-chip 2-D separation of four model proteins using single separation medium. (A) Non-native IEF with focusing order of (1) actin, (2) bovine serum albumin and trypsin inhibitor, and (3) parvalbumin from left to right; (B) electrokinetic transfer of focused proteins; (C) SDS gel electrophoresis. Images were captured at either 90 or 150 s following the initiation of IEF or SDS gel electrophoresis separations, respectively. Images were obtained using either green fluorescence of protein fluorescein conjugates in IEF or red fluorescence of Sypro Red-labeled proteins during electrokinetic transfer and size-based separation. Taken with permission from ref. 56.

described a protein concentration device in PDMS. In this work, the PDMS layer is bonded to a microscope slide and the apexes of the channels pass ~20 μm from each other (57). The application of an electric field across the thin-walled section concentrates on the anodic side. About a 103–106-fold protein concentration was achieved in less than 30 min. Proteins (ovalbumin and BSA) were subsequently separated using the microfluidic platform. The authors hypothesize that an exclusion-enrichment effect (EEE) of the microchannel formed between the PDMS and glass is the reason for the concentration of negatively charged protein molecules near the PDMS thin-walled section on the anodic channel side. This work demonstrates a novel approach to sample concentration and should serve as a foundation for the next generation microfluidic concentrator.

Adsorption of biomolecules is a common occurrence in microfluidics and has been one impediment to commercialization of many chip-based devices. Irrespective of the material used in the fabrication of MDs, protein adsorption, in particular, is a common problem. Two remedies to adsorption include the use of a dynamic coating and permanent surface modification. Neither strategy is perfect. For example, while dynamic coating is convenient and, hence, preferred due to its relative ease of use, it is not permanent and is problematic in some microfluidic applications (mass spectrometry or chemical reactors). Chemical modification can be experimentally difficult, sometimes requiring tedious multistep synthesis complicating the MD. A solution to this problem is to develop novel polymeric materials that are both protein resistant and amenable to fabrication without surface modification. Liu et al. developed a poly(ethylene glycol)-functionalized acrylic copolymer that is resistant to protein and peptide adsorption (58). Fabricated devices were used without surface modification to separate fluorescein isothiocyanate (FITC)-labeled β-lactoglobulin, porcine thyroglobulin, myoglobin, and human serum albumin (HAS) in less than 30 s. Besides protein and peptide separations, the polymer should have wide usage when fabricating MDs for microreactors, biosensors, and other devices.

Multidimensional protein separation techniques on microfluidic chips are still difficult to achieve. Wang et al. coupled IEF and CE or CGE with high-ionic strength buffers in a microfluidic platform (59). In this technique, a high-resolution pH gradient from pH 3 to 10 for IEF was used in a short microchannel (10 mm). Proteins moved into the peak transfer region and one set of microfluidic valves were closed to isolate particular protein peaks. The peak transfer region was connected to the second-dimension separation channel by opening a second set of valves. The microfluidic valves are integral to the success of the technique as they prevent peak interdiffusion and band dispersion during the peak transfer process. In this work, GFP and three fluorescein- or Alexa Fluor 488-labeled proteins were used as samples: ovalbumin, low-density lipoprotein, and trypsin inhibitor. A natural extension to this work involves coupling the microfluidic platform to mass spectrometry (MS), which could help in the area of proteomics and genomics.

Herr et al. developed a microfluidic platform that couples liquid-phase IEF and CE. In this work, GFP and FITC-labeled dextran and ovalbumin were separated by IEF in the first dimension and CE in the second (60). The separation medium was a mixture of ampholytes (focused ampholytes for IEF and unfocused ampholytes for CE) and full-field fluorescence imaging was used for detection. Analysis of each dimension showed time-dependent species displacement and band-broadening behavior consistent with IEF and CE, respectively.

There are a number of available techniques that can be utilized for p*I*-based sample fractionation (multichamber IEF, Gradiflow, etc.) but they require either large sample volumes (10 mL range) or membranes that are unstable at extreme pH conditions. Song et al. developed a novel continuous-flow p*I*-based sorting technique for proteins and peptides in a microfluidic platform (61). What sets their technique apart from other techniques is that the electric field for separation is generated through the diffusion of different buffer species in situ within the microfluidic channel and not from an external power source. For example, when two solutions with different buffer concentrations and molarity meet, a laminar flow boundary is formed and a diffusion potential is developed between the two solutions due to different diffusion coefficients. Although small, this diffusion potential can create a sufficient electric field for p*I*-based sorting. In their work, recombinant GFP and FITC-labeled ovalbumin were used as model proteins. The sorting resolution is estimated to be as high as 0.1 pH unit which is sufficient for sample preparation. This technique has the potential to be an alternative method to IEF and for use in high-throughput applications.

There are a number of excellent studies detailing the use of microfluidics for examining proteolytic digestion (62–65). Given the focus of this review is on protein separations in microchips, rather than comment on them, the reader is encouraged to examine the references just noted for a better understanding of proteolysis, peptide separation, and identification on chips.

Acknowledgments

The authors gratefully acknowledge financial support for this research by grants from the National Science Foundation (CHE-0515363 and DMR-0351848).

References

1. Applegate, R. W., Squier, J., Vestad, T., Oakey, J., Marr, D. W. M., Bado, P., Dugan, M. A. and Said, A. A. (2006) Microfluidic sorting system based on optical waveguide integration and diode laser bar trapping. *Lab Chip.* **6**, 422.
2. Chen, C. C., Zappe, S., Sahin, O., Zhang, X. J., Fish, M., Scott, M. and Solgaard, O. (2004) Design and operation of a microfluidic sorter for drosophila embryos. *Sens. Actuators B.* **102**, 59.
3. Pamme, N. and Wilhelm, C. (2006) Continuous sorting of magnetic cells via on-chip free-flow magnetophoresis. *Lab Chip.* **6**, 974.
4. Sundararajan, N., Kim, D. and Berlin, A. A. (2005) Microfluidic operations using deformable polymer membranes fabricated by single layer soft lithography. *Lab Chip.* **5**, 350.
5. Wang, J., Ibanez, A., Chatrathi, M. P. and Escarpa, A. (2001) Electrochemical enzyme immunoassays on microchip platforms. *Anal. Chem.* **73**, 5323.
6. Seong, G. H., Heo, J. and Crooks, R. M. (2003) Measurement of enzyme kinetics using a continuous-flow microfluidic system. *Anal. Chem.* **75**, 3161.
7. Hadd, A. G., Jacobson, S. C. and Ramsey, J. M. (1999) Microfluidic assays of acetylcholinesterase inhibitors. *Anal. Chem.* **71**, 5206.
8. Bromberg, A. and Mathies, R. A. (2003) Homogeneous immunoassay for detection of TNT and its analogues on a microfabricated capillary electrophoresis chip. *Anal. Chem.* **75**, 1188.
9. Wang, J., Ibanez, A. and Chatrathi, M. P. (2002) Microchip-based amperometric immunoassays using redox tracers. *Electrophoresis.* **23**, 3744.
10. Wang, J., Ibanez, A. and Chatrathi, M. P. (2003) On-chip integration of enzyme and immunoassays: simultaneous measurements of insulin and glucose. *J. Am. Chem. Soc.* **125**, 8444.
11. Johnson Hou, C., Milovic, N., Godin, M., Russo, P. R., Chakrabarti, R. and Manalis, S. R. (2006) Label-free microelectronic PCR quantification. *Anal. Chem.* **78**, 2526.
12. Easley, C. J., Karlinsey, J. M. and Landers, J. P. (2006) On-chip pressure injection for integration of infrared-mediated DNA amplification with electrophoretic separation. *Lab Chip.* **6**, 601.
13. Alex, H. Y. N., Wang, G., Lin, B. and Chan, W. (2005) Microwave plasma treatment of polymer surface for irreversible sealing of microfluidic devices. *Lab Chip.* **5**, 1173.
14. Munce, N. R., Li, J., Herman, P. R. and Lilge, L. (2004) Microfabricated system for parallel single-cell capillary electrophoresis. *Anal. Chem.* **76**, 4983.
15. Lu, H., Koo, L. Y., Wang, W. M., Lauffenburger, D. A., Griffith, L. G. and Jensen, K. F. (2004) Microfluidic shear devices for quantitative analysis of cell adhesion. *Anal. Chem.* **76**, 5257.
16. MacDonald, M. P., Spalding, G. C. and Dholakia, K. (2003) Microfluidic sorting in an optical lattice. *Nature.* **426**, 421.
17. Leach, A. M., Wheeler, A. R. and Zare, R. N. (2003) Flow injection analysis in a microfluidic format. *Anal. Chem.* **75**, 967.
18. McDonald, J. C., Duffy, D. C., Anderson, J. R., Chiu, D. T., Wu, H. K., Schueller, O. J. A. and Whitesides, G. M. (2000) Fabrication of microfluidic systems in poly(dimethylsiloxane). *Electrophoresis.* **21**, 27.
19. Parikesit, G. O. F., Markesteijn, A. P., Kutchoukov, V. G., Piciu, O., Bossche, A., Westerweel, J., Garini, Y. and Young, I. T. (2005) Electroosmotic flow analysis of a branched U-turn nanofluidic device. *Lab Chip.* **5**, 1067.
20. Wang, J., Pumera, M., Chatrathi, M. P., Escarpa, A., Konrad, R., Griebel, A., Dorner, W. and Lowe, H. (2002) Towards disposable lab-on-a-chip: poly(methylmethacrylate) microchip electrophoresis device with electrochemical detection. *Electrophoresis.* **23**, 596.
21. Thorsen, T., Maerkl, S. and Quake, S. R. (2002) Microfluidic large-scale integration. *Science.* **298**, 580.
22. Seong, G. H. and Crooks, R. M. (2002) Efficient mixing and reactions within microfluidic channels using microbead-supported catalysts. *J. Am. Chem. Soc.* **124**, 13360.
23. Quake, S. R. and Scherer, A. (2000) From micro- to nanofabrication with soft materials. *Science.* **290**, 1536.
24. Paegel, B. M., Yeung, S. H. and Mathies, R. A. (2002) Microchip bioprocessor for integrated nanovolume sample purification and DNA sequencing. *Anal. Chem.* **74**, 5092.
25. Melin, J., Johansson, H., Soderberg, O., Nikolajeff, F., Landegren, U., Nilsson, M. and Jarvius, J. (2005) Thermoplastic microfluidic platform for single-molecule detection, cell culture, and actuation. *Anal. Chem.* **77**, 7122.

26. Garcia, A. L., Ista, L. K., Petsev, D. N., O'Brien, M. J., Bisong, P., Mammoli, A. A., Brueck, S. R. J. and López, G. P. (2005) Electrokinetic molecular separation in nanoscale fluidic channels. *Lab Chip.* **5**, 1271.

27. Grover, W. H., Ivester, R. H. C., Jensen, E. C. and Mathies, R. A. (2006) Development and multiplexed control of latching pneumatic valves using microfluidic logical structures. *Lab Chip.* **6**, 623.

28. Lee, J., Park, C. and Whitesides, G. M. (2003) Solvent compatibility of poly(dimethylsiloxane)-based microfluidic devices. *Anal. Chem.* **75**, 6544.

29. Dusseiller, M. R., Niederberger, B., Städler, B., Falconnet, D., Textor, M. and Vörös, J. (2005) A novel crossed microfluidic device for the precise positioning of proteins and vesicles. *Lab Chip.* **5**, 1387.

30. Dong, Y., Phillips, K. S. and Cheng, Q. (2006) Immunosensing of staphylococcus enterotoxin B (SEB) in milk with PDMS microfluidic systems using reinforced supported bilayer membranes (r-SBMs). *Lab Chip.* **6**, 675.

31. Sengupta, S., Battigelli, D. A. and Chang, H. (2006) A micro-scale multi-frequency reactance measurement technique to detect bacterial growth at low bio-particle concentrations. *Lab Chip.* **6**, 682.

32. Huang, B., Wu, H., Kim, S. and Zare, R. N. (2005) Coating of Poly(dimethylsiloxane) With n-dodecyl-D-maltoside to minimize nonspecific protein adsorption. *Lab Chip.* **5**, 1005.

33. Dhopeshwarkar, R., Sun, L. and Crooks, R. M. (2005) Electrokinetic concentration enrichment within a microfluidic device using a hydrogel microplug. *Lab Chip.* **5**, 1148–1154.

34. Gaspar, A., Hernandez, L., Stevens, S. and Gomez, F. A. (2008) Electrochromatography in fritless chromatographic microchips packed with conventional reversed-phase silica particles. *Electrophoresis.* **29**, 1638–1642.

35. Gaspar, A., Piyasena, M., Daroczi, L. and Gomez, F. A. (2008) Magnetically controlled valve for flow manipulation in polymer microfluidic devices. *Microfluid. Nanofluidics.* **6**, 525–531.

36. Pugmire, D. L., Waddell, E. A., Haasch, R., Tarlov, M. J. and Locascio, L. E. (2002) Surface characterization of laser-ablated polymers used for microfluidics. *Anal. Chem.* **74**, 871–878.

37. Lai, S., Wang, S. N., Luo, J., Lee, L. J., Yang, S. T. and Madou, M. J. (2004) Design of a compact disk-like microfluidic platform for enzyme-linked immunosorbent assay. *Anal. Chem.* **76**, 1832–1837.

38. Liu, R. H., Yang, J. N., Lenigk, R., Bonanno, J. and Grodzinski, P. (2004) Self-contained, fully integrated biochip for sample preparation, polymerase chain reaction amplification, and DNA microarray detection. *Anal. Chem.* **76**, 1824–1831.

39. Juang, Y. J., Lee, J. J. and Koelling, K. W. (2002) Hot embossing in microfabrication. Part I: Experimental. *Polym. Eng. Sci.* **42**, 539–550.

40. Fiorini, G. S., Lorenz, R. M., Kuo, J. S. and Chiu, D. T. (2004) Rapid prototyping of thermoset polyester microfluidic devices. *Anal. Chem.* **76**, 4697–4704.

41. Guo, Y., Uchiyama, K., Nakagama, T., Shimosaka, T. and Hobo, T. (2005) An integrated microfluidic device in polyester for electrophoretic analysis of amino acids. *Electrophoresis.* **26**, 1843–1848.

42. Sahlin, E., Beisler, A. T., Woltman, S. J. and Weber, S. G. (2002) Fabrication of microchannel structures in fluorinated ethylene propylene. *Anal. Chem.***74**, 4566–4569.

43. Wu, Z., Jensen, H., Gamby, J., Bai, X. and Girault, H. H. (2004) A flexible sample introduction method for polymer microfluidic chips using a push/pull pressure pump. *Lab Chip.* **4**, 512–515.

44. Das, C., Fredrickson, C. A., Xia, Z. and Fan, Z. H. (2007) Device fabrication and integration with photodefinable microvalves for protein separation. *Sens. Actuators A.* **134**, 271–277

45. Bjorkman, H., Ericson, C., Hjerten, S. and Hjort, K. (2001) Diamond microchips for fast chromatography of proteins. *Sens. Actuators A.* **79**, 71–77.

46. Yin, H., Killeen, K., Brennen, R., Sobek, D., Werlich, M. and Van de Goor, T. (2005) Microfluidic chips for peptide analysis with an integrated HPLC column, sample enrichment column, nanoelectrospray tip. *Anal. Chem.* **77**, 527–533.

47. Yin, H. and Killeen, K. (2007) The fundamental aspects and applications of Agilent HPLC-Chip. *J. Sep. Sci.* **30**, 1427–1434.

48. Pavel, A., Levkina, S., Eeltinka, S., Strattonb, T. R., Brennenc, R., Robotti, K., Yinc, H., Killeenc, K., Svec, F. and Frechet, J. M. J. (2008) Monolithic porous polymer stationary phases in polyimide chips for the fast high-performance liquid chromatography separation of proteins and peptides. *J. Chromatogr. A.* **1200**, 55–61.

49. Emrich, C.A., Medintz, I. L., Chu, W. K. and Mathies, R. A. (2007) Microfabricated two-dimensional electrophoresis device for differential protein expression profiling. *Anal. Chem.* **79**, 7360–7366.

50. Cui, H., Horiuchi, K., Dutta, P. and Ivory, C. F. (2005) Multistage isoelectric focusing in a polymeric microfluidic chips. *Anal. Chem.* 77, 7878–7886.
51. Renzi, R. F., Stamps, J., Horn, B. A., Ferko, S., VanderNoot, V. A., West, J. A. A., Crocker, R., Wiedenman, B., Yee, D. and Fruetel, J. A. (2005) Hand-held microanalytical instrument for chip-based electrophoretic separations of proteins. *Anal. Chem.* 77, 435–441.
52. Zeng, Y. and Harrison, D. J. (2007) Self-assembled colloidal arrays as three-dimensional nanofluidic sieves for separation of biomolecules on microchips. *Anal. Chem.* **79**(6), 2289–2295.
53. Schulze, P., Ludwig, M., Kohler, F. and Belder, D.(2005) Deep UV laser-inducted fluorescence detection of unlabeled drugs and proteins in microchip electrophoresis. *Anal. Chem.* 77, 1325–1329.
54. Vieillard, J., Mazurczyk, R., Morin, C., Hannes, B., Chevolot, Y., Desbene, P. -L. and Krawczyk, S. (2007) Application of microfluidic chip with integrated optics for electrophoretic separations ofproteins. *J. Chromatogr. B.* **845**, 218–225.
55. Shadpour, H., Hupert, M. L., Patterson, D., Liu, C., Galloway, M., Stryewski, W., Goettert, J. and Soper, S. A. (2007) Multichannel microchip electrophoresis device fabricated in polycarbonate with an integrated contact conductivity sensor array. *Anal. Chem.* **79**, 870–878.
56. Li, Y., Buch, J. S., Rosenberger, F., DeVoe, D. L. and Lee, C. S. (2004) Integration of isoelectric focusing with parallel sodium dodecyl sulfate gel electrophoresis for multidimensional protein separations in a plastic microfluidic network. *Anal. Chem.* **76**, 742–748.
57. Kim, S. M., Burns, M. A. and Hasselbrink, E. F. (2006) Electrokinetic protein preconcentration using a simple glass/poly(dimethylsiloxane) microfluidic chip. *Anal. Chem.* **78**, 4779–4785.
58. Liu, J., Sun, X. and Lee, M. L. (2007) Adsorption-resistant acrylic copolymer for prototyping of microfluidic devices for proteins and peptides. *Anal. Chem.* **79**, 1926–1931.
59. Wang, Y. -C., Choi, M. H. and Han, J. (2004) Two-dimensional protein separation with advanced sample and buffer isolation using microfluidic valves. *Anal. Chem.* **76**, 4426–4431.
60. Herr, A. E., Molho, J. I., Drouvalakis, K. A., Mikkelsen, J. C., Utz, P. J., Santiago, J. G. and Keey, T. W. (2003) On-chip coupling of isoelectric focusing and free solution electrophoretic for multidimensional separations. *Anal. Chem.* **75**, 1180–1187.
61. Song, Y. -A., Hsu, S., Stevens, A.L. and Han, J. (2006) Continuous-flow pI-based sorting of proteins and peptides in a microfluidic chip using diffusion potential. *Anal. Chem.* **78**, 3528–3536.
62. Cooper, J. W., Che, J., Li, Y. and Lee, C. S. (2003) Membrane-based nanoscale proteolytic reactor enabling protein digestion, peptide separation, and protein identification using mass spectrometry. *Anal. Chem.* 75, 1067–1074.
63. Hardouin, J., Joubert-Caron, R. and Caron, M. J. (2007) HPLC-chip-mass spectrometry for protein signature identifications. *Sep. Sci.* **30**, 1482–1487.
64. Xie, J., Miao, Y., Shih, J., Tai, Y. -C. and Lee, T. D. (2005) Microfluidic platform for liquid chromatography-tandem mass spectrometry analyses of complex peptides mixtures. *Anal. Chem.* 77, 6947–6953.
65. Ghitun, M., Bonneil, E., Fortier, M. -H., Yin, H., Killeen, K. and Thibault, P. J. (2006) Integrated microfluidic devices with enhanced separation performance: application to phosphoproteome analyses of differentiated cell model systems. *Sep. Sci.* **29**, 1539–1549.

Chapter 9

Tagging Recombinant Proteins to Enhance Solubility and Aid Purification

Dermot Walls and Sinéad T. Loughran

Abstract

Protein fusion technology has enormously facilitated the efficient production and purification of individual recombinant proteins. The use of genetically engineered affinity and solubility-enhancing polypeptide "tags" has increased greatly in recent years and there now exists a considerable repertoire of these that can be used to solve issues related to the expression, stability, solubility, folding, and purification of their fusion partner. In the case of large-scale proteomic studies, the development of purification procedures tailored to individual proteins is not practicable, and affinity tags have therefore become indispensable tools for structural and functional proteomic initiatives that involve the expression of many proteins in parallel. Here, the rationale and applications of a range of established and more recently developed solubility-enhancing and affinity tags are outlined.

Key words: Protein, Tag, Solubility, Affinity chromatography

1. Introduction

The biotechnology industry demands rapid and efficient procedures for expressing and purifying authentic proteins. The advent of genetic engineering has brought with it the ability to express and manipulate any DNA sequence and produce the encoded protein in recombinant form using a heterologous host such as *Escherichia coli* (*E. coli*). Thanks to gene fusion techniques and site-directed mutagenesis, which permit the optimization of transcription and translation regulatory signals including codon content, virtually any polypeptide can be produced in recombinant form in bacterial, fungal, or eukaryotic host cells. Major stumbling blocks remain however, including protein insolubility, incorrect folding, host cell toxicity, and protein instability. Sometimes the result of

Dermot Walls and Sinéad T. Loughran (eds.), *Protein Chromatography: Methods and Protocols*, Methods in Molecular Biology, vol. 681, DOI 10.1007/978-1-60761-913-0_9, © Springer Science+Business Media, LLC 2011

enforced hyperexpression of the desired protein, these impediments can nonetheless be addressed by expressing the protein of interest as a fusion with certain partner "tag" peptides or proteins. Thus, numerous solubility-enhancing or affinity tags of either natural origin or of artificial design have been exploited to enhance protein solubility and yield and facilitate purification. Unlike purification strategies that start with native sources of the desired protein, the chromatography of recombinant proteins is dramatically simpler thanks to the range of tag-associated affinity-based recovery systems that has been developed. Recombinant proteins that are joined to a tag with moderate affinity and high specificity for a particular ligand can often be isolated from crude lysates in a single step following selective binding to matrices on which the ligand has been immobilized. Many tag-fusion expression vectors are now available for the purposes of enhancing solubility, enabling one-step affinity-based purification, and facilitating the detection of the recombinant protein of interest. In structural and functional proteomic studies, it is not practicable to tailor purification procedures to individual proteins, and high-throughput protein purification can only be achieved with the assistance of such affinity tags.

Escherichia coli remains an important host cell for recombinant protein expression because of its easy manipulation and high biomass-to-cost ratio. Some heterologous proteins that are produced in *E. coli* form aggregates of insoluble folding intermediates known as inclusion bodies, which are often the result of insufficient time for correct folding being available under conditions of hyperexpression. Protein denaturation and refolding in vitro is expensive, time-consuming, and can lead to losses in both yield and bioactivity. Inclusion body formation can often be prevented by optimization of host culture conditions or by fusion of the desired protein to some of the larger affinity tags that are available. Affinity tags include enzymes, protein domains, or small polypeptides most of which bind with high specificity to a range of substrates such as carbohydrates, small biomolecules, metal chelates, antibodies, or antibody-binding molecules. Other tags that only possess solubility-enhancing properties are frequently used in tandem with an affinity tag to aid purification.

In order to generate a tagged protein, the DNA sequence that encodes the protein of interest is joined to the expression vector sequence in such a way as to generate an in-frame fusion with the polypeptide tag-coding sequence. The fusion tag may thus be placed at either the N-terminus or C-terminus of its partner depending on the cloning strategy used, and optimal positioning must be elicited as needed. Often, expression vectors are designed so that a short flexible hydrophilic linker or spacer peptide is located between the tag and its fusion partner. This may serve two purposes: (1) to ensure sufficient spatial separation between tag and recombinant protein, so as to maximize tag availability during chromatography but minimize any potential interactions with its partner protein and (2) to facilitate the inclusion of an endoprotease cleavage site that can be subsequently exploited for removal of the tag. The latter is often a primary consideration following chromatography due to the potential impact that the tag may have on the function or structure of the recombinant protein. This is of particular relevance both in the case of therapeutic proteins, where the authenticity of the expressed protein is of paramount importance, and for structural studies where the protein's native conformation may be lost due to the presence of the tag. A limited number of proteases are exploited for the purpose of tag removal and these are discussed in detail elsewhere in this volume along with detailed protocols for their use (see Chapter 19). In general, there are no structural similarities between the various solubility and affinity tags and the choice as to which one might work best in a specific instance must be determined experimentally. Here, we report on the principal solubility and affinity tags that are available for use (see also Table 1).

2. Solubility-Enhancing Fusion Partners

2.1. Maltose-Binding protein

Maltose-binding protein (MBP) was one of the first fusion partners to be used for the purposes of alleviating problems associated with the expression and purification of recombinant proteins. MBP is a part of an *E. coli* pathway that is responsible for the uptake and efficient catabolism of maltodextrins (1). MBP binds strongly to amylose resin (2) and one-step affinity chromatography of MBP-tagged proteins using amylose–agarose columns can typically lead to a fusion protein of 70–90% purity (3). In addition to acting as an affinity tag, the 42 kDa MBP can enhance the expression and folding of fusion proteins in which it is the N-terminal (4–6) or C-terminal partner (7). The *MBP* sequence provides an optimum context for translation initiation, and the expressed protein can be located in the cytoplasm or secreted to the less reducing environment of the periplasm, depending on whether the secreted or non-secreted form of MBP is used as partner (8). There is much evidence of MBP-associated enhancement of the folding/solubility of otherwise insoluble fusion partner (7, 9–11). This enhancement is passive and not completely understood (12), and there is some evidence to suggest that it may be the result of MBP-recruited chaperones being located in the vicinity of the fusion partner (13). Expression vectors for generating proteins fused to MBP are commercially available (the pMAL series, New England Biolabs; pIVEX, Roche), and following purification, the MBP tag

Table 1
Selected tags for solubility enhancement and the affinity purification of recombinant proteins

Tag name	Tag type	Purification	Comments
Thioredoxin (TX)	Solubility enhancement	Phenylarsine oxide matrices	Often poor yield and quality of product. Tandem tagging with polyhistidine (His-patch ThioFusion system, Invitrogen) enables purification by IMAC
Glutathione *S*-transferase (GST)	Solubility enhancement and affinity	Glutathione–sepharose matrices	Elution under mild conditions with free glutathione or IMAC when used with hexa-histidine tag. Vector series (pGEX) available from GE healthcare
Maltose-binding protein (MBP)	Solubility enhancement and affinity	Cross-linked amylose resins	Elution under mild conditions with free maltose or IMAC when used with hexa-histidine tag. pMAL and pIVEX vectors available from New England Biolabs and Roche, respectively
Small ubiquitin-related modifier (SUMO)	Solubility enhancement	None when used alone	Enhanced solubility and expression in *E. coli* and eukaryotic expression systems. SUMO-His_6 fusions purified by IMAC
NusA	Solubility enhancement	None when used alone	IMAC when used with hexa-histidine tag
Z-tag, ZZ-tag, and GB1 tag	Solubility enhancement and affinity	Protein A-sepharose and IgG resins	Mostly tagged in tandem with polyhistidine, enabling IMAC purification to be used
Arg-tag	Affinity	Cation-exchange resins	Eluted with a NaCl linear gradient at alkaline pH. Resins include SP-Sephadex (Sigma-Aldrich); HiTrap SP-sepharose FF columns (GE Healthcare)
FLAG-tag	Affinity	Monoclonal antibodies/ affinity gels for detection and purification (Sigma-Aldrich)	Competitive elution by FLAG octapeptide when using anti-FLAG M2 antibody. Calcium/ EDTA-dependent affinity procedure in the case of M1 antibody. Tag is removable by enterokinase cleavage
Streptavidin-binding peptide (SBP)	Affinity	Immobilized streptavidin	Eluted with biotin (2 mM)

(continued)

Table 1 (continued)

Tag name	Tag type	Purification	Comments
Strep-tag II	Affinity	A modified streptavidin (Strep-tactin; multiple commercial sources)	Can be advantageous when purifying metalloproteins and when NMR is to be used. Eluted with biotin and derivatives, in particular, desthiobiotin
C-myc-tag	Affinity	Monoclonal antibody-based detection	Mab9E10 (multiple suppliers). Physiological conditions followed by elution at low pH
Softag1, 3	Affinity	Monoclonal antibody-based detection	Mild elution conditions (http://www.lucigen.com)
S-tag	Affinity	S-fragment of RNaseA (Novagen)	Elution conditions are harsh: 3 M guanidine thiocyanate, 0.2 M citrate pH 2, 3 M magnesium chloride. Tag removal by enzymatic cleavage is recommended to obtain functional proteins
Calmodulin-binding peptide	Affinity	Immobilized calmodulin (various commercial sources)	Binds calmodulin in the presence of 0.2 mM $CaCl_2$. Eluted in 2 mM EGTA without or with 1 M NaCl. Not to be used for expression in eukaryotic cells
Cellulose-binding domain	Affinity	Cellulose matrices	Tag size varies between 4 and 20 kDa. Eluted with either urea (4 M) and guanidine–HCl (Family I CBDs) or ethylene glycol (Family II/III CBDs)
Chitin-binding domain	Affinity	Chitin matrices (NEB). No need for additional protease step to remove tag	Used in tandem with an intein. Intein self-cleavage is then induced with dithiothreitol (30–50 mM), β-mercaptoethanol, or cysteine
Biotin carboxyl carrier protein (BCCP)	Affinity	Biotinylated proteins purified on avidin/streptavidin resins	PinPointXa series of vectors and SoftLink avidin resin (Promega)
Elastin-like peptides (ELP)	Affinity	Temperature-induced aggregation, separation, and resolubilization	Target protein self-association in high-salt buffers at temperatures above 30°C; intein activity (if present) induced in response to a mild pH shift

(continued)

Table 1 (continued)

Tag name	Tag type	Purification	Comments
His-tag	Affinity	Ni-NTA agarose (QIAgen) Ni-NTA superflow (QIAgen) Co^{2+}-CMA (Talon; Clontech) Ni sepharose (GE Healthcare) Ni sepharose high performance (GE Healthcare) Nickel affinity gel (Sigma-Aldrich) Cobalt affinity gel (Sigma-Aldrich)	Imidazole 20–250 mM or decreasing pH gradient
Tandem affinity purification (TAP)-tag	Affinity (TAP)	IMAC	Imidazole 20–250 mM or low pH
S3S-TAP-tag	TAP	IgG beads (Pharmacia) followed by calmodulin beads (Stratagene)	Calmodulin-binding protein (CBP) fused to two protein-A IgG-(ZZ) binding domains. Contains a TEV cleavage site between CBP and ZZ. There are issues due to large tag size, need for proteolytic removal of tag, and CBP interference with calcium signalling in eukaryotic cells
FF-ZZ TAP-tag	TAP	S-tag agarose (Merck), Strep-tactin beads (QIAgen)	Modified TAP-tag designed for mammalian protein complex isolation. Elution with desthiobiotin
Strep/FLAG-TAP (SF-TAP)	TAP	IgG sepharose resin (GE Healthcare); anti-FLAG M2 affinity resin (Sigma-Aldrich)	Elution with FLAG octapeptide. Improved recovery following purification, due to substitution of CBP with FLAG tags. The tagged protein is produced from a bicistronic vector that permits selection of transfected cells
GS-tag	TAP	Strep-tactin superflow resin (IBA); anti-FLAG M2 affinity matrix (Sigma-Aldrich)	Elution with desthiobiotin or FLAG octapeptide. Purification under native conditions, no need for enzymatic removal of tag
PTP-tag	TAP	HPC4 monoclonal antibody (anti-protein C affinity matrix, Roche) in the presence of calcium	A Strep-II-3× Flag-2× ZZ-tag. Less background but low yield of affinity co-purified proteins. Eluted with EGTA or ProtC peptide. Substitution of CBP with protein C epitope eliminates CBP-associated issues

can be excised by specific protease cleavage in the polypeptide linker region between the fusion partners. The MBP system is widely used in combination with a small affinity tag and vectors for generating combinatorially tagged polyhistidine–MBP fusion proteins have also been described, enabling the user to derive additional benefit from the improved specificity of purification using immobilized metal affinity chromatography (IMAC; see Subheading 3.1) (14–16) and to address the observation that not all MBP fusions can be purified on amylose (17).

2.2. Glutathione S-Transferase

Recombinant glutathione *S*-transferase (GST), a 26 kDa protein whose corresponding gene was originally obtained from the parasitic helminth *Schistosoma japonicum*, is another well-established solubility/affinity tag (18). Fusion proteins are affinity purified from crude lysate preparations on immobilized sepharose–glutathione matrices and recovered under mild non-denaturing elution conditions in the presence of reduced glutathione. Although inexpensive, large-scale production of GST fusions is a lengthy process, a result of the slow binding kinetics of GST to glutathione–sepharose resins (19). Fusion to the GST partner ensures that translation of the recombinant protein initiates efficiently and affords some protection against intracellular protease cleavage. The GST-tag can be fused at the N- or C-terminus and has successfully been used in bacteria, yeast, mammalian cells, and baculovirus-infected insect cells. In many cases, fusion proteins are soluble and form dimers in aqueous solution, but overall GST is a poor solubility enhancer. It is a homodimer, an unwelcome complication making it an unsuitable partner for the isolation of many oligomeric proteins, and has exposed cysteine residues that can promote oxidative aggregation (20). Some insoluble GST fusions can still be purified by affinity chromatography following solubilization with mild detergents (21). Typically, expression vectors for GST fusion proteins (such as the pGEX series from GE Healthcare) encode specific endopeptidase cleavage sites that have been engineered between the tag and partner proteins. GST moieties can therefore be enzymatically cut off and removed by affinity chromatography on a glutathione–sepharose matrix, and the desired fusion partner is then purified to homogeneity by other methods such as ion exchange or gel filtration. Fusing proteins to GST is a popular molecular biology technique for studying biomolecular interactions ("GST pulldowns"; (22, 23)). In high-throughput proteomics, GST fusion proteins have been directionally immobilized onto protein microarrays (24, 25). Successful structure-function studies involving protein–protein and DNA–protein interactions also have been described (22). Affinity chromatography, followed by removal of the GST-tag, has been used to purify numerous proteins prior to structural analyses and crystallography, and crystal structures have also been described for fusion proteins that retain the GST-tag (26, 27). The GST-tag is also widely used in tandem with smaller affinity tags such as polyhistidine thus permitting additional benefit to be had from using IMAC purification systems (see Subheading 3.1).

2.3. Small Ubiquitin-Related Modifier

The addition of small ubiquitin-related modifier (SUMO) protein is a reversible post-translational modification that has been shown to occur in eukaryotic cells. SUMOylation, however, unlike ubiquitination, does not lead to degradation of the target protein at the 26S proteosome, but appears to regulate protein transport to different intracellular compartments, such as the nucleus, and to play a role in transcriptional regulation (28, 29). The SUMO conjugation/deconjugation pathway is highly conserved in eukaryotes, but absent in prokaryotes. Yeast has one SUMO gene (SMT3), while three genes have been identified in vertebrates (SUMO-1, -2, and -3) (30–32). Although the overall sequence identity between SUMOs and Ub is low, they share a common three-dimensional structure (33). The SUMO pathway is similar to those already described for ubiquitin, in that it is an ATP-dependent ligase cascade which serves to couple SUMO through an isopeptide bond to the e-NH2-group of lysine residues of the acceptor protein (34). SUMO itself is a 100 amino acid protein that has been shown to modulate protein structure and function by covalent modification of target proteins (32, 35, 36). SUMO-specific proteases remove and recycle SUMO from target proteins. The SUMO proteases, such as yeast Ulp1, recognize a Gly-Gly- containing motif found at the C-terminus of SUMO, with a strict requirement for tertiary structure elements only present on correctly folded SUMO. Consequently, such cleavage of proteins that are joined to SUMO is restricted to the junction between SUMO and its protein partner. The advantage to this is that erroneous cleavage within the target protein is much less likely to occur (37–40).

In general, recombinant fusion of yeast SUMO (Smt3) to the N-terminus of target proteins for expression in bacteria has been shown to work well, leading to the enhancement of both protein solubility and expression (39–42). The SUMO tag can then be removed by digestion with Ulp1, thus regenerating the native N-terminus of the recombinant fusion partner (39). A SUMO fusion system that facilitates efficient expression of recombinant proteins in *E. coli* has been described (39, 41, 43). Lee et al. used a sticky-end polymerase chain reaction (PCR)-based strategy to design a SUMO fusion protein expression vector that allows directional cloning of any target gene (44). Their approach involved joining the protein of interest to hexahistidine (His6)-tagged Smt3 (Smt3 is the yeast SUMO protein), thus making it possible to carry out both fusion protein purification and SUMO protease cleavage using a Ni^{2+}-resin column.

This kind of one-column strategy will be useful for the development of better high-throughput platforms for the purification of recombinant proteins. One comparative study demonstrated that SUMO was superior to other commonly used fusion tags in enhancing expression and solubility of several fusion partners, with the distinction of generating recombinant proteins with native sequences following tag removal (40).

The *E. coli* protein ElaD, a recently identified ubiquitin (Ub) protease, has been shown to specifically cleaves Ub fusion proteins, but not the SUMO fusion proteins (45), implying that SUMO may be a better choice of fusion partner when *E. coli* is used as host. Although SUMO fusion systems work well in prokaryotes, it is likely that intracellular premature processing by endogenous desumoylases will serve to limit their utility in eukaryotic protein expression systems. One method, known as the "split SUMO" approach, was designed to overcome this problem and involves first fusing the N-terminus of the target protein to the C-terminal half of SUMO (CTHS), which is not cleaved by endogenous desumoylases but does retain the capacity to enhance expression of its fusion partner. Following protein purification, the full SUMO structure is then reconstituted by incubation with the N-terminal half of SUMO (NTHS) which interacts strongly with the CTHS, thus generating a substrate for SUMO protease (42). In a more recent attempt to address the same issue, a double mutant of the yeast Smt3 protease (termed SUMOstar) was recently developed, which is not recognized by native desumoylases. When used as fusion partner, SUMOstar was shown to lead to the enhanced expression of a range of test proteins in several eukaryotic cell lines and a baculovirus system (46–48).

2.4. Other Solubility-Enhancing Fusion Partners

Thioredoxins are universal oxido-reductases that facilitate the reduction of other proteins by cysteine thio-disulfide exchange. *Escherichia coli trxA* is a small protein (109aa, 11,675 kDa) that demonstrates high solubility in the *E. coli* cytoplasm. It also has an inherent thermal stability, and is located on the cytoplasmic side of the adhesion zones between the inner and outer bacterial cell membrane. Though not itself an affinity tag, these two properties of *trxA* may be conferred on *trxA* fusion partners, and can therefore sometimes be exploited to enable the rapid purification of *trxA*-fusion proteins. Release from the *E. coli* cytoplasm is achieved by osmotic shock or freeze/thaw treatments. Thioredoxin can be fused to the amino- or carboxyl-terminus of the protein of interest but typically the *trxA* coding sequence is placed at the 5′-end as it promotes efficient translation initiation of the recombinant gene (49, 50). A variety of expression vectors for this purpose are commercially available, and derivatives of *trxA* have been engineered to possess affinity for immobilized metal ions (e.g. His-patch ThioFusion system from Invitrogen) or avidin/streptavidin (51).

Various cytokines, growth factors, and fluorescent proteins have been shown to retain remarkable solubility in the *E. coli* cytoplasm when expressed as c-terminal fusion proteins with *trxA* (52–54). Other more recent examples, in which the solubility of archaeal proteins and bioactive viscotoxins were considerably enhanced, serve to demonstrate the utility of *trxA* as a versatile fusion partner (55, 56). Protein structural studies are frequently hindered by the difficulty in obtaining diffracting crystals of the target protein, and a fusion partner may sometimes aid the crystallization of difficult targets (57). *TrxA* itself crystallizes in several different forms, and one recent screen of a variety of fusion partners demonstrated significant benefits when *trxA* was joined via the tetrapeptide linker sequence GSAM to proteins that are difficult to crystallize (58).

N-utilization substance A (NusA) is a well-conserved essential transcription elongation and anti-termination factor (59). Fusion with the highly soluble *E. coli* NusA has been shown to enhance the expression and solubility of proteins to which it is joined at the N-terminus in particular (60). One comprehensive study compared the solubility-enhancing capabilities of NusA and MBP using a diverse set of aggregation-prone partner proteins (12). It was concluded that both tags, although very different in terms of their biological functions and physiochemical properties, performed very similarly overall as solubility promoters and played passive roles in the folding of their fusion partners. Another hydrophilic fusion partner is the *E. coli* protein disulfide isomerase I (DsbA), which can increase the cytoplasmic and periplasmic solubility of target proteins. A disadvantage in using these tags alone, however, is that affinity purification is not an option. Vectors enabling the use of NusA, MBP, and GST in combination with a hexa-histidine-tag have been reported, however, permitting IMAC purification of fusions proteins that benefit from the solubility-promoting properties of the larger tag partner (61, 62). Other reported solubility/affinity tags include derivatives of the *Staphylococcus aureus* protein A (Z-tag and ZZ-tag; (63, 64)), a mutated derivative of the *Streptococcus* protein G β1 domain (GB1; (65–67)), disulphide bond C (DsbC; (68)), Phage T7 protein kinase (T7PK; (69)), the 17 kDa protein (Skp; (69)), and the small polypeptide solubility-enhancing tag (SET; (70)).

3. Affinity Tags

3.1. Immobilized Metal Affinity Chromatography

IMAC, a now ubiquitous form of affinity chromatography was first introduced in 1975 by Porath et al. (71) and is currently the most widely used technique for purifying recombinant proteins. IMAC is a selective tool for the separation of metal-binding peptides which is based on the interaction of certain amino acid residues (e.g. accessible His, Ser, Cys, Glu, and Asp residues (72–74)) on the surface of peptides, and the metal ions within an immobilized metal chelate (75). The reversible nature of this interaction means that it can be first exploited for peptide adsorption and then subsequently disrupted under various elution conditions, involving making alterations to pH or salt concentration. The differential affinity of proteins for immobilized metal ions stems from the coordination bonds formed between the metal ions involved and electron donor groups present in some amino acid residues (e.g. His, Cys, Trp, and Arg). In IMAC, these electron donor groups form complexes with transition metal ions such as Cu^{2+}, Co^{2+}, Zn^{2+}, or Ni^{2+} which are typically immobilized on polymeric supports with chelating pendant groups such as iminodiacetic acid (IDA) or nitrilotriacetic acid (NTA). The retention of proteins on IMAC supports is influenced by a number of variables, including the nature of chelating groups, the specific metal ion (75, 76), and the surrounding milieu (salt concentration and pH). Each chelating group exhibits its own selectivity and adsorption capacity toward a specific protein.

Histidine exhibits the strongest interaction with immobilized metal ion chelates and as a result the oligohistidine tag, His_6, is one of the most frequently used fusion tags for protein recovery. Ni(II)-nitrilotriacetic acid (Ni-NTA) exhibits a high affinity for adjacent histidine residues and is one of the most commonly used IMAC matrices for target protein capture by metal-ligand covalent bonding. In general, the order for the adsorption capacity of proteins with accessible histidines and chelating ligands is IDA (tridentate) > NTA (nitrilotriacetic acid, tetradentate) ≥ CM-Asp (carboxymethylated aspartic acid, tetradentate) > TED (pentadentate) (77). Controlled release of captured proteins can be achieved by increasing the concentration of imidazole in the purification buffer, by changing the pH or stripping the metals off the resins with strong metal chelators such as EDTA.

The popularity of histidine (His)-tag affinity purification is due in part to its high affinity with Ni-NTA, a relatively inexpensive matrix that is able to withstand multiple regeneration cycles under stringent conditions, and also due to the ease of controlled release using mild (e.g. non-denaturing) conditions. In addition, as the tertiary structure of the His-tag is not important for purification, an insoluble His-tagged recombinant protein can be purified by IMAC under denaturing conditions and subsequently refolded (for a review see (78)). IMAC binding of a polyhistidine-tagged fusion protein allows the fusion partner to remain free in solution and thus to be able to fold unhindered by any constraints due to the fact that it is bound to a solid support (79). A gradual decrease in denaturant concentration induces protein refolding and elution is achieved by increasing the imidazole concentration or by using a decreasing pH gradient (80, 81). Recently, Dong et al. proposed an artificial chaperone-assisted IMAC (AC-IMAC) procedure for protein refolding and purification from inclusion bodies (82). In this work, the authors first overexpressed His-tagged enhanced green fluorescent protein (EGFP), and then refolded and purified it from solubilized inclusion bodies by AC-IMAC. In a systematic comparison of refolding by AC-IMAC with refolding by IMAC or in bulk solutions, it was shown that in addition to the high purification advantage associated with IMAC, the AC-IMAC method was superior to either IMAC or AC in its high refolding efficiency.

Several further key advantages have established His-tag IMAC as the most widely used method for purifying recombinant proteins for biochemical and especially structural studies. The His-tag combines the advantages of being inert, of low immunogenicity, and of small size (0.84 kDa) (77, 83). In most cases, its small size means the His-tag does not interfere with the biochemical activities of the partner protein (84–90) or with most downstream applications (77). IMAC can be performed in the presence or absence of chaotropic agents, and is compatible with strong denaturing reagents such as urea and guanidinium–HCl, as well as a large number of non-ionic detergents, making it extremely useful in the initial steps of protein purification immediately after the extraction/isolation procedure (77).

His-tagged proteins can be readily detected by Western blotting using anti-His antibodies. Cross-reactivity of the primary antibody with endogenous histidines in mammalian and insect expression systems may be a considerable drawback, however (91). The His-tag IMAC purification process has been adapted and optimized for high throughput on-chip protein purification using protein microarrays with metal chelate-modified glass slide surfaces (92) and has the potential to permit rapid screening of therapeutic reagents leading to the discovery of novel protein functions. High-throughput protein purification using an automated set-up for high-yield IMAC has also been described (93). This procedure allowed fully automated purification from up to 60 cell lysates with milligram yields of pure recombinant protein in 18.5 h.

There are reports of His-tags altering the binding characteristics or structure of their partner protein when compared to the corresponding native protein (94, 95). Hang et al. showed that although His-tagged subunits of the terminase enzyme from bacteriophage-γ formed holoenzymes with wild-type catalytic activity, one of the subunits displayed an altered capacity to interact with DNA that was not seen in its native counterpart (94). The length, composition, and location of the His-tag can therefore require optimization depending on the sequences of the native protein (95–97). The use of His-tags is not recommended for purifying proteins that contain metal ions. Likewise, the presence of cysteine and naturally occurring histidine-rich regions in host proteins may result in unwanted protein binding during IMAC (98).

Despite the universal application of IMAC, there are always difficult tagged proteins that show weak binding to the metal chelating resin. This can be caused by concealment of the His-tag, and may be alleviated by switching its position to the other terminus of the protein (99) or by introducing a linker to separate the His-tag from the partner protein (100). These difficulties can also be overcome by increasing the length of the His-tag to eight or ten histidines (101, 102), and a vector has been described that allows the parallel cloning of target genes with different His-tag lengths using a single insert (44).

IMAC is very sensitive to the presence of metal chelators (77). In *E. coli* expression systems, the cell lysate contains many unspecific weak chelators such as dicarboxylic acids from the citric acid cycle. Under stress conditions, *E. coli* can also produce highly specific metal chelators known as metallophores (83). One report linked the failure to purify low-abundance His-tagged proteins from *E. coli* lysates to metal-ion leakage from purification columns (103). In that study, the authors used His-tagged GFP (His_6-GFP) to examine the effect of *E. coli* lysate on the protein binding capacity of IMAC columns and concluded that low-molecular-weight components of the lysate, associated with the periplasm, severely reduced the binding capacity of the column. By removing the periplasmic material before cell lysis, the authors observed a tenfold increase in the yield of His_6-GFP when it was diluted with *E. coli* lysate before purification to simulate a low-abundance protein. The separation of contaminant proteins from the target protein is often difficult due to similarities in physico-chemical properties. A recent seminal report described the use of proteomics to design a host cell that was tailored for highly efficient protein purification (104). The authors outlined a highly efficient strategy, based on proteomic analysis and elution chromatography, in which a protein of interest may be isolated from co-purifying contaminants. Strains of *E. coli* were first engineered to be deficient in three prevalent host proteins that were found in a strategic fraction of an elution profile of Ni(II)-IMAC. Recombinant GFP was then used as a model protein in the expression system and its elution was directed to the optimized fraction with a His_6-tag, thereby easing its recovery.

His-tag/IMAC-based strategies currently dominate in the area of recombinant protein purification and examples of novel applications continue to emerge. IMAC has recently been deployed for the purification of single-chain variable antibody fragments using an improved hexa-histidine tag phagemid vector (105). Purification by IMAC of whole lentiviral vectors is a promising approach to overcome production issues that limit the widespread availability of clinical grade lentiviral vectors. Histidine-rich protein sequences have been genetically engineered for display at the virus surface and enable retention of whole packaged viral vectors on an IMAC column (106). One study reported a 56% recovery of a histidine-tagged retroviral vector using an agarose-nitrilotriacetic acid-Ni(II) resin (106), while another demonstrated that the commercially available CIM-IDA adsorbent can be used to efficiently concentrate and purify histidine-tagged lentiviral vectors (107).

3.2. Other Affinity Tags

Strep-tag is an eight amino acid peptide that binds to the biotin-binding pocket of streptavidin. Optimization of both partners has led to the development of Strep-tag II peptide and Strep-tactin, the latter being an engineered derivative of streptavidin (108). The Strep-tag II polypeptide binds Strep-tactin with an affinity that is two orders of magnitude greater than its affinity for streptavidin. The tag itself is short, proteolytically stable, and biologically inert, and the Strep-tag II-Strep-tactin affinity system is therefore exploited for the one-step isolation of bioactive and highly pure Strep-tag II-tagged proteins under physiological conditions. The larger 38 amino acid streptavidin-binding peptide (SBP) was selected as a robust high-affinity streptavidin-binding aptamer (109). Tagging recombinant proteins with SBP is of particular interest when intermediate amounts of protein (up to 0.5 mg) need to be produced and purified in a high-throughput manner. An additional advantage here is that a wide variety of streptavidin-derivatized materials is commercially available.

Short oligomers of arginine (Arg-tags) have been used in the past as tags to facilitate protein purification by ion-exchange chromatography. These cationic peptides, such as the R_9 tag which consists of nine arginines, are versatile in that they can also promote cellular internalization and surface immobilization of recombinant proteins to which they are fused (110).

Calmodulin-binding peptide (CBP), derived from human skeletal muscle myosin light chain kinase, binds calmodulin with high affinity (111). Stringent washing conditions coupled with the lack of calmodulin-binding proteins in *E. coli* means that CBP-tagged fusion proteins can be recovered in good yield with high specificity. The tag is not suitable for use in eukaryotic cells as it interacts with endogenous proteins and can interfere with calcium signalling pathways.

Cellulose-binding domains (CBD) are non-catalytic domains that have been identified in many proteins and their size, relative locations, and affinity for their natural substrate varies considerably (112). Cellulose is an attractive matrix for the affinity purification or immobilization of CBD-tagged proteins. It is inert, has low non-specific affinity, is available in many forms, and has received approval for many pharmaceutical uses. CBDs have been used for applications involving the immobilization of enzymes, cytokines, or other ligands (113). Some cellulose–CBD interactions are very strong indeed and often require the use of strong chaotropic agents to promote the release of CBD-tagged protein, thus necessitating protein refolding in vitro. Others can be eluted under milder conditions using ethylene glycol. A method for purifying CBD-tagged molecules has been reported whereby they are induced to self-aggregate leading to their selective pull-down (114). Elsewhere, a novel CDP-antibody-binding domain hybrid tag has been described and used to immobilize antibodies and cells onto regenerated cellulose hollow fibre membranes (115, 116).

Chitin-binding domains have also been exploited as tags for the affinity purification of recombinant proteins. The IMPACT™ system (Intein-Mediated Purification with an Affinity Chitin-binding Tag; NEB/Novagen) is based on fusing the target protein to a chitin-binding domain in tandem with an intein protease. Following affinity selection of the fusion protein on a chitin matrix, the intein undergoes specific cleavage by a thiol reagent or pH and temperature shift which releases the target protein from the chitin-bound tag (117, 118).

FLAG-tag is an eight amino acid peptide that was designed for immunoaffinity chromatography (119). FLAG-tagged proteins may be recovered from crude lysates in a rapid one-step procedure and detected in easy immunoassay formats. Although highly selective and frequently used in research, its binding capacity is low making scale-up a costly undertaking, and being immunogenic, the tag must be removed in the case of proteins intended for therapeutic use. The HA, c-Myc, and T7 epitope tags are often used to tag recombinant proteins (69, 120, 121), but such antibody-based affinity systems are not routinely used for purification purposes. Softagl and Softag3 are small peptides that are recognized by polyol-binding mAb. Elution can be achieved under mild conditions and such "gentle immunoaffinity chromatography" is useful for studying protein interactions within multisubunit protein complexes (122). S-tag, a 15 amino acid polypeptide derived from the N-terminal helix of RNAse A, complements the fold of immobilized truncated RNAse A with very high affinity, thereby providing a system for the efficient purification of S-tag-fused proteins (123).

Elastin-like polypeptides (ELP) have the property of becoming reversibly insoluble at relatively low transition temperatures, and have been used successfully as affinity tags (124). "Inverse transition cycling" (ITC) describes the technique in which an ELP tag is used to reversibly precipitate and purify a genetically fused target protein (125, 126). The method is highly scalable and eliminates the need for expensive affinity resins and apparatus, and when coupled with a self-cleaving intein tag, the requirement for tag removal using an additional protease step is removed (127).

3.3. Tandem Affinity Purification Tagging

Tandem affinity purification (TAP) tagging is a method first described using yeast cells that enables the rapid in situ purification of protein complexes as a result of their co-purification with a "TAP-tagged" recombinant protein. The original TAP tag consisted of two IgG-binding domains of *S. aureus* protein A and a CBP separated by a protease cleavage site (128, 129). TAP-tagging, which involves two sequential purification steps, has been used to identify protein complexes from yeast (130), insect (131), human cells (132), and plants (133–135). The S3S-TAP-tag (a cleavable S-tag combined with Strep-tag II; (136)) is a recently developed universally applicable TAP-tag system suitable for the isolation of mammalian protein complexes. Strep/FLAG-TAP (SF-TAP) tag is a small tag that combines a tandem Strep-tag II and a FLAG-tag, and eliminates the need for an intermediary time-consuming proteolytic cleavage step (137). Another tandem tag is FF-ZZ, which consists of two FLAG tags (FF) followed by two protein-A IgG-binding domains (ZZ). Replacing CBP with FLAG resulted in higher recovery during purification (138). Additional tag combinations have been described whereby the CBP moiety has been replaced with other tags in an effort to address various issues including the yield of TAP-tagged protein and background due to non-specific protein interactions (SF-ZZ, GS-tag, PTP-tag (139–141); see Table 1). TAP-tagging is now an important strategy in the study of protein interaction networks and has recently been used to generate a proteome-wide interaction network in yeast (142).

4. Tagging Proteins and Structural Studies

Protein samples for structural biology projects are increasingly being generated by recombinant DNA technology. NMR spectroscopy, X-ray crystallography, and cryo-electron microscopy all require considerable quantities of protein, usually in the order of 5–50 mg. Research centres such as the Southeast Collaboratory for Structural Genomics (SECSG; http://www.secsg.org) have developed high-throughput (HT) and cost-effective pipelines for protein production, crystallization, and structure determination (143). Virtually all such large structural genomics centres use IMAC as their affinity strategy to purify polyhistidine-tagged recombinant proteins (83). Small tags are favoured for structural studies as they do not substantially increase the size of the target protein and tag removal may not be required in order to generate good quality crystals (144). *Escherichia coli* is the preferred host for recombinant protein expression here, and isotope labelling protocols for NMR spectroscopy and selenomethionine incorporation for X-ray crystallography are now well established.

There are obstacles, however, when using *E. coli* as it cannot reproduce eukaryotic post-translational modifications such as glycosylation, which are often essential for the correct folding and bioactivity of proteins and so protein expression in a eukaryotic host may be necessary (145). Furthermore, some proteins, especially larger proteins and membrane proteins, are either not expressed or are insoluble when produced in *E. coli* although the situation is improving as a result of the development of new vectors and host strains alike. A semi-automated large-scale process for the production of polyhistidine-tagged recombinant proteins in the Baculovirus expression system has been described (146).

Hexa-histidine tags do not enhance the solubility of their fusion partner, unlike MBP or GST tags. Crystal growth is impeded as a result of conformational heterogeneity induced by such large tags, however, creating the need for tag removal by endopeptidase cleavage following expression. If the tag is to be left in place however, it is now clear that a rigid short polypeptide spacer, as opposed to a longer more flexible one, is required between tag and partner protein so as to eliminate such conformational heterogeneity and promote the growth of high quality crystals (27). Not all proteins rendered soluble by fusion to MBP remain soluble after proteolytic removal of the tag, although circumvention of this problem has been reported using a novel modified expression vector (147). In vivo expression systems are increasingly being supplemented by prokaryotic and eukaryotic cell free protein synthesis systems (148–151). This approach, although still be set by problems of inefficiency, enables the production of cDNA-encoded proteins that can seriously interfere with the physiology of their host cell.

5. Concluding Remarks

Protein tagging technology has profoundly impacted on our ability to express and purify recombinant proteins. There is now a considerable repertoire of polypeptide tags that can be exploited to enhance expression, improve solubility, and greatly simplify the purification and detection of their partner protein. Faced with a requirement for soluble correctly folded protein, the investigator now benefits from new rapid cloning and gene fusion systems, the possibility of using of multiple tags simultaneously, the availability of multiple constructs designed to enable direct comparisons to be made between different tags on the same protein, and methods for tag removal following affinity purification. High-throughput protein production, as required by the pharmaceutical industries and structural genomics centres, has additionally benefited from major advances in robotic instrumentation designed to facilitate the expression and purification of many proteins in parallel.

His-tagging is likely to consolidate its position as the method of choice for both pharmaceutical production and as a universal platform for proteomic studies.

It remains unclear as to how the solubility of a given protein may be enhanced by its fusion partner. It has variously been suggested that the tag may act as a nucleation site for the folding of the target protein, or that the tag acts as a magnet for chaperone proteins, or is itself a general molecular chaperone ((12) and references therein). Predicting which solubility-enhancing tag will work best with a given protein can be difficult and it may be best to try out a few of these by expressing the target protein from different plasmid vectors. No affinity tag is ideal in every respect and combinatorial tagging allows the user to simultaneously benefit from the solubility-enhancing property of one tag and the affinity property of another. In addition to those available from commercial sources, the European Molecular Biology Laboratory (EMBL) protein production facility has an available and growing collection of cloning and expression vectors that were generated and published by EMBL co-workers (http://www.pepcore.embl.de/strains_vectors/vectors/index.html). This repository is a useful resource in that one can avail of vectors that enable fusion of the protein of interest to various combinations of affinity and solubility-enhancing tags with different intervening protease cleavage sites.

Acknowledgments

We gratefully acknowledge funding from the Health Research Board (HRB grant RP/2005/212) and Enterprise Ireland (EI grant IP 2008-0530).

References

1. Nikaido, H. (1994) Maltose Transport System of *Escherichia coli*: An ABC-Type Transporter. *FEBS Lett.* **346**, 55–58.
2. di Guan, C., Li, P., Riggs, P. D., and Inouye, H. (1988) Vectors that Facilitate the Expression and Purification of Foreign Peptides in *Escherichia coli* by Fusion to Maltose-Binding Protein. *Gene.* **67**, 21–30.
3. Pattenden, L. K., and Thomas, W. G. (2008) Amylose Affinity Chromatography of Maltose-Binding Protein: Purification by both Native and Novel Matrix-Assisted Dialysis Refolding Methods. *Methods Mol. Biol.* **421**, 169–189.
4. Kapust, R. B., and Waugh, D. S. (1999) *Escherichia coli* Maltose-Binding Protein is Uncommonly Effective at Promoting the Solubility of Polypeptides to Which it is Fused. *Protein Sci.* **8**, 1668–1674.
5. Sachdev, D., and Chirgwin, J. M. (2000) Fusions to Maltose-Binding Protein: Control of Folding and Solubility in Protein Purification. *Methods Enzymol.* **326**, 312–321.
6. Riggs, P. (2000) Expression and Purification of Recombinant Proteins by Fusion to Maltose-Binding Protein. *Mol. Biotechnol.* **15**, 51–63.

7. Dyson, M. R., Shadbolt, S. P., Vincent, K. J., Perera, R. L., and McCafferty, J. (2004) Production of Soluble Mammalian Proteins in *Escherichia coli*: Identification of Protein Features that Correlate with Successful Expression. *BMC Biotechnol.* **4**, 32.
8. Baneyx, F., and Mujacic, M. (2004) Recombinant Protein Folding and Misfolding in *Escherichia coli*. *Nat. Biotechnol.* **22**, 1399–1408.
9. Kataeva, I., Chang, J., Xu, H., Luan, C. H., Zhou, J., Uversky, V. N., Lin, D., Horanyi, P., Liu, Z. J., Ljungdahl, L. G., Rose, J., Luo, M., and Wang, B. C. (2005) Improving Solubility of *Shewanella oneidensis* MR-1 and *Clostridium thermocellum* JW-20 Proteins Expressed into *Escherichia coli*. *J. Proteome Res.* **4**, 1942–1951.
10. Busso, D., Delagoutte-Busso, B., and Moras, D. (2005) Construction of a Set Gateway-Based Destination Vectors for High-Throughput Cloning and Expression Screening in *Escherichia coli*. *Anal. Biochem.* **343**, 313–321.
11. Braud, S., Moutiez, M., Belin, P., Abello, N., Drevet, P., Zinn-Justin, S., Courcon, M., Masson, C., Dassa, J., Charbonnier, J. B., Boulain, J. C., Menez, A., Genet, R., and Gondry, M. (2005) Dual Expression System Suitable for High-Throughput Fluorescence-Based Screening and Production of Soluble Proteins. *J. Proteome Res.* **4**, 2137–2147.
12. Nallamsetty, S., and Waugh, D. S. (2006) Solubility-Enhancing Proteins MBP and NusA Play a Passive Role in the Folding of their Fusion Partners. *Protein Expr. Purif.* **45**, 175–182.
13. Randall, L. L., Hardy, S. J., Topping, T. B., Smith, V. F., Bruce, J. E., and Smith, R. D. (1998) The Interaction Between the Chaperone SecB and its Ligands: Evidence for Multiple Subsites for Binding. *Protein Sci.* 7, 2384–2390.
14. Hamilton, S. R., O'Donnell, J. B., Jr, Hammet, A., Stapleton, D., Habinowski, S. A., Means, A. R., Kemp, B. E., and Witters, L. A. (2002) AMP-Activated Protein Kinase Kinase: Detection with Recombinant AMPK alpha1 Subunit. *Biochem. Biophys. Res. Commun.* **293**, 892–898.
15. Nallamsetty, S., Austin, B. P., Penrose, K. J., and Waugh, D. S. (2005) Gateway Vectors for the Production of Combinatorially-Tagged His6-MBP Fusion Proteins in the Cytoplasm and Periplasm of *Escherichia coli*. *Protein Sci.* **14**, 2964–2971.
16. Nallamsetty, S., and Waugh, D. S. (2007) A Generic Protocol for the Expression and Purification of Recombinant Proteins in *Escherichia coli* using a Combinatorial His6-Maltose Binding Protein Fusion Tag. *Nat. Protoc.* **2**, 383–391.
17. Routzahn, K. M., and Waugh, D. S. (2002) Differential Effects of Supplementary Affinity Tags on the Solubility of MBP Fusion Proteins. *J. Struct. Funct. Genomics.* **2**, 83–92.
18. Smith, D. B., and Johnson, K. S. (1988) Single-Step Purification of Polypeptides Expressed in *Escherichia coli* as Fusions with Glutathione S-Transferase. *Gene.* **67**, 31–40.
19. Hunt, I. (2005) From Gene to Protein: A Review of New and Enabling Technologies for Multi-Parallel Protein Expression. *Protein Expr. Purif.* **40**, 1–22.
20. Kaplan, W., Husler, P., Klump, H., Erhardt, J., Sluis-Cremer, N., and Dirr, H. (1997) Conformational Stability of pGEX-Expressed *Schistosoma japonicum* Glutathione S-Transferase: A Detoxification Enzyme and Fusion-Protein Affinity Tag. *Protein Sci.* **6**, 399–406.
21. Frangioni, J. V., and Neel, B. G. (1993) Use of a General Purpose Mammalian Expression Vector for Studying Intracellular Protein Targeting: Identification of Critical Residues in the Nuclear Lamin A/C Nuclear Localization Signal. *J. Cell Sci.* **105** *(Pt 2)*, 481–488.
22. Vikis, H. G., and Guan, K. L. (2004) Glutathione-S-Transferase-Fusion Based Assays for Studying Protein–Protein Interactions. *Methods Mol. Biol.* **261**, 175–186.
23. Singh, C. R., and Asano, K. (2007) Localization and Characterization of Protein–Protein Interaction Sites. *Methods Enzymol.* **429**, 139–161.
24. Jung, J. W., Jung, S. H., Kim, H. S., Yuk, J. S., Park, J. B., Kim, Y. M., Han, J. A., Kim, P. H., and Ha, K. S. (2006) High-Throughput Analysis of GST-Fusion Protein Expression and Activity-Dependent Protein Interactions on GST-Fusion Protein Arrays with a Spectral Surface Plasmon Resonance Biosensor. *Proteomics.* **6**, 1110–1120.
25. Saaem, I., Papasotiropoulos, V., Wang, T., Soteropoulos, P., and Libera, M. (2007) Hydrogel-Based Protein Nanoarrays. *J. Nanosci. Nanotechnol.* 7, 2623–2632.
26. Zhan, Y., Song, X., and Zhou, G. W. (2001) Structural Analysis of Regulatory Protein Domains using GST-Fusion Proteins. *Gene.* **281**, 1–9.
27. Smyth, D. R., Mrozkiewicz, M. K., McGrath, W. J., Listwan, P., and Kobe, B. (2003) Crystal Structures of Fusion Proteins with Large-Affinity Tags. *Protein Sci.* **12**, 1313–1322.

28. Zheng, G., and Yang, Y. C. (2004) ZNF76, a Novel Transcriptional Repressor Targeting TATA-Binding Protein, is Modulated by Sumoylation. *J. Biol. Chem.* **279**, 42410–42421.

29. Gill, G. (2005) Something About SUMO Inhibits Transcription. *Curr. Opin. Genet. Dev.* **15**, 536–541.

30. Kawabe, Y., Seki, M., Seki, T., Wang, W. S., Imamura, O., Furuichi, Y., Saitoh, H., and Enomoto, T. (2000) Covalent Modification of the Werner's Syndrome Gene Product with the Ubiquitin-Related Protein, SUMO-1. *J. Biol. Chem.* **275**, 20963–20966.

31. Muller, S., Matunis, M. J., and Dejean, A. (1998) Conjugation with the Ubiquitin-Related Modifier SUMO-1 Regulates the Partitioning of PML within the Nucleus. *EMBO J.* **17**, 61–70.

32. Melchior, F. (2000) SUMO – Nonclassical Ubiquitin. *Annu. Rev. Cell Dev. Biol.* **16**, 591–626.

33. Bayer, P., Arndt, A., Metzger, S., Mahajan, R., Melchior, F., Jaenicke, R., and Becker, J. (1998) Structure Determination of the Small Ubiquitin-Related Modifier SUMO-1. *J. Mol. Biol.* **280**, 275–286.

34. Johnson, E. S. (2004) Protein Modification by SUMO. *Annu. Rev. Biochem.* **73**, 355–382.

35. Johnson, E. S., and Blobel, G. (1999) Cell Cycle-Regulated Attachment of the Ubiquitin-Related Protein SUMO to the Yeast Septins. *J. Cell Biol.* **147**, 981–994.

36. Tatham, M. H., Jaffray, E., Vaughan, O. A., Desterro, J. M., Botting, C. H., Naismith, J. H., and Hay, R. T. (2001) Polymeric Chains of SUMO-2 and SUMO-3 are Conjugated to Protein Substrates by SAE1/SAE2 and Ubc9. *J. Biol. Chem.* **276**, 35368–35374.

37. Mossessova, E., and Lima, C. D. (2000) Ulp1-SUMO Crystal Structure and Genetic Analysis Reveal Conserved Interactions and a Regulatory Element Essential for Cell Growth in Yeast. *Mol. Cell.* **5**, 865–876.

38. Catanzariti, A. M., Soboleva, T. A., Jans, D. A., Board, P. G., and Baker, R. T. (2004) An Efficient System for High-Level Expression and Easy Purification of Authentic Recombinant Proteins. *Protein Sci.* **13**, 1331–1339.

39. Malakhov, M. P., Mattern, M. R., Malakhova, O. A., Drinker, M., Weeks, S. D., and Butt, T. R. (2004) SUMO Fusions and SUMO-Specific Protease for Efficient Expression and Purification of Proteins. *J. Struct. Funct. Genomics.* **5**, 75–86.

40. Marblestone, J. G., Edavettal, S. C., Lim, Y., Lim, P., Zuo, X., and Butt, T. R. (2006) Comparison of SUMO Fusion Technology with Traditional Gene Fusion Systems: Enhanced Expression and Solubility with SUMO. *Protein Sci.* **15**, 182–189.

41. Zuo, X., Li, S., Hall, J., Mattern, M. R., Tran, H., Shoo, J., Tan, R., Weiss, S. R., and Butt, T. R. (2005) Enhanced Expression and Purification of Membrane Proteins by SUMO Fusion in *Escherichia coli*. *J. Struct. Funct. Genomics.* **6**, 103–111.

42. Butt, T. R., Edavettal, S. C., Hall, J. P., and Mattern, M. R. (2005) SUMO Fusion Technology for Difficult-to-Express Proteins. *Protein Expr. Purif.* **43**, 1–9.

43. Zuo, X., Mattern, M. R., Tan, R., Li, S., Hall, J., Sterner, D. E., Shoo, J., Tran, H., Lim, P., Sarafianos, S. G., Kazi, L., Navas-Martin, S., Weiss, S. R., and Butt, T. R. (2005) Expression and Purification of SARS Coronavirus Proteins using SUMO-Fusions. *Protein Expr. Purif.* **42**, 100–110.

44. Lee, J., and Kim, S. H. (2009) High-Throughput T7 LIC Vector for Introducing C-Terminal Poly-Histidine Tags with Variable Lengths without Extra Sequences. *Protein Expr. Purif.* **63**, 58–61.

45. Catic, A., Misaghi, S., Korbel, G. A., and Ploegh, H. L. (2007) ElaD, a Deubiquitinating Protease Expressed by *E. Coli*. *PLoS One.* **2**, e381.

46. Liu, L., Spurrier, J., Butt, T. R., and Strickler, J. E. (2008) Enhanced Protein Expression in the baculovirus/insect Cell System using Engineered SUMO Fusions. *Protein Expr. Purif.* **62**, 21–28.

47. Peroutka, R. J., Elshourbagy, N., Piech, T., and Butt, T. R. (2008) Enhanced Protein Expression in Mammalian Cells using Engineered SUMO Fusions: Secreted Phospholipase A2. *Protein Sci.* **17**, 1586–1595.

48. Panavas, T., Sanders, C., and Butt, T. R. (2009) SUMO Fusion Technology for Enhanced Protein Production in Prokaryotic and Eukaryotic Expression Systems. *Methods Mol. Biol.* **497**, 303–317.

49. Katti, S. K., LeMaster, D. M., and Eklund, H. (1990) Crystal Structure of Thioredoxin from *Escherichia coli* at 1.68 A Resolution. *J. Mol. Biol.* **212**, 167–184.

50. LaVallie, E. R., DiBlasio, E. A., Kovacic, S., Grant, K. L., Schendel, P. F., and McCoy, J. M. (1993) A Thioredoxin Gene Fusion Expression System that Circumvents Inclusion Body Formation in the *E. Coli* Cytoplasm. *Biotechnology (N. Y).* **11**, 187–193.

51. Smith, P. A., Tripp, B. C., DiBlasio-Smith, E. A., Lu, Z., LaVallie, E. R., and McCoy, J. M. (1998) A Plasmid Expression System for Quantitative In Vivo Biotinylation of Thioredoxin Fusion Proteins in *Escherichia coli*. *Nucleic Acids Res.* **26**, 1414–1420.
52. LaVallie, E. R., Lu, Z., Diblasio-Smith, E. A., Collins-Racie, L. A., and McCoy, J. M. (2000) Thioredoxin as a Fusion Partner for Production of Soluble Recombinant Proteins in *Escherichia coli*. *Methods Enzymol.* **326**, 322–340.
53. Dummler, A., Lawrence, A. M., and de Marco, A. (2005) Simplified Screening for the Detection of Soluble Fusion Constructs Expressed in *E. Coli* using a Modular Set of Vectors. *Microb. Cell Fact.* **4**, 34.
54. Hammarstrom, M., Hellgren, N., van Den Berg, S., Berglund, H., and Hard, T. (2002) Rapid Screening for Improved Solubility of Small Human Proteins Produced as Fusion Proteins in *Escherichia coli*. *Protein Sci.* **11**, 313–321.
55. Kim, S., and Lee, S. B. (2008) Soluble Expression of Archaeal Proteins in *Escherichia coli* by using Fusion-Partners. *Protein Expr. Purif.* **62**, 116–119.
56. Bogomolovas, J., Simon, B., Sattler, M., and Stier, G. (2009) Screening of Fusion Partners for High Yield Expression and Purification of Bioactive Viscotoxins. *Protein Expr. Purif.* **64**, 16–23.
57. Derewenda, Z. S. (2004) The use of Recombinant Methods and Molecular Engineering in Protein Crystallization. *Methods.* **34**, 354–363.
58. Corsini, L., Hothorn, M., Scheffzek, K., Sattler, M., and Stier, G. (2008) Thioredoxin as a Fusion Tag for Carrier-Driven Crystallization. *Protein Sci.* **17**, 2070–2079.
59. Gusarov, I., and Nudler, E. (2001) Control of Intrinsic Transcription Termination by N and NusA: The Basic Mechanisms. *Cell.* **107**, 437–449.
60. Davis, G. D., Elisee, C., Newham, D. M., and Harrison, R. G. (1999) New Fusion Protein Systems Designed to Give Soluble Expression in *Escherichia coli*. *Biotechnol. Bioeng.* **65**, 382–388.
61. Harrison, R. G. (2000) Expression of Soluble Heterologous Proteins via Fusion with NusA Protein. *inNovation.* **11**, 4–7.
62. Cabrita, L. D., Dai, W., and Bottomley, S. P. (2006) A Family of *E. Coli* Expression Vectors for Laboratory Scale and High Throughput Soluble Protein Production. *BMC Biotechnol.* **6**, 12.
63. Graslund, S., Eklund, M., Falk, R., Uhlen, M., Nygren, P. A., and Stahl, S. (2002) A Novel Affinity Gene Fusion System Allowing Protein A-Based Recovery of Non-Immunoglobulin Gene Products. *J. Biotechnol.* **99**, 41–50.
64. Zhao, Y., Benita, Y., Lok, M., Kuipers, B., van der Ley, P., Jiskoot, W., Hennink, W. E., Crommelin, D. J., and Oosting, R. S. (2005) Multi-Antigen Immunization using IgG Binding Domain ZZ as Carrier. *Vaccine.* **23**, 5082–5090.
65. Cheng, Y., and Patel, D. J. (2004) An Efficient System for Small Protein Expression and Refolding. *Biochem. Biophys. Res. Commun.* **317**, 401–405.
66. Card, P. B., and Gardner, K. H. (2005) Identification and Optimization of Protein Domains for NMR Studies. *Methods Enzymol.* **394**, 3–16.
67. Bao, W. J., Gao, Y. G., Chang, Y. G., Zhang, T. Y., Lin, X. J., Yan, X. Z., and Hu, H. Y. (2006) Highly Efficient Expression and Purification System of Small-Size Protein Domains in *Escherichia coli* for Biochemical Characterization. *Protein Expr. Purif.* **47**, 599–606.
68. Zhang, Z., Li, Z. H., Wang, F., Fang, M., Yin, C. C., Zhou, Z. Y., Lin, Q., and Huang, H. L. (2002) Overexpression of DsbC and DsbG Markedly Improves Soluble and Functional Expression of Single-Chain Fv Antibodies in *Escherichia coli*. *Protein Expr. Purif.* **26**, 218–228.
69. Chatterjee, D. K., and Esposito, D. (2006) Enhanced Soluble Protein Expression using Two New Fusion Tags. *Protein Expr. Purif.* **46**, 122–129.
70. Zhang, Y. B., Howitt, J., McCorkle, S., Lawrence, P., Springer, K., and Freimuth, P. (2004) Protein Aggregation during Overexpression Limited by Peptide Extensions with Large Net Negative Charge. *Protein Expr. Purif.* **36**, 207–216.
71. Porath, J., Carlsson, J., Olsson, I., and Belfrage, G. (1975) Metal Chelate Affinity Chromatography, a New Approach to Protein Fractionation. *Nature.* **258**, 598–599.
72. Storcksdieck genannt Bonsmann, S., and Hurrell, R. F. (2007) Iron-Binding Properties, Amino Acid Composition, and Structure of Muscle Tissue Peptides from In Vitro Digestion of Different Meat Sources. *J. Food Sci.* **72**, S019–S029.
73. Swain, J. H., Tabatabai, L. B., and Reddy, M. B. (2002) Histidine Content of Low-

Molecular-Weight Beef Proteins Influences Nonheme Iron Bioavailability in Caco-2 Cells. *J. Nutr.* **132**, 245–251.

74. Taylor, P. G., Martinez-Torres, C., Romano, E. L., and Layrisse, M. (1986) The Effect of Cysteine-Containing Peptides Released during Meat Digestion on Iron Absorption in Humans. *Am. J. Clin. Nutr.* **43**, 68–71.
75. Porath, J. (1992) Immobilized Metal Ion Affinity Chromatography. *Protein Expr. Purif.* **3**, 263–281.
76. Ueda, E. K., Gout, P. W., and Morganti, L. (2003) Current and Prospective Applications of Metal Ion-Protein Binding. *J. Chromatogr. A.* **988**, 1–23.
77. Chaga, G. S. (2001) Twenty-Five Years of Immobilized Metal Ion Affinity Chromatography: Past, Present and Future. *J. Biochem. Biophys. Methods.* **49**, 313–334.
78. Li, M., Su, Z. G., and Janson, J. C. (2004) In Vitro Protein Refolding by Chromatographic Procedures. *Protein Expr. Purif.* **33**, 1–10.
79. Hutchinson, M. H., and Chase, H. A. (2006) Adsorptive Refolding of Histidine-Tagged Glutathione S-Transferase using Metal Affinity Chromatography. *J. Chromatogr. A.* **1128**, 125–132.
80. Rogl, H., Kosemund, K., Kuhlbrandt, W., and Collinson, I. (1998) Refolding of *Escherichia coli* Produced Membrane Protein Inclusion Bodies Immobilised by Nickel Chelating Chromatography. *FEBS Lett.* **432**, 21–26.
81. Zouhar, J., Nanak, E., and Brzobohaty, B. (1999) Expression, Single-Step Purification, and Matrix-Assisted Refolding of a Maize Cytokinin Glucoside-Specific Beta-Glucosidase. *Protein Expr. Purif.* **17**, 153–162.
82. Dong, X. Y., Chen, L. J., and Sun, Y. (2009) Refolding and Purification of Histidine-Tagged Protein by Artificial Chaperone-Assisted Metal Affinity Chromatography. *J. Chromatogr. A.* **1216**, 5207–5213.
83. Manjasetty, B. A., Turnbull, A. P., Panjikar, S., Bussow, K., and Chance, M. R. (2008) Automated Technologies and Novel Techniques to Accelerate Protein Crystallography for Structural Genomics. *Proteomics.* **8**, 612–625.
84. Sharma, S. K., Evans, D. B., Vosters, A. F., McQuade, T. J., and Tarpley, W. G. (1991) Metal Affinity Chromatography of Recombinant HIV-1 Reverse Transcriptase Containing a Human Renin Cleavable Metal Binding Domain. *Biotechnol. Appl. Biochem.* **14**, 69–81.
85. Zhang, Z., Tong, K. T., Belew, M., Pettersson, T., and Janson, J. C. (1992) Production, Purification and Characterization of Recombinant Human Interferon Gamma. *J. Chromatogr.* **604**, 143–155.
86. Franke, C. A., and Hruby, D. E. (1993) Expression and Single-Step Purification of Enzymatically Active Vaccinia Virus Thymidine Kinase Containing an Engineered Oligohistidine Domain by Immobilized Metal Affinity Chromatography. *Protein Expr. Purif.* **4**, 101–109.
87. Kipriyanov, S. M., Dubel, S., Breitling, F., Kontermann, R. E., Heymann, S., and Little, M. (1995) Bacterial Expression and Refolding of Single-Chain Fv Fragments with C-Terminal Cysteines. *Cell Biophys.* **26**, 187–204.
88. Vaughan, T. J., Williams, A. J., Pritchard, K., Osbourn, J. K., Pope, A. R., Earnshaw, J. C., McCafferty, J., Hodits, R. A., Wilton, J., and Johnson, K. S. (1996) Human Antibodies with Sub-Nanomolar Affinities Isolated from a Large Non-Immunized Phage Display Library. *Nat. Biotechnol.* **14**, 309–314.
89. Eldin, P., Pauza, M. E., Hieda, Y., Lin, G., Murtaugh, M. P., Pentel, P. R., and Pennell, C. A. (1997) High-Level Secretion of Two Antibody Single Chain Fv Fragments by *Pichia pastoris. J. Immunol. Methods.* **201**, 67–75.
90. Passafiume, M., Vulliez-le Normand, B., Riottot, M. M., and Bentley, G. A. (1998) Sequence Analysis of a Monoclonal Antibody Specific for the preS2 Region of Hepatitis B Surface Antigen, and the Cloning, Expression and Characterisation of its Single-Chain Fv Construction. *FEBS Lett.* **441**, 407–412.
91. Kimple, M. E., and Sondek, J. (2004) Overview of Affinity Tags for Protein Purification. *Curr. Protoc. Protein Sci. Chapter 9*, Unit 9.9.
92. Kwon, K., Grose, C., Pieper, R., Pandya, G. A., Fleischmann, R. D., and Peterson, S. N. (2009) High Quality Protein Microarray Using In Situ Protein Purification. *BMC Biotechnol.* **9**, 72.
93. Steen, J., Uhlen, M., Hober, S., and Ottosson, J. (2006) High-Throughput Protein Purification using an Automated Set-Up for High-Yield Affinity Chromatography. *Protein Expr. Purif.* **46**, 173–178.
94. Hang, Q., Woods, L., Feiss, M., and Catalano, C. E. (1999) Cloning, Expression, and Biochemical Characterization of Hexahistidine-Tagged Terminase Proteins. *J. Biol. Chem.* **274**, 15305–15314.

95. Gaberc-Porekar, V., Menart, V., Jevsevar, S., Vidensek, A., and Stalc, A. (1999) Histidines in Affinity Tags and Surface Clusters for Immobilized Metal-Ion Affinity Chromatography of Trimeric Tumor Necrosis Factor Alpha. *J. Chromatogr. A.* **852**, 117–128.
96. Chant, A., Kraemer-Pecore, C. M., Watkin, R., and Kneale, G. G. (2005) Attachment of a Histidine Tag to the Minimal Zinc Finger Protein of the *Aspergillus Nidulans* Gene Regulatory Protein AreA Causes a Conformational Change at the DNA-Binding Site. *Protein Expr. Purif.* **39**, 152–159.
97. Chaga, G., Bochkariov, D. E., Jokhadze, G. G., Hopp, J., and Nelson, P. (1999) Natural Poly-Histidine Affinity Tag for Purification of Recombinant Proteins on Cobalt(II)–Carboxymethylaspartate Crosslinked Agarose. *J. Chromatogr. A.* **864**, 247–256.
98. de Vries, E. G., de Hooge, M. N., Gietema, J. A., and de Jong, S., Correspondence Re, Ferreira, C. G., et al. Apoptosis: Target of Cancer Therapy. *Clin. Cancer Res.* **8**, 2024–2034, 2002. *Clin. Cancer Res.* **9**, 912; author reply 913.
99. Xu, C. G., Fan, X. J., Fu, Y. J., and Liang, A. H. (2008) Effect of Location of the His-Tag on the Production of Soluble and Functional Buthus Martensii Karsch Insect Toxin. *Protein Expr. Purif.* **59**, 103–109.
100. Loughran, S. T., Loughran, N. B., Ryan, B. J., D'Souza, B. N., and Walls, D. (2006) Modified His-Tag Fusion Vector for Enhanced Protein Purification by Immobilized Metal Affinity Chromatography. *Anal. Biochem.* **355**, 148–150.
101. Grisshammer, R., White, J. F., Trinh, L. B., and Shiloach, J. (2005) Large-Scale Expression and Purification of a G-Protein-Coupled Receptor for Structure Determination – an Overview. *J. Struct. Funct. Genomics.* **6**, 159–163.
102. Yeliseev, A. A., Wong, K. K., Soubias, O., and Gawrisch, K. (2005) Expression of Human Peripheral Cannabinoid Receptor for Structural Studies. *Protein Sci.* **14**, 2638–2653.
103. Magnusdottir, A., Johansson, I., Dahlgren, L. G., Nordlund, P., and Berglund, H. (2009) Enabling IMAC Purification of Low Abundance Recombinant Proteins from *E. Coli* Lysates. *Nat. Methods.* **6**, 477–478.
104. Liu, Z., Bartlow, P., Varakala, R., Beitle, R., Koepsel, R., and Ataai, M. M. (2009) Use of Proteomics for Design of a Tailored Host Cell for Highly Efficient Protein Purification. *J. Chromatogr. A.* **1216**, 2433–2438.
105. Zhao, Q., Chan, Y. W., Lee, S. S., and Cheung, W. T. (2009) One-Step Expression and Purification of Single-Chain Variable Antibody Fragment using an Improved Hexahistidine Tag Phagemid Vector. *Protein Expr. Purif.* **68**, 190–195.
106. Ye, K., Jin, S., Ataai, M. M., Schultz, J. S., and Ibeh, J. (2004) Tagging Retrovirus Vectors with a Metal Binding Peptide and One-Step Purification by Immobilized Metal Affinity Chromatography. *J. Virol.* **78**, 9820–9827.
107. Cheeks, M. C., Kamal, N., Sorrell, A., Darling, D., Farzaneh, F., and Slater, N. K. (2009) Immobilized Metal Affinity Chromatography of Histidine-Tagged Lentiviral Vectors using Monolithic Adsorbents. *J. Chromatogr. A.* **1216**, 2705–2711.
108. Schmidt, T. G., and Skerra, A. (2007) The Strep-Tag System for One-Step Purification and High-Affinity Detection or Capturing of Proteins. *Nat. Protoc.* **2**, 1528–1535.
109. Keefe, A. D., Wilson, D. S., Seelig, B., and Szostak, J. W. (2001) One-Step Purification of Recombinant Proteins using a Nanomolar-Affinity Streptavidin-Binding Peptide, the SBP-Tag. *Protein Expr. Purif.* **23**, 440–446.
110. Fuchs, S. M., and Raines, R. T. (2005) Polyarginine as a Multifunctional Fusion Tag. *Protein Sci.* **14**, 1538–1544.
111. Stofko-Hahn, R. E., Carr, D. W., and Scott, J. D. (1992) A Single Step Purification for Recombinant Proteins. Characterization of a Microtubule Associated Protein (MAP 2) Fragment which Associates with the Type II cAMP-Dependent Protein Kinase. *FEBS Lett.* **302**, 274–278.
112. Carrard, G., Koivula, A., Soderlund, H., and Beguin, P. (2000) Cellulose-Binding Domains Promote Hydrolysis of Different Sites on Crystalline Cellulose. *Proc. Natl. Acad. Sci. U.S.A.* **97**, 10342–10347.
113. Shoseyov, O., Shani, Z., and Levy, I. (2006) Carbohydrate Binding Modules: Biochemical Properties and Novel Applications. *Microbiol. Mol. Biol. Rev.* **70**, 283–295.
114. Nahalka, J., and Nidetzky, B. (2007) Fusion to a Pull-Down Domain: A Novel Approach of Producing Trigonopsis variabilisD-Amino Acid Oxidase as Insoluble Enzyme Aggregates. *Biotechnol. Bioeng.* **97**, 454–461.
115. Xu, Y., and Foong, F. C. (2008) Characterization of a Cellulose Binding Domain from *Clostridium Cellulovorans* Endoglucanase–Xylanase D and its use as a Fusion Partner for Soluble Protein Expression in *Escherichia coli*. *J. Biotechnol.* **135**, 319–325.

116. Craig, S. J., Shu, A., Xu, Y., Foong, F. C., and Nordon, R. (2007) Chimeric Protein for Selective Cell Attachment onto Cellulosic Substrates. *Protein Eng. Des. Sel.* **20**, 235–241.

117. Evans, T. C., Jr, and Xu, M. Q. (1999) Intein-Mediated Protein Ligation: Harnessing Nature's Escape Artists. *Biopolymers.* **51**, 333–342.

118. Xu, M. Q., Paulus, H., and Chong, S. (2000) Fusions to Self-Splicing Inteins for Protein Purification. *Methods Enzymol.* **326**, 376–418.

119. Einhauer, A., and Jungbauer, A. (2001) The FLAG Peptide, a Versatile Fusion Tag for the Purification of Recombinant Proteins. *J. Biochem. Biophys. Methods.* **49**, 455–465.

120. Hage, D. S. (1999) Affinity Chromatography: A Review of Clinical Applications. *Clin. Chem.* **45**, 593–615.

121. Munro, S., and Pelham, H. R. (1986) An Hsp70-Like Protein in the ER: Identity with the 78 kD Glucose-Regulated Protein and Immunoglobulin Heavy Chain Binding Protein. *Cell.* **46**, 291–300.

122. Thompson, N. E., Arthur, T. M., and Burgess, R. R. (2003) Development of an Epitope Tag for the Gentle Purification of Proteins by Immunoaffinity Chromatography: Application to Epitope-Tagged Green Fluorescent Protein. *Anal. Biochem.* **323**, 171–179.

123. Kim, J. S., and Raines, R. T. (1993) Ribonuclease S-Peptide as a Carrier in Fusion Proteins. *Protein Sci.* **2**, 348–356.

124. Banki, M. R., Feng, L., and Wood, D. W. (2005) Simple Bioseparations using Self-Cleaving Elastin-Like Polypeptide Tags. *Nat. Methods.* **2**, 659–661.

125. Meyer, D. E., and Chilkoti, A. (1999) Purification of Recombinant Proteins by Fusion with Thermally-Responsive Polypeptides. *Nat. Biotechnol.* **17**, 1112–1115.

126. Lim, D. W., Trabbic-Carlson, K., Mackay, J. A., and Chilkoti, A. (2007) Improved Non-Chromatographic Purification of a Recombinant Protein by Cationic Elastin-Like Polypeptides. *Biomacromolecules.* **8**, 1417–1424.

127. Wu, W. Y., Mee, C., Califano, F., Banki, R., and Wood, D. W. (2006) Recombinant Protein Purification by Self-Cleaving Aggregation Tag. *Nat. Protoc.* **1**, 2257–2262.

128. Rigaut, G., Shevchenko, A., Rutz, B., Wilm, M., Mann, M., and Seraphin, B. (1999) A Generic Protein Purification Method for Protein Complex Characterization and Proteome Exploration. *Nat. Biotechnol.* **17**, 1030–1032.

129. Puig, O., Caspary, F., Rigaut, G., Rutz, B., Bouveret, E., Bragado-Nilsson, E., Wilm, M., and Seraphin, B. (2001) The Tandem Affinity Purification (TAP) Method: A General Procedure of Protein Complex Purification. *Methods.* **24**, 218–229.

130. Gavin, A. C., Bosche, M., Krause, R., Grandi, P., Marzioch, M., Bauer, A., Schultz, J., Rick, J. M., Michon, A. M., Cruciat, C. M., Remor, M., Hofert, C., Schelder, M., Brajenovic, M., Ruffner, H., Merino, A., Klein, K., Hudak, M., Dickson, D., Rudi, T., Gnau, V., Bauch, A., Bastuck, S., Huhse, B., Leutwein, C., Heurtier, M. A., Copley, R. R., Edelmann, A., Querfurth, E., Rybin, V., Drewes, G., Raida, M., Bouwmeester, T., Bork, P., Seraphin, B., Kuster, B., Neubauer, G., and Superti-Furga, G. (2002) Functional Organization of the Yeast Proteome by Systematic Analysis of Protein Complexes. *Nature.* **415**, 141–147.

131. Forler, D., Kocher, T., Rode, M., Gentzel, M., Izaurralde, E., and Wilm, M. (2003) An Efficient Protein Complex Purification Method for Functional Proteomics in Higher Eukaryotes. *Nat. Biotechnol.* **21**, 89–92.

132. Gingras, A. C., Aebersold, R., and Raught, B. (2005) Advances in Protein Complex Analysis Using Mass Spectrometry. *J. Physiol.* **563**, 11–21.

133. Rohila, J. S., Chen, M., Chen, S., Chen, J., Cerny, R., Dardick, C., Canlas, P., Xu, X., Gribskov, M., Kanrar, S., Zhu, J. K., Ronald, P., and Fromm, M. E. (2006) Protein–Protein Interactions of Tandem Affinity Purification-Tagged Protein Kinases in Rice. *Plant J.* **46**, 1–13.

134. Rubio, V., Shen, Y., Saijo, Y., Liu, Y., Gusmaroli, G., Dinesh-Kumar, S. P., and Deng, X. W. (2005) An Alternative Tandem Affinity Purification Strategy Applied to Arabidopsis Protein Complex Isolation. *Plant J.* **41**, 767–778.

135. Van Leene, J., Stals, H., Eeckhout, D., Persiau, G., Van De Slijke, E., Van Isterdael, G., De Clercq, A., Bonnet, E., Laukens, K., Remmerie, N., Henderickx, K., De Vijlder, T., Abdelkrim, A., Pharazyn, A., Van Onckelen, H., Inze, D., Witters, E., and De Jaeger, G. (2007) A Tandem Affinity Purification-Based Technology Platform to Study the Cell Cycle Interactome in *Arabidopsis thaliana*. *Mol. Cell. Proteomics.* **6**, 1226–1238.

136. Lehmann, R., Meyer, J., Schuemann, M., Krause, E., and Freund, C. (2009) A Novel S3S-TAP-Tag for the Isolation of T Cell

Interaction Partners of Adhesion and Degranulation Promoting Adaptor Protein (ADAP). *Proteomics.***9**, 5288–5295.

137. Gloeckner, C. J., Boldt, K., Schumacher, A., and Ueffing, M. (2009) Tandem Affinity Purification of Protein Complexes from Mammalian Cells by the Strep/FLAG (SF)-TAP Tag. *Methods Mol. Biol.* **564**, 359–372.
138. Tsai, A., and Carstens, R. P. (2006) An Optimized Protocol for Protein Purification in Cultured Mammalian Cells using a Tandem Affinity Purification Approach. *Nat. Protoc.* **1**, 2820–2827.
139. Van Leene, J., Witters, E., Inze, D., and De Jaeger, G. (2008) Boosting Tandem Affinity Purification of Plant Protein Complexes. *Trends Plant Sci.* **13**, 517–520.
140. Burckstummer, T., Bennett, K. L., Preradovic, A., Schutze, G., Hantschel, O., Superti-Furga, G., and Bauch, A. (2006) An Efficient Tandem Affinity Purification Procedure for Interaction Proteomics in Mammalian Cells. *Nat. Methods.* **3**, 1013–1019.
141. Schimanski, B., Nguyen, T. N., and Gunzl, A. (2005) Highly Efficient Tandem Affinity Purification of Trypanosome Protein Complexes Based on a Novel Epitope Combination. *Eukaryot. Cell.* **4**, 1942–1950.
142. Collins, S. R., Miller, K. M., Maas, N. L., Roguev, A., Fillingham, J., Chu, C. S., Schuldiner, M., Gebbia, M., Recht, J., Shales, M., Ding, H., Xu, H., Han, J., Ingvarsdottir, K., Cheng, B., Andrews, B., Boone, C., Berger, S. L., Hieter, P., Zhang, Z., Brown, G. W., Ingles, C. J., Emili, A., Allis, C. D., Toczyski, D. P., Weissman, J. S., Greenblatt, J. F., and Krogan, N. J. (2007) Functional Dissection of Protein Complexes Involved in Yeast Chromosome Biology using a Genetic Interaction Map. *Nature.* **446**, 806–810.
143. Wang, B. C., Adams, M. W., Dailey, H., DeLucas, L., Luo, M., Rose, J., Bunzel, R., Dailey, T., Habel, J., Horanyi, P., Jenney, F. E., Jr, Kataeva, I., Lee, H. S., Li, S., Li, T., Lin, D., Liu, Z. J., Luan, C. H., Mayer, M., Nagy, L., Newton, M. G., Ng, J., Poole, F. L.,II, Shah, A., Shah, C., Sugar, F. J., and Xu, H. (2005) Protein Production and Crystallization at SECSG – an Overview. *J. Struct. Funct. Genomics.* **6**, 233–243.
144. Bucher, M. H., Evdokimov, A. G., and Waugh, D. S. (2002) Differential Effects of Short Affinity Tags on the Crystallization of *Pyrococcus furiosus* Maltodextrin-Binding Protein. *Acta Crystallogr. D Biol. Crystallogr.* **58**, 392–397.
145. Lee, J. E., Fusco, M. L., and Ollmann Saphire, E. (2009) An Efficient Platform for Screening Expression and Crystallization of Glycoproteins Produced in Human Cells. *Nat. Protoc.* **4**, 592–604.
146. Schlaeppi, J. M., Henke, M., Mahnke, M., Hartmann, S., Schmitz, R., Pouliquen, Y., Kerins, B., Weber, E., Kolbinger, F., and Kocher, H. P. (2006) A Semi-Automated Large-Scale Process for the Production of Recombinant Tagged Proteins in the Baculovirus Expression System. *Protein Expr. Purif.* **50**, 185–195.
147. Donnelly, M. I., Zhou, M., Millard, C. S., Clancy, S., Stols, L., Eschenfeldt, W. H., Collart, F. R., and Joachimiak, A. (2006) An Expression Vector Tailored for Large-Scale, High-ThroughputPurificationofRecombinant Proteins. *Protein Expr. Purif.* **47**, 446–454.
148. Vinarov, D. A., Lytle, B. L., Peterson, F. C., Tyler, E. M., Volkman, B. F., and Markley, J. L. (2004) Cell-Free Protein Production and Labeling Protocol for NMR-Based Structural Proteomics. *Nat. Methods.* **1**, 149–153.
149. Endo, Y., and Sawasaki, T. (2006) Cell-Free Expression Systems for Eukaryotic Protein Production. *Curr. Opin. Biotechnol.* **17**, 373–380.
150. Sawasaki, T., Ogasawara, T., Morishita, R., and Endo, Y. (2002) A Cell-Free Protein Synthesis System for High-Throughput Proteomics. *Proc. Natl. Acad. Sci. U.S.A.* **99**, 14652–14657.
151. Liguori, L., Marques, B., Villegas-Mendez, A., Rothe, R., and Lenormand, J. L. (2007) Production of Membrane Proteins Using Cell-Free Expression Systems. *Expert Rev. Proteomics.* **4**, 79–90.

Part II

Protocols

Chapter 10

Storage and Lyophilisation of Pure Proteins

Ciarán Ó'Fágáin

Abstract

This article outlines empirical procedures for the storage of pure proteins with preservation of high levels of biological activity. It describes simple and workable means of preventing microbial contamination and proteolytic degradation, and the use of various types of stabilising additives. It sets out the principles of lyophilisation (otherwise known as freeze-drying, a complex process comprising freezing, primary drying and secondary drying stages), outlines a general procedure for the use of lyophiliser apparatus and mentions notable pitfalls to be avoided.

Key words: Protein stability, Storage of pure proteins, Stabilising additives, Excipients, Freezing, Freeze-drying, Lyophilisation

1. Introduction

There is often a need to store an isolated or purified protein for varying periods of time. If the protein is to be studied, it will take some time to characterise the properties of interest. If the protein is an end product or is for use as a tool in some procedure, it will likely be used in small quantities over an extended period. It is vital, therefore, that the protein retains as much as possible of its original, post-purification, biological (or functional) activity throughout an extended storage period. This "shelf life" may vary from a few days to more than 1 year. Shelf life can depend on the nature of the protein and on the storage conditions. This article outlines the means by which activity losses occur on storage and discusses a range of measures to prevent or lessen these protein-inactivating events.

Extremes of temperature and pH will, naturally, be avoided as conditions for routine or long-term storage. Various other factors, however, can lead to loss or deterioration of a protein's biological

Dermot Walls and Sinéad T. Loughran (eds.), *Protein Chromatography: Methods and Protocols*, Methods in Molecular Biology, vol. 681, DOI 10.1007/978-1-60761-913-0_10, © Springer Science+Business Media, LLC 2011

activity. Typical physical phenomena are aggregation (1–3) and precipitation. Adsorption to surfaces may also lead to inactivation (1, 2). Biological deterioration can result from the loss of an essential cofactor, from the dissociation of subunits (in the case of an oligomeric protein), or from the action of proteolytic enzymes (either endogenous or arising from microbial contamination) (1).

The purely chemical reactions are few and well defined. Deamidation of glutamine and asparagine can occur at neutral-to-alkaline pH values, while peptide bonds involving aspartic acid undergo cleavage under acidic conditions. Cysteine is prone to oxidation, as are tryptophan and methionine. Alkaline conditions lead to the reduction of disulphide bonds and this is often followed by β-elimination or thiol-disulphide exchange reactions (1, 3, 4). Where reducing sugars are present with free protein amino groups (N-termini or lysine residues), there may be destructive glycation of amino functional groups by the reactive aldehyde or keto groups of the sugar (the Maillard reaction) (5).

Aggregation and deleterious chemical reactions can occur even at moderate temperatures. In one study, virtually complete aggregation of lyophilised bovine serum albumin (BSA) occurred over 24 h at 37°C following addition of just 3 μl of physiological saline. The degree of aggregation was less, but still significant, at lower temperatures. Ovalbumin, glucose oxidase, and β-lactoglobulin underwent similar aggregation under the same conditions (3) (see Note 1).

Certain proteins are more stable at room temperature than in the refrigerator and are said to be *cold labile*. This cold denaturation has been well characterised for myoglobin and a few other proteins (6). It is a property of the protein itself and is distinct from freezing inactivation (see Subheading 3.7). This phenomenon arises from the fact that it is thermodynamically possible for a protein to unfold at low as well as at high temperatures (see (7) for a summary of the notable features of cold denaturation). Happily, cold denaturation is reversible (7). These findings underline how important it is to ascertain correct storage conditions for the protein of interest. Simple reliance on the laboratory refrigerator to minimise activity losses may not suffice.

Storage concerns a protein's *long-term* or *kinetic* protein stability, which is distinct from (and need not correspond with) *thermodynamic* (*folding* or *conformational*) stability. Kinetic stability measures the persistence of activity with time (or, put another way, the progressive loss of function). It can be represented by the scheme

$$\mathrm{N} \xrightarrow{k_{in}} \mathrm{I}$$

where N is the native, functional protein, I is an irreversibly inactivated form, and k_{in} is the rate constant for the inactivation process. The equation $V_{in} = -d[N]/dt = k_{in}[N]$ describes the process mathematically, where V_{in} is the experimentally observed rate of disappearance of the native form (1). Often the activity loss will be first order, although more complex inactivation patterns are well documented (8). Note that an apparently unimolecular first order time course of inactivation may mask a more complex set of inactivating molecular events (8).

Water participates directly in many of the above deleterious chemical reactions (including proteolysis) and also provides a medium for molecular movement and interactions. Thus, removal of water can effectively prevent deterioration of the protein. Lyophilisation, or freeze-drying, is a method for the preservation of labile materials in a dehydrated form. It can be particularly suitable for high-value biomolecules such as proteins. The process involves the removal of bulk water from a frozen protein solution by sublimation under vacuum with gentle heating (primary drying). This may be followed by controlled heating to higher temperatures for removal of the remaining "bound" water from the protein preparation (secondary drying). Residual moisture levels are often lower than 1% (9). If the freeze-drying operation is carried out correctly (Subheadings 3.6–3.9 below), the protein will preserve much of its initial biological activity in a dry, less bulky state, which offers many advantages for long-term storage. When one wishes to use the protein preparation, one can rehydrate it simply by the addition of an appropriate volume of pure water or suitable buffer solution.

At very low temperatures, a liquid may behave in one of two ways. *Eutectic* solutions undergo a sharp liquid–solid freezing transition over a narrow temperature range, while *amorphous* liquids show a *glass transition* where viscosity increases dramatically with cooling and the solution takes on the macroscopic properties of a solid, even though it has not crystallised. Below this glass transition temperature (T_g') virtually no adverse chemical or biological reaction can take place. Above T_g', however, the viscous, rubbery material is very prone to deterioration (10). Understanding these low-temperature features of liquids is important for effective freeze-drying. Eutectic and glass transitions will greatly influence the freeze-drying protocol (the way the freeze dryer is run) and the choice of substances used as preservatives or excipients in the product formulation subjected to lyophilisation. The process of freeze-drying has been described in detail (10–12), as are the effects of additives (13, 14). Reference (15) contains a useful treatment which touches on the underlying theory. Franks (11) outlines principles for process design, while Oetjen has published a detailed monograph (16).

Effective freeze-drying cannot be hurried. Process times of 72 h or longer are not unusual, depending on the nature of the product formulation and the properties of its constituents.

Complex lyophiliser equipment may have high capital and running costs. Freeze-drying is usually reserved, therefore, for high-value proteins or is used in cases where alternative product presentations (such as ammonium sulphate precipitates) are unsuitable or give insufficient shelf lives.

A complex interplay of chemical and physical phenomena takes place during freeze-drying. The product yield (that is, the percentage recovery of the initial active protein) depends on the formulation in which the protein is placed prior to lyophilisation (10, 12), while its ease of rehydration and its stability on long-term storage (or "shelf life") are influenced by the processing regime (10). Subheading 3.6 describes, in broad terms, the operation of a typical freeze dryer apparatus. Some critical factors concerning the operations of freezing, primary and secondary drying are outlined in Subheadings 3.7–3.9 below.

Good Manufacturing Practice, ISO 9000 disciplines or regulatory concerns are beyond the scope of this article.

2. Materials

Mention of suppliers' names does not imply endorsement of particular product(s).

2.1. Buffers and Chemicals

The choice and preparation of an appropriate buffer system (composition, pH, molarity, ionic strength) for a given protein will vary with that protein's characteristics, so it is not possible to give precise directions in a chapter such as this. Consult reference works and/or the scientific literature for guidance regarding a protein of interest (see also Subheading 3.7). Useful chemicals and additives are discussed in Subheadings 3.1, 3.2, and especially 3.3. Highly purified forms of chemicals are preferable.

2.2. Containers

All containers used for storage of pure proteins should be of good quality and should tolerate temperatures as low as −20 or even −80°C if freezer storage is desired or necessary. A number of manufacturers (such as Sarstedt, Germany, or Nunc, Denmark; there are many others) supply pre-sterilised screwcap plasticware with good mechanical and low-temperature properties. Clean glassware exhaustively and sterilise it by dry heat. Autoclave screw caps or rubber stoppers that cannot withstand dry heat (see Note 2). Ensure that vials (usually glass) or any other vessels (e.g., round-bottomed flasks) used for lyophilisation are of sufficient quality to withstand the temperatures and pressures associated with freeze-drying. Vial characteristics are important (see Note 3). Use the same model of vial consistently (10). Also choose stoppers with care (see Note 2).

2.3. Filters

Membrane or cartridge filters, of pore size 0.22 μm for sterile filtration, are available from companies such as Gelman or Sartorius, with or without a Luer lock for extra secure attachment to a hand-held syringe.

2.4. Low-Temperature Storage

Use an ordinary domestic refrigerator for storage at temperatures of 4–6°C. Modern machines are often combined with a freezer unit, which can easily maintain temperatures as low as –20°C. Storage at –70°C or below will require a specialised low-temperature freezer.

2.5. Lyophiliser Equipment

A typical freeze dryer configuration comprises a sample container, which can be cooled to about –40°C or heated above ambient temperature, a condenser usually cooled to –60°C, and a high-performance vacuum pump (see also Note 4). The reduced pressure within the system allows sublimation of the bulk water (primary drying). The condenser acts as a low-temperature trap for the sublimed water, which collects in the form of ice; i.e., the condenser provides a temperature gradient within the apparatus. This explains why the condenser must be colder than the sample container. Regardless of the type of freeze dryer, one must use it at all times in accordance with manufacturer's instructions. Subheading 3.6 below assumes that the reader studies this article together with the pertinent user's handbook.

3. Methods

3.1. Prevention of Bacterial Contamination

3.1.1. Antimicrobials

Microbial contamination can lead to significant losses of a pure protein by proteolysis. Even if one can eliminate or remove contaminating microorganisms, the protein of interest may already have lost some activity or may have deteriorated in ways that are difficult to detect. Where permissible and feasible, add an antimicrobial compound such as sodium azide (to a final concentration of 0.1% w/v) or thiomersal (sodium merthiolate, a mercury-containing compound, to a final concentration of 0.01% w/v) in order to prevent microbial growth (Both of these compounds are poisonous: handle them with care; see also Note 5). Note that azide will inactivate haem-containing proteins such as peroxidase.

3.1.2. Filtration

1. It is good practice to perform filtration operations in a class 2 laminar flow microbiological safety cabinet, the design of which prevents contamination of the sample. Following manufacturer's instructions closely, turn on the cabinet's fans and allow to run for at least 10 min to allow adequate filtration of cabinet air. Open and remove the front door. Swab the cabinet's internal surfaces, and the outer surfaces of storage

containers brought inside the cabinet, with 70% v/v alcohol and allow to evaporate.

2. Sterilise labile materials, which cannot be autoclaved or irradiated, by filtration. A filter of pore size 0.22 μm will exclude all bacteria. Disposable filter cartridges are widely available in various configurations; often with low protein-binding capacities.
3. Draw the solution to be sterilised into a syringe and then remove the needle or tube.
4. Connect the filter to the syringe nozzle, ensuring it is firmly mounted. Uncap a suitable sterile storage container directly beneath the filter outlet (using standard aseptic manipulations to avoid contamination of container or cap) and depress the syringe plunger to force the protein solution through the sterilising filter into the container. Recap immediately. It is not always possible to use a filter as fine as 0.22 μm directly (see Note 6); neither is it possible to "flame" plastic containers in a bunsen burner as part of aseptic technique.
5. Upon completion of the filtration manoeuvres, remove the storage containers and dispose of waste materials appropriately. Swab the internal surfaces of the cabinet with alcohol once again, replace the front cover, and allow to run for 10 min (or according to user's handbook) before shutting down.

3.2. Avoidance of Proteolysis

It can be difficult to remove proteases completely during protein purification. Unless the object protein is completely pure (homogeneous), even tiny amounts of contaminating proteolytic enzymes can cause serious losses of activity during extended storage periods. The molecular diversity of proteases complicates matters: there are exo-acting (remove amino acid residues from the N- or C-termini) and endo-acting (cleave internal peptide bonds) serine, cysteine (or thiol), acid, and metalloproteases (17). Use EDTA in the concentration range 2–5 mM to complex the divalent metal ions essential for metalloprotease action. Pepstatin A is a potent but reversible inhibitor of acid proteases; use at concentrations of around 0.1 μM, as with similar protease inhibitors. Phenylmethylsulphonyl fluoride (PMSF) reacts irreversibly with the essential serine in the active site of serine proteases, inactivating them (it can also act on some thiol proteases). Use it at a final concentration of 0.5–1 mM, having dissolved it first in a solvent such as acetone (it is poorly soluble in water) (18). Ensure before addition that none of these compounds will adversely affect the protein of interest (see Note 7; see also Chapter 4 for a detailed discussion on how to avoid proteolysis during protein purification procedures).

If the protein of interest is itself a proteolytic enzyme, one may need to store it in dried form (Subheading 3.5 below) or as a freeze-dried preparation (Subheadings 3.6–3.9 below). In some cases, it may be possible simply to place the enzyme in a solution having a pH value far removed from the protease's optimum. Trypsin, for example, is most active at mildly alkaline pH values. Stock solutions of trypsin for daily use are often prepared in 1 mM HCl, where the very acid pH value renders the enzyme effectively incapable of catalysis. This helps prevent autolysis during the course of the experiment. The trypsin molecule does not inactivate under these conditions and is fully active on dilution into a suitable assay solution (19). Of course, one must ascertain by experiment whether the protease will tolerate such storage conditions.

3.3. Use of Stabilising Additives

Timasheff and Arakawa have shown that substances such as glycerol or sucrose are preferentially excluded from the vicinity of protein molecules: their binding to protein is thermodynamically unfavourable and the protein becomes preferentially hydrated (20). This preferential hydration of the protein molecule arises from a polyol-induced increase in the surface tension of the solvent water (21). Interaction between the protecting additive and the peptide backbone is unfavourable (22). Loss of the protein's compact folded structure (denaturation) would increase the protein–solvent interface and, thus, the extent of thermodynamically unfavourable interaction between the additive and the protein. The result is that the protein molecule is stabilised. Naturally occurring stabilising additives increase a protein's T_m (the temperature at which 50% of the protein molecules are unfolded) but do not affect the protein's denaturation Gibbs energy (ΔG_D°) (23). The intrinsic conformational stability of the protein molecule itself is not increased but its unfolding is greatly disfavoured by virtue of the stabilising additives in the medium.

Note that the additives discussed below are *generally* applicable as stabilising agents for proteins but a given substance may not be effective for a particular protein. Both sucrose and polyethylene glycol (PEG), for instance, are good stabilisers of invertase but have denaturing effects on lysozyme (24).

3.3.1. Addition of Salts

Certain salts can significantly stabilise proteins in solution. The effect varies with the constituent ions' positions in the Hofmeister lyotropic series, which relates to ionic effects on protein solubility (25, 26). This series ranks both cations and anions in order of their stabilising effects. Below, the most stabilising ions are on the left while those on the right are destabilising.

$$(CH_3)_4N^+ > NH_4^+ > K^+, Na^+ > Mg^{2+} > Ca^{2+} > Ba^{2+}$$

$$SO_4^{2-} > Cl^- > Br^- > NO_3^- > ClO_4^- > SCN^-$$

The "stabilising" ions force protein molecules to adopt a tightly packed, compact structure by "salting out" hydrophobic residues. This prevents the initial unfolding event of any protein deterioration process. Most stabilising ions seem to act via a surface tension effect (20). Ammonium sulphate, widely used as a stabilising additive and as a non-inactivating precipitant, comprises two of the most stabilising ions in the above list, the NH_4^+ cation and the SO_4^- anion. To stabilise proteins in solution while avoiding precipitation, add ammonium sulphate to a final concentration in the range 20–400 mM (27). Do this by adding a minimal volume of a stock solution of ammonium sulphate of known molarity or by careful addition of solid ammonium sulphate. Sprinkle the solid salt, a few grains at a time, into the protein solution. Ensure that each portion of ammonium sulphate added dissolves fully before the addition of the next lot. This will prevent accumulation of undesirable high local salt concentrations (ensure that the protein-containing buffer is of sufficient molarity to resist a possible pH decrease upon addition of ammonium sulphate, the salt of a strong acid and a weak base). Salts containing citrate, sulphate, acetate, phosphate, and quaternary ammonium ions are also useful (27). Note that the nature of the counterion will influence the overall effect on protein stability (20).

Polyethyleneimine is a cationic polymer with numerous uses, including protein stabilisation (28). Both high- and low-molecular weight fractions of polyethyleneimine, at 0.01–1%(w/v) concentrations, greatly increased the shelf lives of dehydrogenases and hydrolases stored at 36°C. The effect seems to be kinetic rather than thermodynamic, as the denaturation temperature of lactate dehydrogenase was unaffected by the presence of polyethyleneimine (28) (see Note 8). The cationic surfactant benzalkonium chloride (0.01 or 0.1% w/v; 0.3 or 3 mM) maintained the activity of bovine lactoperoxidase stored at 37°C, pH 7 for much longer than that of a control sample (but not at pH 6) (29).

It is assumed that ions (or other substances) used/added are not substrates, activators, or inhibitors of the enzyme/protein under study (ammonium ion, for instance, is a substrate for glutamate dehydrogenase) and that added ions do not interfere with, or precipitate, essential ions already in solution.

3.3.2. Use of Osmolytes

Osmolytes are a diverse group of substances comprising such compounds as polyols (e.g., glycerol, xylitol), mono- (e.g., glucose), oligo- (e.g., sucrose, trehalose), and poly-saccharides, neutral polymers (such as PEG), amino acids (and their derivatives), and methylamines (such as sarcosine and trimethylamine *N*-oxide or TMAO) (22). They are not strongly charged and have little effect on enzyme activity below 1 M concentration. In general, they affect water's bulk solution properties and do not interact directly with the protein (27).

Use polyols and sugars at high final concentrations: typical figures range from 10 to 40% (w/v) (27). Sugars are reckoned to be the best stabilisers (but reducing sugars can react with protein amino groups, leading to inactivation (5); this can be avoided by using non-reducing sugars or sugar alcohols). Glycerol is a widely used low molecular weight polyol. It is easily removed by dialysis and it does not interfere with ion exchange chromatography (27). Glycerol has two significant disadvantages, however: it is a good bacterial substrate (27) and it greatly lowers the glass transition temperature (T_g') of materials to be preserved by lyophilisation (see Subheading 3.7) or drying (see Subheading 3.5). Xylitol, a 5-carbon sugar alcohol, can often replace glycerol, can be recycled from buffers, and is not a convenient food source for bacteria (27).

Polymers such as PEG are generally added to a final concentration of 1–15% (w/v). They increase the viscosity of the single-phase solvent system and so help prevent aggregation. Higher polymer concentrations, however, will promote the development of a two-phase system. The protein may concentrate in one of these phases and, possibly, aggregate (27).

Amino acids with no net charge, notably glycine and alanine, can be stabilising in the range 20–500 mM, as can related compounds, such as gamma-amino butyric acid (GABA) and TMAO, at similar concentrations (27).

3.3.3. Substrates and Specific Ligands

Addition of specific substrates, cofactors, or competitive (reversible) inhibitors to purified proteins can be very stabilising (indeed, they may be necessary where an essential metal ion or coenzyme is only loosely bound to the apoprotein). Occupation of the protein's binding/active site(s) leads to minor but significant conformational changes in the polypeptide backbone. The protein adopts a more tightly folded conformation, reducing any tendency to unfold (30) and (sometimes) rendering it less prone to proteolytic degradation. Occlusion of the protein's active site(s) by a bound substrate molecule or reversible inhibitor will protect those amino acid side chains that are critical for function. A starch degrading amyloglucosidase enzyme (from an *Aspergillus* species) stored in the presence of 14% (w/v) partial starch hydrolysate was 80% more stable over a 24-week period at ambient temperature than the corresponding enzyme preparation stored in the hydrolysate's absence (31).

Note that dialysis (or some other procedure for the removal of low molecular mass substances) may be necessary to avoid carry-over effects of the substrate or inhibitor when the protein is removed from storage for use in a particular situation where maximal activity is desired.

3.3.4. Use of Reducing Agents and Prevention of Oxidation Reactions

The thiol group of cysteine is prone to destructive oxidative reactions. One can prevent or minimise these by using reducing agents such as 2-(formerly β-) mercaptoethanol (a liquid with an unpleasant smell) or dithiothreitol ("Cleland's reagent", or DTT, a solid with little odour). Add 2-mercaptoethanol to reach a final concentration of 5–20 mM, then keep the solution under anaerobic conditions, achieved by gently bubbling an inert gas such as nitrogen through the solution before adding the reducing agent. Fill the solution to the brim of a screwcap container to minimise headspace and the chances of gaseous exchange. DTT is effective at lower concentrations: usually 0.5–1 mM will suffice (18); Schein has advised that the DTT concentration should not exceed 1 mM, as it can act as a denaturant at higher temperatures and is not very soluble in high salt (27) (note that reducing agents are themselves prone to oxidation: solutions containing them must be stored so as to eliminate, or minimise, contact with air). DTT oxidises to form an internal disulphide which is no longer effective but which will not interfere with protein molecules (18). On the other hand, 2-mercaptoethanol participates in intermolecular reactions and can form disulphides with protein thiol groups (18). Such thiol-disulphide exchanges are highly undesirable and may actually lead to inactivation or aggregation. It is probably best to use reducing agents only where they are known (or demonstrated) to be beneficial (25).

Oxidation of thiol groups may be mediated by divalent metal ions, which activate molecular oxygen. Complexation of free metal ions (where they are not essential for activity) can prevent destructive oxidation of thiols. Polyethyleneimine at 1% (w/v) concentration protected the -SH groups of lactate dehydrogenase against oxidation and prevented the consequential aggregation of the protein, even in the presence of Cu^{2+} ions; the protecting effect was ascribed to metal chelation by polyethyleneimine (28). See Subheading 3.2 above regarding the use of EDTA to complex metal ions.

3.3.5. Extremely Dilute Solutions

Very dilute protein solutions are highly prone to inactivation. This is especially true of oligomeric proteins, where dissociation of subunits can occur at low protein concentration. Protein solutions of concentration less than 1–2 mg/mL should be concentrated as rapidly as possible (18) by ultrafiltration (see Note 9) or sucrose concentration (see Note 10).

Where rapid concentration is not possible, inactivation may be prevented by the addition of an exogenous protein such as BSA, typically to a final concentration of 1 mg/mL. Scopes has discussed possible reasons for the undoubted benefits of BSA addition (18). It may seem foolish to deliberately add an exogenous, contaminating protein such as BSA to a pure protein preparation but occasionally this may be a price to pay in order to avoid inactivation.

3.4. Low Temperature Storage

Refrigeration at 4–6°C often suffices to preserve a protein provided the hints in Subheadings 3.1–3.3 of this chapter are followed.

Many proteins are supplied commercially in 50% glycerol or as slurries in approximately 3 M ammonium sulphate. Freezing of such preparations is not necessary and should be avoided.

Some proteins can deteriorate at "refrigerator" temperatures and require storage at temperatures <0°C. Usually, temperatures between −18 and −20°C (domestic freezer) will permit stable storage (see Note 11). Sometimes, however, a low temperature laboratory freezer, designed to maintain temperatures in the range −70 to −80°C, may be needed (see Note 12).

Most protein solutions will freeze to a solid at temperatures <0°C (mixtures containing high concentrations of glycerol will remain liquid at −20°C: see Subheading 3.3.2 above and Subheading 3.7). Much more complex events occur on freezing of a protein-containing mixture or biological system than the simple phase change would suggest. Differential freezing of particular components of the mixture can lead to enormous concentration effects and to dramatic changes of pH at low temperatures. These processes can lead, in turn, to protein inactivation. Freezing damage and its avoidance are discussed in Subheading 3.7. This problem can often be minimised by judicious choice of stabilising additives (Subheading 3.3 above).

Prevention of freezing will, of course, avoid freezing damage. It is possible to undercool liquids without freezing by preventing the nucleation of ice crystals. This means that proteins can be stored well below 0°C in the liquid phase. The preparation of protein-containing aqueous-organic emulsions that can maintain complete biological activity in the liquid state over extended periods at −20°C has been described (32). The method is very useful for small volumes of valuable proteins, avoids the need to use additives and is more economical than freeze-drying. The same process is used for many different proteins and one can remove portions of a sample without effect on the activity of the remainder. The actual storage temperature matters little, provided the upper temperature is <4°C and the lower temperature remains above −40°C, the nucleation temperature for ice crystal formation (33).

3.5. Drying for Stable Storage

The advantages of water removal as a protein storage/stabilisation strategy have been set out in Subheading 1 above. Lyophilisation can remove >95% of water from a protein preparation but there is the risk of freezing damage (Subheading 3.7). One can design protein-compatible formulations with glass transition temperatures (T_g') typically as high as 37°C (15). With these high T_g' values, controlled evaporative drying can be used in place of lyophilisation to stabilise proteins in the solid state. Worthwhile evaporation rates will occur below these high T_g' values at reduced pressure. Evaporation is faster, less costly, and more easily controlled than freeze-drying (15, 33). The high T_g' values also mean that one can sometimes store the resulting dried product at ambient temperature: as long as room temperature does not exceed the glass transition temperature, the protein formulation will not undergo a glass/rubber transition during storage at room temperature. The glass-forming compounds are typically carbohydrates; maltose and maltohexose are particularly valuable (15). Reconstitution of the solid protein preparation is accomplished simply by rehydration with added water or buffer. The method has been patented and is described elsewhere (34). An alternative formulation for vacuum drying of proteins, involving the use of a cationic, soluble polymer (e.g., diethylaminoethyl dextran) and the sugar alcohol, lactitol, has also been patented (35) and published (36).

3.6. Lyophilisation: Operation of Freeze Dryer

The directions below for freeze-drying are very general. Exact details will depend on the material and apparatus in question and on user's requirements. Useful practical guidelines are given elsewhere (37) and a fuller treatment of lyophilisation, with emphasis on instability, stabilisation, and formulation of protein preparations, is also available (38).

3.6.1. Start Up

Ensure that the valve connecting the vacuum to the drying chamber is closed. Start the vacuum pump and allow it to evacuate. Observe the decrease in pressure on the vacuum indicator. It is important to start the pump first so as to reach a steady-state high vacuum long before evacuation of the main chamber. It is also important that the pump warms up thoroughly before the condenser or shelves are cooled: water in any form reaching the pump can cause damage. A 30-min warm up time will usually suffice. Close the condenser drain outlet, which should always be left open when the freeze dryer is not in use (if not, open it, allow any water to drain completely and then close tightly once again). Switch on the condenser and allow it to cool to −60°C.

3.6.2. Filling and Loading

Fill only minimal amounts of material into each sample container to ensure a high ratio of surface area to volume: this will aid effective freeze-drying. Fill depths should not exceed 20 mm (11). The product matrix will tend to inhibit sublimation of water vapour from the surface of the ice crystal. This resistance depends on the depth of liquid and on the solids content of the product (12). A solid content of about 10% w/w is usually best (11) (see Note 13). Also, a good head space in the vial or ampoule will allow easier and better gaseous movement. One can conveniently load filled vials into metal trays at the bench and then place the trays into the drying chamber. The trays should have level bottoms to make good contact with the freeze dryer shelves. Many vials have narrow necks, which match with special rubber stoppers. The design of these stoppers allows one to cap the vials partially while maintaining contact between their contents and the atmosphere (sealing of the vials takes place later under vacuum). If one wishes to use such closures, one should partially stopper the vials before loading them into the freeze dryer.

If the freeze dryer has flexible temperature probes (carefully clean each one), arrange them in different places throughout the chamber. Conditions will not be homogeneous across all vials and monitoring should be as complete as equipment will allow. If possible, dip some probes directly into the material to be lyophilised, right to the bottom of the vials. A temperature record of the actual solution (instead of the shelf underneath it) is well worth the loss of a small amount of product.

3.6.3. The Freeze-Drying Operation

1. Bring the material well below its freezing temperature (or glass transition temperature) as quickly as possible. This is particularly important (Subheading 3.7 below). If possible, let the product reach a steady –40°C before drawing the vacuum.
2. Check that the door of the drying chamber is properly closed and sealed (or, for simpler devices, attach the flasks of previously frozen material to the instrument manifold, checking for a good seal), then open the vacuum valve to evacuate the chamber. The vacuum gauge may take a few minutes to register a vacuum as the air within the drying chamber evacuates. Soon, however, the pressure within the drying chamber will decrease and a steady-state high vacuum will result.

There now exists a high vacuum within the drying chamber together with a temperature gradient from the shelves/product (–40°C) to the condenser (–60°C). These conditions permit sublimation of the bulk water in the product over a period of hours. The sublimed water will collect as ice on the condenser. Sublimation, or primary drying, removes only the bulk water in the system; it is insufficient to remove the "bound" water closely associated with the protein molecules. Removal of this bound water (secondary drying) requires heating which may be applied through the shelves on which the product rests. One can programme the shelves to heat to a particular temperature at a defined rate appropriate for the product in question. One must select the heating regime with particular care (Subheadings 3.8 and 3.9 below).

3.6.4. Termination of Run and Removal of Product

1. Terminate the freeze-drying process when the cake has a good appearance and the product has reached a sufficiently low steady-state percentage moisture (determined in a separate series of experiments).
2. Seal vials, partially capped with rubber stoppers before insertion into the chamber, by operating the special screw press. This

operation will insert the stoppers fully before air is admitted into the drying chamber. Sealing of vials while still under vacuum has an advantage over the use of an inert, moisture-free gas such as nitrogen. Air will rush in when a vial, sealed under vacuum, is uncapped. The sound of the inrushing air will be absent from a defective vial which has failed to seal or where the seal has broken down. In this way, the user will immediately know of the defect and be aware that the vial contents may have deteriorated (9).

3. Close the valve to the vacuum pump firmly. Slowly release the vacuum within the drying chamber by a minimal opening of the air inlet valve: air admission should be as gentle as possible to prevent undue sudden stresses on sealed vials and also to prevent disturbance or upset of chamber contents by a strong jet of incoming air. The vacuum/pressure gauge will show a rise in pressure; this will eventually equalise to atmospheric and one can then open the chamber door. Vials of freeze-dried product can be removed for storage (preferably at refrigerated temperatures).

3.6.5. Shutting Down

If the freeze dryer is to be re-used immediately, one must be certain that the condenser's ice capacity will withstand the accumulation of ice from two runs (see Note 13). If there is no more material for lyophilisation, shut down the apparatus carefully. Remove any material spilled in the drying chamber and clean the apparatus carefully according to manufacturer's instructions. Leave the chamber door slightly ajar to allow circulation of air and to prevent sticking and compression of the door seals. Switch off the condenser and open its drain outlet (the drain should remain open until the freeze dryer is next used). Over a period of hours, the ice on the condenser surfaces will melt and drain away. Once the condenser temperature has returned to ambient and all the melted ice has drained away, allow the vacuum pump to run for a further 3 h before shutting it down. This is to prevent any damage to the pump due to occurrence or accumulation of condensation. To maintain good pump performance, change the oil often according to handbook's directions.

3.6.6. Using Simpler Freeze Dryers

Use of simpler apparatus with manifold or centrifuge accessories is carried out in much the same stepwise fashion as above. In these cases, aliquots of the product are frozen in individual open-necked flasks or tubes. Switch on the vacuum pump and condenser and allow to run as noted above. Freeze the flask contents by immersion in an alcohol cooling bath (follow the normal precautions). Swirl or rotate the flask during the freezing step to effect even distribution of the product over the widest possible surface area. This will minimise the depth of material through which water loss must occur (see Subheading 3.6.2 above and (12)).

Connect the frozen material directly to the manifold assembly (or load into the centrifugal tube dryer) and draw a vacuum in the chamber as quickly as possible so as to minimise back-melting of ice, while the sample is still under atmospheric pressure. The rapid reduction of pressure by the vacuum pump will aid sublimation and help minimise melting to water. While control of temperature and heating rates are more problematic with such accessories, visual inspection of the material and of the dried cake is much easier than in an enclosed chamber.

3.7. Lyophilisation: Freezing

In freezing, liquid water crystallises to yield solid ice. Water, like many other substances, is a *eutectic* material with a sharp transition between the liquid and solid states. Non-crystalline materials, in contrast, undergo a *glass transition* and become rigid at a certain low temperature (T_g') due to a notable increase in viscosity with decreasing temperature. The material becomes increasingly rubbery until rigidification occurs at T_g' and liquid movement effectively ceases (see Note 14). Even for eutectics, ice formation usually does not occur at the thermodynamic freezing point: the actual temperature of freezing is generally 10–15°C lower than this, so that supercooling (or undercooling) is required (10, 12). It is generally satisfactory to cool the product to just below 0°C initially, but not so low as to induce crystallisation. When the entire product has cooled to this set temperature, reduce the temperature sharply to crystallise bulk water to ice. Typical temperatures lie in the range −20 to −30°C. Allow time for ice to form in all containers before proceeding to the next stage, which involves further cooling below the lowest eutectic temperature (for crystalline materials) or below the glass transition temperature (T_g', for amorphous substances; see Note 15). A common final temperature is −40°C. The product is reliably solidified only below these temperatures and only now may primary drying (Subheading 3.8) begin. It is vital to maintain the freezing temperature below T_g' before and during primary drying. If the temperature rises above this value, the material ceases to be a solid and becomes a viscous rubber very prone to deleterious reactions, with resulting losses of activity. The rubber will also undergo a mechanical collapse, take on a different appearance, and be difficult to rehydrate (10). Thus, accurate measurement of T_g' is of great importance; in most cases (where the protein of interest is in dilute solution) T_g' depends on the types and proportions of the excipients and salts in the product formulation (10, 38).

Other changes, many of which can lead to significant protein denaturation, may occur in the product during freezing (39). Altered secondary structures have been detected in lyophilised proteins (4, 39, 40). The protein may undergo inactivation, as indicated by poor recovery of pre-lyophilisation activity. As the bulk water freezes to ice, the amount of liquid water remaining naturally decreases. This leads to great freeze concentration of solutes such as salts, perhaps with far-reaching effects (see Note 16). Concentration increases the rates of unwanted chemical reactions such as oxidations. Buffer components may crystallise differentially, leading to pronounced pH shifts; also, pK_a values are temperature dependent. For phosphate buffers, one should use potassium or mixed salts in preference to sodium salts (10, 41). For all of these reasons, one should accomplish the freezing steps (crystallisation of bulk water and cooling below the eutectic or T_g' temperatures) in any lyophilisation process as quickly as possible.

Suitable excipients/protective substances can lessen/overcome some of these damaging effects: especially, they can influence melting or collapse temperatures (38). Excipients may be classed as bulking agents, tonicity modifiers, buffers, and cryoprotectants/lyoprotectants (38). Bulking agents help to ensure the development of a plug of dried material and are used where the product's solids content is low. Bulking agents can also prevent blowout (loss of the dried product from the container along with the vapour (14, 38)). Typical bulking agents include polyols and certain sugars, mostly non-reducing.

Buffers must be chosen with great care, considering possible pK_a variations with temperature, solubility and compatibility with the protein(s) and other constituents. Useful information can be found in ref. (42) and in monographs and articles dealing with the handling of proteins. Note that the most stable pH for a protein in solution may not be appropriate for the solid state (38).

Cryoprotectants and lyoprotectants stabilise a protein against the effects of freezing and of lyophilisation/storage, respectively. These protecting additives are preferentially excluded from the protein surface, strengthening of the water "shell" surrounding the protein (20). Sugars, certain salts, and polyols can each be beneficial (although PEG–dextran mixtures may give undesirable and damaging phase separations at low temperatures (42)). In general, non-reducing sugars are preferable since these cannot participate in Maillard reactions. Xylitol's T_g' is −46.5°C in a freeze concentrate of 42.9 weight % water, while T_g' values for sorbitol, sucrose, and trehalose are −43.5, −32, and −29.5°C at water weight % values of 18.7, 35.9, and 16.7, respectively (10, 38). Mannitol and lactose may separate as crystals from a frozen solution under certain conditions (11, 38), so these compounds should be used with caution. Glycerol is not an ideal lyophilisation excipient: it has a notably low T_g' value of −65°C at 46 weight % water (10), meaning that glycerol-containing formulations become glassy only at very low temperatures. Volatile excipient compounds, such as ammonium bicarbonate, will be removed with the subliming ice and therefore will not occur within the final product (11). Sucrose or glycine in combination can be a useful starting point for the formulation of a solid protein product (38). T_g of protein formulations increases with the protein:excipient ratio (38). Useful discussions and examples of cryoprotectants occur in ref. (13, 20, 27, 38); see also Note 14 and Subheading 3.8 below.

3.8. Lyophilisation: Primary Drying

Primary drying is the sublimation under vacuum of bulk ice from the product to the much colder condenser typically held at −60°C. Never allow the sample temperature to equal or exceed T_g': if it does, collapse and product deterioration may occur. Sublimation, however, will be faster at higher temperatures so, if possible, heat the samples to a few degrees below T_g' to quicken the process (see Note 14). A usual safety margin is 2–5°C below the eutectic or collapse temperature (12, 38). "Collapse temperature" is usually equivalent to T_g' for an amorphous substance. The sublimation rate depends on the vacuum and on the condenser temperature and so depends on the characteristics of the lyophiliser (10). It is important that the temperature of the actual product remains constant (and, therefore, is monitored) throughout primary drying. Drying often becomes easier as the temperature approaches T_g' or the eutectic temperature: the sublimation rate can increase by about 13% for each 1°C rise in temperature (12). Resistance to drying also decreases with decreasing product thickness (i.e., filling height) and with increasing vial diameter (which influences the area of the drying surface) (12). The sublimation rate will decrease as primary drying proceeds and, therefore, the degree of product cooling due to sublimation will decrease also. Be sure to adjust any heat input to the containers as the product dries to prevent a net rise in product temperature; an unchecked rise could lead to collapse. Even under uniform conditions, primary drying times may vary by up to 10%. Make sure to include a delay period (ascertained empirically and occupying a period of hours) at the end of the primary drying cycle, to ensure that all ice has sublimed satisfactorily and to avoid collapse.

3.9. Lyophilisation: Secondary Drying

Secondary drying removes the remaining unfrozen, bound water from the lyophilising material to yield a final product, low in residual moisture, which will be stable for an extended period without deterioration. The unfrozen water is removed by heating the containers in which the primarily dried product rests. The rule of thumb for initiation of secondary drying is the equivalence of product and container temperatures. For safety, however, one should include a delay period at the end of primary drying in any freeze-drying protocol to prevent collapse of any vials that have not quite finished sublimation (12): see (37, 38) and Subheading 3.8 above. The partial pressure of water within the drying chamber drops at the end of primary drying as the last of the ice sublimes. If it can be monitored, this drop in water partial pressure can be a good indicator of the completion of primary drying. Even during secondary drying, with much of the original water gone, the preparation's temperature should never rise above T_g' (10). T_g', however, rises as the residual water content drops (12). One can, therefore, increase the temperature (within limits) during secondary drying (overheating during primary and/or secondary drying will likely be deleterious, however). The vacuum need not be exhaustive during secondary drying; indeed, one should use a pressure in the region of 0.2 Torr for secondary drying (12, 37).

3.10. Lyophilisation: Quality Indices

Some freeze dryers have a built-in chart recorder to provide a profile of shelf temperatures (and perhaps other parameters such as vacuum) during the lyophilisation run. Study this carefully and retain with other batch information. Note the cake shape and texture and any variations between vials, especially those located at different positions within the chamber. Choose a representative number of vials from different chamber locations for scrupulous moisture determination (see Note 17). Measure the yield or recovery (%) of the initial biological activity by appropriate assay following rehydration of a representative number of vials. While waiting to assay, note the time required for complete product rehydration. Also note whether any turbidity remains on rehydration, or after what interval turbidity appears in a clear sample (10). Note that the formulation used greatly influences yield, while the process parameters affect ease of rehydration and shelf life (10). Persistence of the rehydrated biological activity can be measured at suitable or convenient time intervals. Accelerated degradation methods can predict shelf lives of the lyophilised preparation at temperatures of interest for long-term storage (Subheading 3.11 below).

3.11. Stability Analysis and Accelerated Degradation Testing

Kinetic stability, defined in Subheading 1 above, is usually measured at elevated temperatures (1) but the inactivating event(s) at high temperatures may not mirror that/those at the much lower temperatures used for storage. It is not feasible, however, to monitor stability in real time at the actual storage temperature: the experiment would take too long. Inaccuracy may result over shorter intervals, since only minimal losses, scarcely distinguishable from the starting activity, would be apparent.

Accelerated degradation (accelerated storage) protocols can be of value in these situations *provided* the activity decay is first order at each of the temperatures tested and all data are scrupulously accurate and precise. Accelerated degradation involves the periodic assay of samples incubated at different temperatures and use of the Arrhenius equation ($\ln k = -E_a/RT + \ln A$ where k is the first-order activity decay constant, E_a the activation energy, R the gas constant, and T the temperature in Kelvins) to predict "shelf lives" at temperatures of interest. Extrapolation of the

Arrhenius plot (ln k versus $1/T$, slope $-E_a/R$) can give the rate constant (and hence the useful life) at a particular temperature. Kirkwood (43) gives guidelines for the proper use of accelerated methods. Experimental data undergo transformations before use in the Arrhenius plot (conversion to natural log or reciprocal values), perhaps affecting error relationships. To minimise such errors, always use a computer for statistical fitting of data. Use of good quality replicate results is very important (see Note 18). Accelerated storage testing has been used as a practical means of quality assurance for biological standards (44) and has been employed in some scientific investigations (45).

4. Notes

1. Loss or decrease of the protein's biological or functional activity will be the main and most important index of deterioration. Often, however, the degree or time course of activity loss will not give any indication of the underlying molecular cause (although aggregation may be readily visible). Mozhaev (1) provides a useful table of methods to identify the molecular changes leading to inactivation of the protein.
2. The type of stopper used can influence the residual moisture contents of freeze-dried materials. Such effects can arise from the nature of the stopper material itself and also from its prehistory (16). Such considerations will likely apply also to dry preparations for storage, so stoppers should be chosen with care.
3. The vial diameter, glass type and bottom shape, and thickness will all affect the rate of heat transfer from the shelves to product. The vials should withstand freezing and pressure changes and should be uniform with respect to internal diameter and bottom thickness. The vial bottoms should be completely flat to make good contact with the shelves (10).
4. The most basic lyophiliser equipment will have a unit housing a condenser and vacuum pump, to which one may attach a centrifugal test tube holder or a manifold for the drying of multiple product containers (often round-bottomed flasks). Freezing is accomplished in a separate cooling bath, usually filled with an alcohol. Such equipment can be used successfully for small-volume samples but fine and reproducible control of the overall process may not be possible. Higher-grade equipment, with temperature-programmable shelves and a number of temperature probes, is preferable. Shelf-equipped freeze dryers are especially suitable for use with rubber-capped pharmaceutical vials. An externally operated screw press may be present, allowing one to seal vials (partially closed with

rubber stoppers so as not to restrict gaseous movement) under vacuum before releasing air into the chamber.

5. Thiomersal and azide are totally unacceptable in any product for internal administration. Do not discard azide compounds or azide-containing solutions down laboratory sinks. Not only is azide toxic but it can accumulate in old lead piping, leading to the formation of potentially explosive compounds.
6. Some biological matrices, particularly sera, will not filter effectively through a 0.22 μm filter alone. One may need to prefilter the material initially through a coarser 0.45 μm filter to which the desired 0.22 μm filter is connected in series. Alternatively, one can accomplish the finer filtration as a separate operation. One can best filter larger volumes (hundreds of millilitres or litres) using a stack of filters clamped in a special filtration unit. A filter as coarse as 1 μm may be used directly in contact with the solution of interest, the stack comprising progressively finer filters until the sterilising 0.22 μm filter is encountered at the bottom of the stack. Technical representatives of filtration manufacturers can give specialist advice for individual cases.
7. Many suppliers offer specific inhibitors of proteases or classes of protease. These inhibitors are often peptides or proteins, e.g., aprotinin, soybean trypsin inhibitor. A cocktail of protease inhibitors is available in tablet form from Roche Applied Science under the trade name "Complete". The product is stated to give effective inhibition of serine, cysteine, and metalloproteases during protein extractions from a variety of tissues and sources.
8. Curiously, however, lactate dehydrogenase activity levels at pH 5 decreased with increasing concentrations of polyethyleneimine. In contrast, the polymer-stimulated activity at pH 7.2 and 9 (28).
9. Many different vessels and membranes for laboratory-scale ultrafiltration, with a range of defined molecular weight cutoffs, are commercially available. These may comprise permanent stirred pressure cells with replaceable membranes (for volumes in the range 10–500 mL) or disposable centrifugal concentrators (for volumes up to 10 mL) (27). Schein gives some useful observations on ultrafiltration and suggests some other means of achieving protein concentration (27).
10. Sucrose concentration is an effective and rapid means of concentrating a dilute protein solution. Place the solution of interest into a suitably treated, softened dialysis tube, and secure the ends tightly. Tear off a piece of aluminium foil such that the dialysis tube will rest on the foil with roughly 5–6 cm to spare all round. Shake some solid sucrose onto the foil,

then rest the dialysis tube on top of the sucrose. Shake more sucrose on top of the dialysis tube, wrap the foil around the sucrose and dialysis tube to form a parcel, and place in the refrigerator. Water from the dilute protein solution will move by osmosis through the pores of the dialysis tube to the surrounding solid sucrose, leading to concentration of the protein. Examine the dialysis tubing every 15–20 min. The sucrose surrounding the dialysis tubing will gradually form a viscous liquid which can be removed periodically and replaced with fresh solid sucrose. Volume reduction can take place quite quickly. The method has the drawback that sucrose will enter the dialysis tube in amounts not readily calculable (the sucrose will likely help to stabilise the protein, of course). If the presence of sucrose is undesirable, gently pull the dialysis tube between finger and thumb to force its contents into one end. Knot or clamp the dialysis tube tightly as close as possible to the concentrated solution and then dialyse the shortened dialysis tube against a suitable buffer to remove the sucrose. Note that the dialysis tube will swell in dilute buffer as water moves by osmosis into the protein solution which will have a high sucrose concentration. The dialysis tube must be clamped very tightly and as short as possible to prevent undue "re-dilution" of the sucrose-concentrated protein solution.

11. It can be a good idea to place a maximum/minimum thermometer inside the refrigerator, freezer, or incubator(s) close to the containers of interest in order to record any significant variations of temperature which may occur over an extended period (ensure first that the thermometer will withstand the low or high temperatures).
12. Low temperature freezers typically function at –70 to –80°C. These temperatures are extremely cold and can inflict a "cold burn" on exposed skin. Always wear insulating or autoclave gloves when handling low temperature items: latex or nitrile laboratory gloves are not sufficient.
13. The volumes of bulk liquid subjected to freeze-drying must never exceed manufacturer's recommendations. If the condenser's ice capacity is reached or exceeded, the degree of product drying will be insufficient and many problems can result.
14. The glass transition temperature, T_g', and the product water content are critical parameters in the freeze-drying process. It is essential to prevent or minimise damage to the protein of interest during freezing. Inclusion of excipients (additives) with high T_g' values in the protein formulation to be freeze dried can be very useful. The mixture will form a glass at relatively low temperatures, minimising freezing damage. A high

T_g' will also allow the use of higher temperatures during primary drying with less danger of product collapse. Any constituent that will lower the unbound water content of the freeze concentrate will help shorten the secondary drying operation (10) but uncrystallised salts will decrease T_g', since any salt will bring about a depression of freezing point. Thus the salts content of the product formulation should be as low as is practicable (10). The optimum temperature for freezing and primary drying depends on the ratio protein:protectant additive:salt in the freeze concentrate rather than in the initial solution (10, 14). The ratio protein:other solids in the freeze concentrate influences T_g' (10).

15. The glass transition temperature can be determined by differential scanning calorimetry, although a sensitive instrument is required (10). Electrical resistance measurements can sometimes be useful but are not suitable for non-electrolyte mixtures (11). However, microscopic observation of freeze-drying over a range of temperatures is arguably the most direct, sensitive, and unambiguous method for the determination of the collapse temperature (12).
16. There have been reports of protein damage due to mechanical stresses at the interfaces of separated liquid phases arising from freeze concentration effects ((40) and references therein). Sucrose and trehalose exerted little protection against this phenomenon, despite being good glass formers. Rapid cooling below the glass transition temperature appeared to minimise damage from this cause, since the protein spends less time in a freeze-concentrated solution before attainment of the glassy state (40).
17. The residual moisture content of, and its distribution throughout, the lyophilised preparation will dictate its long-term stability. Uneven moisture distribution between vials often leads to biphasic activity-loss profiles on extended storage (10). Each 1% of moisture can depress T_g' by more than 10°C (15). Significant aggregation of lyophilised recombinant human serum albumin occurred within hours upon incubation at 37°C and 96% relative humidity (4), indicating just how critical the residual moisture content can be.
18. Amorphous solid preparations will follow Arrhenius kinetics provided they remain in the glassy state. However, if any of the elevated temperatures used exceeds the glass transition temperature (T_g'), the product will become rubbery and will no longer obey the Arrhenius equation. Other situations in which deviations from Arrhenius kinetics may occur are outlined in ref. (46); see also Note 13 and (38).

Acknowledgment

Dr. Felix Franks indicated numerous useful references.

References

1. Mozhaev, V.V. (1993) Mechanism-based strategies for protein thermostabilisation. *Trends Biotech.* **11**, 88–95.
2. Sluzky, V., Klibanov, A.M. and Langer, R. (1992) Mechanism of insulin aggregation and stabilisation in agitated aqueous solutions. *Biotech. Bioeng.* **40**, 895–903.
3. Liu, W.R., Langer, R. and Klibanov, A.M. (1991) Moisture-induced aggregation of lyophilized proteins in the solid state. *Biotech. Bioeng.* **37**, 177–184.
4. Costantino, H.R., Langer, R. and Klibanov, A.M. (1995) Aggregation of a lyophilized pharmaceutical protein, recombinant human albumin. *Bio/Technology* **13**, 493–496.
5. Quax, W.J. (1993) Thermostable glucose isomerases. *Trends Food Sci. Tech.* **4**, 31–34.
6. Privalov, P.L. (1990) Cold denaturation of proteins. *Crit. Rev. Biochem. Mol. Biol.* **25**, 281–305.
7. Franks, F. (1993) Conformational stability of proteins, in *Protein biotechnology: isolation, characterisation and stabilisation* (Franks, F., ed), Humana, Totowa, NJ. pp. 395–436.
8. Sadana, A. (1988) Enzyme deactivation. *Biotech. Adv.* **6**, 349–446.
9. Thuma, R.S., Giegel, J.L. and Posner, A.H. (1987) Manufacture of quality control materials, in *Laboratory quality assurance* (Howanitz, P.J. and Howanitz, J.H., eds) McGraw-Hill, NY. pp. 101–123.
10. Franks, F. (1990) Freeze-drying: from empiricism to predictability. *Cryo Letters* **11**, 93–110.
11. Franks, F. (1998) Freeze-drying of bioproducts: putting principles into practice. *Eur. J. Pharma. Biopharm.* **45**, 221–229.
12. Pikal, M.J. (1990) Freeze-drying of proteins. Part 1: process design. *BioPharm* **3**, 18–27.
13. Carpenter, J.F. and Crowe, J.H. (1988) The mechanism of cryoprotection of proteins by solutes. *Cryobiology* **25**, 244–255.
14. Pikal, M.J. (1990) Freeze-drying of proteins. Part 2: formulation selection. *BioPharm* **3**, 26–30.
15. Franks, F., Hatley, R.H.M. and Mathias, S.F. (1991) Materials Science and the production of shelf-stable biologicals. *Pharm. Technol. Int.* **3**, 24–34.
16. Oetjen, G.-W. (1999) *Freeze-drying.* Wiley-VCH, Weinheim, Germany.
17. Beynon, R.J. & Bond, J.S. (1987) *Proteolytic enzymes: a practical approach.* Oxford University Press, Oxford, UK (2nd edition 2001)
18. Scopes, R.K. (1994) *Protein purification: principles and practice*, 2nd edn, Springer Verlag, Berlin. pp. 317–324.
19. Erlanger, B.F., Kokowsky, N. and Cohen, W. (1961) The preparation and properties of two new chromogenic substrates of trypsin. *Arch. Biochem. Biophys.* **95**, 271–278.
20. Timasheff, S.N. and Arakawa, T. (1997) Stabilisation of protein structure by solvents, in *Protein structure: a practical approach* (Creighton, T.E., ed; 2nd edition) IRL Press, Oxford, UK. pp. 349–364.
21. Kaushik, J.K. and Bhat, R. (1998) Thermal stability of proteins in aqueous polyol solutions. *J. Phys. Chem. B* **102**, 7058-7066.
22. Qu, Y., Bolen, C.L. and Bolen, D.W. (1998) Osmolyte-driven contraction of a random coil protein. *Proc. Natl Acad. Sci. USA* **95**, 9268–9273.
23. Anjum, F., Rishi, V. and Ahmad, F. (2000) Compatibility of osmolytes with Gibbs energy of stabilisation of proteins. *Biochim. Biophys. Acta* **1476**, 75–84.
24. Combes, D., Yoovidhya, T., Girbal, E., Willemot, R.-M. and Monsan, P. (1987) Mechanism of enzyme stabilisation. *Ann. N.Y. Acad. Sci.* **501**, 59–62.
25. Tombs, M.P. (1985) Stability of enzymes. *J. Appl. Biochem.* **7**, 3–24.
26. Klibanov, A.M. (1983) Stabilisation of enzymes against thermal inactivation. *Adv. Appl. Microbiol.* **29**, 1–25.
27. Schein, C.H. (1990) Solubility as a function of protein structure and solvent components. *Bio/Technology* **8**, 308–317.
28. Andersson, M.M. and Hatti-Kaul, R. (1999) Protein stablising effect of polyethyleneimine. *J. Biotech.* **72**, 21–31.
29. Marcozzi, G., Di Domenico, C. and Spreti, N. (1998) Effects of surfactants on the stabilisation of the bovine lactoperoxidase activity. *Biotech. Prog.* **14**, 653–656.

30. Volkin, D.B. and Klibanov, A.M. (1989) Minimising protein inactivation, in *Protein function: a practical approach* (Creighton, T.E., ed; 1st edition), IRL Press, Oxford, UK. pp. 1–24.
31. Shah, N.K., Shah, D.N., Upadhyay, C.M., Nehete, P.N., Kothari, R.M. and Hegde, M.V. (1989) An economical, upgraded, stabilised and efficient preparation of amyloglucosidase. *J. Biotech.* **10**, 267–276.
32. Hatley, R.H.M., Franks, F. and Mathias, S.F. (1987) The stabilisation of labile biomolecules by undercooling. *Process Biochem.* **22**, 169–172.
33. Franks, F. (1993) Storage stabilisation of proteins, in *Protein biotechnology: isolation, characterisation and stabilisation* (Franks, F., ed), Humana, Totowa, NJ. pp. 489–531.
34. Franks, F. and Hatley, R.H.M. (1992) Storage of materials. US Patent no. 5,098,893.
35. Gibson, T.D. and Woodward, J.R. (1991) Enzyme stabilisation. Patent application PCT/GB91/00443, Publ. No. Wo91/14773.
36. Gibson, T.D., Higgins, I.J. and Woodward, J.R. (1992) Stabilisation of analytical enzymes using a novel polymer–carbohydrate system and the production of a stabilised, single reagent for alcohol analysis. *Analyst* **117**, 1293–1297.
37. Tang, X.C. and Pikal, M.J. (2004) Design of freeze-drying processes for pharmaceuticals: practical advice. *Pharmaceut. Res.* **21**, 191–200.
38. Wang, W. (2000) Lyophilisation and development of solid protein pharmaceuticals. *Int. J. Pharmaceut.* **203**, 1–60.
39. Bhatnagar, B. S., Bogner, R. H. and Pikal, M. J. (2007) Protein stability during freezing: separation of stresses and mechanisms of protein stabilization. *Pharmaceut. Dev. Technol.* **12**, 505–523.
40. Heller, M.C., Carpenter, J.C. and Randolph, T.W. (1999) Protein formulation and lyophilization cycle design. *Biotech. Bioeng.* **63**, 166–174.
41. Franks, F. and Murase, N. (1992) Nucleation and crystallisation in aqueous systems during drying: theory and practice. *Pure Appl. Chem.* **64**, 1667–1672.
42. Blanchard, J.S. (1984) Buffers for enzymes. *Meths. Enzymol.* **104**, 404–415.
43. Kirkwood, T.B.L. (1984) Design and analysis of accelerated degradation tests for the stability of biological standards III. Principles of design. *J. Biol. Stand.* **12**, 215–224.
44. Jerne, N.K. and Perry, W.L.M. (1956) The stability of biological standards. *Bull. World Health Organ.* **14**, 167–182.
45. Malcolm, B.A., Wilson, K.P., Matthews, B.W., Kirsch, J.F. and Wilson, A.C. (1990) Ancestral lysozymes reconstructed, neutrality tested, and thermostability linked to hydrocarbon packing. *Nature* **345**, 86–89.
46. Franks, F. (1994) Accelerated stability testing of bioproducts: attractions and pitfalls. *Trends Biotech.* **12**, 114–117.

Chapter 11

Differential Precipitation and Solubilization of Proteins

Barry J. Ryan

Abstract

Differential protein precipitation is a rapid and economical step in protein purification and is based on exploiting the inherent physico-chemical properties of the polypeptide. Precipitation of recombinant proteins, lysed from the host cell, is commonly used to concentrate the protein of choice before further polishing steps with more selective purification columns (e.g. His-Tag, Size Exclusion, etc.). Recombinant proteins can also precipitate naturally as inclusion bodies due to various influences during over-expression in the host cell. Although this phenomenon permits easier initial separation from native proteins, these inclusion bodies must carefully be differentially solubilised so as to reform functional, correctly folded proteins. Here, a typical protein extraction, precipitation, and selective resolubilisation procedure is outlined, based on a recombinantly expressed protein.

Key words: Ammonium sulphate precipitation, Trichloroacetic acid precipitation, Inclusion body solubilisation, Protein refolding

1. Introduction

Protein precipitation can be caused by the differential solubility between a protein-rich soluble phase and a solid chemical precipitant. Soluble proteins can be insolubilised by interaction with a suitable precipitant that decreases the protein's attraction to the solvent and increases the protein's attraction to other protein molecules, resulting in protein accumulation and eventually precipitation. The addition of low molecular weight substances, such as glycerol, polyethylene glycol, and sucrose, and high molecular weight substances such as serum albumin, can have significant effects on protein structure and stability. Preferential hydration of a protein molecule caused by the presence of these additives can increase the protein's stability. Certain salts can also exert a stabilizing effect by "salting out" hydrophobic residues of

Dermot Walls and Sinéad T. Loughran (eds.), *Protein Chromatography: Methods and protocols*, Methods in Molecular Biology, vol. 681, DOI 10.1007/978-1-60761-913-0_11, © Springer Science+Business Media, LLC 2011

a protein, causing the molecule to adapt a more compact, stable structure (1) frequently resulting in precipitation. The use of such protein precipitating molecules is an empirical process, the effects of any given substance on a protein must be determined experimentally. The use of additives can not only be used as a simple approach to increase the stability of a given protein, but also to actively effect protein precipitation. Protein precipitation can be used as a crude protein clean-up method from cell lysates, readily employed after bacterial over-expression of recombinant proteins.

Differential solubilisation of proteins is often employed for proteomic analyses (2, 3), but it too can offer an alternative purification technique for non-soluble recombinant proteins expressed in heterologous hosts. Recombinant proteins expressed as inclusion bodies can be readily separated from the host cell protein matrix; however, careful solubilisation and refolding are critical for obtaining suitable recombinant proteins for further downstream processes. Here a typical recombinant protein precipitation and resolubilisation procedure is outlined.

2. Materials

Note: All consumables may be sourced from Sigma-Aldrich unless otherwise stated.

2.1. Recombinant/ Native Protein Extraction

1. Centrifuge (Bucket Type, e.g. J2-21, Beckman and microfuge, e.g. 5415D Eppendorf).
2. pH Meter (e.g. M-240, Corning).
3. Sonicator (e.g. Vibra Cell, Sonics Scientific).
4. Water bath (temperature controlled).
5. Vortex.
6. Resuspension Buffer One: 50 mM Na_2HPO_4–NaH_2PO_4, pH 8.0, 0.3 M NaCl.
7. –80°C freezer.

2.2. Protein Precipitation Using Ammonium Sulphate

1. Resuspension Buffer One: 50 mM Na_2HPO_4–NaH_2PO_4, pH 8.0, 0.3 M NaCl.
2. Saturated ammonium sulphate: Add 750 g of $(NH_4)_2SO_4$ to 1 L of double-distilled water in a beaker or flask. Stir the solution at room temperature with a magnetic stirrer for 15 min or until saturation. Gently decant the clear supernatant solution after the undissolved solids settle on the bottom of the flask.
3. Graduated pipette (10 mL).

2.3. Protein Precipitation Using Trichloroacetic Acid

1. 2% Deoxycholate (DOC): Add 2 g of DOC to 100 mL of double-distilled H_2O, mix well.
2. 100% Trichloroacetic acid (TCA): Add 1 g of TCA to 454 μL double-distilled H_2O and mix carefully. Store in a light proof bottle at 4°C until required for use. TCA is a harmful skin and eye irritant. Always use correct personal protective equipment when handling it.
3. Acetone (ice cold). Store acetone at –20°C. Use directly from –20°C.
4. Vacuum Dryer (e.g. DNA 110 Speed Vac, Savant).
5. Resuspension Buffer: 50 mM Na_2HPO_4–NaH_2PO_4, pH 8.0, 0.3 M NaCl.
6. Centrifuge (Bucket Type, e.g. J2-21, Beckman and microfuge, e.g. 5415D Eppendorf).

2.4. Protein Solubilisation

1. DNase I (100 U/mL).
2. Resuspension Buffer Two: 50 mM Na_2HPO_4–NaH_2PO_4, pH 8.0, 0.3 M NaCl, 5 mM DTT, 0.35 mg/mL lysozyme, Proteinase Inhibitor Cocktail (see also Chapter 4). Make up as fresh prior to use.
3. Triton X-100.
4. Solubilisation Buffer: 50 mM Na_2HPO_4–NaH_2PO_4, pH 8.0, 0.3 M NaCl, 25 mM DTT, 6 M guanidine HCl.
5. PBS-T: Phosphate buffered saline (PBS, 1×) containing 1% Triton X-100.
6. Centrifuge (Bucket Type, e.g. J2-21, Beckman and microfuge, e.g. 5415D Eppendorf).
7. Vacuum Concentrator (Speed Vac, Savant).

2.5. Protein Refolding

1. Refolding Buffer: 50 mM Na_2HPO_4–NaH_2PO_4, pH 8.0, 0.3 M NaCl, 2.5 mM reduced glutathione, 0.25 mM oxidized glutathione, 0.2 M arginine.
2. Dialysis Buffer: 50 mM Na_2HPO_4–NaH_2PO_4, pH 8.0, 0.3 M NaCl.
3. Dialysis apparatus (dialysis tubing and clips, magnetic mixer, large clean container).
4. Guanidine hydrochloride: 6 M stock, made in double-distilled H_2O.
5. Amicon protein concentration device (e.g. Ultra-15 Centrifugal Filter Units, Amicon).
6. Gradient maker apparatus (see Fig. 1).

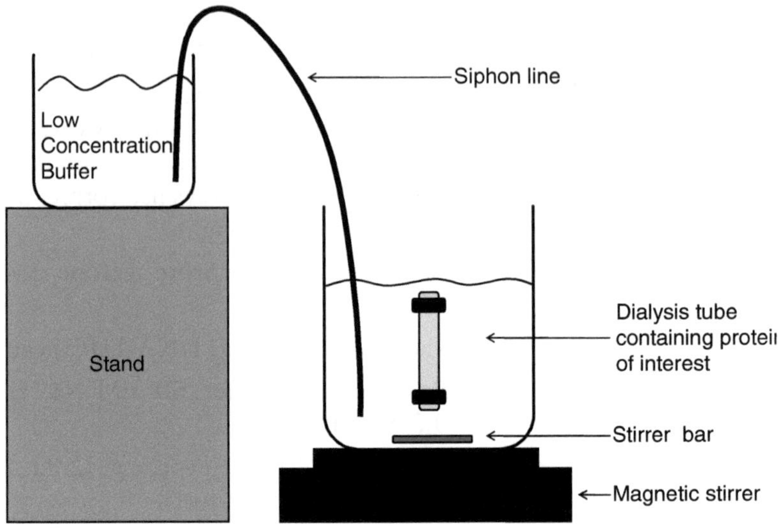

Fig. 1. Schematic diagram of a simple gradient maker.

3. Methods

3.1. Recombinant/ Native Protein Extraction

The source of the protein will determine the optimal technique to release the protein from the tissue or cells in which it is contained. The typical freeze-thaw cell lysis procedure (below) is generally sufficient to lyse most bacterial cell types, although other options are available (see Notes 1–3).

1. Collect the bacterial cells by transferring the bacterial culture to a pre-chilled sterile centrifuge tube and centrifuge at low speed (5 min, 800×*g*) in a previously cooled centrifuge (4°C).
2. Carefully remove the culture media from the bacterial cell pellet, ensuring the pellet is not disturbed.
3. Resuspend the cell pellet in Resuspension Buffer One, in 10% of the original culture volume.
4. Freeze the resuspended cells to −80°C by placing the resuspension solution (still in the plastic centrifuge tube) into a pre-equilibrated −80°C freezer, then warm the cells to 37°C (using a pre-equilibrated water bath) for 10 min. Repeat this freeze-thaw process three times.

   Sonication can also be used if the protein is not released during the freeze-thaw steps. It is crucial to maintain the cell suspension on ice during the sonication process (see Notes 4–6).
5. Sonicate at ten amplitude microns for 10–20 s.
6. Allow the cell suspension to stand on ice for 30 s.

7. Repeat steps 5 and 6 three more times.
8. Check the recombinant protein induction/expression by loading and analyzing a representative sample (typically 50 μg protein) onto a SDS–PAGE gel.

3.2. Protein Precipitation Using Ammonium Sulphate

A common and inexpensive first step to isolate proteins during protein purification is precipitation with an external additive. This additive alters the physico-chemical properties of the protein causing it to fall out of solution. Ammonium sulphate (see Note 7) is commonly used for large-scale precipitations.

1. Gently stir the protein mixture with the aid of a magnetic stirring bar at 4°C. Add, using a graduated pipette, the saturated ammonium sulphate solution drop-wise to the protein solution until precipitates start to form (see Notes 8–10).
2. Once sufficient saturated salt solution has been added to cause precipitation of the protein of interest (indicated by collection of precipitate at the bottom of the container), centrifuge the mixture at 10,000 × *g* for 15 min. Collect the precipitate by carefully discarding as much supernatant as possible (see Note 10).
3. Resuspend the protein pellet at 4°C in Resuspension Buffer One for further downstream processes.

3.3. Protein Precipitation Using Trichloroacetic Acid

Trichloroacetic acid (TCA) is routinely employed for small-scale operations or precipitations of protein preparations that are at low concentration; however, it should be noted that this procedure is protein denaturing, and *caution* must be exercised when working with TCA (see Notes 11 and 12).

1. To one volume of protein solution, add 1/100 volume of 2% DOC (sodium deoxycholate).
2. Vortex and incubate for 30 min at 4°C.
3. Add 1/10 volume of 100% TCA. Vortex the solution and incubate statically overnight at 4°C.
4. Centrifuge the sample for 15 min at 4°C (10,400 × *g*). Gently remove the supernatant and retain the pellet. Carefully dry the tube by inversion on tissue paper (Note: the pellet may be difficult to see).
5. Optional: Wash the pellet twice with one volume of ice-cold acetone. Vortex and re-pellet the samples by centrifugation 10,400 × *g* for 5 min at 4°C between washes (see Note 13).
6. Dry the samples under vacuum (e.g. Speed Vac, Savant) or allow to air dry.
7. Resuspend the protein pellet in a buffer of choice for further downstream processes.

3.4. Protein Solubilisation

Recombinant protein expression in a heterologous host frequently results in insoluble and inactive proteins. Regularly, protein overexpression results in the production of inclusion bodies, which are insoluble aggregates of misfolded protein. Although these inclusion bodies can easily be purified, further characterisation of this protein mass is often impossible without solubilisation of the protein of interest and refolding into an active form (see Note 14). A typical inclusion body solubilisation and refolding protocol is outlined below (see also Note 15).

1. Carry out steps 1–3 as outlined in Subheading 3.1, except resuspend the cell pellet in 10% of the original culture volume of Resuspension Buffer Two.
2. Slowly add Triton X-100 (to a final concentration of 1% v/v), and mix gently.
3. Carry out the sonication procedure as outlined in Subheading 3.1, steps 5–7.
4. Incubate the cell debris with DNase I (100 U/mL) for 1 h at 37°C.
5. Collect the inclusion bodies by centrifugation at 30,000 × *g* for 30 min at 4°C.
6. Wash the inclusion body pellet twice with PBS-T, followed by centrifugation at 30,000 × *g* for 30 min at 4°C.
7. Solubilise the pelleted inclusion bodies in the solubilisation buffer and allow total solubilisation to occur at 4°C for 1 h, with occasional gentle mixing.
8. After 1 h solubilisation, remove all remaining insoluble material by centrifugation 30,000 × *g* for 10 min at 4°C (see Note 16).
9. Determine the protein concentration and adjust to 1 mg/mL by dilution in solubilisation buffer and proceed directly to refolding at 4°C (Subheading 3.5).

3.5. Protein Refolding

1. Dilute the solubilised proteins as quickly as possible (to yield a final protein concentration of 0.1 mg/mL, see Note 17) into pre-chilled Refolding Buffer (see Note 18).
2. Dialyze the diluted solubilised protein overnight against a 200-fold volume of dialysis buffer with slowly decreasing concentrations of GuHCl (typically decrease GuHCl concentration as follows: 6, 4, 2, 1, 0.5 and then 0 M in a continual dialysis approach; see Fig. 1 and Notes 19 and 20).
3. Centrifuge the dialysate at 4°C for 30 min at 30,000 × *g*.
4. Carefully remove the liquid protein-rich layer, concentrate (e.g. Amicon filtration) and store at an appropriate temperature (see Note 21).

4. Notes

1. There are several methods to achieve this, including repeated freezing and thawing, sonication, homogenization at high pressures, enzymatic lysis, or permeabilization by organic solvents. The method of choice depends on how fragile the protein is and how robust the host cell is.
2. Proteins can also be selectively released from the various compartments of a bacterial host, for example proteins expressed in the periplasmic envelope can be selectively lysed by a method similar to that described by French et al. (4). Pellet the bacterial cells to be disrupted by centrifugation at 800 × *g* for 3 min. Resuspend the pellet in Fractionation Buffer (F1) in 20% of the original culture volume. The F1 comprises (final concentrations): 500 μg/mL lysozyme, 20% w/v sucrose, 1 mM EDTA, pH 8.0, 200 mM guanidine hydrochloride, and 200 mM Tris–HCl, pH 8.0, at room temperature. Statically incubate the resuspended cells at room temperature for 15 min, after which add an equal volume of ice-cold water. Stand the mixture at room temperature for 15 min. Remove the cell debris by centrifugation at 10,400 × *g* for 10 min. Transfer the supernatant (containing the periplasmic fraction) to a clean container for further purification.
3. Proteins expressed and transported to the culture supernatant can also be conveniently concentrated by a method outlined by Caldwell and Lattemann (5). In brief, this method involves adding an equal volume of a PRMM solution (0.05 mM pyrogallol red, 0.16 mM sodium molybdate, 1.0 mM sodium oxalate, 50.0 mM succinic acid, 20% v/v methanol in H_2O, adjusted to pH 2.0 with HCl) to cleared (0.22 μm filtered) culture supernatant. Adjust the pH of the solution to 2.8 (±0.1), and allow the proteins to precipitate for 1–2 h at room temperature, followed by an overnight incubation at 4°C. Sediment the precipitate by centrifugation at 10,000 × *g* for 1 h, and carefully remove the supernatant. Repeatedly rinse the precipitate with 1 mL of acetone. Remove all traces of acetone by evaporation at room temperature. Solubilise the precipitate by adding 100 μL of 2× sodium dodecyl sulphate–polyacrylamide gel electrophoresis (SDS–PAGE) sample buffer (25% glycerol, 8% SDS, 4% β-mercaptoethanol, 0.02% bromophenol blue, 100 mM Tris–HCl, pH 6.8). These samples can be applied directly to an SDS–PAGE gel, if this type of analysis is required.
4. The sonication process can generate large amounts of heat, which is why pulses are limited to ~20 s. In between pulses, cool the tube in ice or ice-water slurry for 30 s. If a large volume is required to be sonicated, split the cell suspension into

two tubes, and alternate the sonication and cooling steps. Sonicate on ice where possible.

5. The extraction process also releases proteases, which will digest all proteins in the solution. If the protein is sensitive to proteolysis, it is desirable to employ a protease inhibitor (see Chapter 4), to proceed quickly, and to keep the extract cooled to minimize proteolysis.
6. Lysozyme (500 μg/mL, to assist cell-wall degradation) and DNase I (100 U/mL, to degrade genomic DNA) can be added to the lysis buffer.
7. The addition of high concentrations of salt to a protein solution causes precipitation by removing water from hydrophobic patches on the protein's surface, resulting in these patches aggregating together causing the protein to come out of solution. A number of salts can be used for this process; NaCl, Na_2SO_4, KCl, $CaCl_2$, and $MgSO_4$, however, $(NH_4)_2SO_4$ is by far the most commonly used additive. This is due to several advantageous characteristics of the salt, including the fact that it has a high solubility in water (4 M saturation) and it has a low density at saturation allowing precipitated proteins to be collected by centrifugation. Hence, the first proteins to be purified during ammonium sulphate precipitation are water-soluble proteins (6).
8. Final concentrations of ammonium sulphate must be calculated using standard nomograms or with online tools ("Ammonium Sulphate Calculator", available at http://www.encorbio.com/protocols/AM-SO4.htm). Adding increasing amounts of ammonium sulphate causes the different fractions of a protein mixture to precipitate at different rates. One advantage of this method is that it can be performed inexpensively with very large volumes. Additionally, the high salt content of the precipitated protein permits its direct addition onto a hydrophobic interaction chromatography (HIC) purification column, thus speeding up the overall purification process.
9. Purification of integral membrane proteins requires the addition of a detergent such as sodium dodecyl sulphate (1% SDS) to dissolve cell membranes and keep membrane proteins in solution during purification. It should be noted that SDS causes protein denaturation, hence milder detergents such as 1% Triton X-100 or 1% CHAPS can be used to retain the protein's native conformation during cell-membrane dissolution.
10. Ammonium sulphate salt can be added either in saturated solution (as described above) or directly as salt crystals. It may be advantageous to add ammonium sulphate directly

into the protein mixture as powdered solids during large-scale purification processes so that the effect of dilution by the salt solution is minimized. If a saturated salt solution is employed the amount of ammonium sulphate solution added must be recorded accurately, often this is achieved by dispensing from a graduated pipette. It is critical to avoid the spatial non-uniformity in the salt concentration during the addition of the salt solution. Localized concentration "*hot-spots*" will prematurely initiate the precipitation of other proteins and inadvertently affect the precipitation process. Record the volume of the saturated ammonium sulphate solution required to precipitate the protein of interest. Also note that protein precipitation is not instantaneous; it may require 15–20 min to equilibrate.

11. TCA is a harmful skin and eye irritant. Always use correct personal protective equipment when handling it.
12. There are numerous options to effect other types of protein precipitation including (but not limited to) acetone precipitation (useful to simultaneously eliminate acetone soluble components and increase protein concentration), ethanol precipitation (useful to simultaneously concentrate proteins and remove traces of GuHCl prior to SDS–PAGE analysis), acidified acetone/methanol (50/50 v/v; useful to simultaneously remove acetone and methanol soluble interferences such as SDS prior to IEF analysis), and chloroform/methanol (50/50 v/v; useful to simultaneously remove salt and detergents).
13. The presence of trace amounts of TCA, carried through from the precipitation, can acidify the resuspension sample buffer. If further downstream processes are pH sensitive, the sample buffer should be titrated with 1 M NaOH or 1 M Tris–HCl, pH 8.5, to obtain the desired pH for the required process. Acidified SDS–PAGE sample buffer, for example, can give a yellow colour. Correct titration will result in reversion to the typical blue sample buffer colour. Hint: A simple method to overcome this is to resuspend the samples in a slightly basic SDS–PAGE loading buffer (e.g. pH 9.0). Hence, any residual TCA left it will be neutralized by the basic buffer allowing direct addition onto the SDS–PAGE gel. Additionally, excess TCA traces will cause the Coomassie dye to precipitate during SDS–PAGE. If this is a problem repeat the optional wash steps outlined.
14. The interactions between solvents and proteins, and also proteins and proteins, determine the solubility of any given protein. Interactions can be classified as either attractive or repulsive. A protein will be soluble in a particular solvent, if the net-free energy of the proteins interactions is adequately

negative (i.e. attractive). Additionally, protein solubility is improved if protein–protein interactions have sufficiently positive net-free energy (i.e. repulsive), although it should be noted that protein–protein interaction is modulated by the chemical nature of the solvent of choice. Conversely, insolubility typically results from net attractive forces between proteins and net repulsive forces between the solvent of choice and the protein of interest. Furthermore, a soluble protein can be insolubilised by a change in its free energy state in relation to the proteins, or the solvent, it interacts with and, hence, additions/subtractions to a protein solution should be carefully assessed on a small scale (7).

15. The separation of one protein, or family of proteins, from other proteins by means of differential solubility with chemical reagents is based on the differential solubility between a liquid phase and a solid phase. The optimisation of this procedure is empirical but Lindwall et al. (8) outline an optimisation procedure based on a sparse matrix approach; solubilisation buffers are composed based on "*solubility space*" which is related to accepted protein solubilisation theories. This method assists in identifying suitable solubilisation conditions for most over-expressed proteins.
16. It is important to remove existing aggregates that can act as nuclei to trigger aggregation during folding.
17. The final protein concentration should not exceed 0.05–0.1 mg/mL as dilute protein mixtures refold optimally at this concentration. A rapid and efficient mix is essential at this step.
18. The addition of a mild solubilising agent [e.g. 1 M 3(1-pyridinio)-1-propane sulfonate] during the refolding steps limits re-aggregation of re-folding proteins.
19. Continual dialysis can be set up by using a gradient maker. In its simplest form, this consists of two containers of the same shape connected by a syphon. One container contains the low-concentration buffer, and the other contains high-concentration buffer. The buffer is withdrawn from the low-concentration container to the high-concentration container. This will produce a linear gradient from high to low buffer concentrations over the total volume of the gradient. Once the "*low-concentration buffer*" supply has been depleted, the dialysis tubing is removed from the larger vessel and placed in a similar, clean vessel containing fresh buffer at the same concentration as the original "*low-concentration buffer*". The "*low-concentration buffer*" vessel is replaced with a vessel containing buffer at the next lower concentration level and the process is allowed to continue until the "*low-concentration buffer*" supply is depleted again. This process is repeated until

the vessel containing the dialysis tubing has reached the desired final concentration, typically 0 M GuHCl.

20. See Rudolph and Lilie (9) for a comprehensive overview of protein refolding and (10) for an industrial viewpoint on that topic.

21. Most proteins can be stored at 4°C, without significant denaturation, for up to 24 h. For intermediate storage times (24 h to 1 week), the protein should be filter sterilized (through a 0.22 μm filter) and stored at 4°C. Additional supplements, such as a bacteriostatic agent (e.g. 0.1% sodium azide) can be included to avoid bacterial growth. For storage times greater than 1 week (up to several months), it is advisable to freeze the protein preparation. Rapid freezing helps reduce protein denaturation. It is useful to freeze the solution in small aliquots to avoid repeated freeze/thaw cycles, which may reduce the biological activity of the protein. Additional stabilizing agents can also be added prior to freezing, such as glycerol (5–50% w/v), serum albumin (10 mg/mL), reducing agents (such as 1 mM DTT), and ligands/co-factors (depending on the nature of the target protein). Extended protein storage (several months to years) should be carried out at −80°C or in liquid nitrogen. The addition of 50% (w/v) glycerol is recommended for storage at this temperature. Alternative strategies include storing the protein as an ammonium sulphate precipitate at 4°C, or at lower temperatures in a lyophilized form (*see also* Chapter 10 for protocols and discussion regarding the storage and lyophilisation of proteins).

References

1. O'Fágáin, C. (1997) Protein stability and its measurement, in *Stabilising Protein Function* (O'Fágáin, C., ed.), Springer, Berlin, pp 69–75.
2. Ramos, Y., García, Y., Llopiz, A., and Castellanos-Serra, L. (2008) Selectivity of bacterial proteome fractionation based on differential solubility: A mass spectrometry evaluation. *Anal. Biochem.* **377**, 134–140.
3. Leimgruber, R. M. (2005) Extraction and solubilisation of proteins for proteomic studies, in *The Proteomics Protocols Handbook* (Walker J. M., ed.), Humana, Totowa, NJ, pp 1–18.
4. French, C.; Keshavarz-Moore, E., Ward, J.M. (1996) Development of a simple method for the recovery of recombinant proteins from the *E.coli* periplasm. *Enzyme Microb. Technol.* **19**, 332–8.
5. Caldwell R.B., and Lattemann C.T. (2004) Simple and reliable method to precipitate proteins from bacterial culture supernatant. *Appl. Environ. Microb.* **70**, 610–612.
6. Doonan, S. (2004) Bulk purification by fractional precipitation, in *Protein purification protocols, methods in molecular biology* (Cutler, P., ed.), Humana, Totowa, NJ, **244**, pp 117–125.
7. Rothstein F. (1994) Differential precipitation of proteins, in *Protein Purification Process Engineering* (Harrison, R. G. ed), Marcel Dekker Inc., NY, pp 115–116.
8. Lindwall, G., Chau, M.-F., Gardner, S.R., and Kohlstaedt, L.A. (2000) A sparse matrix approach to the solubilisation of overexpressed proteins. *Protein Eng.* **13**, 67-71.
9. Rudolph, R., and Lilie, H. (1996) In vitro folding of inclusion body proteins. *FASEB J.* **10**, 49–56.
10. Jungbauer, A., and Kaar, W. (2007) Current status of technical protein refolding. *J. Biotechnol.* **128**, 587–96.

Chapter 12

Ion-Exchange Chromatography: Basic Principles and Application to the Partial Purification of Soluble Mammalian Prolyl Oligopeptidase

Philip M. Cummins, Oonagh Dowling, and Brendan F. O'Connor

Abstract

Ion-exchange chromatography (IEC) allows for the separation of ionizable molecules on the basis of differences in charge properties. Its large sample-handling capacity, broad applicability (particularly to proteins and enzymes), moderate cost, powerful resolving ability, and ease of scale-up and automation have led to it becoming one of the most versatile and widely used of all liquid chromatography (LC) techniques. In this chapter, we review the basic principles of IEC, as well as the broader criteria for selecting IEC conditions. By way of further illustration, we outline protocols necessary to partially purify a serine peptidase from bovine whole brain cytosolic fraction, covering crude tissue extract preparation through to partial purification of the target enzyme using anion-exchange chromatography. Protocols for assaying total protein and enzyme activity in both pre- and post-IEC fractions are also described. The target serine peptidase, prolyl oligopeptidase (POP, EC3.4.21.26), is an 80-kDa enzyme with endopeptidase activity towards peptide substrates of ≤30 amino acids. POP is a ubiquitous post-proline cleaving enzyme with particularly high expression levels in the mammalian brain, where it participates in the metabolism of neuroactive peptides and peptide-like hormones (e.g. thyroliberin, gonadotropin-releasing hormone).

Key words: Liquid chromatography, Anion-exchange, Cation-exchange, Matrix, pH, Ionic strength, DEAE-Sepharose® Fast Flow, Prolyl oligopeptidase, Bovine brain

1. Introduction

Bioseparation involves resolution of the components in complex mixtures encountered in biological and biochemical systems, thereby enabling scientists to determine both the identity and concentration of each component and, if necessary, to isolate a desired component from other contaminating molecules for further analysis or application. Bioseparation processes are frequently dominated by liquid chromatography (LC) steps. Resolution of

Dermot Walls and Sinéad T. Loughran (eds.), *Protein Chromatography: Methods and protocols*, Methods in Molecular Biology, vol. 681, DOI 10.1007/978-1-60761-913-0_12, © Springer Science+Business Media, LLC 2011

mixtures by LC is based on the principle that, under a given set of conditions, individual solutes dissolved in a ***mobile phase*** will differentially interact with a chemically modified ***stationary phase*** as a function of differences in individual solute distribution coefficients (K). In this way, LC exploits inherent differences between biomolecules (e.g. molecular size, hydrophobicity, binding specificity, charge) in order to achieve their separation from one another.

1.1. Basic Principles of Ion-Exchange Chromatography

With its origins dating back to the 1940s, ion-exchange chromatography (IEC) was designed specifically for the separation of differentially charged or ionizable molecules (1, 2). Both chemists and biochemists have routinely employed this technique for the purification of proteins (3, 4), enzymes (4–6), antibodies (3, 7), peptides (8), amino acids, and nucleic acids (9, 10), as well as simpler carbohydrates (11) and organic compounds (12). Its large sample-handling capacity, broad applicability (including high performance and high-throughput application formats), moderate cost, powerful resolving ability, and ease of scale-up and automation have led to it becoming one of the most versatile and widely used of all LC techniques. Like other forms of column-based LC (e.g. gel-permeation, affinity, hydrophobic interaction, etc.), this technique comprises both mobile and stationary phases, the former typically an aqueous buffer system into which the mixture to be resolved is introduced, and the latter usually an inert organic matrix chemically derivatized with ionizable functional groups that carry a displaceable oppositely charged counterion. These counterions exist in a state of equilibrium between the mobile and stationary phases, giving rise to two possible IEC formats, namely anion- and cation-exchange (see Fig.1). Exchangeable matrix counterions may include protons (H^+), hydroxide groups (OH^-), single charged monoatomic ions (Na^+, K^+, Cl^-), double charged monoatomic ions (Ca^{2+}, Mg^{2+}), and polyatomic inorganic ions (SO_4^{2-}, PO_4^{3-}), as well as organic bases (NR_2H^+) and acids (COO^-).

The pH-dependent ionization of electrolyte groups (weak acids or bases) can impart a net positive or negative charge on biomolecules, subsequently enabling their separation from one another via IEC. This can be explained by taking as an example the separation of a mixture of proteins. During IEC, the mobile phase pH will determine the net charge on both the matrix functional group and on individual proteins within the sample mixture. The polyampholyte nature of proteins mean they can carry both positive and negative charges, the former largely attributable to the ionization of lysine and arginine side chains, and the latter to aspartate and glutamate side chain ionizations. As a general rule, a protein will have a net negative charge above its isoelectric point or p*I* (i.e. pH at which a protein has zero net charge), and vice-versa. Naturally, different proteins may have different p*I* values,

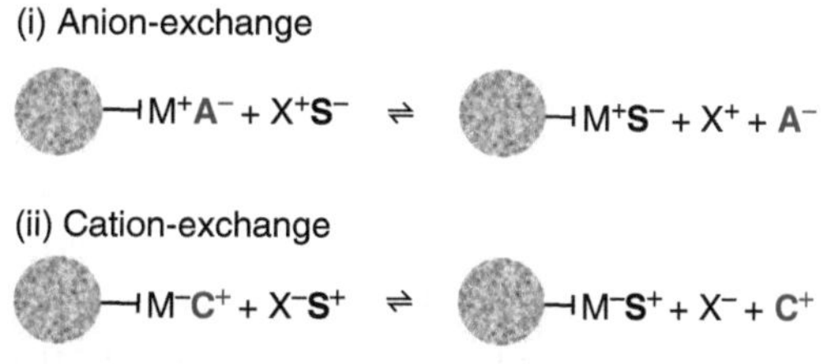

Fig. 1. Ion-exchange chromatography schematic. It is the nature of the counterions displaced from the matrix functional groups (M^+, M^-) which determines the IEC format. Thus, with anion-exchange chromatography, the stationary phase (usually a porous bead) displays a positively charged functional group with counter anion (A^-) that can be displaced by an anionic solute (S^-). By contrast, with cation-exchange chromatography, the stationary phase displays a negatively charged functional group with counter cation (C^+) that can be displaced by a cationic solute (S^+).

and therefore a different net charge at any given pH. A mobile phase pH can therefore be selected to ensure that the net charge on a protein of interest within a mixture is opposite to that of the matrix functional group, ensuring that it will displace the functional group counterion and bind the matrix (adsorption). Conversely, oppositely charged "contaminant" proteins will not be retained.

Bound protein analytes can be eluted (desorption) in one of two possible ways: (1) pH and (2) ionic strength. Changing the mobile phase pH alters the net charge of the bound protein, and thus its matrix-binding capacity. More commonly, increasing the concentration of a similarly charged species within the mobile phase can compete with and ultimately displace the bound ionic species. During anion-exchange chromatography, for example, negatively charged protein analytes can be competitively displaced by the addition of negatively charged chloride ions (e.g. from sodium chloride). By gradually increasing the salt concentration in the mobile phase, the affinity of interaction between the salt ions and the functional groups will eventually exceed that which exists between the protein charges and the functional groups, resulting in protein displacement and elution.

1.2. Criteria for Selection of IEC Conditions

In the following Subheadings (1.2.1–1.2.4), the broader criteria for selecting IEC conditions is discussed, thereby allowing one to better comprehend and adapt the described protocols (13). The ensuing materials and methods sections (Subheadings 2 and 3) outline the protocols necessary to partially purify a serine peptidase from bovine whole brain soluble fraction. Specifically, we describe the preparation of a crude tissue extract using homogenization, centrifugation, and ammonium sulphate precipitation, followed by partial purification using anion-exchange LC. To allow one to accurately monitor purification efficiency, protocols for assaying total protein and enzyme activity in both pre- and post-IEC fractions are also described. Our focus for these protocols is prolyl oligopeptidase (POP, prolyl endopeptidase, EC3.4.21.26), an 80-kDa serine peptidase with endopeptidase activity towards peptide substrates of ≤30 amino acids. POP is a ubiquitous post-proline cleaving enzyme with particularly high expression levels in the mammalian brain, where it participates in the metabolism of neuroactive peptides and peptide-like hormones (e.g. thyroliberin, gonadotropin-releasing hormone, substance P, Arg-vasopressin). Several studies indicate putative roles for POP in regulation of the central nervous system at both the physiological (memory, learning, mood) and pathological (Alzheimer's, Huntington's and Parkinson's diseases) levels (14, 15).

1.2.1. Selection of Ion-Exchange Matrix

The choice of a suitable ion-exchange matrix is probably the single most important aspect of any ion-exchange protocol and is based on various factors, which include the following: (1) desired ion-exchanger charge/strength, (2) linear flow rate/sample volume, and (3) sample properties. As mentioned above, ion-exchange functional groups fall into two charge categories. Positively charged diethylaminoethyl (DEAE) and quarternary ammonium (Q) functional groups, for example, are routinely employed in anion-exchange chromatography, whilst negatively charged carboxymethyl (CM), sulphomethyl (S), and sulphopropyl (SP) groups are typical cation-exchangers. Both exchanger types can be further categorized as either "strong" or "weak." Strong ion-exchangers are fully ionized over a broad working pH range (i.e. show no loss or gain of charge with varying pH), whilst weak ion-exchangers are only partially ionized over a narrow pH range (i.e. charge can vary significantly with pH). Consequently, with strong ion-exchangers, individual proteins can adsorb to several exchanger sites, often necessitating harsh elution conditions (up to 1 M NaCl) that may compromise sample stability and resolution. Strong ion-exchangers are therefore often used for initial development and optimization of purification protocols (and for binding proteins with p*I* values lying towards the more extreme ends of the pH scale). By contrast, weak ion-exchangers are more flexible in terms of selectivity, and are a more common choice for the separation of proteins that retain their functionality over the pH 6–9 range, as well as for labile proteins that may require mild elution conditions (Table 1 highlights a range of commercially available ion-exchanger resins categorized on the basis of charge and strength).

The size, porosity, and binding capacity of resin particles are also important when selecting an ion-exchanger matrix. These resin properties are normally based on chromatographic column size and dimensions, in conjunction with sample volume and concentration. Commercially available resins range in size from 10 to 400 μm. Larger particles are frequently used in initial protein purification stages that require fast elution rates and high capacity

Table 1
Overview of commercial ion-exchanger properties

Exchange type	Ion-exchange group	Functional group	Buffer counter ions	Commercial resin	Range
Strong anion	Quarternary ammonium(Q)	$CH_2N^+(CH_3)_3$	Cl^-, $HCOO^-$, CH_3COO^-, SO_4^{2-}	Q Sepharose® Capto®Q Dowex®1X2 Amberlite®/ Amberjet® QAE Sephadex®	pH 2–12
Weak anion	Diethylaminoethyl (DEAE)	$O(CH_2)_2N^+H(C_2H_5)_2$	Cl^-, $HCOO^-$, CH_3COO^-, SO_4^{2-}	DEAE-Sepharose® Capto®DEAE DEAE Cellulose	pH 2–9
Strong cation	Sulfopropyl(SP)	$(CH_2)_3SO_3^-$	Na^+, H^+, Li^+	Capto®S SP Sepharose® SP Sephadex® TSKgel SP-5PW	pH 4–13
Weak cation	Carboxymethyl (CM)	OCH_2COO^-	Na^+, H^+, Li^+	CM Cellulose CM Sepharose® CM Sephadex® CM Sepharose® CL6B TSKgel CM-5PW	pH 6–10

but low-to-intermediate resolution, whilst smaller particles are ideal for final purification stages requiring higher resolution. Moreover, commercial resins have binding capacities ranging from less than 2 mg/mL to more than 150 mg/mL.

Selection of the most suitable exchanger functional group for a purification will also be dictated by the target protein biochemical properties such as p*I* and pH stability. For example, if a protein has a low p*I* (<5.0), but is more stable at pH values above this, then an anion-exchanger should be chosen and vice-versa. The purification of POP by IEC is a good example. With a p*I* in the region of 4.8 (16, 17), POP could potentially adsorb to either a strong cation-exchanger or a weak-anion exchanger. However, with a pH optimum from pH 7.4 to 8.0, coupled with rapid

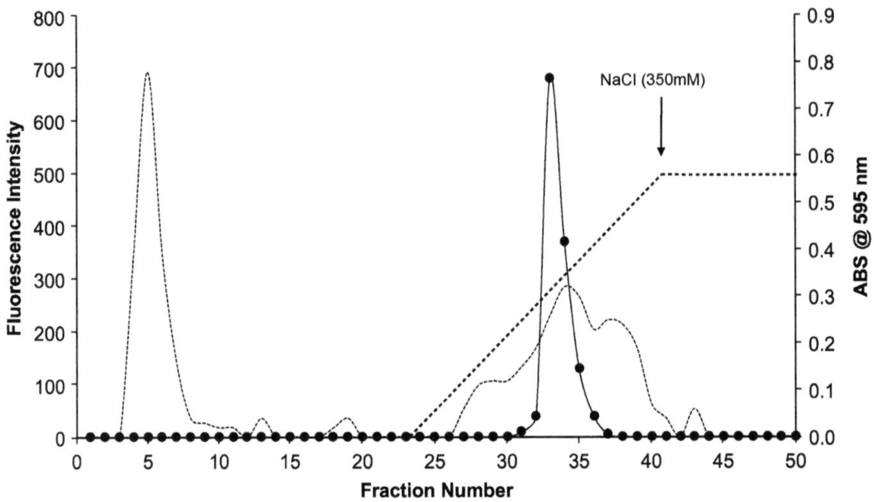

Fig. 2. Partial purification of prolyl oligopeptidase (POP) from bovine whole brain cytosol via DEAE-Sepharose® Fast Flow anion-exchange chromatography. Dialysed post-ammonium sulphate POP fraction (7.4 mL) was applied to a pre-equilibrated 20-mL DEAE-Sepharose® column (50 mM Tris–HCl pH 8.0, 5 mM DTT, 0.5 mM EDTA). Following removal of unbound contaminants, bound POP was eluted with a linear NaCl gradient (*dashed lines*) from 0 to 350 mM. Five-milliliter fractions were collected and assayed for POP activity (*lines with filled circles*) via fluorimetric assay at excitation and emission wavelengths of 370 and 440 nm, respectively (fluroescence intensity values plotted). Protein (*dotted lines*) was monitored by BCA microplate assay at 595 nm. Peak POP elution occurred at 180 mM NaCl. All data courtesy of Dr. Brendan O'Connor (unpublished observations). *DTT* dithiothreitol, *EDTA* ethylene diamine tetraacetic acid, *NaCl* sodium chloride.

Table 2
Partial purification of prolyl oligopeptidase (POP) from bovine whole brain cytosolic fraction

Sample	Total protein (mg)	Total activity (units)	Specific activity (units/mg)	Purification factor	Recovery (%)
Brain cytosol	122.26	1,835,292	15,011	1	100
Ammonium sulphate	24.95	1,578,408	63,263	4.2	86
DEAE anion exchange	10.69	927,209	86,736	5.8	51

Enzyme activity (units) expressed as picomoles of MCA released (i.e. from *Z*-Gly-Pro-MCA) per minute at 37°C. Protein (mg) was determined by BCA microplate assay. All data courtesy of Dr. Brendan O'Connor (unpublished observations)

destabilization of the enzyme below pH 5.0 (Dr. Oonagh Dowling – unpublished observations), the anion-exchange option is favoured. To illustrate this, Fig. 2 and Table 2 demonstrate how anion-exchange chromatography using DEAE-Sepharose® Fast Flow (GE Healthcare) can be used for the initial-stage purification of POP from bovine whole brain soluble tissue extract. Of relevance, the broad pH stability previously reported for the "serum" forms of POP and its *Z*-Pro-Prolinal-insensitive homolog (ZIP) (18, 19) has previously been exploited in a cation-exchange protocol to resolve one isoform from the other (20).

1.2.2. Selection of Buffer Conditions

In order to prevent any variation in matrix and protein net charge, maintenance of a constant mobile phase pH during IEC is essential to avoid pH fluctuations which can occur when both protein and exchanger counterions (particularly if these are H^+ or OH^- ions) are released into the mobile phase. A number of important factors dictate choice of the mobile phase buffer, which include: (1) buffer charge; (2) buffer strength; and (3) buffer pH. The buffering ion should not interact with the ion-exchanger functional groups (i.e. positively charged buffers should be used in anion-exchange, and vice-versa). For example, Tris buffers are often used with DEAE exchangers, whilst phosphate and acetate buffers are frequently used with CM exchangers. Moreover, the minimum buffering strength recommended for ion-exchange is approximately 10 mM within 0.3 pH units of the buffer dissociation constant or pKa (i.e. the pH at which buffering capacity is strongest). Finally, a buffer pH should be selected that permits the protein of interest to remain stable, whilst allowing it to bind reversibly to the matrix. It should also be close enough to the pH at which the protein begins to dissociate from the column to prevent the need to adjust the pH or ionic strength during elution to levels that would destabilize the protein.

1.2.3. Selection of Adsorption and Elution Conditions (pH and Ionic Strength)

Mobile phase pH can be altered to favour either adsorption or elution of proteins. In general, a pH is chosen which will just permit binding of the target protein. This is usually about 1 pH unit above or below the target protein p*I*. A greater difference in pH would lead to stronger protein binding, the need for stronger elution conditions, and decreases in sample resolution and recovery of target protein activity. A pH change can also be used to induce desorption of the target protein (a pH decrease in the case of anion-exchangers and vice-versa). As with pH, mobile phase ionic strength can also be used to control target protein adsorption and elution. As a general rule, the highest ionic strength that will allow adsorption (e.g. 20–50 mM NaCl) and the lowest ionic strength that will allow elution are recommended.

1.2.4. Selection of Elution Format

Two elution formats can be identified as follows: (1) isocratic elution and (2) gradient elution. With isocratic elution, a single buffer is used throughout the entire separation. Sample components (including the target protein) are only loosely adsorbed to the column matrix. Since individual proteins will have different distribution coefficients, separation is achieved by their relative speeds of migration over the column. To achieve optimum resolution of sample components therefore, a small sample volume (1–5% of the bed volume) and a long exchanger column (1:20 diameter:length ratio) are necessary. Although this technique is time consuming and the desired protein invariably elutes in a large volume, no gradient-forming apparatus is required and column regeneration is usually unnecessary. More commonly, conditions are selected which result in the complete adsorption of the desired protein to the column matrix, necessitating an alteration of eluent conditions to achieve its desorption. With gradient elution, continuous or discontinuous (stepwise) variations in the ionic strength and/or pH of the eluent are used to promote target protein desorption. Whilst stepwise gradients are technically simpler, continuous gradients generally give better resolution.

2. Materials

Unless otherwise indicated, all chemicals can be purchased from Sigma-Aldrich.

2.1. Preparation of Bovine Whole Brain Cytosolic Extract

1. Buffer A: 100 mM potassium phosphate pH 7.4, 5 mM DTT (dithiothreitol), and 0.5 mM EDTA (ethylenediamine tetra acetic acid) (see Note 1).
2. Bovine whole brain should be obtained from a freshly slaughtered animal. The brain tissue can be sectioned and frozen at −80°C for long-term storage.
3. Container with crushed ice.
4. Homogenizer (e.g. Sorvall Omni Mixer, standard food blender).
5. Refrigerated centrifuge and rotor (e.g. Beckman J2-MC/JA-20 rotor; 36,000 × *g*).
6. Refrigerated ultracentrifuge (e.g. Beckman L8-M/70Ti rotor; 100,000 × *g*).

2.2. Ammonium Sulphate Precipitation

1. Buffer B: 50 mM Tris–HCl pH 8.0, 5 mM DTT, and 0.5 mM EDTA (see Note 2).
2. Solid $(NH_4)_2SO_4$ (ammonium sulphate).
3. NaOH (sodium hydroxide) at 1 M.
4. Container with crushed ice.
5. Dialysis tubing and sealing clips (Sigma-Aldrich).
6. Glass beakers for both "salting out" and dialysis tube preparation steps.
7. Magnetic stirrer and Bunsen burner.

2.3. POP Partial Purification by IEC

1. Buffer B (see Subheading 2.2).
2. NaCl (sodium chloride) at 350 mM prepared in Buffer B.
3. DEAE-Sepharose® Fast Flow anion-exchange resin (GE Healthcare, Sigma-Aldrich), usually supplied in as pre-swollen beads (see Note 3).
4. Test tubes for the fraction collector.
5. Glass column, low pressure pump (Model EP-1 Econo Pump), fraction collector (Model 2110), silicone tubing and luer-lock fittings (Bio-Rad Laboratories are recommended for all of the aforementioned LC hardware) (see Note 4).
6. Gradient-forming device with 100–200 mL capacity (Sigma-Aldrich) (see Note 5).

2.4. Assay of Post-DEAE Fractions

1. Buffer A (see Subheading 2.1).
2. Bradford and BCA (bicinchoninic acid) protein assay reagents (Bio-Rad Laboratories and Pierce Protein Research Products).
3. BSA protein assay standard at 1 mg/mL (in ultra pure water).
4. Both fluorescent standard (1 mM 7-amino-4-methyl-coumarin, MCA) and substrate stock (10 mM *Z*-Gly-Pro-MCA) can be prepared in 100% dimethyl sulphoxide (DMSO) and stored as frozen aliquots at −20°C (Bachem).
5. Acetic acid at 1.5 M (BDH).
6. Waterbath at 37°C.
7. Fluorescence spectrophotometer (Perkin-Elmer LS50 is recommended).
8. Glass and quartz microcuvettes, the latter for monitoring absorbance at ≤280 nm.

3. Methods

3.1. Preparation of Bovine Whole Brain Cytosolic Extract

All steps to be conducted at 4°C. Latex gloves should also be worn.

1. Extract preparation has been described previously (5). Briefly, homogenize fully a 50 g whole brain slice in 200 mL of ice-cold Buffer A (see Subheading 2.1) (see Note 6).
2. Centrifuge homogenate for 45 min at 36,000 × *g* to yield a supernatant (S1) and pellet (P1).
3. Resuspend P1 in 100 mL of ice-cold distilled water (osmotic shock step to release occluded POP activity) and recentrifuge as above to yield a second supernatant (S2) and pellet (P2). Discard the P2 pellet.

4. Combine S1 and S2 fractions and ultracentrifuge for 45 min at 100,000 × *g* to yield a whole brain supernatant (S3) for storage as 40 mL aliquots at –20°C/–80°C. Discard the pellet (P3).

3.2. Ammonium Sulphate Precipitation

All steps to be conducted at 4°C. Latex gloves should also be worn.

1. Add solid ammonium sulphate to 40 mL of S3 with constant stirring to give 45% (w/v) saturation (10.67 g at 4°C) and adjust to pH 7.4 using 1 M NaOH (see Note 7).
2. Stir for 1 h in a constant gentle manner, and then remove the precipitated ("salted-out") contaminants by refrigerated centrifugation for 45 min at 36,000 × *g*. Following centrifugation, retain the supernatant (S4) and discard the pellet (P4).
3. Add solid ammonium sulphate to S4 with constant stirring to give 75% saturation (9.39 g at 4°C) and adjust to pH 7.4 using 1 M NaOH.
4. Repeat step 2 and retain the pellet (P5). Resuspend P5 in 5 mL of Buffer B to create a "post-ammonium sulphate extract."
5. Dialyse the post-ammonium sulphate extract for 12 h against Buffer B (see Note 8).

3.3. POP Partial Purification by IEC

All steps to be conducted at 4°C. Latex gloves should also be worn. A column flow rate of 1 mL/min and a fraction collection volume of 5 mL should be used throughout.

1. Equilibrate a 20-mL DEAE-Sepharose® Fast Flow column (diameter: 2.5 cm, height: 3.0 cm) with 100 mL of Buffer B.
2. Apply all of the dialysed post-ammonium sulphate extract (from step 5 above) to the column and wash through the unbound contaminants with 100 mL of Buffer B.
3. Elute bound POP using a 100 mL linear NaCl gradient (0–350 mM) prepared in Buffer B (see Note 9).
4. Regenerate the DEAE column with 60 mL of 350 mM NaCl in Buffer B, followed by 100 mL of NaCl-free Buffer B (see Note 10).
5. Assay (as soon as possible) eluted fractions for total protein and POP activity. Following assay (described below), eluted fractions containing the highest levels of POP activity should be pooled for storage (–20°C) and further purification.

3.4. Assay of Post-DEAE Fractions

Determination of total protein in pre- and post-IEC fractions can be done by monitoring fraction absorbance at 280 nm or using widely available Bio-Rad or BCA standard assay protocols based on the methods of Bradford (21) and Smith et al. (22), respectively (see also Chapter 13). Determination of POP activity is based on a modification of the original method of Yoshimoto et al. using *Z*-Gly-Pro-MCA as substrate (23) and is described below:

1. Prepare the substrate stock (10 mM *Z*-Gly-Pro-MCA) in 100% DMSO. To 200 μL of substrate stock, add 600 μL of DMSO, followed by Buffer A to a final volume of 10 mL. This will yield a final substrate concentration of 200 μM in 8% DMSO (see Note 11).
2. Add 400 μL of 200 μM substrate to 100 μL of post-DEAE fraction and incubate at 37°C for 30 min. Fractions should be assayed in triplicate (see Note 12).
3. Terminate assay reactions after 30 min with 1 mL of 1.5 M acetic acid.
4. For blanking purposes, prepare a negative control by addition of 1 mL of 1.5 M acetic acid to a 100 μL aliquot of Fraction 1 "prior" to addition of substrate (see Note 13).
5. Monitor liberated MCA by fluorescence spectrophotometry at excitation and emission wavelengths of 370 and 440 nm, respectively (e.g. Perkin-Elmer LS50). The fluorimeter excitation slit width can be set at 10 nm, whilst the emission slit width can be adjusted (from 2.5 to 20 nm) to obtain fluorimetric readings within range of a relevant MCA standard curve (0–10 or 0–100 μM) (see Note 14).

4. Notes

1. Potassium phosphate buffer can be prepared from the 100 mM "acid" (K_2HPO_4) and "base" (KH_2PO_4) forms of potassium phosphate. The base form can then be adjusted to pH 7.4 using the acid form. Moreover, as DTT loses much of its reducing potency within 12–24 h in solution, it should be prepared (in distilled water) as a concentrated stock (100×) and stored (−20°C), only to be thawed and added to buffers immediately prior to use.
2. Tris–HCl buffer can be prepared from 100 mM trizma base and adjusted down to the desired pH with concentrated HCl (hydrochloric acid). Moreover, as Tris buffers are temperature sensitive, they should be adjusted to pH 7.8 when being prepared at room temperature (pH will rise to 8.0 when the buffer is equilibrated to 4°C).
3. Various materials ranging from silica and complex polysaccharides (e.g. dextran, agarose, and cellulose) to more complex organic polymers (e.g. polyacrylate, polyvinyl, polyether, and polystyrene-divinyl benzene) have been used in the manufacture

of ion-exchange resins. Resins are usually porous beads (although fibrous, microgranular, and composite matrices are also available), supplied either as dried preparations or in a pre-swollen state to be used in LC applications ranging from bio-analytical monitoring and research to process-scale protein separations. Vendors also provide ion-exchange resins in pre-packed IEC columns and microplates for use with standard LC setups, HPLC systems, and high-throughput applications.

4. Column pouring should be performed at 4°C. Prior to pouring, the column exit valve and tubing should be purged of air with distilled water. A pre-measured volume of the suspended pre-swollen ion-exchange resin can then be poured into the column in "one" pour (this is essential if one is to avoid "gaps" in the resin bed, which can reduce column performance and resolution). The buffer volume or "headspace" above the resin bed should be kept to a minimum (10–20% of bed volume) in order to ensure accurate delivery of a chosen elution gradient. Moreover, if there is airspace above the buffer level covering the top of the resin bed (dependent on column dimensions and bed volume), a small piece of parafilm (~1 cm^2) can be placed floating on top of the buffer over the resin. This will act as a "shock absorber" to prevent fluid turbulence (which can cause disruption of the resin bed surface as the buffer is pumped down through the column). Once poured, the column should be washed with several volumes of distilled water to remove preservative.
5. When setting up the gradient maker, ensure that the narrow fluid channel connecting the two gradient compartments is properly purged of air. This can be achieved using a long needle syringe to draw fluid through the channel, or by briefly exerting downward pressure on one of the gradient buffers to force fluid through the channel. Moreover, only gentle magnetic stirring should be used during gradient elution to prevent bubble formation and possible blockage of the gradient fluid channel.
6. Tissue homogenization should be conducted using short, repeated bursts of the homogenizer/blender (i.e. 2–3 s). This will minimize "shearing" and "foaming," both of which can reduce recovery levels of active enzyme.
7. The "salting out" procedure should be conducted on ice in a small glass beaker placed at the bottom of a small polystyrene dry-ice container (the container can subsequently be placed directly onto a magnetic stirrer).
8. Prior to ion-exchange, samples must be dialysed into the starting buffer (in this case, Buffer B). At least 200 volumes

of dialysis buffer are recommended (i.e. relative to the sample volume), with a buffer change after 3–4 h. Dialysis tubing can be prepared by placing a relevant length into a beaker of boiling water containing a large spatula of EDTA (disodium form). This will soften the tubing and remove heavy metal ions. After 5 min in boiling water, the tubing should be rinsed thoroughly in cold distilled water.

9. Linear ionic strength gradients are very reproducible and can be prepared by mixing two buffers of differing ionic strengths in linear volume ratios. Linear pH gradients cannot be prepared by mixing two buffers of different pH values in linear volume ratios due to differences in the buffering capacities of the two buffers being mixed. In addition, the mixed buffer then has to titrate the buffering action of the ion-exchanger and the adsorbed proteins. Consequently, pH gradients are less frequently employed.

10. Avoid leaving high salt buffers on the column for extended periods following regeneration as this may lead to salt crystallization (which necessitates column repouring). Moreover, for medium-to-long-term storage, the column should be regenerated in the normal fashion, washed in several volumes of distilled water, followed by several volumes of either 0.02% (v/v) sodium azide or 20% (v/v) ethanol.

11. When diluting the 10-mM *Z*-Gly-Pro-MCA stock, Buffer A should be pre-warmed and added very slowly to the final volume of 10 mL. Moreover, the DMSO should be added in 100 μL increments in parallel with the buffer to prevent the substrate precipitating out of solution. The 200 μM substrate should be prepared from stock only as required.

12. Prior to assay, both fraction triplicates and substrate should be pre-equilibrated to 37°C. At $t=0$, the substrate should be added to sequentially numbered fractions at exactly 15-s intervals. At $t=30$, the acetic acid should be added to fractions at exactly 15-s intervals. In this way, all fractions receive the exact 30 min assay time.

13. When assaying any sample for POP activity, the appropriate negative control should always be; sample (crude or purified), followed sequentially (at $t=30$ min) by stopping agent (1.5 M acetic acid) + substrate (200 μM *Z*-Gly-Pro-MCA).

14. The MCA standard curve must be prepared under identical assay conditions and read at the same fluorimeter settings as those used for the assay in order to properly quantitate MCA release for use in the calculation of enzyme activity expression. Units of POP activity are defined as picomoles of MCA released per minute at 37°C.

References

1. Fritz, J. S. (2004) Early milestones in the development of ion-exchange chromatography: a personal account. *J Chromatogr A* **1039,** 3–12.
2. Lucy, C. A. (2003) Evolution of ion-exchange: from Moses to the Manhattan Project to modern times. *J Chromatogr A* **1000,** 711–24.
3. Kent, U. M. (1999) Purification of antibodies using ion-exchange chromatography. *Methods Mol Biol* **115,** 19–22.
4. Levison, P. R. (2003) Large-scale ion-exchange column chromatography of proteins: comparison of different formats. *J Chromatogr B Analyt Technol Biomed Life Sci* **790,** 17–33.
5. Cummins, P. M. and O'Connor, B. (1996) Bovine brain pyroglutamyl aminopeptidase (type-I): purification and characterization of a neuropeptide-inactivating peptidase. *Int J Biochem Cell Biol* **28,** 883–93.
6. Wu, B., Wu, L., Chen, D., Yang, Z., and Luo, M. (2009) Purification and characterization of a novel fibrinolytic protease from *Fusarium* sp. CPCC 480097. *J Ind Microbiol Biotechnol* **36,** 451–59.
7. Knudsen, H. L., Fahrner, R. L., Xu, Y, Norling, L. A., and Blank, G. S. (2001) Membrane ion-exchange chromatography for process-scale antibody purification. *J Chromatogr A* **907,** 145–54.
8. Mant, C. T. and Hodges, R. S. (2008) Mixed-mode hydrophilic interaction/cation-exchange chromatography: separation of complex mixtures of peptides of varying charge and hydrophobicity. *J Sep Sci* **31,** 1573–84.
9. Fekkes, D., Voskuilen-Kooyman, A., Jankie, R., and Huijmans, J. (2000) Precise analysis of primary amino acids in urine by an automated high-performance liquid chromatography method: comparison with ion-exchange chromatography. *J Chromatogr B Biomed Sci Appl* **744,** 183–88.
10. Yang, Y., Hebron, H. R., and Hang, J. (2008) High performance DNA purification using a novel ion-exchange matrix. *J Biomol Tech* **19,** 205–10.
11. Bonn, G. (1987) High-performance liquid chromatographic isolation of ^{14}C-labelled gluco-oligosaccharides, monosaccharides and sugar degradation products on ion-exchange resins. *J Chromatogr* **387,** 393–98.
12. Hajós, P. and Nagy, L. (1998) Retention behaviours and separation of carboxylic acids by ion-exchange chromatography. *J Chromatogr B Biomed Sci Appl* **717,** 27–38.
13. Ahmed, H. (ed.) (2004) Principals and reactions of protein extraction, purification and characterization. CRC Press, Boca Raton, FL.
14. Männisto, P. T., Venäläinen, J., Jalkanen, A., and García-Horsman, J. A. (2007) Prolyl oligopeptidase: a potential target for the treatment of cognitive disorders. *Drug News Perspect* **20,** 293–305.
15. Polgár, L. (2002) The prolyl oligopeptidase family. *Cell Mol Life Sci* **59,** 349–62.
16. Kalwant, S. and Porter, A. G. (1991) Purification and characterization of human brain prolyl endopeptidase. *Biochem J* **276,** 237–44.
17. Sharma, K. K. and Ortwerth, B. J. (1994) Purification and characterization of prolyl oligopeptidase from bovine lens. *Exp Eye Res* **59,** 107–15.
18. Cunningham, D. F. and O'Connor, B. (1998) A study of prolyl endopeptidase in bovine serum and its relevance to the tissue enzyme. *Int J Biochem Cell Biol* **30,** 99–114.
19. Birney, Y. A. and O'Connor, B. F. (2001) Purification and characterization of a *Z*-pro-prolinal-insensitive *Z*-Gly-Pro-7-amino-4-methyl coumarin-hydrolyzing peptidase from bovine serum: a new proline-specific peptidase. *Protein Expr Purif* **22,** 286–98.
20. Cunningham, D. F. and O'Connor, B. (1996) Identification and initial characterization of a *N*-benyloxycarbonyl-prolyl-prolinal (*Z*-Pro-Prolinal)-insensitive 7-(*N*-benzyloxycarbonyl-glycyl-prolyl-amido)-4-methylcoumarin (*Z*-Gly-Pro-NH-Mec)-hydrolyzing peptidase in bovine serum. *Eur J Biochem* **244,** 900–3.
21. Bradford, M. (1976) A rapid and sensitive method for the quantitation of microgram quantities of protein utilizing the principle of protein dye-binding. *Anal Biochem* **72,** 248–54.
22. Smith, P. K., Krohn, R. I., Hermanson, G. T., Mallia, A. K., Gartner, F. H., Provenzano, M. D., et al. (1985) Measurement of protein using bicinchoninic acid. *Anal Biochem* **150,** 76–85.
23. Yoshimoto, T., Ogita, K., Walter, R., Koida, M., and Tsuru, D. (1979) Post-proline cleaving enzyme: synthesis of a new fluorogenic substrate and distribution of the endopeptidase in rat tissues and body fluids of man. *Biochim Biophys Acta* **569,** 184–92.

Chapter 13

Protein Quantitation and Analysis of Purity

Eva M. Campion, Sinéad T. Loughran, and Dermot Walls

Abstract

The accurate quantitation of proteins and an analysis of their purity are essential in numerous areas of scientific research, and are a critical factor in many clinical applications. The large and varied number of techniques employed for this purpose is therefore not surprising. The selection of a suitable assay is dependent on factors such as the level of sensitivity required, the presence of interfering agents, and the composition of the protein itself. Here, protocols for the most commonly used protein determination methodologies are outlined, as well as for the more recently adapted technique of quantitative immuno-polymerase chain reaction.

Key words: Protein, Lowry, Bradford, BCA, Quantitative immuno-PCR (qIPCR)

1. Introduction

The ability to easily and accurately quantitate total protein content in a given sample is a fundamental requirement of many biological studies. Indeed, the routine measurement of total protein content is a well-established essential step in many areas of basic biochemical research and routine clinical practise (1).

Numerous and varied methods to assay total protein content have been described in the literature. The most commonly utilized methods rely on (a) the intrinsic ability of protein molecules to absorb ultraviolet (UV) light (UV absorption) (2), (b) the use of protein-binding dyes (Bradford Assay, Silver staining) (3, 4), and (c) the reduction of copper in the presence of a chromogenic reagent [Lowry and Bicinchoninic acid (BCA) assays] (5, 6). Since each of these methods has its own strengths and weaknesses (see Note 1) (7), none of these assays should be employed without first considering its suitability for the application in question (see Table 1). There is in fact no absolute method that produces

Dermot Walls and Sinéad T. Loughran (eds.), *Protein Chromatography: Methods and Protocols*, Methods in Molecular Biology, vol. 681, DOI 10.1007/978-1-60761-913-0_13, © Springer Science+Business Media, LLC 2011

Table 1
A comparison of UV absorption, Bradford, Lowry, and BCA protein assay methods

Method name	Advantages	Disadvantages
UV absorption	• Simple and fast • Inexpensive • Sample is recoverable	• Many buffer components absorb strongly in this region • The presence of nucleic acid can greatly influence the absorption • Least sensitive method
Bradford assay	• Simple and fast • Inexpensive • Very sensitive • Compatible with a wide range of buffers	• Non-linear standard curve over wide ranges • Response to different proteins can vary widely: choice of standard is very important
Lowry assay	• Sensitive • Commonly referenced procedure • Easily adapted to microplate format	• Time-consuming • Susceptible to many interfering compounds • Variation in the content of tyrosine and tryptophan residues will influence the assay
BCA assay	• Very sensitive • Rapid • Compatible with a wide range of buffers • Little variation in response between different proteins	• The reaction does not go to completion when performed at room temperature or 37°C (difficult when prepping large numbers of samples)

accurate results in every instance and often it is necessary to employ more than one type of protein assay (8). In addition to total protein concentration, the specific activity of a particular target protein in a sample is of significance when proteins are being purified or when different protein samples are being compared (4). In broad terms, measurement of the level of a specific protein of interest may be undertaken by one of two methods, a specific biological assay (or bioassay) or an immunoassay. The specificity of an immunoassay relies on the interaction between the protein and an antibody, and determining the quantity of bound antibody in an immunoassay is routinely achieved by virtue of either using a labelled primary antibody or detecting the latter using a secondary antibody that is itself labelled (9). There are many immunoassay formats possible, and labels that are used include enzymes [e.g. enzyme linked immunosorbent assay (ELISA)], radioactive labels [radioimmunoassay (RIA)], magnetic labels [e.g. magnetic immunoassay (MIA)], fluorescent tags [as used in flow cytometry/fluorescence-activated cell sorting (FACS) analysis and fluorescence microscopy], or a piece of DNA [as in real-time quantitative immuno-polymerase chain reaction (qIPCR)].

A pure protein is one that is free from any quantifiable amounts of impurities (10). Any purity determination is only as reliable as the analytical methods used, and factors such as the structural properties of the protein itself, the amount of protein available, the nature of potential contaminants in the sample, and the accuracy of the estimate required should always be considered when selecting the method of analysis (11). In reality, it may only be necessary to ensure the sample is free of contaminating products that may affect the application in question, and thus aspects of the process such as the intended use (e.g. bulk enzyme preparations, protein crystallography, primary sequence analysis, or therapeutic applications), the source of the protein (animal tissue, human serum, recombinant micro-organisms, or hybridomas), and the purification processes employed should all accordingly dictate the extent of analysis required (4). Here, general protocols for the following methods of protein estimation and purity are described: ultraviolet (UV) protein absorption assays, the Bradford and Lowry assays, Macro- and Micro-BCA assays, ELISA, sodium dodecyl sulphate polyacrylamide gel electrophoresis (SDS-PAGE) and associated staining methods, Western immunoblotting, and several formats for qIPCR.

2. Materials

2.1. Ultraviolet (UV) Protein Absorption Assay Using a Spectrophotometer

1. Cuvettes (Quartz for wavelengths <215 nm) (Sarstedt).
2. Buffer solution in which the protein sample is dissolved (for blanking, see Note 2).

2.2. Preparation of a Standard Curve

1. Stock solution: prepare a stock solution of a standard protein, e.g. bovine serum albumin (BSA), lysozyme, albumin, or γ-globulin (see Note 3) at a suitable concentration (1–5 mg/mL see Note 4), and in the same buffer as the protein of unknown concentration.
2. Buffer (for blanking).

2.3. UV Protein Absorption Assay Using the Nanodrop® ND-1000

1. Distilled H_2O (dH_2O).
2. Lab-wipes (Thermo-Scientific).
3. Buffer for blanking (see Note 5).
4. Nanodrop® ND-1000 Spectrophotometer (NanoDrop Technologies).

2.4. Bradford Assay

1. Protein standard solutions (e.g. 1 mg/mL BSA): dilute in the range 20–100 μg/mL in a total volume of 100 μL. Aliquot and store at −20°C.

2. Buffer (for blanking, see Note 6).
3. 1 M sodium hydroxide (NaOH) (see Note 7).
4. Bradford reagent (available from Sigma-Aldrich), or prepare oneself by adding 100 mg Coomassie brilliant blue G-250 (Thermo-Scientific) to 50 mL 95% (v/v) ethanol. When Coomassie brilliant blue G-250 has dissolved add 100 mL 85% (w/v) phosphoric acid and stir overnight. Dilute to 1 L with dH_2O. Store for up to 3 months at 4°C.
5. Cuvettes (see Note 8) (Sarstedt).

2.5. Lowry Protein Assay

1. Lowry solution A: 2% (w/v) sodium carbonate (Na_2CO_3) in 0.1 M sodium hydroxide (NaOH). Store for up to 3 months at room temperature.
2. Lowry solution B: 1% (w/v) copper sulphate pentahydrate ($CuSO_4 \cdot 5H_2O$, Sigma-Aldrich) in dH_2O. Store for up to 1 year at room temperature.
3. Lowry solution C: 2% (w/v) sodium potassium tartrate ($NaKC_4H_4O_6 \cdot 4H_2O$) in dH_2O. Store for up to 3 months at room temperature.
4. Lowry working solution: prepare immediately before use; Lowry solution A: Lowry solution B: Lowry solution C in the ratio 100:1:1 (v:v:v), respectively.
5. Folin-Ciocalteu reagent: available as 2 N reagent (Sigma-Aldrich). Dilute 1:1 in dH_2O. This solution is light sensitive and should be prepared just prior to use and kept in a light-protected container.
6. Protein standards: prepare a dilution series of standard protein, e.g. albumin in the range 0–100 μg/mL in a total volume of 1 mL.
7. Buffer for blanking (see Note 9).

2.6. Bicinchoninic Acid Assay

2.6.1. Macro-BCA Assay

1. Macro-BCA reagent A: dissolve 1 g sodium bicinchoninate (BCA, Pierce), 2 g sodium carbonate (Sigma-Aldrich), 0.16 g sodium tartrate (Sigma-Aldrich), 0.4 g NaOH, and 0.95 g sodium bicarbonate in dH_2O. Adjust the pH to 11.25 with 10 M NaOH and bring to 100 mL with dH_2O. Stable for 1 year at room temperature.
2. Macro-BCA reagent B: dissolve 0.4 g $CuSO_4 \cdot 5H_2O$ in 10 mL distilled water. Stable for 1 year at room temperature.
3. Macro-BCA working solution: mix 50 volumes of Macro-BCA reagent A with 1 volume of Macro-BCA reagent B (prepare fresh before use). The working solution should be green in colour.
4. Glass or disposable polystyrene cuvettes (Sarstedt).

5. Buffer for blanking (see Note 10).

2.6.2. Micro-BCA Assay

The Micro-BCA assay uses three reagents whose concentrations are significantly higher than the two reagents used in the Macro-BCA format. Here, BCA is prepared as a separate reagent in order to avoid its precipitation.

1. Micro-BCA assay reagent A: dissolve 8 g sodium carbonate monohydrate, 1.6 g sodium tartrate in dH_2O, adjust the pH to 11.25 with 10 M NaOH and bring to 100 mL with dH_2O. Stable for 1 year at room temperature.
2. Micro-BCA assay reagent B: dissolve 4 g BCA in 100 mL dH_2O. Stable for 1 year at room temperature.
3. Micro-BCA assay reagent C: dissolve 0.4 g $CuSO_4{\cdot}5H_2O$ in 10 mL dH_2O. Stable for 1 year at room temperature.
4. Micro-BCA assay solution (prepare fresh): 25:25:1 (v/v/v) Micro-BCA assay reagent A/Micro-BCA assay reagent B/Micro-BCA assay reagent C.
5. Buffer for blanking (see Note 10).

2.7. Indirect ELISA

1. Microtitre plates (Biosciences).
2. Antigen-coating buffer: 0.1 M Na_2CO_3/sodium bicarbonate ($NaHCO_3$) pH 9.6 (adjust the pH if necessary using HCl).
3. Phosphate-buffered saline (PBS).
4. Assay buffer: 20 mM PBS pH 7.4 (adjust pH to 7.4 by adding dilute HCl or NaOH if necessary), 0.05% (v/v) Tween 20.
5. Blocking buffer: 20 mM PBS, 5% (w/v) non-fat dried milk (or other suitable blocking agent see Note 11).
6. Primary antibody (diluted to the optimal concentration in blocking buffer immediately before use. This concentration should be experimentally determined, for most applications dilution in the range 1:200 to 1:1,000 is sufficient, consult manufacturer's guidelines).
7. Conjugated secondary antibodies directed towards the primary antibody (diluted to the optimal concentration in blocking buffer immediately before use).
8. Substrate, e.g. for peroxidase system: add 100 μL of 0.8 mg/mL *o*-phenylenediamine dihydrochloride (OPD; Sigma-Aldrich) dissolved in 0.1 M phosphate-citrate buffer pH 5.0, containing 0.4 mg/mL urea hydrogen peroxide (UHP; Sigma-Aldrich) and incubate at room temperature for 30 min.
9. Stop solution: alkaline phosphatase –3 M NaOH; peroxidase –3 M HCl or 3 M H_2SO_4.

2.8. SDS-PAGE

1. Leupeptin (Sigma-Aldrich): dissolve 2 mg/mL leupeptin in dH_2O and store at −20°C.
2. Aprotinin (Sigma-Aldrich): make 0.1 M stock solution of aprotinin in dH_2O and store at −20°C.
3. Phenylmethanesulphonylfluoride (PMSF) (Sigma-Aldrich): make 100 mg/mL PMSF in isopropanol and store at −20°C in the dark.
4. Suspension buffer: 0.1 M NaCl, 0.01 M Tris–HCl (pH 7.6), 0.001 M NaOH-EDTA (pH 8.0), 1 μg/mL leupeptin, 1 μg/mL aprotinin, 100 μg/mL PMSF. Store at 4°C.
5. 2× SDS loading buffer (sample buffer): 100 mM Tris–HCl (pH 7.6), 4% (w/v) SDS, 20% (v/v) glycerol, 10% (v/v) 2-mercaptoethanol, 0.2% (w/v) bromophenol blue. Store at room temperature.
6. Acrylagel (National Diagnostics). Acrylagel is toxic, and a known carcinogen. Consult the corresponding material safety data sheet (MSDS) before use.
7. Bis-acrylagel (National Diagnostics). Bis-acrylagel is an irritant. Consult the corresponding MSDS before use.
8. 1.5 M Tris–HCl (pH 8.8).
9. 1 M Tris–HCl (pH 6.8).
10. dH_2O.
11. 10% (w/v) sodium dodecyl sulphate (SDS) in dH_2O.
12. 10% (w/v) ammonium persulphate (APS) dH_2O (prepare fresh) (Sigma-Aldrich). APS is a strong and harmful oxidizing agent. Consult the corresponding MSDS before use.
13. TEMED (*N*,*N*,*N*′,*N*′-tetramethylethylenediamine).
14. 5× Tris-glycine running buffer (electrode buffer): 15.1 g Tris base, 95.4 g glycine, 50 mL 10% (w/v) SDS. Make up to 1 L with dH_2O and store at room temperature.
15. 1× Tris-glycine running buffer: 200 mL 5× Tris-glycine running buffer, 800 mL dH_2O. Store at room temperature.
16. Molecular weight marker (pre-stained) (GE Healthcare).
17. ATTO protein gel electrophoresis system.

2.9. Coomassie Brilliant Blue Staining of SDS-PAGE Gels

1. Coomassie brilliant blue G-250 solution (Sigma-Aldrich).
2. Destain: 450 mL methanol, 450 mL dH_2O, 100 mL glacial acetic acid. Store at room temperature.

2.10. Silver Staining of SDS-PAGE Gels

1. Fixing solution: methanol, acetic acid, and formalin (40:10:0.05, by volume).
2. Wash solution: ethanol, acetic acid, and water (10:5:85, by volume).

3. Silver nitrate solution (Thermo-Scientific): 0.2% (w/v) $AgNO_3$, 0.076% formalin (prepare fresh). Silver Nitrate is harmful. Consult the corresponding MSDS before use.
4. Oxidizing solution: 3.4 mM potassium dichromate and 3.2 mM nitric acid.
5. Developing solution: 0.28 mM sodium carbonate and 1.9% (v/v) formaldehyde. Formaldehyde is toxic, consult the corresponding MSDS prior to use.
6. Stop solution: methanol, acetic acid (50:12).
7. 1% (v/v) acetic acid.

2.11. Western Blotting

1. Bio-Rad Trans-Blot® SD semi-dry electrophoretic transfer cell.
2. Transfer buffer: 750 mL dH_2O, 2.9 g glycine, 5.8 g Tris base, 3.7 mL 10% (w/v) SDS, 200 mL methanol. Adjust volume to 1 L with dH_2O and store at 4°C (storing at this temperature is critical).
3. 1× Tris-buffered saline (TBS): 6.1 g Tris base, 8.8 g NaCl, 800 mL dH_2O. Adjust the pH to 7.5 with HCl and adjust the volume to 1 L with dH_2O. Store at room temperature.
4. TBS-T: 1 L 1× TBS, 1 mL Tween 20 (Sigma-Aldrich). Store at room temperature.
5. Blocking buffer: 5 g non-fat dry milk powder (or other appropriate blocking agent see Note 11), 100 mL TBS-T. Store at 4°C.
6. Nitrocellulose blotting membrane (Labkem).
7. 3MM filter paper (Whatman).
8. Scalpel blade (Swann-Morton).
9. Ponceau S (Sigma-Aldrich).
10. dH_2O.
11. Primary antibody (diluted to optimal working concentration in blocking buffer immediately before use. This concentration should be experimentally determined, for most applications dilution in the range 1:200 to 1:1,000 is sufficient, consult manufacturers' guidelines).
12. Conjugated secondary antibody directed towards the primary antibody (diluted to an optimal working concentration in blocking buffer immediately before use).
13. Substrate (e.g. 5-bromo-4-chloro-3-indolyl phosphate/Nitro Blue Tetrazolium, BCIP/NBT, Sigma-Aldrich) or 3,3′,5,5′-tetramethylbenzidine (TMB, Sigma-Aldrich).

2.12. qIPCR

2.12.1. Preparation of Biotinylated DNA-Label

1. Template DNA (e.g. linearized pUC19 plasmid DNA).
2. Template-specific oligonucleotide primers: 5′-biotinylated forward primer and unmodified reverse primer.
3. PCR purification kit, e.g. QIAquick PCR purification kit (Qiagen).
4. Agarose powder (Sigma-Aldrich).
5. 50× Tris-acetate/EDTA electrophoresis buffer (TAE): 242 g Tris base, 57.1 mL glacial acetic acid, 100 mL 0.5 M NaOH-EDTA (pH 8.0). Adjust to 1 L with dH_2O and store at room temperature.
6. 1× TAE buffer: 20 mL 50× TAE, 980 mL dH_2O. Store at room temperature.
7. Loading dye: 40% (w/v) sucrose, 0.25% (w/v) bromophenol blue. Store at room temperature.
8. SYBR Safe DNA gel stain (Invitrogen).
9. Horizontal agarose gel electrophoresis system (Sigma-Aldrich).
10. DNA size markers (e.g. 100 bp ladder from Invitrogen).
11. PCR mixture: 5 μL 10× PCR buffer, 1 μL of each dNTP (200 μM), 1 μL of each primer (100 μmol/L), 1.25 U *Taq* polymerase, 50 ng template DNA, bring to 50 μL with molecular grade H_2O.
12. Prepare a 1.5% (w/v) agarose gel: dissolve the agarose in 1× TAE by boiling, with intermittent mixing until completely dissolved. Add SYBR Safe DNA gel stain according to the manufacturer's instructions. Cast the gel according to the instructions accompanying the apparatus being used.
13. G-50 sephadex column (Roche).
14. Cuvettes (Sarstedt).

2.12.2. Preparation of Antibody/DNA-Label Conjugate

1. Detection antibody [in PBS, 1 mM EDTA, pH 8.0 (adjust the pH of buffer by adding dilute HCl/NaOH if necessary)].
2. 2-Iminothiolane (Traut's reagent, Sigma-Aldrich).
3. 1 M glycine-NaOH pH 7.3.
4. 5′ Amino-modified DNA (in PBS, 1 mM EDTA, pH 7.2, adjust the pH of the buffer by adding dilute HCl or NaOH if necessary).
5. Sulpho-succinimidyl4-[*N*-maleimidomethyl]cyclohexane-1- carboxylate (SMCC).
6. Zeba 2 mL desalt spin column (Thermo-Scientific).
7. Ion exchange column (Resource Q 1 mL, Biosciences).
8. 10 mM Tris–HCl pH 8.0.
9. 1.5 M NaCl.
10. Centricon YM-100 (Millipore).

2.12.3. Biotinylation of Detection Antibody

1. Detection antibody (in amine-free buffer, pH 7.2–8.0).
2. Biotinamido-caproate-*N*-hydroxysuccinimide ester (BNHS) 1 mg/mL: dissolve 1 mg of BNHS in 1 mL dimethyl sulphoxide (DMSO) immediately before use.
3. 1 M $NaHCO_3$ pH 8.5 (adjust the pH if necessary using NaOH or H_2SO_4).
4. PD-10 gel filtration column (Sephadex G-25 M), GE Healthcare *or* dialysis tubing (Sigma-Aldrich).
5. PBS pH 7.4 (adjust pH to 7.4 by adding dilute HCl or NaOH if necessary).

2.12.4. qIPCR Assay (Common Components)

1. Wash buffer: 0.154 M NaCl, 5 mM Tris–HCl pH 7.75, 0.02% (w/v) sodium azide (NaN_3) (see Note 12).
2. Blocking buffer: PBS, 0.1–1% (w/v) BSA (or other suitable blocking agent, see Note 13) and 0.05% (v/v) Tween 20.
3. Antigen/unknown samples: dilute the antigen standard, preparing a suitable concentration range (see Note 14).
4. Ultrapure water (see Note 15).

2.12.5. Additional Components Assay Format I

1. Capture antibody (diluted to optimal working concentration in 0.2 M NaH_2PO_4; see Note 16).
2. Polypropylene PCR plate (see Note 17).
3. Biotinylated detection antibody (diluted to optimal working concentration in blocking buffer).
4. Streptavidin (5 nM in blocking buffer).
5. Biotinylated DNA-label (0.7 pM in blocking buffer) (see Subheading 3.12.1).

2.12.6. Additional Components Assay Format II

1. Capture antibody (diluted at optimal concentration in 0.2 M NaH_2PO_4).
2. Polypropylene PCR plate.
3. DNA-conjugated detection antibody (see Subheading 3.12.2) (diluted to optimal working concentration).

2.12.7. Additional Components Assay Format III

1. Streptavidin-coated microtitre plates (Roche).
2. DNA-conjugated detection antibody (see Subheading 3.12.2) (diluted to optimal working concentration).
3. Biotinylated capture antibody (diluted to optimal working concentration).

2.12.8. Additional Components Assay Format IV

1. Polypropylene PCR plate.

2. DNA-conjugated detection antibody (see Subheading 3.12.2) (diluted at optimal concentration).
3. Antigen-coating buffer: 0.1 M Na_2CO_3-$NaHCO_3$ pH 9.6 (adjust the pH if necessary using HCl).

2.12.9. Real-Time PCR (qPCR) Assay

Real-Time PCR mixture (assemble on ice):

1. 2 μL 10× PCR buffer (4 mM $MgCl_2$) (Promega).
2. 200 μM dNTPs (dUTP, dGTP, dCTP, dATP) (Invitrogen).
3. Hot start *Taq* polymerase (Promega).
4. 200–400 nM primers [specific for DNA-label (see Subheading 3.12.1)] (Eurofins MWG operon).
5. 0.5 μL 50× SYBR green solution (vortex for 30 s before use) (Quantace).
6. Uracil-*N*-glycosylase (UNG) (Bioline) (see Note 18).
7. Bring to 20 μL with molecular grade H_2O (Sigma-Aldrich).
8. Filter/barrier tips (Anachem).

3. Methods

3.1. Protein Determination by Ultraviolet (UV) Light Absorption

Simple and rapid estimations of protein concentration can be made by monitoring the absorbance of ultraviolet light. Absorbance of near-UV light at 280 nm depends largely on the presence of aromatic amino acids, in particular tryptophan and tyrosine, and to a much more minor extent on phenylalanine and disulphide bonds. Absorption is affected by pH and ionic strength. In addition, strong interference from nucleic acids is a particular problem at this wavelength. As a method for determining protein concentration, UV absorption gives no more than a quick and rough estimate unless the protein preparation is pure and its extinction coefficient is known. The sample is recoverable, however, and the user should proceed to one of the other assays below when a more accurate determination of protein concentration is required.

1. Switch on the UV spectrophotometer and set the wavelength at 280 nm. Leave the instrument to stabilize for 15–20 min.
2. Calibrate the instrument to zero absorbance using a water blank. Ensure to use suitable cuvettes (quartz or other cuvettes known to be transparent at the given wavelength).
3. Measure the A_{280} of the buffer used to prepare the sample. The cuvette should be filled with a sufficient volume to cover the aperture through which the light beam passes (do not allow any bubbles to inhibit the path of the light).

Table 2
Preparation of protein standards for standard curve generation

Final concentration (e.g. BSA μg/mL)	Volume of stock (μL) (BSA 0.1 mg/mL)	Volume of buffer (μL)
0	0	100
10	10	90
20	20	80
30	30	70
40	40	60
50	50	50

4. Measure the absorbance of the protein sample by replacing the buffer blank with a cuvette containing the protein sample.
5. If the value of A_{280} exceeds 2, dilute the sample using buffer and read the absorbance again.

3.2. Preparation of a Standard Curve

1. Prepare a stock solution of a protein standard (e.g. BSA 0.1 mg/mL) and set up dilutions for the preparation of a standard curve (see Table 2). The concentration of protein in the samples must fall within the linear range of the standard curve.
2. Pipette duplicate aliquots of protein standards into microfuge tubes and dilute appropriately (see Note 19) with a suitable buffer, e.g. 0.15 M NaCl.
3. Mix well by inversion or using a vortex.
4. Carry out the assay as per Subheading 3.1.
5. Prepare a standard curve by plotting absorbance versus protein concentration (see Note 20).
6. Use the equation of the line to calculate the protein concentrations of the unknown samples.

3.3. Ultraviolet (UV) Protein Absorption Assay Using a NanoDrop® ND-1000 Apparatus

NanoDrop® Spectrophotometers were developed for microvolume quantitation and analysis. These instruments employ fibre optic technology and use the inherent surface tension of the liquids being analyzed to create a column between the ends of the optical fibres (the sensors). In this way, the measurement optical path is formed. Very small protein sample volumes (1–2 μL) can therefore be rapidly analyzed spectrophotometrically without the need for cuvettes or capillaries. The procedure for use with the Nanodrop® ND-1000 instrument is outlined below.

1. Clean the upper and lower sensor surfaces with dH_2O water and a lab-wipe to eliminate any dried/leftover sample that might be present (cleaning in this manner is sufficient to prevent any sample carryover).
2. Open the NanoDrop® software program and select the appropriate component (see Note 21).
3. Pipette 1.2–2 μL of dH_2O onto the sensor. Carefully bring down the arm.
4. Follow the onscreen prompts to initialize/calibrate the instrument.
5. Clean the sensors as before and pipette 1.2–2 μL of the buffer (for blanking) onto the sample surface. Lower the arm.
6. Follow the onscreen prompts to blank the instrument.
7. Clean the sample surfaces and pipette 1.2–2 μL of the sample onto the sensor and lower the arm.
8. Click "measure" and record the concentration of the sample, now shown onscreen.
9. In order to analyze multiple samples, clean the sensor between measurements. (recalibration or reblanking is not necessary).
10. Once finished, clean the sensors and switch off the instrument.

3.4. The Bradford Protein Assay

The Bradford Assay is based on the formation of a complex between Brilliant Blue G-250 and the proteins in a solution. The absorption maximum of the dye shifts from 465 to 595 nm when it complexes with protein and the amount of absorption observed is proportional to the quantity of protein present. This is a simple, rapid, inexpensive assay, and unlike other protein assay procedures such as Lowry and BCA, the Bradford assay is compatible with reducing agents that are often used for the purposes of stabilizing proteins in solution. The Bradford assay is not suitable, however, if even low concentrations of detergents are present in the sample, and in that case the BCA protein determination procedure is to be recommended. The linear concentration range of the Bradford assay when using BSA as standard is 0.1–1.5 mg/mL of protein.

1. Prepare the protein standards in the range 20–100 μg (see Note 22) diluted in the same buffer as the unknown (e.g. H_2O/0.15 M NaCl) in a final volume of 100 μL as per Subheading 3.2 (see Note 23).
2. Prepare two blanks by pipetting 100 μL of buffer into two microfuge tubes.
3. Pipette 20–100 μg of the protein sample of unknown concentration into microfuge tubes in a total volume of 100 μL (if desired, add an equal volume of 1 M NaOH to samples

and standards to prevent precipitation upon the addition of Bradford reagent).

4. Add 5 mL Bradford reagent and mix well by inversion or using a vortex. Allow the samples to stand for 2–60 min at room temperature (avoid foaming as this will lead to poor reproducibility).
5. Meanwhile, switch on the UV spectrophotometer and set the wavelength at 595 nm. Leave the instrument to stabilize for 15–20 min.
6. Calibrate the instrument to zero absorbance using air as a blank.
7. Transfer samples and standards to cuvettes and determine the absorption at 595 nm.
8. Make a standard curve by plotting absorbance at 595 nm against protein concentration. Use the standard curve to determine the concentration of protein in the unknown sample (see Note 24).

3.5. The Lowry Assay

The Lowry assay involves two reactions, the first resulting in the formation of a copper ion complex with amide bonds, which forms reduced copper in alkaline solution. The second reaction is the reduction of the Folin-Ciocalteu reagent, mainly by the reduced copper-amide bond complex, but also by tyrosine and tryptophan residues. The reduced reagent is blue and is thus detectable using a spectrophotometer, detectable in the region of 500–750 nm. The assay has a dynamic range of 1–100 μg of protein. A major limitation of the Lowry assay is the fact that it is sensitive to a considerable range of agents that are frequently found in many lysis buffers commonly used during cell lysis. In addition, this assay is strongly biased to those proteins that are rich in tyrosine and tryptophan.

1. Prepare the protein standards in the range 0–100 μg as per Subheading 3.2 in a total volume of 1 mL (see Note 25).
2. Pipette 200 μL of each standard and the samples of unknown protein concentration into microfuge tubes. Prepare two blank tubes using 200 μL water/buffer.
3. To 200 μL of sample, blank or standard, add 1 mL of freshly prepared Lowry working solution. Let the solution stand at room temperature for 10–30 min (see Note 26).
4. Add 100 μL of diluted Folin-Ciocalteu reagent, vortex immediately (essential for obtaining reproducible results), and stand for 30–60 min (do not exceed 60 min) at room temperature in the dark (complete mixing of the reagent must be accomplished quickly to avoid decomposition of the reagent before it reacts with protein).

5. Transfer the samples to cuvettes.
6. Set the spectrophotometer to 660 nm and leave to stabilize for 15–20 min (see Note 27). Zero the instrument using the blank sample. Measure the absorbance of all the samples in turn.
7. Plot a standard curve of absorbance as a function of initial protein concentration and use it to determine the unknown protein concentrations (see Note 28).

3.6. Bicinchoninic Acid Assay

The principle of the BCA assay is similar to that of the Lowry procedure, in that both rely on the formation of a Cu^{2+}–protein complex under alkaline conditions, followed by reduction of the Cu^{2+} to Cu^{1+}. Here, BCA reagent replaces the Folin-Ciocalteu reagent, and the amount of reduction is proportional to the protein present as before. BCA forms a blue-purple complex with Cu^{1+} in alkaline solutions, the appearance of which can be monitored by absorbance at 562 nm. Unlike the Lowry assay, BCA does not interact with detergents and is less susceptible to interference from other compounds that may be found in buffers used for cell lysis and protein preparation. Some reducing or chelating agents such as DTT and EGTA are best avoided, however, as they interfere by either reducing or sequestering Cu^{2+}.

3.6.1. Macro-BCA Assay

1. Prepare the protein standards in the range 0–100 μg diluted in dH_2O as per Subheading 3.2 in a final volume of 100 μL. Prepare a blank tube using 100 μL water/buffer.
2. Pipette 100 μL of sample/blank/standard into test tubes, add 2 mL of the working solution and mix thoroughly.
3. Incubate for 15 min at 60°C.
4. Cool the samples to room temperature (see Note 29) (the colour is stable up to 1 h following incubation at 60°C).
5. Mix the samples using a vortex and transfer to cuvettes.
6. Set the spectrophotometer to read A_{562} and leave to stabilize for 15–20 min. Zero the instrument, blank, and measure the absorbance of every sample in turn.
7. Plot a standard curve of absorbance as a function of initial protein concentration and use it to determine the protein concentrations in the samples (see Subheading 3.2).

3.6.2. Micro-BCA Assay

1. Set up microfuge tubes containing samples and known amounts of standard protein in the range of 0.5–20 μg with a final sample volume of 500 μL. Prepare two blank tubes using 500 μL water/buffer.
2. Add 500 μL of micro-BCA assay solution to each tube, vortex, and incubate the sample for 15 min at 60°C (see Note 30).
3. Cool samples to room temperature (see Note 29).

4. Vortex the samples and read the absorbance at 562 nm as described in Subheading 3.1.

3.7. Immunoassay: Indirect ELISA

ELISAs are common antibody-based tests that are used to detect antigens or other antibodies in a sample. It is imperative that a robust antibody directed against the protein of interest (i.e. the antigen) be available to the investigator for the purpose of target protein quantitation. There are many direct and indirect formats available, and the specific detection antibody can be covalently linked to a reporter enzyme or can alternatively be itself detected using an enzyme-labelled secondary antibody. Many ELISAs utilize chromogenic substrates, though newer assays employ fluorogenic markers enabling much higher sensitivities to be achieved. The amount of target protein in a sample is inferred from the level of activity/signal from the reporter. Here, the format of a typical indirect ELISA is outlined.

3.7.1. Antigen Coating to Microtitre Plates

1. Dilute the antigen standard and the unknown sample in the antigen-coating buffer to set up a concentration range of 0.1–10 μg/mL.
2. Add 50 μL of these to separate wells of the microtitre plate. Set up control wells containing no antigen and standard wells containing serially diluted antigen to provide data to construct a calibration curve. It is advisable to prepare triplicate wells in all cases.
3. Incubate the plates at room temperature for 5–6 h on a microtitre plate shaker, or alternatively overnight at 4°C.
4. Wash the wells three times with assay buffer followed by three further washes with PBS (see Note 31 for washing procedure). Ensure the wells are washed sufficiently by completely filling them with wash buffer. It is important not to let the wells dry out or enzyme activity will be lost.
5. Add 200 μL of blocking buffer and incubate overnight.
6. Thoroughly wash the plate again three times with the assay buffer followed by three washes with PBS as before (see Note 31).

3.7.2. Primary Antibody Binding Step

7. Add 100 μL of primary antibody (at the experimentally determined optimal working concentration) to each well and incubate for 5– 6 h at room temperature.
8. Wash the plates as before (see Note 31).

3.7.3. Secondary Antibody Binding Step

9. Add 100 μL of the diluted secondary antibody (at the experimentally determined optimal working concentration) to each well and incubate at room temperature for 1 h.

10. Thoroughly wash the plates with assay buffer, as before (see Note 31).

3.7.4. Detection Step

11. Detect bound antibody by the addition of a suitable substrate solution to the wells (see Note 32).
12. As soon as the colour develops add stop solution.
13. Read the plates immediately at A_{450} nm on a microtitre plate reader.
14. Draw the standard curve and estimate the protein concentration in the unknown sample.

3.8. Sodium Dodecyl Sulphate Polyacrylamide Gel Electrophoresis of Proteins

SDS-PAGE is a widely used method for fractionating proteins based on their mobility in an electric field. Binding of the detergent SDS confers different proteins with similar charge per unit mass ratios, and separation by PAGE is therefore based on protein size. Instructions for use with an ATTO protein gel electrophoresis system are laid out here and can easily be adapted to other equivalent systems.

1. Wash glass plates with detergent, rinse first with tap water, then with dH_2O, and finally wipe in one direction with tissue soaked in 100% ethanol.
2. Place the gasket around the ridged plate; assemble the plates and secure with clamps.
3. Prepare a 10% resolving gel by mixing 3.3 mL acrylagel, 1.35 mL bis-acrylagel, 2.5 mL 1.5 M Tris–HCl (pH 8.8), 2.61 mL dH_2O, 0.10 mL 10% (w/v) SDS, 0.10 mL 10% (w/v) APS (freshly prepared), and 0.01 mL TEMED.
4. Pour the gel to within 2 cm of the top of the larger plate (allowing space for the stacking gel) and overlay with 100% ethanol. The resolving gel should be fully polymerized within 30 min.
5. Prepare the 5% stacking gel by mixing 0.42 mL of acrylagel with 0.168 mL bis-acrylagel, 0.312 mL 1 M Tris–HCl (pH 6.8), 1.55 mL dH_2O, 0.025 mL 10% (w/v) SDS, 0.025 mL 10% (w/v) APS, and 0.0025 mL TEMED.
6. Remove the ethanol, pour the stacking gel, and immediately insert a clean comb (wiped with 100% ethanol). Allow the gel to polymerize for at least 20 min.
7. Fill the electrophoresis tank with 1× Tris-glycine running buffer to a level of about 5 cm deep.
8. After polymerization, remove the clamps and gaskets and lower the pre-poured gels into the buffer (place the gels into the buffer at an angle to exclude air bubbles from the gel interface) notched plate facing inwards.

9. Fix the gel plates firmly in place using the pressure plates. Completely fill the tank (including the chamber formed by the inner plates) with 1× running buffer and carefully remove the comb from the gel (see Note 33).
10. Load the sample into the wells. Include at least one well with a pre-stained molecular weight marker.
11. Complete the assembly of the gel unit and connect to a power supply.
12. Electrophoresis at a constant current of 30 mA per gel until the blue dye front has reached the bottom of the gel.
13. When completed remove the plates, separate and process the gel as required (see Subheadings 3.9–3.11).

3.9. Coomassie Brilliant Blue Staining of SDS-PAGE Gels

The dye Coomassie brilliant blue G-250 binds non-specifically to virtually all proteins and is commonly used as a convenient stain for proteins following fractionation by PAGE. Gels are first soaked in a solution of the dye, and any dye that is not bound to protein diffuses out of the gel during the destain steps. This stain binds to proteins more or less stoichiometrically, and so this method can be used when relative amounts of protein needs to be established by densitometry.

1. After electrophoresis, disconnect the gel unit from the power supply and disassemble the apparatus. Carefully separate the plates and remove and discard the stacking gel. Use the sharp blade to cut one corner from the resolving gel to allow the orientation of the gel to be followed.
2. Immerse the gel in Coomassie blue solution for 30 min with constant gentle agitation.
3. After staining, immerse the gel in destain with constant agitation. Refresh the destain four or five times at 1 h intervals until all the background staining has been removed from the gel (see Note 34).
4. An image of the gel in black and white can be captured using a UV trans-illuminator (switched to white light only and using a white tray), or alternatively the gel can be photographed or scanned using a computer (see Note 35).

3.10. Silver Staining of SDS-PAGE Gels

Silver staining is a highly sensitive method for visualizing proteins following electrophoresis. Detection of 0.3–10 ng of protein is possible, making it up to 100 times more sensitive that Coomassie blue staining. Although sensitive, not all proteins stain equally using this method, and the linear dynamic range is rather limited. Silver staining is therefore not favoured when quantitative protein expression analysis is required. Commercially available silver staining kits that offer improved compatibility with subsequent mass spectrometric work include SilverQuest™ Silver Staining Kit (Invitrogen) and PlusOne Silver Staining Kit (GE Healthcare).

1. Following electrophoresis (see Note 36), disconnect the gel unit from the power supply and disassemble the apparatus. Carefully separate the plates and remove and discard the stacking gel. Cut one corner from the resolving gel to allow the orientation of the gel to be followed.
2. Place the gel in a clean container in fixing solution for 1–2 h with gentle agitation at room temperature (for gels previously stained with Coomassie blue, wash overnight with dH_2O, and start from point 3).
3. Discard the fixing solution and wash the gel three times for 20 min each in wash solution.
4. Incubate the gel in oxidizing solution for 10 min.
5. Remove the oxidizing solution and wash the gel for 10 min in copious amounts of water. Repeat the washing procedure until the pale yellow colour has completely faded.
6. Soak the gel in silver nitrate solution for 30 min to stain.
7. Following staining, wash the gel twice for 2 min each using plenty of water.
8. Place the gel in developing solution for 1 min. Replenish the developing solution and incubate for a further 5 min. Repeat this process until bands are stained satisfactorily.
9. Immerse the gel in stop solution for 5 min.
10. Store the gel at 4°C in 1% (v/v) acetic acid. See Note 35 for analysis.

3.11. Western Blotting

Western blotting (also known as immunoblotting) is a method used to detect specific proteins, for which an antibody is available, in a cell culture or tissue extract. Total cellular proteins are first fractionated by gel electrophoresis under denaturing or non-denaturing/native conditions. The proteins are then transferred to a nitrocellulose or nylon membrane onto which they are then immobilized (the blot). Primary antibodies (either monoclonal or polyclonal) that are specific to the protein under investigation are then used to "probe" the blot. When bound to the blot, these antibodies are then in turn located using a secondary antibody that is labelled with a reporter enzyme such as a peroxidase (POD) or alkaline phosphatase. Reporter enzyme activity is then detected by incubation of the blot with an appropriate substrate, yielding a detectable product that maps to the location of the protein of interest. Specific proteins can thus be detected, even when present at very low levels in cell extracts. The outcome of a Western blot experiment depends heavily on the quality of the primary antibody available for the protein under study, and there are now many companies that offer specific antibodies against a vast array of protein targets, enzyme-conjugated secondary antibodies, and detection kits. The user should adhere to any recommendations made by the manufacturers of the apparatus, antibodies, transfer membrane, and immunodetection kits being employed. One should also survey any published literature on the protein under study and take note of the reagents and immunoblotting conditions that are most frequently described. The directions given here are generic and assume the use of a Bio-Rad Trans-Blot® SD semi-dry electrophoretic transfer cell.

3.11.1. Electrophoretic Transfer of Proteins from PAGE Gels to Nitrocellulose Filters

1. Carry out SDS-PAGE as previously described (see Subheading 3.8). The use of commercially available pre-stained protein size markers is recommended.
2. Disconnect the gel unit from the power supply and disassemble the apparatus. Carefully separate the plates and remove and discard the stacking gel. Cut one corner from the resolving gel to allow the orientation of the gel to be followed.
3. Soak the gel in transfer buffer for 15 min to equilibrate the gel removing salts and detergents.
4. Cut the transfer membrane (e.g. nitrocellulose) to the same dimensions as the gel, along with six pieces of 3MM filter paper required to complete the gel/membrane sandwich.
5. Place a pre-soaked sheet of filter paper onto the platinum anode. Roll a pipette over the surface of the filter paper to exclude all air bubbles. Repeat this step with two more sheets of filter paper.
6. Place pre-wetted blotting membrane on top of the filter paper and exclude all air bubbles as before (not all types of membrane require pre-wetting).
7. Place the equilibrated gel carefully on top of the nitrocellulose membrane, align the gel on the centre of the membrane, and ensure all the air bubbles removed.
8. Place another three sheets of pre-wetted filter paper on top of the gel, again remove air bubbles (see Note 37).
9. Put the cathode on top of the stack and place the safety cover on the transfer unit. Transfer the gel for 30 min at 17 V (transfer conditions should be experimentally determined for optimal results).
10. Once the transfer is complete, disassemble the apparatus. Leave the gel in place on top of the nitrocellulose membrane so that the membrane can be cut to the shape of the gel (including the cut corner for orientation) using a sharp blade. The gel and any excess membrane can then be discarded.

11. The coloured molecular weight markers should be clearly visible on the membrane.

3.11.2. Staining of Proteins Immobilized on Nitrocellulose Filters

1. Following electrophoretic transfer, immerse the nitrocellulose membrane in 20 mL Ponceau S solution, and stain for 5 min with constant agitation.
2. After proteins are visualized, wash the membrane with several changes in dH_2O until all the excess stain has been washed away. The loading of equal quantities of sample across all lanes – or otherwise as the case may be – will be apparent. The membrane can now be used for immunological probing.

3.11.3. Immunological Detection

1. Following Ponceau S staining, incubate the membrane in blocking buffer (see Subheading 2.11) for 3 h at room temperature with constant agitation.
2. Discard the blocking buffer and quickly rinse the membrane prior to the addition of the primary antibody (diluted appropriately in blocking buffer) at 4°C overnight on a rocking platform.
3. Remove the primary antibody (see Note 38) and wash the membrane three times for 15 min each with 50 mL TBS-T [if using conjugated primary antibodies then proceed to develop the blot after these steps (i.e. go to step 6)].
4. Add the secondary antibody (diluted in blocking buffer) to the membrane and incubate for 90 min at room temperature with constant agitation.
5. Remove the secondary antibody and wash the membrane three times for 15 min each with 50 mL TBS-T.
6. During the final wash, bring 2 mL aliquots of the appropriate substrate (e.g. BCIP/NBT, or TMB) to room temperature.
7. Place the membrane in a clean container and cover with the substrate (if using a chemiluminescent assay then follow the manufacturer's instructions accordingly).
8. When sufficient chromogenic development has occurred (usually within 10 min, but in some cases it may take several hours), revealing bands of satisfactory intensity, rinse the membrane in dH_2O to stop the reaction.
9. Photograph or scan the blot.

3.12. qIPCR Assay

Nucleic acid amplification methods can be used for signal generation in antibody-based immunoassays (12). qIPCR is based on the use of specific antibodies that have been conjugated to a nucleic acid molecule that is targeted for amplification by real-time PCR, the method of choice for quantitative determinations of low levels of DNA. The benefits that are brought to bear include dramatically improved limits of detection (by a factor of up to 10,000), the capacity to work with small sample volumes, and assay formats that are compatible with complex biological matrices (13).

Various synthetic DNA sequences and the corresponding PCR primers may be used as a target region (14). Several strategies are available for linking antibody recognition with nucleic acid amplification. *Assay Format (I)*: The capture antibody is first adsorbed to the PCR tube surface and streptavidin is used to couple biotinylated DNA to biotinylated detection antibody. *Assay Format (II)*: The capture antibody is first adsorbed to the PCR tube surface and DNA-antibody conjugate is mixed with the antigen or sample before addition to the well. *Assay Format (III)*: Streptavidin-coated microtitre plates are used, and capture antibody, DNA-antibody conjugate, and antigen or sample are premixed before addition to the well. *Assay Format (IV)*: Antigen and samples are adsorbed to the surface of the PCR well before the DNA-antibody conjugate is added to the well. All methods require some prior knowledge of real-time PCR and access to the the appropriate instrumentation and software [The procedure described below employs the ABI 7500 and the corresponding 7500 software (Applied Biosystems)]. Here, we describe a number of different ways to assemble a qIPCR detection system.

3.12.1. Preparation of Biotinylated DNA-Label

1. Use a 5′-biotinylated forward primer and an unmodified reverse primer when using PCR to prepare biotinylated DNA.
2. Assemble the PCR component mixture in a total volume of 50 μL (see Note 39).
3. Perform PCR using the following temperature and time profile: hold at 95°C for 5 min (1 cycle); denature at 95°C for 30 s, anneal at *X*°C for 30 s (the annealing temperature *X* is specific to the primer set), extend at 72°C for 40 s (40 cycles); then perform a final extension step at 72°C for 5 min.
4. Analyze the PCR product by agarose gel electrophoresis [constant voltage (5 V/cm, usually 100 V), for 1–2 h] and view under UV illumination. Loading dye can be used as a tracking marker, and a 1 kb or 100 bp ladder should be loaded as a size marker. Locate the PCR product on the gel.
5. Excise and purify the PCR product using a PCR product Clean-Up kit according to the manufacturer's instructions.
6. Concentrate the purified biotinylated DNA-label using a G-50 Sephadex Column according to the manufacturer's instructions.
7. Determine the concentration of the DNA-label using UV-visible spectrophotometry.

8. Aliquot appropriately and store at –20°C. Biotinylated DNA-labels are stable at –20°C for up to 1 year.

3.12.2. Preparation of Antibody/DNA-Label Conjugate

1. Activate the antibody with thiol groups by mixing ~5 mg of antibody (in PBS, 1 mM EDTA, pH 8.0) with 2-iminothiolane Traut's reagent in tenfold molar excess in a total volume of 1 mL.
2. Incubate the reaction for 1 h at room temperature.
3. Terminate the reaction with 30 μL of 1 M glycine-NaOH pH 7.3.
4. Meanwhile, mix 40–50 nanomoles of amino-modified DNA (in PBS, 1 mM EDTA, pH 7.2) with 2 mM of sulpho-SMCC in a total volume of 500 μL.
5. Incubate for 30 min at room temperature.
6. Terminate the reaction with 13 μL of 1 M glycine-NaOH pH 7.3.
7. Purify the activated antibody and the activated DNA-label using a 2 mL Zeba desalt spin column (Thermo-Scientific) according to the manufacturer's instructions.
8. Mix together and incubate on a shaker for 3 h at room temperature.
9. Incubate overnight at 4°C on a shaker.
10. Purify the conjugate using an ion exchange column (Resource Q 1 mL), according to the manufacturer's instructions, using 10 mM Tris–HCl pH 8.0 and elute with a 0–1.5 M NaCl gradient.
11. Pool fractions that contain the antibody/DNA conjugate and concentrate to 0.3 mL using a Centricon YM-100.
12. Test the concentrated fraction by ELISA and qPCR (see Subheadings 3.7 and 3.12.9).

3.12.3. Biotinylation of Detection Antibody

1. Pipette 2.5 mg/mL of detection antibody into a microfuge tube.
2. Add BNHS (1 mg/mL in DMSO) in five to ten times molar excess and mix gently but thoroughly.
3. Add one-tenth volume of 1 M $NaHCO_{3,}$ pH 8.5 to the antibody solution.
4. Incubate at room temperature for 1–4 h.
5. Purify the biotinylated antibody either by dialysis (for 24 h with three changes of PBS pH 7.4) or using a PD-10 gel filtration column according to the manufacturer's instructions.

6. Store aliquots of biotinylated antibody at −20°C (or under conditions consistent with the stability properties of the antibody).

3.12.4. qIPCR Assay Set Up

General comments: (a) In all assay formats, set up a no-antigen control (to control for unspecific binding of antibody and DNA to the surface), a negative sample control (to determine that the sample matrix does not give rise to a signal), and a no antibody control (to check for any unspecific binding between antibodies). (b) Prepare a standard curve by making a dilution series of antigen (0.1–10 μg/mL) including six to seven concentrations. Vortex and spin between every dilution. (c) Set up two or three replicates of all standards and unknowns that are to be assayed.

3.12.5. Assay Format I

1. Add 25 μL of diluted capture antibody to a microtitre plate and incubate overnight on microtitre plate shaker at 4°C.
2. Wash the wells three times with wash buffer (see Note 31). Ensure the wells are washed sufficiently by filling the wells with wash buffer.
3. Add 200 μL of blocking buffer and incubate for 1 h at 37°C.
4. Wash the wells three times with wash buffer.
5. Set up dilutions of the antigen standard and the unknown sample across a concentration range of 0.1–10 μg/mL in the antigen-coating buffer.
6. Add 25 μL of antigen or unknown sample to each well as appropriate and incubate at room temperature for 1 h.
7. Wash the wells three times with wash buffer.
8. Add 25 μL diluted biotinylated detection antibody and incubate for 1 h at room temperature.
9. Wash the wells six times with wash buffer.
10. Incubate with 25 μL streptavidin (5 nM) for 30 min at room temperature.
11. Add 25 μL of the diluted biotinylated DNA-label and incubate at room temperature for 1 h.
12. Wash the wells six times with wash buffer.
13. Wash six times with ultrapure water.
14. Quantify using qPCR (see Subheading 3.12.9).

3.12.6. Assay Format II

1. Add 25 μL of the diluted capture antibody to a microtitre plate and incubate overnight on a microtitre plate shaker at 4°C.
2. Wash the wells three times with wash buffer (see Note 31). Ensure the wells are washed sufficiently by filling the wells with wash buffer.

3. Add 200 μL of blocking buffer and incubate for 1 h at 37°C.
4. Wash the wells three times with wash buffer.
5. At the same time, dilute the antigen standard and the unknown sample in a concentration range of 0.1–10 μg/mL in the antigen-coating buffer.
6. Mix the antigen or sample together with the diluted chemically conjugated antibody and DNA-label. Incubate for 1 h at 37°C.
7. Add 25 μL of this mix to the wells and incubate for 30 min at room temperature.
8. Wash the wells six times with wash buffer.
9. Wash six times with ultrapure water.
10. Quantify using qPCR (see Subheading 3.12.9).

3.12.7. Assay Format III

1. Prepare dilutions of the antigen standard and the unknown sample in the antigen-coating buffer across a concentration range of 0.1–10 μg/mL.
2. Block the streptavidin-coated microtitre plates with 200 μL blocking buffer for 1 h at 37°C.
3. At the same time, mix 5.5 μL antigen or sample with 11 μL of diluted chemically conjugated antibody and DNA-label, and 11 μL of the diluted biotinylated capture antibody. Incubate this mix at room temperature for 1 h.
4. Wash the wells three times with wash buffer (see Note 31).
5. Add 25 μL of the mix to the wells and incubate for 30 min at room temperature.
6. Wash the wells six times with wash buffer.
7. Wash six times with ultrapure water.
8. Quantify using qPCR (see Subheading 3.12.9).

3.12.8. Assay Format IV

1. Dilute the antigen standard and the unknown sample in a concentration range of 0.1–10 μg/mL in the antigen-coating buffer.
2. Add 50 μL to each well.
3. Incubate the plates on microtitre plate shaker, overnight at 4°C.
4. Wash the wells three times with wash buffer (see Note 31).
5. Add 200 μL of blocking buffer and incubate for 1 h at 37°C.
6. Wash the wells three times with wash buffer.
7. At the same time, mix the antigen or sample together with the chemically conjugated antibody and DNA-label. Incubate for 1 h at 37°C.

8. Add 25 μL of the mix to the wells and incubate for 30 min at room temperature.
9. Wash the wells six times with wash buffer.
10. Wash six times with ultrapure water.
11. Quantify using qPCR (see Subheading 3.12.9).

3.12.9. Real-Time PCR (qPCR)

1. Prepare the reagents for the PCR as a master-mix (make up 10% more than the volume required so as to account for potential pipetting error).
2. Include three "no template controls" (NTC, a minus sample control) in each reaction plate. Set up a positive control sample using DNA/conjugate as template.
3. Aliquot 20 μL of PCR master-mix per reaction into the individual wells of the PCR plate containing the immobilized DNA.
4. Immediately seal the plate with an optical adhesive cover and protect from light. Centrifuge at 500×*g* for 2 min to eliminate air bubbles. It is important to check to ensure that no bubbles remain in any well.
5. Perform amplification and detection under the following cycling conditions: hold at 95°C for 5 min (1 cycle); denature at 95°C for 20 s, anneal at *X*°C for 20 s (the annealing temperature *X* is specific to the primer set), extend at 72°C for 25 s (45 cycles).
6. Record fluorescence data during the 72°C step.
7. Include a dissociation stage to analyze PCR product melting temperature.
8. Analyze the qPCR data obtained with the appropriate software. Check for bimodal dissociation curves or an abnormal amplification plot.

4. Notes

1. When selecting the method to be used for total protein determination, care should be taken to consider sample and buffer properties and the objective of the analysis being undertaken. The advantages and limitations of the most commonly employed methods are compared in Table 2, as an aid to the user in the selection of the most suitable assay for the sample in question.
2. All solutions should be at room temperature prior to the assay (cold solutions can cause condensation on the surface of the

cuvette, whereas warm solutions often lead to bubbles causing inaccuracy in the readings).

3. The selection of a protein standard is an important step in every protein assay as the standard and sample should be of a similar molecular weight and should respond in the same way to the assay in order to minimize errors in estimations.
4. Protein concentrations are measured in milligrams/millilitre (mg/mL) and micrograms/microlitre (μg/μL). Concentrations can also be recorded in micrograms/millilitre (μg/mL) for very small concentrations.
5. Common buffer components such as acetate, Brij 35 detergent, deoxycholate, SDS, Triton X-100, Tween 20, and urea are known to absorb strongly at 280 nm and should be avoided. Nucleic acid contamination of samples can also greatly influence the absorption.
6. Common buffer components such as acetic acid, ammonium sulphate, 2-glycerol, mercaptoethanol, Tris, and SDS are known not to be compatible with the Bradford Assay.
7. NaOH is added to ensure that the sample does not precipitate upon addition of Bradford reagent.
8. Quartz cuvettes cannot be used as the Coomassie brilliant blue G-250 dye binds to this material.
9. Many substances are known to interfere with the Lowry protein assay including CAPS, barbital, cesium chloride, EDTA, citrate, cysteine, diethanolamine, dithiothreitol, EGTA, HEPES, mercaptoethanol, nonidet P-40, phenol, sodium deoxycholate, polyvinyl pyrrolidone, sodium salicylate, thimerosol, tricine, Tris, and Triton X-100.
10. Substances known to interfere with the BCA assay include Tris, ammonium sulphate, EDTA, DTT, EGTA, and SDS.
11. Commonly used blocking agents include BSA, serum, non-fat dry milk, casein, and gelatin. The selection of blocking agent is dependent on the antibody being used (see the manufacturers' guidelines).
12. Allow wash buffer to equilibrate to room temperature before use.
13. Non-fat dried milk is not compatible for use with streptavidin/biotin.
14. For accuracy, prepare standards that span the full range of the assay.
15. Use ultrapure water for preparation of all buffers.
16. Optimal concentrations are determined experimentally or following the manufacturers' guidelines.

17. A plate material that can bind biological materials should be used. For samples containing low concentrations of target antigens, the adsorption to reaction vessels can be minimized using silanized cups.
18. UNG is included in the PCR to prevent the reamplification of carryover PCR product. Through the use of dUTP rather than dTTP, the uracil incorporated into amplicons during PCR can be destroyed by UNG, thus destroying any contaminating PCR product present at the outset. The UNG incubation step involves incubating reactions at 50°C for 2 min and precedes the HotStart *Taq* DNA polymerase activation step (95°C for 5 min). UNG is then thermally inactivated during the first denaturation step of the PCR.
19. Concentrated protein solutions should be diluted so they are within the linear range of the instrument being used. If the approximate sample concentration is unknown, a range of dilutions (1, 1:10, 1:100, 1:1,000) should be assayed. When working with dilute samples, the addition of a non-ionic detergent to the buffer may help to prevent the loss of protein through adsorption of protein on the cuvette.
20. If a clear linear relationship is observed, a standard curve is not necessary. Amounts can be determined using interpolation. However, a standard curve should be prepared to check for accuracy and linearity the first time that an assay is performed.
21. Choose the appropriate protein reference (e.g. BSA, IgG, lysozyme, etc.) for calculations of protein concentration in order to obtain the desired molar extinction coefficient.
22. A microassay format can be used for protein concentrations between 1 and 20 μg/mL. The protocol followed is the same as that of the basic protocol except with a reduced total volume of 1 mL. The assay responds primarily to arginine residues, and an arginine-rich standard may be preferable if the sample is rich in that amino acid.
23. The sample can be diluted in the case of a high protein concentration.
24. The standard curve is not linear, and the precise absorbance varies depending on the age of the assay reagent. As a result, it is important to construct a calibration curve for each assay performed.
25. The amount of colour yielded in this assay is dependent on the amino acid composition of the sample. Thus, two different proteins at the same concentration can give significantly different colour yields. Hence, the choice of a standard similar in composition to that of the sample is an important consideration in the design of this assay.

26. This incubation period is not a critical parameter and can vary from 10 min to several hours with no effect on final absorbance values.
27. If the A_{660} values are low then reread the samples at A_{750}, which may increase the sensitivity of the assay, or at A_{550} if the sample concentration is between 100 and 2,000 μg/mL.
28. Since this assay is not linear at high concentrations, ensure that the sample falls on the linear portion of the standard curve.
29. This can easily be achieved by placing the samples in a water bath at room temperature.
30. Increase the incubation time in order to improve the sensitivity of this assay. The incubation temperature should be lowered to ambient room temperature to obtain a decrease in sensitivity.
31. Washing procedure:
    Wash Step: use a squirt bottle/multi-channel pipette to fill the wells with wash solution. Do not touch the inside surface of the wells with the pipette tips or bottle nozzle. *Expulsion of wash solution:* holding the microtitre plate firmly (from the underneath) over a sink, turn the plate rapidly and directly upside down causing the liquid to be forced out of the wells and into the sink. Repeat this "dumping motion" a second time. *Blotting and banging the plate*: immediately blot the upside down plate (wells face down) on blotting/tissue paper. Move the plate to an unused section of the blotting paper and allow to drain for 5–10 s. Repeatedly "bang" the plate very hard four to five times over unused areas of the paper. *Wash step*: Use a squirt bottle/multi-channel pipette to refill the wells with wash solution as before and immediately dump and bang the plate again as described above. Following the last wash, leave the plate upside down for 30 s to drain. Bang again three to four times, rotating the plate between each bang.
32. Do not add substrate near the sink as the washing procedure can create aerosols which could recontaminate the wells or substrate.
33. Use a 3 mL syringe fitted with a 22-gauge needle to wash the wells with running buffer and remove any unpolymerized gel. Straighten the wells using a loading tip.
34. Extensive destaining can lead to the loss of some low molecular weight bands and the fading of other bands. Note: when staining IEF gels with Coomassie blue, the gel must first be fixed in a solution of trichloroacetic acid. This leaches out the carrier ampholytes, which would otherwise produce background staining. When staining small peptides (<10 kDa),

the gel is first fixed in a solution containing glutaraldehyde to cross-link the peptides and prevent them from diffusing out of the gel during subsequent staining steps.

35. *Analysis of SDS-PAGE gels*: determine the correct orientation of the gel, using the cut corner to identify the top/bottom and left/right ends. Locate the lanes corresponding to each sample loaded on the gel. Identify the polypeptide bands of interest. Estimate the approximate molecular mass or relative molecular mass for each band of interest using the molecular weight markers as a guide. Note differences in the intensity of band staining. This may be indicative of disparity in abundance between individual polypeptides.

    Identify unusual patterns that might reflect incomplete denaturation or degradation of the proteins being analyzed (overloading or underloading, distortion of lanes, smears, and streaks can all limit the interpretation of results).

36. Maintain a high level of cleanliness during electrophoresis and silver staining to minimize spurious artefacts. Wear gloves to prevent staining of the gel due to finger marks.
37. Ensure that the nitrocellulose membrane is between the gel and the anode or the proteins will be lost from the gel into the buffer rather than transferred to the nitrocellulose.
38. Diluted antibody can usually be retained and stored at −20°C for a further four to five uses.
39. Use filter/barrier tips to avoid cross-contamination during pipetting. Do not touch the microplate wells with the tips during pipetting.

References

1. Rosenburg, I. M. (2005) Getting started with protein purification, in *Protein analysis and purification: benchtop techniques*, 2nd Edition. Birkhäuser, Boston, MA, pp. 118–152.
2. Lovrien, R., and Matulis, D. (2005) Assays for total protein, in *Current protocols in microbiology* (Coico, R., Kowalik, T., Quarles, J., Stevenson, B., Taylor, R., eds.). John Wiley & Sons, Inc., Hoboken, NJ, pp. A.3.A.1–A.3.A.14.
3. Bradford, M. M. (1976) A rapid and sensitive method for the quantitation of microgram quantities of protein utilizing the principle of protein-dye binding. *Anal. Biochem.* **72**: 248–254.
4. Baines, D. (2001) Analysis of purity, in *Protein purification techniques: practical approach*, 2nd Edition (Roe, S., ed.). Oxford University Press, Oxford, pp. 27–46.
5. Lowry, O. H., Rosebrough, N. J., Farr, A. L., and Randall, R. J. (1951) Protein measurement with the Folin phenol reagent. *J. Biol. Chem.* **193**:265–275.
6. Smith, P. K., Krohn, R. I., Hermanson, G. T., Mallia, A. K., Gartner, F. H., Provenzano, M. D., et al. (1985) Measurement of protein using bicinchoninic acid. *Anal. Biochem.* **150**:76–85.
7. Olson, B. J. S. C., and Markwell, J. (2007) Assays for determination of protein concentration, in *Current protocols in protein science* (Coligan, J. E., Dunn, B. E., Speicher, D. W., Wingfield, P. T., Ploegh, H. L., eds.). John Wiley & Sons, Inc., Hoboken, NJ, pp. 3.4.1–3.4.29.
8. Okutucu, B., Dınçer, A., Habib, O., and Zihnioglu, F. (2007) Comparison of five methods for determination of total plasma protein concentration. *J. Biochem. Biophys. Methods* **70(5)**:709–711.
9. Davies, C. (2005) Introduction to immunoassay principles, in *The immunoassay hand-*

book, 3rd Edition (Wild, D., ed.). Elsevier, London, pp. 3–40.

10. Walsh, G. (2002) Protein purification and characterization, in *Proteins: biochemistry and biotechnology.* John Wiley & Sons, Inc., Hoboken, NJ, pp. 89–178.
11. Anicetti, V. R., Keytand, B. A., and Hancock, W. S. (1989) Purity analysis of protein pharmaceuticals produced by recombinant DNA technology. *Trends Biotechnol.* **7(12)**: 342–349.
12. Sano, T., Smith, C. L., and Cantor, C. R. (1992) Immuno-PCR: very sensitive antigen detection by means of specific antibody–DNA conjugates. *Science* **258**:120–122.
13. Niemeyer, C. M., Adler, M., and Wacker, R. (2007) Detecting antigens by quantitative immuno-PCR. *Nat. Protoc.* **2**:1918–1930.
14. Lind, K., and Kubista, M. (2005) Development and evaluation of three real-time immuno-PCR assemblages for quantification of PSA. *J. Immunol. Methods* **304**:107–116.

Chapter 14

Purification of Proteins Fused to Glutathione S-Transferase

Sandra Harper and David W. Speicher

Abstract

This chapter describes the use of glutathione S-transferase (GST) gene fusion proteins as a method for inducible, high-level protein expression and purification from bacterial cell lysates. The protein is expressed in a pGEX vector, with the GST moiety located at the N terminus followed by the target protein. The use of GST as a fusion tag is desirable because it can act as a chaperone to facilitate protein folding, and frequently the fusion protein can be expressed as a soluble protein rather than in inclusion bodies. Additionally, the GST fusion protein can be affinity purified facilely without denaturation or use of mild detergents. The fusion protein is captured by immobilized glutathione and impurities are washed away. The fusion protein then is eluted under mild, non-denaturing conditions using reduced glutathione. If desired, the removal of the GST affinity tag is accomplished by using a site-specific protease recognition sequence located between the GST moiety and the target protein. Purified proteins have been used successfully in immunological studies, structure determinations, vaccine production, protein–protein, and protein–DNA interaction studies and other biochemical analyses.

Key words: Glutathione S-transferase, pGEX, Protein expression, Protein purification, Thrombin, Factor Xa, Fusion tags

1. Introduction

Glutathione S-transferase (GST) is a naturally occurring 26-KDa protein found in eukaryotic cells. The gene from the parasitic helminth *Schistosoma japonicum* was used in the development of the pGEX vectors (1). This chapter describes the use of a GST affinity tag to aid in the purification of recombinant proteins. The 26-KDa GST moiety binds with high affinity to glutathione coupled to a Sepharose matrix. This binding is reversible and the protein can be eluted under mild, non-denaturing conditions by the addition of reduced glutathione to the elution buffer. A specific protease site engineered between the GST moiety and the protein of interest allows removal of the GST moiety from the target

Dermot Walls and Sinéad T. Loughran (eds.), *Protein Chromatography: Methods and Protocols*, Methods in Molecular Biology, vol. 681, DOI 10.1007/978-1-60761-913-0_14, © Springer Science+Business Media, LLC 2011

recombinant protein. The GST can be removed from the sample by re-chromatography on a glutathione column, and the protein of interest purified to homogeneity by other techniques such as gel filtration or ion exchange. The most commonly used vectors are available from GE Healthcare (and are discussed here), although variations from other suppliers also are available.

Successful purification of GST fusion proteins requires several strategic decisions and may require optimization of methods and conditions for specific proteins. A flow diagram highlighting the basic steps of the vector design, expression and purification processes, and some of the key decisions to be made is shown in Fig. 1. Each step involves multiple options that are often interrelated and

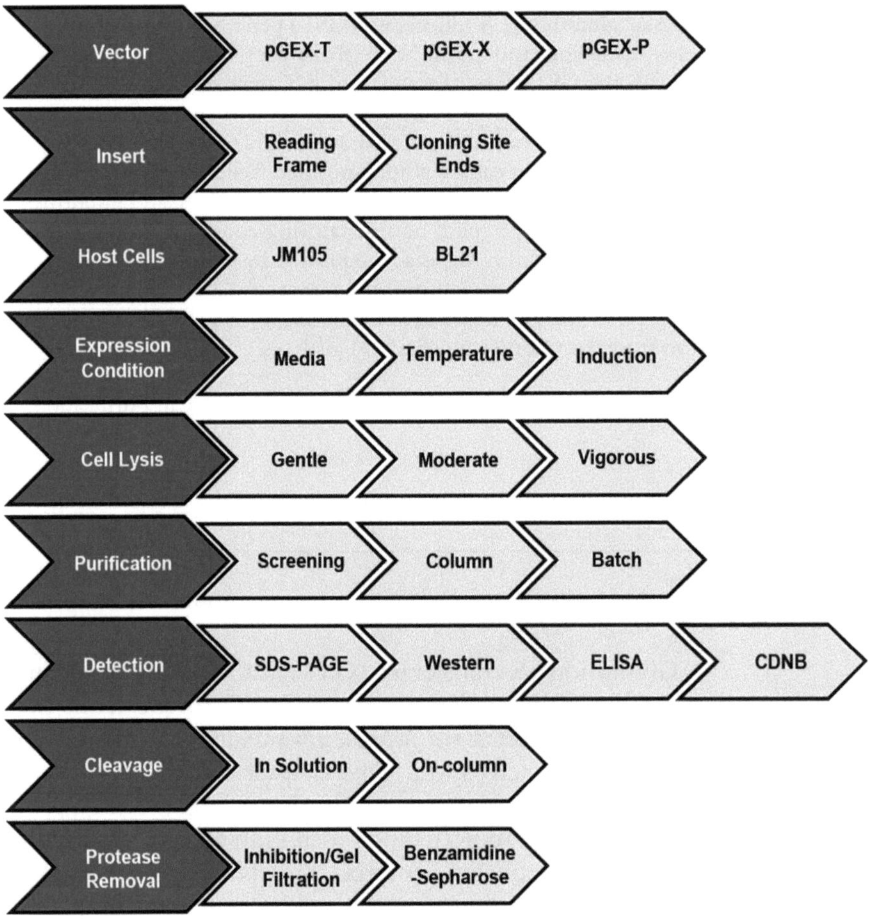

Fig. 1. Flow diagram illustrating the key decision-making steps for designing and executing a successful GST fusion protein purification. Selection of the vector depends on whether the GST moiety ultimately will be cleaved away and the enzyme that will be used. The insert should be compatible with the vector chosen. A variety of host cells are available; the use of *E. coli* BL21 strains and their derivatives are recommended. Expression conditions should be optimized to yield the highest level of soluble fusion protein. Exact compositions of lysis and purification buffers depend largely on target protein characteristics and should be modified accordingly. Conditions for removal of the GST moiety also will need to be optimized. Overall, each step involves multiple options that should be tailored to the characteristics of the target protein and intended purpose of the purified protein.

could affect the final product yield and purity. For instance, choice of vector will be influenced by whether the GST moiety ultimately will be cleaved away from the target protein and the desired protease to be used. Something to consider when evaluating whether or not to cleave the GST moiety away is that GST is a homodimeric protein. If it is suspected that oligomeric state will influence properties of the target protein, the GST moiety should be cleaved away. Once a vector has been chosen and successful cloning has occurred, optimization of protein expression conditions such as *Escherichia coli* host cell strain, temperature, concentration of isopropyl-1-thio-β-d-galactopyranoside (IPTG), and length of induction should be optimized. After sample extraction, immobilized glutathione Sepharose media is used for purification. Cleavage of the target protein from the GST moiety will depend on the vector chosen and can be performed either in solution or while the protein is bound to the column. The enzyme should be inactivated and/or removed from the sample after cleavage. Depending on the sample purity required for ultimate use, the protein should be further purified after removal of the GST moiety using gel filtration, ion exchange, or other purification schemes.

Here we provide a detailed description of the protocols required for successful purification of GST fusion proteins, with an emphasis on maintaining the solubility of the fusion protein and avoiding denaturants or mild detergents. The protocols include basic information on the choice of expression vectors and host *E. coli* cells, detailed procedures for protein expression and purification of expressed protein, enzymatic cleavage of the GST moiety from the target protein, and subsequent steps to purify the target protein to homogeneity.

2. Materials

Use Milli-Q-purified water or equivalent for the preparation of all buffers.

2.1. Expression of GST Fusion Protein

1. Glycerol stock of transformed *E. coli* cells expressing the GST fusion protein of interest in a pGEX vector (see Notes 1–3).
2. Luria Bertani (LB) medium per liter: 10 g tryptone, 5 g yeast, 5 g NaCl. Adjust to pH 7.2 with NaOH. Autoclave.
3. Ampicillin, 5 mg/mL stock, filter sterilize.
4. Refrigerated environmental shaker (refrigeration is necessary if protein expression is to be performed at temperatures below room temperature; e.g., New Brunswick Scientific Innova 4330 Refrigerated Environmental Shaker).
5. IPTG, 100 mM stock.

6. Large centrifuge bottles (e.g., 1-L capacity).
7. 50-mL Centrifuge tubes.

2.2. Affinity Purification of GST Fusion Protein

1. Glutathione Sepharose 4B bulk matrix (GE Healthcare or Sigma) (see Note 4).
2. Column (BioRad 2.5 × 10 cm Econo-Column).
3. Peristaltic pump.
4. 10 mM glutathione buffer: 50 mM Tris–HCl, 10 mM reduced glutathione, pH 8.0 (make fresh daily).
5. PBS/EDTA/phenylmethanesulfonyl fluoride (PMSF): 1× PBS, 5 mM EDTA, 0.15 mM PMSF, adjust to pH 7.4 with NaOH.
6. Pelleted *E. coli* containing expressed target fusion protein.
7. 60-mL Centrifuge tube (capable of handling force 48,000 × *g*).
8. Dounce homogenizer.
9. PBS/EDTA: 1× PBS, 5 mM EDTA, adjust to pH 7.4 with NaOH.
10. Lysis buffer: 50 mM Tris–HCl, 50 mM NaCl, 5 mM EDTA, 1 μg/mL leupeptin, 1 μg/mL pepstatin, 0.15 mM PMSF, 1 mM diisopropyl fluorophosphates (DFP), 1 mM 2-ME, pH 8.0. *Caution: DFP is a very dangerous neurotoxin;* see Note 5 *and supplied precautions* (see Note 6).
11. Wash buffer: 50 mM Tris–HCl, 5 mM EDTA, 1 μg/mL leupeptin, 1 μg/mL pepstatin, 0.15 mM PMSF, pH 8.0.
12. U-buffer: 5 M urea, 50 mM Tris–HCl, 5 mM EDTA, 5 mM 2-ME, 1 μg/mL leupeptin, 1 μg/mL pepstatin, 0.15 mM PMSF, 1 mM DFP pH 8.0.
13. Triton X-100.
14. PBS/glycerol: 1× PBS, 20% glycerol, 1% Triton X-100, 5 mM 2-ME, 5 mM EDTA, 5 mM 2-ME, 1 μg/mL leupeptin, 1 μg/mL pepstatin, 0.15 mM PMSF, 1 mM DFP, adjust to pH 7.4 with NaOH.

2.3. Enzymatic Cleavage to Remove GST Affinity Tag

1. Affinity-purified fusion protein.
2. Vector-specific enzyme [thrombin (Sigma), factor Xa (Roche Diagnostics), or PreScission protease (GE Healthcare)] (see Note 7).
3. 0.15 mM PMSF in isopropanol.
4. PBS/EDTA/PMSF: 1× PBS, 5 mM EDTA, 0.15 mM PMSF, adjust to pH 7.4 with NaOH.

2.4. Removal of GST Affinity Tag After Protease Cleavage

1. Cleaved fusion protein dialysed into PBS/EDTA/PMSF.
2. Glutathione Sepharose column.
3. PBS/EDTA/PMSF.

2.5. Further Purification of Cleaved Recombinant Protein

1. Cleaved fusion protein.
2. PBS.
3. Concentrator (e.g., Amicon Ultra, Millipore).
4. HPLC system with in-line detector and fraction collector.
5. Gel filtration column compatible with the molecular weight range of the sample to be purified.

3. Methods

3.1. Expression of GST Fusion Protein

One of the driving forces behind the development of the GST fusion system is that the GST protein accumulates in the cytoplasm (1) and, due to the large size of this affinity tag, the GST moiety has the propensity to confer solubility to the fusion partner – thereby eliminating the need for lengthy purification schemes for insoluble proteins that usually involve denaturants. Unfortunately, in practice it is unpredictable which fusion proteins will remain soluble and which will accumulate in occlusion bodies. As a general rule of thumb, the larger the protein and the more complex folding it must undergo, the more likely it will accumulate in inclusion bodies. In most cases, where the fusion protein is initially observed primarily in inclusion bodies, it is desirable to modify culture conditions – such as lowering growth temperature, increasing aeration, or altering induction conditions – to attempt to obtain soluble protein, rather than to solubilize inclusion bodies with denaturants and refold the protein after extraction. Hence, before conducting a large-scale protein expression it is recommended to complete a pilot experiment to check solubility. Since the main determinant of successful purification of functional GST fusion proteins in good yields is protein solubility, it is worth the effort to optimize culture conditions prior to large-scale purification.

3.1.1. Screening Protocol for Protein Expression

Following is a screening protocol to check protein expression levels of fusion protein in mini-cultures. The presence of an increase in protein levels at the expected molecular weight on an SDS-PAGE gel after induction with IPTG is a good indication of successful protein expression. After induction, a prominent band at the expected molecular weight (molecular weight of target protein + ~26 KDa for the GST moiety) should be visible. If it is suspected that the protein is expressed at a low level, verification of correct expression can be accomplished using Western blotting with an antibody specific for either the target protein or GST. If the samples will not immediately be analyzed by SDS-PAGE, they may be stored overnight at 4°C or at –20°C for longer-term storage. Additionally, samples grown in mini-cultures can be used to evaluate solubility of a particular fusion protein (see Notes 8 and 9).

1. Inoculate several isolated colonies into separate 50-mL centrifuge tubes containing 10 mL LB supplemented with 100 μg/mL ampicillin. Grow the inoculated culture in an environmental shaker at 250–300 rpm at 37°C overnight.
2. The next morning, add 0.5 mL of the overnight culture to 4.5 mL LB with 100 μg/mL ampicillin. Grow for 1 h at 37°C with shaking at 250–300 rpm.
3. Remove 1 mL of the un-induced sample. Centrifuge at 16,000 × *g* for 2 min and remove the supernatant. Freeze the pelleted sample or store on ice until ready to run SDS-PAGE.
4. To the mini-culture add IPTG to a final concentration of 1 mM and grow for an additional 2–3 h.
5. Remove 1 mL of the induced sample. Centrifuge at 16,000 × *g* for 2 min and remove the supernatant. Freeze the pelleted sample or store on ice until ready to run SDS-PAGE.
6. Re-suspend the cell pellets in 200 μl SDS-PAGE sample buffer; heat 3–5 min at 90°C.
7. Analyze 5–10 μL using SDS-PAGE followed by Coomassie blue staining (or 1–2 μL for Western blot) (Note 10).

3.1.2. Large-Scale Protein Expression

Yields of GST fusion proteins using expression in *E. coli* are highly variable, often ranging from 10 to 50 mg/L, but potentially could be much lower in cases where the fusion protein is toxic to the cells or unstable (see Note 11). Since expression levels are typically high, adequate amounts of protein usually can be obtained from several liters or less of bacterial culture grown in shaker flasks. The following protocol describes protein production at 37°C; however, exact temperature and induction conditions for each construct should be modified to improve the yield of soluble protein. The growth of the *E. coli* culture can be monitored by reading the optical density at 600 nm (A_{600}) (see Note 12), and protein expression levels can be analyzed using SDS-PAGE of un-induced and induced samples (see Note 13). Once the culture has been induced with IPTG, do not allow the cells to grow for extended periods of time, as cell lysis can occur and release proteases into the cytoplasm that could degrade the fusion protein.

1. Using aseptic technique, transfer some of a glycerol culture or isolated colonies grown on a streaked plate to 100 mL LB with 100 μg/mL ampicillin.
2. Incubate the inoculated culture overnight in an environmental shaker at 37°C at 250–300 rpm.
3. The next morning remove a 1 mL aliquot of cells and check the OD_{600} using a spectrophotometer (see Note 12).

4. Using aseptic technique, dilute the overnight starter culture 1:20 into 600 mL fresh LB supplemented with 100 μg/mL ampicillin (see Note 14).
5. Incubate the culture at 37°C at 250–300 rpm until the OD_{600} is 0.5–0.7 (see Note 12).
6. Remove a 1 mL aliquot of cells to be saved for gel analysis (see Note 13).
7. Add IPTG to1.0 mM final concentration.
8. Incubate at 37°C at 250–300 rpm for an additional 3 h (see Note 12), while monitoring the growth at OD_{600}. At saturation they will stop dividing. Remove a 1-mL aliquot of cells for gel analysis (see Note 13).
9. Harvest the cells by centrifugation at 4,000 ×*g* for 20 min at 4°C.
10. Carefully decant the supernatant, leaving ~15–50 mL in the centrifuge bottle.
11. Re-suspend the cells and transfer to a 50-mL centrifuge tube.
12. Centrifuge for 20 min at 4,000 ×*g*, at 4°C.
13. Decant the supernatant.
14. The cell pellet may be used immediately or frozen for up to several months at −80°C.
15. Analyze un-induced and induced samples by SDS-PAGE to check protein expression levels (see Note 13).

3.2. Affinity Purification of GST Fusion Protein

3.2.1. Column Purification of GST Fusion Proteins

Soluble GST fusion proteins are purified easily using an immobilized glutathione Sepharose column. There are several options of immobilized glutathione chromatography media available to purify soluble GST fusion proteins from bacterial cell lysates (see Note 4). The protocol described below is an adaptation of the manufacturer's recommendation using glutathione Sepharose 4B poured into a column and using a peristaltic pump to control flow rates. Protease inhibitors and reducing agents should be added to the buffers, as required, to minimize proteolysis of the fusion protein. An exception is that serine protease inhibitors must be removed from the glutathione buffer prior to enzymatic removal of the GST moiety, as they will inhibit enzyme activity (see Note 15). Save a small aliquot from each step of the purification for analysis by SDS-PAGE to monitor the location of the fusion protein throughout the purification (see Notes 16 and 17). A given column or batch of resin should be used exclusively with a single fusion protein to minimize potential cross contamination.

As an alternative to column purification, a protocol describing batch purifications is described in Subheading 3.2.2. Batch purifications are quick and simple, but frequently the yield and purity of the protein obtained will be somewhat lower than that obtained through chromatographic separations. In order to minimize proteolysis, all steps of the protein purification should be carried out at 4°C, unless otherwise noted.

1. Re-suspend and pour 20 mL glutathione Sepharose 4B resin into a 2.5 × 8 cm column (see Notes 18 and 19).
2. Thoroughly wash the glutathione Sepharose with 5–10 bed volumes (see Note 18) PBS at 1.5 mL/min to remove the ethanol storage solution.
3. Re-suspend the pelleted *E. coli* cells in 15 mL cold lysis buffer (cells might be freshly prepared or thawed frozen cell pellets) (see Note 20).
4. Lyse the cells by sonication on ice (~10 times for 10 s each with 1 min rest between bursts to minimize sample heating). Save 50 μL of the lysate for gel analysis and transfer the remainder to a 60-mL centrifuge tube (see Note 21).
5. Centrifuge the lysate at 48,000 × *g* for 20 min at 4°C.
6. Decant the supernatant into a clean, 50-mL centrifuge tube.
7. Re-suspend the pellet in 15 mL PBS buffer using a dounce homogenizer (see Note 20).
8. Run 5–10 μL each of lysate, supernatant, and re-suspended pellet on an SDS-PAGE gel to verify that the fusion protein is in the supernatant fraction (see Note 8). If the fusion protein is located in the pellet fraction, see Note 9 for tips to improve soluble protein expression or Subheading 3.2.3.
9. Load the soluble fusion protein to the equilibrated glutathione Sepharose column using a flow rate of 0.1 mL/min (see Notes 22 and 23). Collect fractions and run gels to verify that fusion protein is binding to the column and that the capacity has not been exceeded (see Note 24). If the fusion protein binds poorly to the resin, see Note 25 for several possible remedies.
10. Wash the column with 5–10 bed volumes of PBS/EDTA/PMSF using a flow rate of 1.5 mL/min.
11. Wash the column with 10 bed volumes of PBS/EDTA using a flow rate of 1.5 mL/min (see Notes 22 and 23).
12. Elute the fusion protein with glutathione buffer using a flow rate of 0.3 mL/min. Fractions may be monitored using A_{280} and analysis by SDS-PAGE (see Note 26). Pool the fractions that contain the GST fusion protein. The purified protein may be stored at 4°C and should be ~90% pure at this stage. If problems are encountered eluting the fusion protein, see Note 27. If high levels of contamination are present, see Note 28 for troubleshooting tips.

3.2.2. Batch Purification of GST Fusion Proteins

GST proteins also can be purified using a batch purification method. A batch purification method is quick, easy, and requires little equipment; however, the resulting protein may contain more impurities and have a lower yield than a chromatography-based purification. Batch purifications are best utilized when screening purification conditions. The batch purification outlined below is generally performed at room temperature. To minimize the risk of proteolytic degradation, the procedure can be performed at 4°C by increasing the incubation time two- to fourfold.

1. Lyse cells by sonication on ice (~10 times for 10 s each, with 1 min rest between bursts to minimize sample heating). Save 50 μL of the lysate for gel analysis and transfer the remainder to a 60-mL centrifuge tube (see Note 21).
2. Centrifuge the lysate at 48,000 × *g* for 20 min at 4°C.
3. Decant the supernatant into a clean, 50-mL centrifuge tube.
4. Re-suspend the pellet in 15 mL PBS buffer using a dounce homogenizer (see Note 20).
5. Run 5–10 μL each of lysate, supernatant, and re-suspended pellet on an SDS-PAGE gel to verify that the fusion protein is in the supernatant fraction (see Note 8). If the fusion protein is located in the pellet fraction, see Note 9 for tips to improve soluble protein expression or Subheading 3.2.3.
6. Add 2 mL 50% slurry glutathione Sepharose bulk matrix (1 mL bed volume) per 100 mL bacterial lysate supernatant obtained in step 3 (see Note 18 for preparation of a 50% slurry). Incubate for 30 min at room temperature with gentle agitation.
7. Centrifuge at 500 × *g* for 5 min at room temperature. Remove the supernatant and save for analysis by SDS-PAGE to determine binding efficiency.
8. Wash the resin with 10 bed volumes of PBS. Invert the tube several times to mix, then centrifuge at 500 × *g* for 5 min at room temperature. Remove the supernatant. Repeat wash and centrifugation steps for a total of three washes of 10 bed volumes each.
9. Elute the fusion protein by re-suspending the resin with 1.0 mL glutathione elution buffer per milliliter bed volume. Incubate the sample with gentle agitation for 10 min at room temperature. Centrifuge at 500 × *g* for 5 min at room temperature. Transfer the fusion protein-containing supernatant to a separate tube. Repeat elution and collection steps a total of three times. The supernatants may be pooled or analyzed separately by SDS-PAGE to monitor protein content (see Notes 16 and 17).

3.2.3. Extraction of GST Fusion Proteins from Inclusion Bodies

If attempts to shift protein expression from inclusion bodies to the soluble fraction fail, insoluble GST fusion proteins sometimes can be purified in the presence of denaturants such as urea, followed by refolding. Solubilization using detergents such as sarkosyl (*N*-laurylsarosine) also has been successfully employed (2, 3). Although the following protocol has been used successfully to denature and re-nature GST fusion proteins, each fusion protein construct is unique and the exact denaturation and re-naturation conditions must be determined empirically. Common denaturants that have been successfully employed include guanidine HCl, urea, Tween 20, cetrimonium bromide (CTAB), and SDS (4, 5). The denaturants then must be completely removed to allow proper refolding of the protein. Conditions that should be optimized to facilitate refolding include: type of denaturant, pH, presence of reducing agent, speed of denaturant removal, and protein concentration. Once the protein has been re-natured, verify that the protein has regained its native conformation and function and remove any improperly folded protein.

1. Pre-equilibrate the glutathione Sepharose column; sonicate the pelleted *E. coli* cells, and separate the lysate supernatant and pellet fractions by centrifugation (see Subheading 3.2.1, steps 1–8).
2. Re-suspend the lysate pellet that includes the inclusion bodies, in 15 mL wash buffer. Centrifuge the sample for 20 min at 48,000 × *g* at 4°C. Decant the supernatant and re-suspend the washed pellet in 12 mL U-buffer (re-suspend in 20 μL U-buffer per milliliter of original culture). Incubate for 2 h on ice.
3. Centrifuge the sample for 20 min at 48,000 × *g* at 4°C. Transfer the supernatant that contains the denatured fusion protein to a clean centrifuge tube.
4. Add Triton X-100 to the supernatant to a final concentration of 1%.
5. Dialyze the sample into PBS/glycerol for 2–3 h. The volume of the dialysis buffer should be a minimum of 20 times the sample volume.
6. Dialyze the sample overnight vs. PBS with protease inhibitors using a buffer volume that is >100 times the sample volume.
7. Remove the sample from dialysis and centrifuge for 20 min at 4,000 × *g* at 4°C.
8. The extracted and re-natured protein from the inclusion bodies now can be applied to glutathione Sepharose columns and purified by following steps 9–12 of Subheading 3.2.1.

3.3. Enzymatic Cleavage to Remove GST Affinity Tag

Depending on the vector chosen, the GST affinity tag can be removed with thrombin, factor Xa, or PreScission protease, either in solution or while still bound to the column matrix. Cleavage in solution offers the advantage of more control over optimization of the cleavage conditions such as temperature, enzyme-to-substrate ratio, length of incubation, and buffer conditions. An advantage of on-column cleavage is the high level of purity obtained, but this comes at the expense of generally low yield due to less efficient protease cleavage and decreased control of the digestion conditions. Digestion can be performed in the glutathione buffer used to elute the protein from the affinity matrix provided there are no serine protease inhibitors in this buffer. After incubation, the enzyme can be inhibited using a variety of protease inhibitors or removed using a HiTrap benzamidine column. Separation of the target protein and the GST moiety can be achieved by re-chromatography on the glutathione Sepharose column (after dialysis to PBS buffer) to remove the GST and any un-cleaved fusion protein. See Notes 29 and 30 for digestion of GST fusion proteins while bound to the column matrix (this is recommended for use with PreScission protease).

1. Add the appropriate amount of thrombin or factor Xa to the affinity-purified fusion protein and incubate at 37°C (thrombin) or 25°C (factor Xa) for the desired length of time (see Note 31).
2. Inactivate the enzyme by adding 0.3 mM PMSF (final concentration) to the sample. To ensure complete inhibition, incubate the sample for 15 min at 37°C for thrombin or for 30 min at 25°C for factor Xa (see Note 33).
3. Dialyze the sample against PBS/EDTA/PMSF twice using 2 L per dialysis for a minimum of 4 h for each dialysis (see Note 34).
4. Centrifuge the dialyzed sample for 20 min at 4,000 × *g* at 4°C to remove any precipitated material that may have formed during the digestion or dialysis steps. At this point, the sample can be reapplied to the glutathione Sepharose column to remove the GST moiety and any undigested fusion protein.

3.3.1. Removal of GST Affinity Tag After Protease Cleavage in Solution

If the protein was digested in solution, the target protein can be purified further by re-chromatography on the glutathione Sepharose column to remove any un-cleaved fusion protein and the GST moiety. After a dialysis step to remove the glutathione present from the initial isolation, the sample is reapplied to the column and the sample of interest and thrombin is located in the column flow through. Protein purification should be performed at 4°C to minimize degradation of the target protein.

1. If reusing a glutathione Sepharose column, wash with >3 bed volumes of glutathione buffer at 1.5 mL/min (see Notes 35 and 36).
2. Wash the glutathione Sepharose column with 10 bed volumes of PBS at 1.5 mL/min.

3. Load the dialyzed, cleaved, fusion protein onto the column at 0.1 mL/min (see Note 23). Collect fractions for analysis by SDS-PAGE (see Note 37).
4. Wash the column with 2–3 bed volumes of PBS/EDTA/PMSF at 1.5 mL/min.
5. Elute the bound GST and any un-cleaved fusion protein with glutathione buffer at 0.3 mL/min for 5 bed volumes. Collect fractions for analysis by SDS-PAGE.
6. Analyze all unbound and bound fractions by SDS-PAGE. Pool fractions that contain the cleaved target protein and store at 4°C.

3.4. Further Purification of Cleaved Recombinant Protein

Unless a benzamidine column has been used to remove the protease, the cleaved sample contains the inactivated protease. The sample also may contain small amounts of residual GST or the GST fusion protein that did not rebind to the column, as well as aggregates or other minor contaminants such as proteolytic fragments or host cell proteins. While the protein is usually greater than 90% pure at this stage, many applications require an even higher level of purity. Therefore, HPLC gel filtration is recommended as a final polishing step that will isolate properly folded protein from aggregates or other contaminating species. As an alternative to HPLC, ion exchange chromatography also could be performed.

1. Concentrate the cleaved unbound protein to a final volume that is 0.5–1% of the column volume to be used for the separation using an ultrafiltration device, such as Amicon Ultra (Millipore), according to the manufacturer's instructions (see Note 38).
2. Remove particulates from the concentrated sample using a 0.22 μm filter unit, or centrifuge the sample at 4,000 × *g* for 20 min at 4°C.
3. Inject the concentrated, filtered sample onto an appropriate gel filtration column equilibrated in PBS or other compatible sample buffer.
4. Monitor the A_{280} with an online HPLC detector and collect fractions. Analyze the fractions by SDS-PAGE and pool the sample based on sample purity or concentration.
5. Store the pooled sample at 0–4°C or freeze appropriately until further use.

4. Notes

1. There are multiple pGEX gene fusion vectors available from GE Healthcare. A shared characteristic of all the vectors includes the presence of a *tac* promoter for chemically inducible,

high-level protein expression with IPTG. An internal *laqI*^{q} gene helps to maintain tight control over expression of the insert by binding to the *tac* promoter until induction with IPTG. The available fusion vectors are designed so that the polypeptide of interest can be inserted immediately after the GST gene using a polylinker site (see Fig. 2). A protease cleavage site is located between the GST sequence and the polylinker site so that the GST moiety can be enzymatically removed, if desired (see Note 2). Although not depicted in the vector diagram for all the different vectors, the polylinker sites are followed by stop codons in each reading frame. Although expression in *E. coli* is simple, economical, and efficient, it should be noted *E. coli* does not posttranslationally modify most proteins. If posttranslational modifications are required for function, expression in baculovirus (6) and yeast (7) have also been successful.

2. Choosing an appropriate vector is based on the ultimate use of the protein. If the GST moiety is to be cleaved away from the protein of interest, the pGEX-T series contains a protease cleavage site for thrombin, the pGEX-X series of vectors contain protease cleavage sites for factor Xa, and the pGEX-P series contains a cleavage site for PreScission protease (see Fig. 2). When selecting the protease cleavage site, be sure that the protein of interest does not contain an internal recognition sequence for this protease (see Table 1). One of the advantages of using the thrombin recognition site is that it is generally cost effective, as relatively small amounts of thrombin and short incubation times at 37°C are sufficient to cleave the protein with high efficiency. Factor Xa has very high specificity but is expensive and generally requires high enzyme-to-substrate ratios for efficient cleavage. PreScission protease is exclusive to GE Healthcare but has several advantageous characteristics, namely it is effective at low temperature (5°C) and it is also a GST fusion protein, thereby facilitating removal after cleavage. An additional factor to consider when selecting a vector and the restriction sites that will be utilized is the sequence that remains at the N terminus of the protein after removal of the GST moiety.

3. The use of JM105 or a similar strain is recommended for cloning and maintenance of the vector. Strains of BL21 and its derivatives, or another protease deficient strain, are recommended for maximal protein expression. Strains deficient in cytoplasmic protease gene products such as *lon, ompT, degP, or htpR* may minimize the effects of proteolytic degradation of overexpressed protein by the host. Do not use a strain of *E. coli* that carries the *rec1A* allele to propagate pGEX plasmids, as deletions or rearrangements of plasmid DNA have been reported.

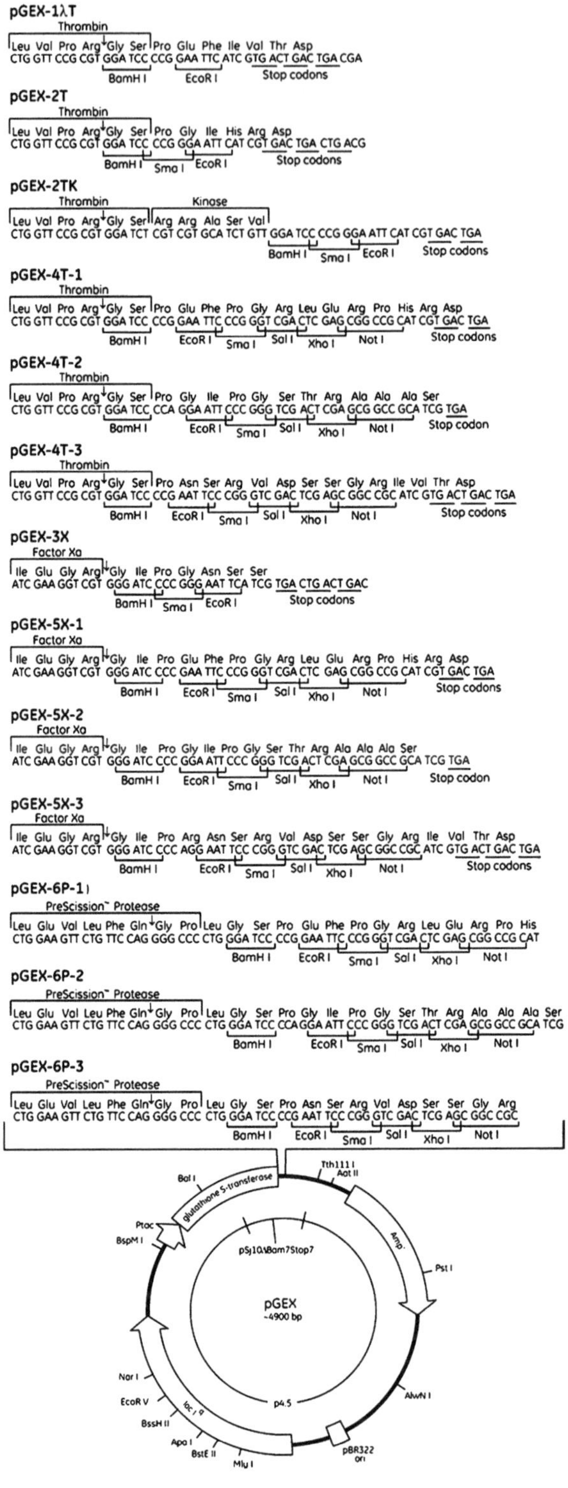

Fig. 2. GST fusion proteins may be constructed using one of ten different pGEX vectors. Shared characteristics of the vectors include a gene for ampicillin resistance; the *lacI*^q gene product is a repressor protein which binds to the operator region of the P*tac* promoter preventing expression of protein product until induction with IPTG. The gene for the protein of interest can be inserted in the polylinker site immediately after the GST gene. The protease cleavage sites are located between the GST moiety and the protein of interest. Although not shown for all vectors, stop codons in all three reading frames follow the polylinker site. A consideration to be made when selecting the vector is to minimize the number of amino acids remaining after proteolytic cleavage (© 2009 General Electric Company – reproduced by permission of the owner).

Table 1
Recognition sequences for selected endoproteases

Enzyme	Optimal recognition sequence
PreScission protease	LEVLFN↓GP
Factor Xa	IEGR↓X, X≠R,P
Thrombin	LVPR↓GS[a] IRPK↓LK[b] MYPR↓GN[b] AR↓G[c] GK↓L[c]

[a] The recognition sequence found in the pGEX plasmids
[b] Thrombin has the general recognition sequence P4-P3-P-R/K↓P1′-P2′ where P3 and P4 are hydrophobic amino acids and P1′ and P 2′ are non-acidic amino acids; two examples are shown
[c] Thrombin also recognizes the general scheme P2-R/K↓P1′ where either P2 or P1′ is G; two examples are shown

4. Glutathione Sepharose 4B bulk media is used in this purification protocol. This chromatography media is ideal for screening conditions and allows for easy scale-up. Purifications are performed easily using either gravity flow or a low-pressure chromatography system, or may be used in batch purifications based on centrifugation. GSTrap FF affinity columns are 1 or 5 mL prepacked columns that also are available and may be connected in series for scale-up. These columns are for use with a liquid chromatographic system, a pump, or a syringe. In addition to these formats, glutathione Sepharose also is available in spin columns and 96-well plate formats for rapid screening of GST fusion protein expression and purification conditions.
5. DFP is an extremely dangerous neurotoxin. Follow the precautions supplied by the manufacturer and only handle the neat reagent in a chemical fume hood.
6. Addition of protease inhibitors to the lysis buffer will help to prevent proteolytic degradation of the target protein during extraction. However, the exact cocktail of protease inhibitors added to the lysis buffer should be tailored to the characteristics of the target protein. Reducing agents such as 2-ME, DTT, or TCEP at concentrations of 1–10 mM should be added to the buffer as needed. Nonionic detergent, such as 1% Triton X-100, also may be added to the buffer to aid in extraction.
7. Thrombin and factor Xa are commercially available from a wide variety of vendors. Each vendor's product may differ

slightly in performance such that once a product is chosen that should be used exclusively. Preparations that are not as pure may contain contaminating proteolytic activities that would cause undesirable proteolysis of the fusion protein.

8. At the end of the induction period, harvest the cells by centrifugation. Re-suspend the pellet in 200 μL of cold PBS. Lyse the cells using a sonicator with an appropriate probe for 10 s; keep the sample on ice. Save a small aliquot of the lysate for analysis by SDS-PAGE. Centrifuge the lysed sample at 16,000 × *g* for 15 min. Remove the supernatant to a separate tube. Re-suspend the pellet in 200 μL PBS. Analyze some of the crude lysate, the supernatant, and the re-suspended pellet by SDS-PAGE or Western blotting. If the recombinant protein is observed in both the lysate and supernatant fraction, the sample is adequately soluble. If the recombinant protein is present primarily in the lysate and re-suspended pellet, the expressed protein is most likely in occlusion bodies. In cases where the protein is located in inclusion bodies, explore different expression conditions to shift expression to a soluble form (see Note 9).
9. If the fusion protein is expressed primarily in inclusion bodies (e.g., >80% insoluble), various strategies may be employed to improve solubility. Often, induction at a lower temperature (15–25°C) is effective at shifting expression from inclusion bodies to the soluble fraction. Cells induced at this lower temperature will grow more slowly and will require overnight induction periods. Use of alternative host strains or modifications to the plasmid construct may be necessary in some cases. If the fusion protein remains primarily in inclusion bodies even after attempts to obtain soluble protein, it may be worthwhile to scale up the production of *E. coli* and purify enough protein from the small amount of soluble protein that is available. In some cases, other fusion protein tags should be explored.
10. Glycerol stocks of the bacterial culture should be prepared by mixing equal volumes of overnight culture and 70% glycerol, invert several times to mix then store at −80°C. Once positive expression is confirmed, DNA sequencing should be performed to ensure no errors were introduced during the PCR amplification or ligation.
11. Low protein expression levels could be the result of rare codon usage. In this case, switching to another strain of *E. coli* that contains extra transcripts for rare codon tRNA can improve protein production significantly. Different media formulations or the addition of up to 2% glucose also may improve expression (8, 9). If it is suspected that the expressed protein is toxic to the cells (usually they will stop growing after induction with IPTG), try inducing at a later time

point for a shorter amount of time. Decreasing the IPTG concentration is another option that should be explored.

12. Monitor cell growth by reading the optical density at 600 nm (A_{600}). An overnight culture should reach an A_{600} ~ 1–1.2. Cells should be induced at an early phase of the logarithmic growth curve for *E. coli*, A_{600} ~0.5–0.7. At 37°C, it will take approximately 2 h for cultures to reach an early log stage of growth. If cells are grown at a lower temperature, the incubation time must be increased, since cells grow more slowly at lower temperatures. Generally, the greatest yield of fusion protein will be obtained when the cells are induced at A_{600} = ~ 0.5. After induction with IPTG, a general guideline for incubation time is as follows: 3 h at 37°C, 5 h at 30°C, and overnight at 25°C or lower. Harvest the cells prior to saturation, A_{600} ~ 1.0–1.2.

13. Protein expression levels can be monitored by analyzing pre- and post-induction samples by SDS-PAGE. Remove a 1-mL aliquot of cells prior to induction and a second 1-mL sample after the 3 h induction period. Centrifuge the aliquots for 2 min at 16,000 × *g* in a bench top centrifuge for 2 min. Remove the supernatant with a pipette. Re-suspend the cell pellet in 200 μL SDS sample buffer and heat for 5 min at 90°C. Load 5–10 μL of the sample on an SDS-PAGE gel. A band at the appropriate molecular weight should be visible on the gel.

14. To ensure adequate aeration, flasks should be filled only to 20–30% of their capacity.

15. It is important that no serine protease inhibitors be present in the sample prior to cleavage with thrombin or factor Xa. The following protease inhibitors must be removed prior to thrombin or factor Xa cleavage: AEBSF, APMSF, antithrombin III, antipain, α_1-antitrypsin, aprotinin, chymostatin, hirudin, leupeptin, and PMSF. In addition, Pefabloc TH benzamidine is specific for thrombin, and Pefabloc FXa is specific to factor Xa. The following protease inhibitors should be avoided with PreScission Protease: 100 mM $ZnCl_2$, 100 μM chymostatin, and 4 mM Pefabloc.

16. There are a number of alternative methods for the detection of GST fusion proteins. For proteins that express well at a high level, the simplest method of monitoring the purification is via SDS-PAGE gels stained with Coomassie blue or silver stain. An alternative for proteins that express at low levels is to use Western blotting using an antibody directed at either GST or the target protein. An alternative for small-scale expressions is to monitor for the presence of GST with an ELISA or CDNB enzyme assay.

17. Save a small amount of protein sample from each step of the purification – including lysis, supernatant, pellet, unbound fractions from sample loading, wash steps, and elution steps – to be analyzed by SDS-PAGE or Western blotting. After all samples have been analyzed and pooled, discard unwanted fractions.
18. A bed volume is equal to one half of the 50% slurry of glutathione Sepharose 4B used for purification. To prepare a 50% slurry of glutathione Sepharose (it is supplied as ~75% slurry): Determine the bed volume required for the purification scale. Re-suspend the glutathione Sepharose 4B. For each mL of bed volume required, pipette 1.33 mL of the 75% slurry to an appropriate size centrifuge tube. Centrifuge at 500×*g* for 5 min. Decant the supernatant. Wash with 10 bed volumes (10 mL per 1.33 mL original slurry) of PBS by gently inverting the tube several times to mix, then centrifuge at 500×*g* for 5 min. Decant the supernatant. Failure to remove the ethanol storage solution may interfere with subsequent binding steps. Add 1 mL of PBS for each 1.33 mL of the original slurry. This results in a 50% slurry.
19. Glutathione Sepharose has an advertised minimum binding capacity of 8 mg/mL. A 20 mL column should be adequate to purify protein from 3×600 mL *E. coli* cultures that contain ~20–40 mg fusion protein per culture. Both the amount of resin used and the column size can be scaled up or down depending on the amount of protein to be purified.
20. Re-suspend the pellet using 25–50 μL wash buffer per milliliter original culture.
21. Cell disruption is complete when the suspension partially clears and turns a slightly darker color. Avoid frothing during sonication, as this can lead to denaturation of the fusion protein. Avoid oversonication, as this can lead to co-purification of host *E. coli* proteins along with the GST fusion protein. High viscosity due to the release of nucleic acids during sonication can be reduced by adding DNase or benzoase to the lysis buffer. Lysozyme up to a concentration of 0.2 mg/mL may be added as an aid to cell lysis. An alternative to sonication is the use of commercially available cell-extraction formulations (e.g., BugBuster from Novagen). However, these products could contain proprietary components that interfere with downstream applications.
22. Use of a peristaltic pump is advised to evenly control flow rates. If compression of the resin occurs, the pressure is too high and the flow rate should be reduced.
23. The binding kinetics between glutathione and GST are relatively slow. Therefore it is important to use low flow rates to

achieve maximum binding capacity, e.g., 0.1 mL/min for a 2.5-cm I.D. column. Using flow rates that are too fast may decrease the amount of bound fusion protein due to the slow kinetics of association. Washing and elution steps may be performed at faster flow rates to save time (1.5 and 0.3 mL/min, respectively). For large volume samples, binding is most conveniently performed overnight, with adjustments to flow rates so that the column does not run dry.

24. Analysis of the unbound fractions will indicate whether the fusion protein is binding or not. If the protein is not detected in early fractions but appears in later fractions, it indicates that the column capacity has been exceeded. Reduction in protein load or increase in column size will alleviate this condition.

25. If the fusion protein binds poorly to the glutathione Sepharose, there are several alternatives to try increasing binding efficiency. Try a milder lysis method. If the method used is too harsh, the protein may become denatured and, therefore, unable to bind to the column. Remove the ethanol storage solution from the glutathione Sepharose; reduce the column with fresh glutathione buffer followed by a wash with PBS immediately prior to sample loading. Increase the amount of resin and/or decrease the flow rate used for loading the sample. Try adding 1–20 mM DTT to the sample. Also, if re-naturing the protein from inclusion bodies, be sure that all denaturants have been removed from the buffer, either through exhaustive dialysis or application of the sample to a desalting column, before applying to the glutathione column. If a problem still persists, clean the column according to the manufacturer's recommendation. If binding still is not restored, try using fresh resin.

26. The yield of fusion protein also can be estimated by measuring absorbance at 280 nm (A_{280}). The extinction coefficient for the GST moiety alone is ~1.5, e.g., 1.00 A_{280} = ~0.6 mg/mL protein, although the extinction coefficient for the fusion protein will depend partially on the absorbance characteristics of the target protein.

27. If there is a problem eluting the protein from the glutathione Sepharose column, try decreasing the flow rate and increasing the volume of elution buffer that is used. Be sure to use fresh reduced glutathione buffer (make it the day that it will be used). Increase the concentration of glutathione up to 40 mM and/or raise the pH of the buffer to pH 8.0–9.0 using NaOH. Add 1–20 mM DTT (or other reducing agent) to the buffer. Addition of a nonionic detergent also may improve solubilization and minimize any aggregation that may be occurring.

28. Contamination of the purified fusion protein with *E. coli* host cell proteins is an indication that sonication has been too

severe. If degraded fragments of the fusion protein are present, try adding additional protease inhibitors to the lysis buffer. Keep all samples, buffers, and collection tubes cold to minimize proteolysis. If degradation is occurring during protein expression, try inducing the sample late (~0.8 OD_{600}) and decrease the length of the induction period. Switching to an alternative host strain also may help.

29. An alternative to in-solution digestion is protease cleavage of fusion protein while bound to the glutathione Sepharose matrix. The fusion protein is extracted and loaded onto the glutathione Sepharose followed by washing with PBS (see Subheading 3.2, steps 1–11). Rather than elute the protein with glutathione buffer, a PBS solution containing the enzyme is loaded onto the column and incubated for several hours at room temperature (4°C for PreScission protease). The cleaved protein is then washed out of the column with several column volumes of PBS. Residual fusion protein and the GST moiety can be removed from the column by washing the column with reduced glutathione buffer. The amount of enzyme and incubation time must be determined empirically for each fusion protein. Analyze all samples by SDS-PAGE to determine cleavage efficiency and protein purity.

30. In a batch mode protease cleavage, add an empirically determined amount of enzyme to the fusion protein bound to the resin. Use 1 mL of PBS-containing enzyme per milliliter of bed volume. Incubate at room temperature for several hours with gentle agitation. Centrifuge at 500 × *g* for 5 min at room temperature to sediment the resin. Remove the supernatant which contains the target protein to a separate tube. Wash the glutathione Sepharose with PBS up to three times to recover the cleaved fusion protein. Incubate the resin with glutathione buffer to remove GST and residual fusion protein. Use 1 mL glutathione buffer per milliliter of resin. Centrifuge at 500 × *g* for 5 min and recover the eluted GST. Repeat up to three times. Analyze samples by SDS-PAGE.

31. If using thrombin protease, digestion can be carried out in the glutathione buffer used to elute the protein from the column. If using factor Xa, it is recommended to dialyze the protein into either a Tris buffer or PBS buffer prior to digestion; the glutathione present in the elution buffer can disrupt the disulfide bridges present in factor Xa leading to inefficient digestion of the fusion protein. An empirical determination of digestion conditions for each fusion protein must be determined in a pilot digestion experiment. A convenient method is to digest 100 μg of fusion protein over a range of enzyme-to-substrate ratios and vary the incubation time. Typical incubation times range from 2 to 8 h. Recommended enzyme-to-substrate

ratios for thrombin are 1:100, 1:350, 1:1,000, and 1:3,000 (units of enzyme per microgram of fusion protein), and for factor Xa are 1:10, 1:25, 1:50, 1:100, 1:300 (microgram enzyme per microgram fusion protein). At desired time points, remove 2 μg of protein and stop digestion by adding the sample aliquot to boiling SDS sample buffer. Analyze the samples by SDS-PAGE to determine the optimal cleavage conditions. In addition to enzyme-to-substrate ratio and time, consider altering buffer conditions, such as increasing or decreasing NaCl concentration or adding Ca^{2+} to the buffer (see Note 32 for troubleshooting tips).

32. If multiple bands are observed on SDS-PAGE for the target protein after digestion, check the target protein sequence for potential secondary protease recognition sites (see Table 1). If secondary cleavage sites exist, re-clone into a different vector. If no cleavage is observed, check the DNA sequence to verify the presence and integrity of the expected protease cleavage site. Make sure that protease inhibitors have been removed completely from the buffer. Add more enzyme and/or increase incubation times to overnight. If the digestion remains incomplete at the highest enzyme-to-substrate ratios and longest time points, consider reengineering the protease cleavage site to include several glycines between the GST moiety and the target protein to decrease the likelihood of steric hindrance interfering with the cleavage (10, 11).

33. Benzamidine Sepharose is a specific media designed for the removal of trypsin-like serine proteases such as thrombin or factor Xa. After digestion is complete, if inhibition of the enzyme using serine protease inhibitors is undesirable, the sample may be applied to a HiTrap benzamidine column to remove the enzyme from the sample.

34. If the sample is to be re-chromatographed on the glutathione Sepharose column to remove GST and any undigested fusion protein, it is critical that the reduced glutathione be removed completely from the sample. Glutathione equilibrates very slowly across dialysis membranes; therefore, if dialysis tubing with a molecular weight cutoff (MWCO) <12,000 is used, three or more changes of buffer may be required for complete removal. Increased dialysis time (overnight) and a larger volume of buffer also may be required for large sample volumes.

35. The same glutathione Sepharose column may be used for both initial isolation of the fusion protein and for re-purification after enzymatic cleavage. It is recommended to dedicate a single column to an individual protein construct to avoid potential cross contamination of different recombinant proteins. Columns may be reused multiple times. If binding efficiency decreases over time, clean the column according to the

manufacturer's recommendations. If binding activity is not restored, a new column should be used.

36. Maximal binding occurs if the column is fully reduced; however, if the glutathione Sepharose column has been washed less than 48 h prior to this step, this step may be skipped.
37. The target protein will be in the unbound fraction and the GST and any un-cleaved fusion protein will bind to the column.
38. Small sample volumes are recommended for good resolution of target protein vs. other contaminants. If it is not feasible to concentrate the sample into a small volume, multiple injections of the sample may be performed.

Acknowledgments

This work was supported in part by the National Institutes of Health Grant HL038794 and institutional grants to The Wistar Institute, including an NCI Cancer Core Grant (CA10815) and grants from the Pennsylvania Department of Health.

References

1. Smith, D. B. and Johnson, K. S. (1988) Single-step purification of polypeptides expressed in *Escherichia coli* as fusions with glutathione S-transferase. *Gene* 67, 31–40.
2. Frangioni, J. V. and Neel, B. G. (1993) Solubilization and purification of enzymatically active glutathione S-transferase (pGEX) fusion proteins. *Anal Biochem* 210, 179–187.
3. Grieco, F., Hay, J. M., and Hull, R. (1992) An improved procedure for the purification of protein fused with glutathione S-transferase. *Biotechniques* 13, 856–858.
4. Singh, S. M. and Panda, A. K. (2005) Solubilization and refolding of bacterial inclusion body proteins. *J Biosci Bioeng* 99, 303–310.
5. Singh, S. M., Eshwari, A. N., Garg, L. C., and Panda, A. K. (2005) Isolation, solubilization, refolding, and chromatographic purification of human growth hormone from inclusion bodies of *Escherichia coli* cells: a case study. *Methods Mol Biol* 308, 163–176.
6. Davies, A. H., Jowett, J. B., and Jones, I. M. (1993) Recombinant baculovirus vectors expressing glutathione-S-transferase fusion proteins. *Biotechnology (N Y)* 11, 933–936.
7. Mitchell, D. A., Marshall, T. K., and Deschenes, R. J. (1993) Vectors for the inducible overexpression of glutathione S-transferase fusion proteins in yeast. *Yeast* 9, 715–722.
8. Grossman, T. H., Kawasaki, E. S., Punreddy, S. R., and Osburne, M. S. (1998) Spontaneous cAMP-dependent derepression of gene expression in stationary phase plays a role in recombinant expression instability. *Gene* 209, 95–103.
9. Pan, S. H. and Malcolm, B. A. (2000) Reduced background expression and improved plasmid stability with pET vectors in BL21 (DE3). *Biotechniques* 29, 1234–1238.
10. Guan, K. L. and Dixon, J. E. (1991) Eukaryotic proteins expressed in *Escherichia coli*: an improved thrombin cleavage and purification procedure of fusion proteins with glutathione S-transferase. *Anal Biochem* 192, 262–267.
11. Hakes, D. J. and Dixon, J. E. (1992) New vectors for high level expression of recombinant proteins in bacteria. *Anal Biochem* 202, 293–298.

Chapter 15

Purification of Proteins Fused to Maltose-Binding Protein

Mario Lebendiker and Tsafi Danieli

Abstract

Maltose-binding protein (MBP) is one of the most popular fusion partners being used for producing recombinant proteins in bacterial cells. MBP allows one to use a simple capture affinity step on amylose–agarose columns, resulting in a protein that is often 70–90% pure. In addition to protein-isolation applications, MBP provides a high degree of translation and facilitates the proper folding and solubility of the target protein. This chapter describes efficient procedures for isolating highly purified MBP-target proteins. Special attention is given to considerations for downstream applications such as structural determination studies, protein activity assays, and assessing the chemical characteristics of the target protein.

Key words: Maltose-binding protein, Protein expression and purification, Protein solubility, Protein aggregation and soluble aggregates, Fusing protein tags, Folding, Purification techniques, Amylose–agarose, TEV protease

1. Introduction

MBP is one of the oldest and most popular fusion partners being used for producing recombinant proteins in bacterial cells. It is the product of the *malE* gene in *Escherichia coli*, part of the maltose/maltodextrin system of that organism, and it acts as a receptor for chemotaxis and gene regulation (1). An advantage of MBP is that it can be expressed in bacterial cells in both secreted and nonsecreted forms. Expression levels are higher when the protein is produced in the cytoplasm, however fusing the target protein to the secreted form of MBP delivers the complex into the periplasm, and this can facilitate the folding of proteins with disulfide bonds (2) (see Note 1).

MBP enhances both the production and solubility of its fusion partner by a mechanism that is still not completely understood. Studies have suggested that MBP functions as a "chaperone magnet"

Dermot Walls and Sinéad T. Loughran (eds.), *Protein Chromatography: Methods and protocols*, Methods in Molecular Biology, vol. 681, DOI 10.1007/978-1-60761-913-0_15, © Springer Science+Business Media, LLC 2011

by recruiting chaperones that normally associate with MBP to the vicinity of the target protein (3) or that form large micelle-like aggregates with incompletely folded passenger proteins held inside (4, 5). Recent studies also show that the MBP plays a passive role in the folding of its target fusion partner and works similarly to the solubility-enhancing protein NusA (6).

Despite the high-metabolic burden on the host cell, owing to the considerable size of the protein (approximately 42 kDa), MBP is still considered to be one of the best choices for circumventing heterologous expression problems. The isolation and purification of a protein tagged with MBP can be achieved by using a cheap and convenient affinity column that can yield tagged protein that is 70–90% pure following a single-capture step. In order to achieve a higher degree of purification, which is often required for downstream applications such as structural studies, one should add additional purification steps such as ion exchange, hydrophobic exchange, and size exclusion chromatography.

Ion exchange chromatography is essential as an intermediate step for separating target proteins from protein contaminants such as chaperons and other host cell proteins. It also allows one to separate the target protein from heterogeneously folded forms that are a consequence of the expression and purification conditions used and from heterogeneity in posttranslational modifications. Sometimes purification techniques that separate proteins according to their charge are insufficient, and other approaches based on different principles, such as hydrophobic exchange chromatography or hydroxyapatite, should be used. As a final polishing step, it is often recommended to use size-exclusion chromatography, not only to eliminate protein contaminants and low molecular weight molecules but also to obtain a homogeneous oligomeric form. An added value of the gel filtration step is that the protein will elute in the final desired buffer.

Following purification, the MBP tag can be removed from the target protein by a specific protease (see Notes 2 and 3). However, structural studies of the proteins, and crystallography in particular, may gain a huge advantage from using the uncleaved protein because the structure of MBP has already been solved, and the rather straightforward procedure of molecular replacement phasing can be employed, instead of the exhaustive time-consuming procedure of heavy atom derivative phasing (see Note 4).

Advantages of the MBP fusion system include enhanced expression, improved solubility, ease of purification, and mild elution conditions. MBP purification procedures are highly efficient and compatible with most downstream applications, making MBP one of the most desirable choices of fusion partner for recombinant protein expression. Here, efficient procedures for isolating highly purified MBP-target proteins are described.

2. Materials

All materials may be sourced from Sigma-Aldrich unless otherwise stated.

2.1. Protein Expression Conditions

1. Vectors: the *Rack1* expression plasmid lpMAL-c2MBP-Rack1 was a kind gift from Prof. Daria Mochly-Rosen.
2. *E. coli* BL21(DE3) strain (Novagen).
3. Luria-Bertani (LB) medium: prepare 1 L using 10 g bactotryptone, 5 g bacto-yeast extract, 10 g NaCl, 166 μL NaOH (10 N), and 10 mL MgSO4 (1 M).
4. Isopropyl-β-d-thiogalactopyranoside.
5. Incubator-Shaker, e.g., Innova 43 (New Brunswick Scientific).

2.2. Amylose–Agarose Column

Homemade amylose–agarose column.

1. Sepharose 6B.
2. Vinyl sulfonic acid.
3. Amylose (Type III from potato, Sigma, St. Louis, MO).
4. 1 M sodium carbonate pH 11.
5. 0.9% NaCl.
6. 20% Ethanol.
7. Sinter glass.

Commercial amylose–agarose column: amylose resin high flow (New England Biolabs).

2.3. Protein Purification

2.3.1. Lysis and Clarification

1. Micro-fluidizer (M-110 EHIS; Microfluidics Corp., Newton, MA).
2. Buffer A (20 mM Tris–Cl pH 7.4, 200 mM NaCl, 1 mM EDTA, 0.02% NaN_3).
3. Protease inhibitor cocktail.
4. DNase 1.
5. Lysozyme (Thermo Scientific).
6. Filter GF/D (Whatman) and 0.45 μm filter (Whatman).

2.3.2. Capture: Affinity Chromatography: Amylose Resin

1. ÄKTAexplorer system (GE Healthcare).
2. Maltose.
3. A 12% sodium dodecyl sulfate–polyacrylamide (SDS–PAGE) gel.

2.3.3. Intermediate Purification: Ion Exchange Chromatography

1. Resource 30Q column (GE Healthcare) 7× 1.6 cm.
2. Buffer A1: 20 mM Tris–Cl pH 8.0, 0.02% NaN_3.

3. Buffer B: buffer A1 + 1 M NaCl.
4. Centriplus cut-off 30 kDa (Amicon, Millipore).

2.3.4. Final Polishing: Size Exclusion Chromatography

1. Sephacryl S100 FF column (GE Healthcare) 92 × 2.6 cm.
2. Buffer A2 (20 mM Tris–Cl pH 8.0, 0.1 M NaCl, 0.02% NaN_3).

2.3.5. Column Regeneration and Storage

1. 0.1% SDS.
2. 0.5 M NaOH.
3. 20% Ethanol.

3. Methods

3.1. Expression Conditions

1. *E. coli* BL21(DE3) competent cells are transformed with 10 ng of pMAL-c2MBP-Rack1 and plated on LB agar plates containing 100 μg/mL ampicilin and incubated for 16 h at 37°C, (see Note 1).
2. A single colony is used to inoculate a tube containing 10 mL LB with 100 μg/mL ampicilin.
3. The cells are grown in a shaker incubator for 16 h, transferred into 1 L of LB medium at an inoculum to medium ratio of 1:100 and placed in a 37°C incubator shaker.
4. IPTG is added to a final concentration of 0.3 mM when the OD600 reaches 0.6.
5. The cells are harvested after 6 h of incubation at 30°C.
6. Pellets are kept at −80°C until further processing.

3.2. Preparation of an Amylose–Agarose Column

As an alternative to commercially available columns, home-made amylose–agarose columns may be prepared using a procedure similar to that used to prepare lactose–sepharose beads (7).

1. Twenty-five milliliters of Sepharose 6B is washed with water in a Sinter glass and with 1 M sodium carbonate pH 11.
2. The resin is resuspended in 25 mL of 1 M sodium carbonate pH 11 and allowed to react by mixing for 70 min at room temperature with 5 mL vinyl sulfonic acid.
3. After washing with 500 mL of water, the resin is resuspended in a 25 mL solution of 2.6 g amylose in 1 M sodium carbonate pH 11, with continuous stirring overnight.
4. After washing again with water, 0.9% NaCl, and water again, the resin is maintained in a solution of 20% ethanol/80% water at 4°C. The Amylose-agarose column can be purchased from New England Biolab (Amylose Resin High Flow #E8022L) or it can be prepared at home (See Note 5).

3.3. Protein Purification

3.3.1. Lysis and Clarification

All procedures should be performed at 4°C.

1. The frozen cell pellet from a 1 L culture is thawed on ice and resuspended in 70 mL of buffer A supplemented with protease inhibitor cocktail 1:200, 50 μg/mL DNase I, and 0.2 mg/mL lysozyme.
2. The cells are lysed mechanically using a Micro-fluidizer at 21,000 psi.
3. Insoluble cell debris is removed from the cell lysate by centrifugation at 4°C for 20 min (15,000 × *g*); subsequently, the cleared lysate is first filtered through a GF/D filter and then a 0.45 μm filter.

3.3.2. Capture: Affinity Chromatography: Amylose Resin (See Fig. 1)

1. An amylose–agarose column 9.2 × 2.6 cm (49 mL) is equilibrated, prior to the lysis steps, with buffer A using an ÄKTAexplorer system at 4°C. Equilibration is confirmed by measuring pH and conductivity. Pressure limit: 0.5 MPa.
2. The column is loaded with filtered lysate at 1.7 mL/min and washed with buffer A at 2.5 mL/min up to low optical density (~5 cv, column volume).
3. Protein is eluted with elution buffer (buffer A + 20 mM maltose) at 1.5 mL/min, collecting fractions of 9 mL during 4 cv.

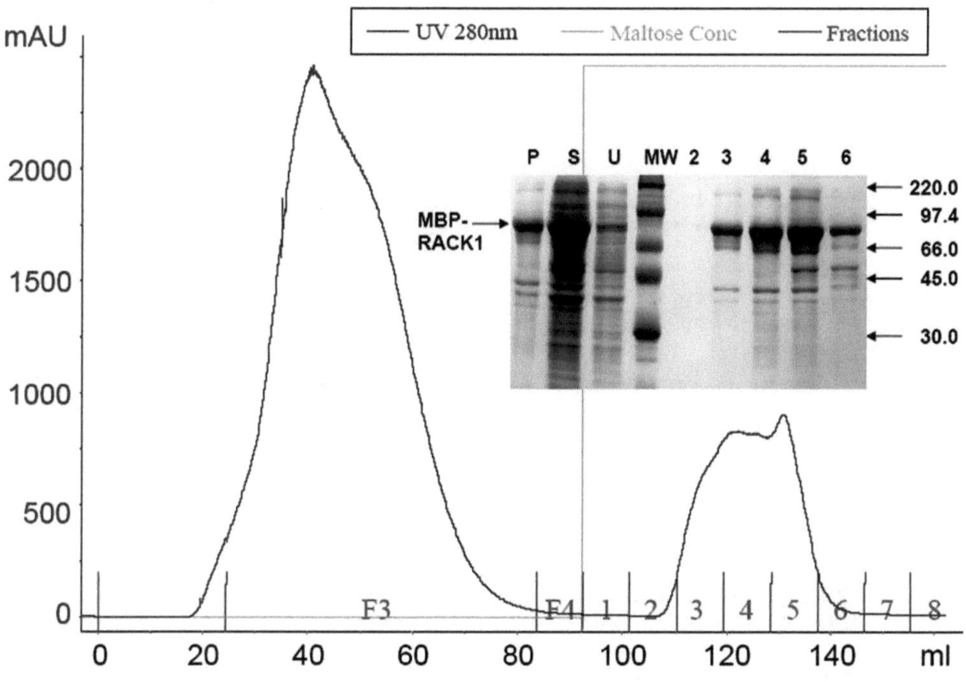

Fig. 1. Capture – affinity chromatography – amylose resin. A cell pellet from a 1 L culture was lysed, clarified, and purified on a homemade amylose resin column, as described in Subheading 3.3.2. Fractions were analyzed by SDS–PAGE. *P* pellet, *S* supernatant after lysis and centrifugation, *U* unbound to amylose resin, *MW* molecular weight markers.

4. Samples from each fraction are analyzed for protein content by SDS–PAGE. Protein-containing fractions are then pooled according to the profile obtained.

3.3.3. Intermediate Purification: Ion Exchange Chromatography (See Fig. 2)

1. A Resource 30Q column (7 × 1.6 cm, 14 mL) is equilibrated with buffer A1, and equilibration is confirmed by measuring pH and conductivity as before.
2. Pooled protein from the affinity step is diluted 1:4 with buffer A1 to reduce conductivity, filtered with a 0.45 μm filter and loaded at 6 mL/min.
3. The column is washed with 4% buffer B at 4 mL/min up to low optical density (~3 cv).
4. Protein is eluted with a 15 cv gradient (4–15% buffer B at 4 mL/min) collecting fractions of 4 mL, and then 4 cv 15–30% buffer B, and 4 cv 30–100% buffer B at 6 mL/min, collecting fractions of 9 mL.
5. Samples from each fraction are analyzed for protein content by SDS–PAGE. Protein-containing fractions are then pooled according to the profile obtained. The main peak elutes at around 11.5% buffer B.

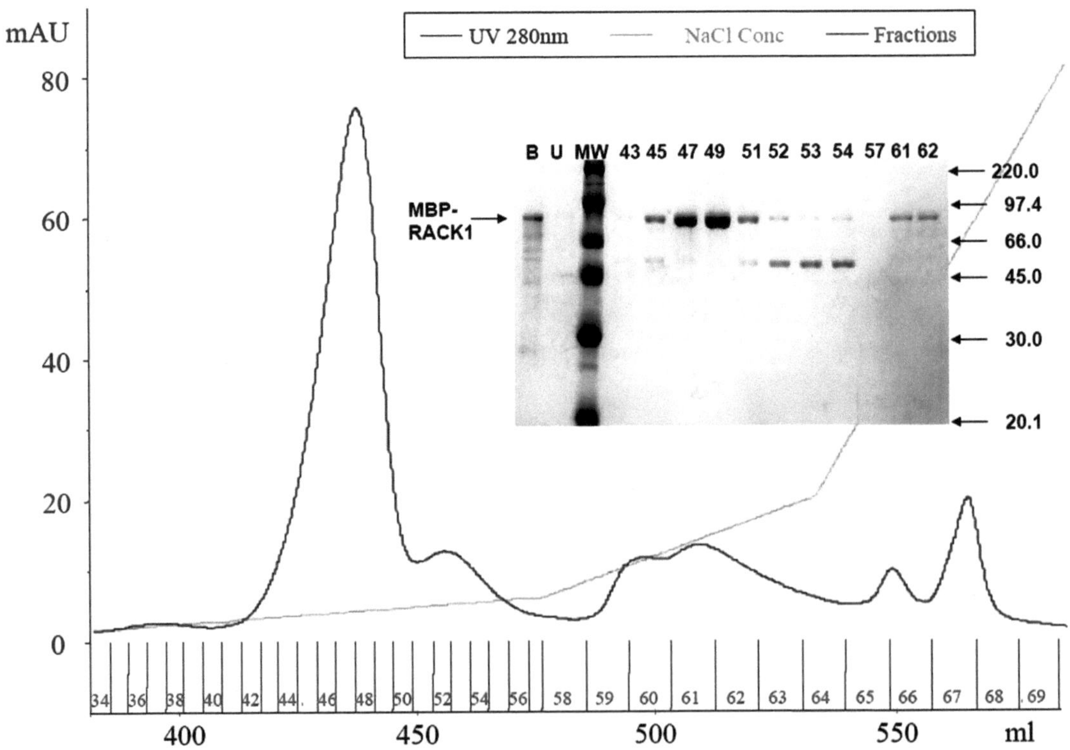

Fig. 2. Intermediate purification: ion exchange chromatography. Affinity purified protein was further purified by anion exchange chromatography, as described in Subheading 3.3.3. Fractions were analyzed by SDS–PAGE. *B* before binding, *U* unbound to anion exchange resin, *MW* molecular weight markers.

6. The pooled sample is concentrated to 10 mL with a Centriplus unit (cut-off 30 kDa).

3.3.4. Final Polishing: Size Exclusion Chromatography (See Fig. 3)

1. A Sephacryl S100 FF column (92 × 2.6 cm, 489 mL) is equilibrated with buffer A2. Equilibration is confirmed by measuring pH and conductivity as before.
2. The concentrated peak from Subheading 3.3.3, step 6 is loaded and run (isocratic elution) at 2 mL/min, collecting fractions of 4 mL.
3. Samples from each fraction are analyzed for protein content by SDS–PAGE. The main peak elutes at around 0.47 cv (see Note 6).

3.3.5. Column Regeneration and Storage

1. Amylose–agarose columns are regenerated with 0.1% SDS at room temperature, then water, and maintained in 20% ethanol at 4°C (see Note 7) (consult manufacturer's instructions if using purchased columns).
2. Resource 30Q columns are regenerated with 0.5 M NaOH (up-flow direction), then circulate buffer until the pH is neutral, then circulate water and finally 20% ethanol. Maintain the columns in 20% ethanol at 4°C.
3. Sephacryl S100 FF columns are regenerated with 0.5 M NaOH (up-flow direction), then circulate buffer until the pH is neutral, and maintain the columns in buffer containing 0.02% NaN_3 at room temperature.

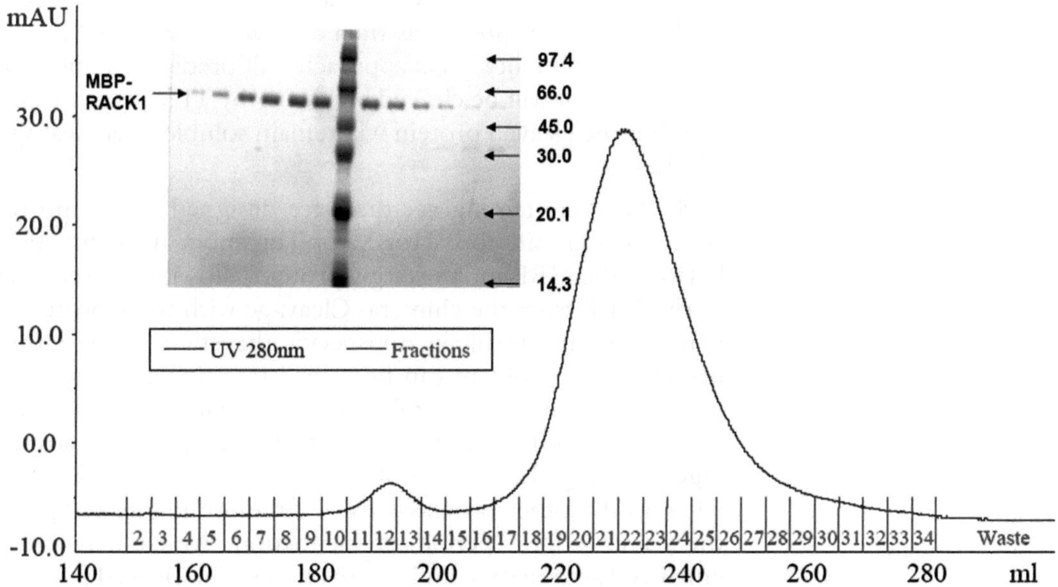

Fig. 3. Final polishing: size exclusion chromatography. The pooled eluate obtained following anion exchange was further purified by gel filtration chromatography as described in Subheading 3.3.4. Fractions were analyzed by SDS–PAGE (*MW* molecular weight markers).

4. Notes

1. The most popular commercial vectors are the pMAL vectors, available from NEB. These vectors allow the expression of a secreted or cytosolic form of MBP, fused to a target protein, under the regulation of an IPTG-induced *tac* promoter. The use of this promoter allows pMAL vectors to be used in a wide variety of bacterial hosts, since the *tac* promoter utilizes the bacterial RNA polymerase for transcription. Removal of the MBP fusion protein is also an option in many vectors, where protease recognition sites such as factor Xa and enterokinase are inserted between the MBP and the target gene multiple cloning site (see Chapter 19 for more details on the removal of affinity tags). The more recent pMAL vectors from NEB contain mutated MBP that allows improved affinity binding to amylose resins. There are also several options for noncommercial vectors, available from the Addgene repository (www.addgene.org). In some noncommercial vectors, the MBP was cloned under the control of a T7 promoter, suitable for tighter regulation of expression using *E. coli* DE3 strains.
2. A noteworthy system was developed by Waugh's group to assess the solubility of the target protein after the removal of the MBP. This system requires the coexpression of the MBP-target protein with a compatible vector containing the TEV protease gene. Both plasmids are cotransformed into the same cell and induced by IPTG and anhydrotetracycline, respectively. The target protein is then analyzed by SDS–PAGE to determine solubility. This approach will predict whether the fusion protein will be cleaved efficiently by TEV protease and whether the cleaved protein will remain soluble after cleavage (8, 9).
3. Early MBP-containing vectors were designed with a proteolytic cleavage site for factor Xa or Thrombin at the junction between the MBP and the target protein, allowing the removal of the MBP from the chimera. Cleavage with these proteases may sometimes result in nonspecific digestion of the target protein. This problem can be solved by using more specific proteases such as Enterokinase, Rhinovirus 3C protease (Precision™/GE Healthcare) or TEV protease. The advantage of using the TEV protease is that it is active at 4°C, whereas the other proteases usually require higher temperatures and a long period of incubation, which may cause enhanced aggregation and inactivation of the cleaved protein. Another advantage of using this protease is its resistance to detergents that are often essential in the preparation of membrane proteins and other hydrophobic proteins (10). There are

several commercial and noncommercial vectors that express a TEV protease that enables one to remove the MBP from the target protein (11–13). TEV protease does not have to be purchased and can be produced using a simple expression and purification procedure in a bacterial system developed by David Waugh's laboratory (14). This option considerably reduces the cost of preparation for downstream applications especially in large-scale production.

4. Large-affinity tags such as MBP may offer some advantages for structural biology applications, since they can facilitate the crystallization of problematic proteins. For this purpose, the target protein must be rigidly fused to the MBP by a short spacer, such as three to five alanines, to reduce the conformational heterogeneity introduced by a flexible linker. Moreover, fusion of membrane proteins to MBP can increase the size of the hydrophilic domain, and eventually facilitate crystallization (15).

5. Tris–HCl, MOPS, HEPES, and phosphate buffers at pH values between 6.5 and 8.5 are all compatible buffers for MBP binding. Since MBP binds to amylose primarily via hydrogen bonds, high ionic strengths such as 1 M NaCl can be used in order to reduce nonspecific adsorption of proteins to the resin (13). Optional additives that can be added are 0.02% sodium azide to avoid bacterial contamination in the medium and reducing agents such as 10 mM β-mercaptoethanol or 1 mM DTT that serve to maintain reduced cysteines and to avoid the formation of nonspecific disulfide bridges that can cause aggregation.

6. Troubleshooting tips

   Inadequate binding of MBP-tagged proteins to the amylose resin:

   (a) The presence of endogenous amylases during bacterial growth may competitively inhibit binding to the amylose column. This problem can be partially overcome by using 0.2% glucose in the growth medium, in order to repress the endogenous amylase expression (see Instruction Manual from pMAL™ Protein Fusion and Purification System, NEB).

   (b) The presence of nonionic detergents such as Triton X-100 and Tween-20 can interfere with binding. If detergents are essential to the target protein, use less than 0.05% in order to solubilize the extract (13). However, if this concentration is too low, you might need to consider improving binding by screening alternative detergents.

   (c) The oligomeric state of the molecule (soluble aggregates) can affect its binding to affinity columns. The presence of soluble aggregates can be analyzed by gel filtration.

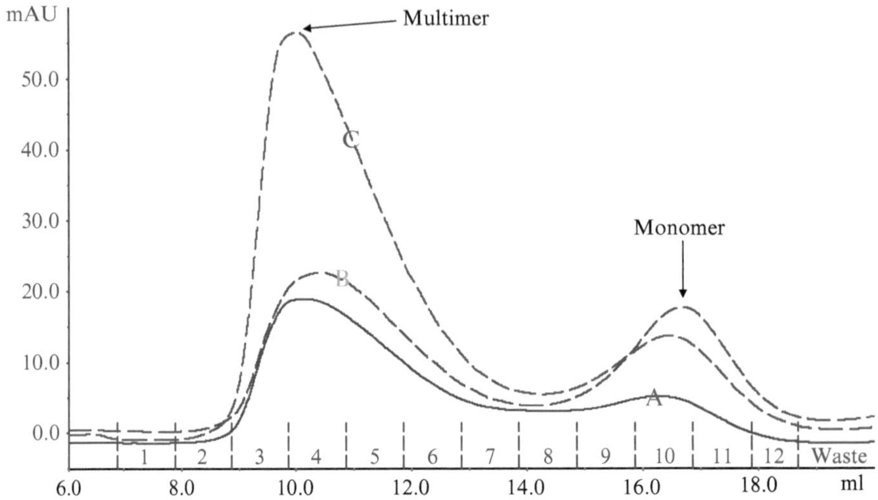

Fig. 4. Gel filtration analysis of the oligomeric state of an MBP fusion protein. A 30 kDa cytosolic protein, prone to aggregation, was fused to MBP in order to increase its solubility. The construct was transformed into *E. coli* BL21 cells. Upon reaching an OD600 of 0.7, the sample was induced with 0.4 mM IPTG and incubated for 4 h at 37°C (**a**). Other transformed cells were grown at 37°C until the OD600 reached 0.3. At this point, 0.1% glycerol and 0.1 mM potassium glutamate were added directly to the medium (we used SIGMA G-1501 L-Glutamic acid monopotassium salt). Next, the sample was subjected to a 20 min heat-shock treatment at 42°C, after which the temperature was reduced to 37°C, and the sample was induced with 0.4 mM IPTG at an OD of 0.7. One sample was harvested at 4 h postinduction at 37°C (**b**), and another was harvested at 16 h postinduction at 37°C (**c**). All samples were lysed, purified on an amylose–agarose column, and loaded on an analytical Superdex 200 gel filtration column to check the oligomeric state of the protein.

The formation of oligomers can be reduced by changing the expression conditions or the purification procedure and by screening different buffers and additives. In some cases, heat-shock treatment can greatly enhance the monomeric fraction of the expressed protein. In Fig. 4, MBP-fused protein was expressed under three different conditions: 37°C for 4 h, heat shock treatment followed by 4 h incubation, and heat shock treatment followed by 16 h induction. These results indicate that when the samples were subjected to heat shock treatment, a larger fraction of the protein shifted to the monomeric state. Although overnight induction resulted in an increased yield of the total protein, the yield of the monomeric form decreased (see Fig. 4).

(d) In some cases, however, the efficiency of binding and purification using amylose resin is not satisfactory. Here, one might consider adding a polyhistidine tag (His6) to the N terminus of MBP. This addition does not interfere with the ability of MBP to promote the solubility and proper folding of its fusion partners, and it can be used for binding to the more commonly used immobilized metal affinity chromatography systems (IMAC) (9, 14,

16, 17) (see also Chapter 17). Alternatively, NEB has recently developed improved MBP mutants with higher abilities to bind amylose resin (NEB pMAL-p4 and c4 series).

Protein is not eluted efficiently from the column.

(e) If the kinetics of the elution is too low, the protein is not completely eluted from the resin or is not eluted in a sharply concentrated peak. The following parameters should be considered as a means of improving the situation: (i) decreasing the elution flow rate, (ii) overnight incubation in the elution buffer, when performing batch purification, and (iii) increasing the concentration of maltose in the elution buffer by using from 20 to 100 mM maltose.

(f) The oligomeric state of the protein can change as a result of the high protein concentration in the column. Here, changes in the buffer can prevent aggregation and the following options should be considered: (i) increasing ionic strength up to 1 M NaCl or KCl, (ii) adding detergents or additives such as glycerol to the buffers, and (iii) performing batch binding instead of column binding.

(g) If multiple protein bands are present after elution, then protein degradation is to be suspected. Western blot analysis can be performed to verify if proteolysis is occurring. Conducting all purification steps at 4°C, reducing the overall time taken to carry out the procedure, and using protease inhibitors during the cell disruption process, can all help to reduce proteolysis.

(h) If the additional bands visible on SDS–PAGE are not the result of target protein degradation, there are two main reasons that usually explain the presence of cellular protein contaminants: (i) contaminating proteins are binding nonspecifically to the resin, (ii) contaminants are sticking to the target protein. If contaminants are bound nonspecifically to the resin, consider decreasing the resin volume to increase competition, or increasing the ionic strength of the buffers (up to 1 M NaCl or KCl), to reduce hydrophobic interactions with the resin. If contaminants stick to the target protein, increasing the washing step is the first option that should be considered. If this does not work, consider increasing the ionic strength of the buffers (up to 1 M NaCl or KCl), adding additives such as glycerol, adding reducing agents in order to disrupt nonspecific intermolecular disulfide bonds, or adding detergents that might reduce hydrophobic interactions. If taking these options does not reduce the presence of contaminants, additional purification steps should be performed before or after affinity purification.

(i) Maltose should be removed completely after the two last purification steps (ion exchange and gel filtration). When MBP fusion proteins are purified by affinity chromatography without further purification columns, dialysis after affinity purification is not enough to eliminate maltose from the protein solution. Maltose can be completely removed by binding the fusion protein to hydroxyapatite, or by ion exchange, or hydrophobic exchange, or any other resin that can bind the fusion protein and not the sugar. The resin is then washed extensively before protein elution (see pMAL™ Protein Fusion and Purification System manual from NEB).

(7) Amylose–agarose columns can be regenerated with 0.1% SDS at room temperature, water (according to New England Biolabs) or 0.1 M NaOH for a very short time and then neutralized. Alternatively, they can be regenerated with 50 mM HEPES pH 7.4, 4 M urea, 0.5% w/v SDS and then 50 mM HEPES pH 7.4, 150 mM $(NH4)_2SO_4$, 2 mM EDTA, 2 mM EGTA and water (18), and kept in 20% ethanol at 4°C. GE Healthcare developed MBPTrap™ HP, a ready to use and very successful new column for purifying recombinant proteins tagged with maltose-binding protein (MBP). MBPTrap™ HP can be easily regenerated using 0.5 M NaOH, so columns can be used for repeated runs, with reproducible results (GE Healthcare Date file 28-9136-33 AA MBPTrap HP 1 ml and 5 ml).

Acknowledgments

We thank Dr. Erez Podoly from the Biological Chemistry Department of The Hebrew University of Jerusalem, now at the Structural Biology Department of Stanford University School of Medicine, for expressing the fusion protein, for his assistance with purification and his helpful suggestions. We thank Dr. Daria Mochly-Rosen from the Department of Chemical and Systems Biology, School of Medicine, Stanford University for kindly providing us with pMAL-c2MBP-Rack1.

References

1. Nikaido, H. (1994) Maltose transport system of *Escherichia coli*: an ABC-type transporter. *FEBS Lett.* **346**, 55–58.
2. Baneyx, F., and Mujacic, M. (2004) Recombinant protein folding and misfolding in *Escherichia coli*. *Nat. Biotech.* **22**, 1399–1408.
3. Randall, L.L., Topping, T.B., Smith, V.F., Diamond, D.L., and Hardy, S.J. (1998). SecB: A chaperone from *Escherichia coli*. *Meth. Enzymol.* 290, 444–459.
4. Nomine, Y., Ristriani, T., Laurent, C., Lefevre, J.F., Weiss, E., and Trave, G. (2001).A strategy for optimizing the monodispersity of fusion proteins: application to purification of recombinant HPV E6 oncoprotein. *Prot. Eng.* **14**, 297–305.
5. Sachdev, D., and Chirgwin, J.M. (1998). Order of fusions between bacterial and mammalian proteins can determine solubility in *Escherichia coli*. *Biochem. Biophys. Res. Comm.* **244**, 933–937.
6. Nallamsetty, S. and Waugh, D.S. (2006) Solubility-enhancing proteins MBP and NusA play a passive role in the folding of their fusion partners. *Protein Expr. Purif.* **45**, 175–182.
7. Lee, R.T., Ichikawa, Y., Allen, H.J., and Lee, Y.C. (1990) Binding characteristics of galactoside-binding lectin (galaptin) from human spleen. *J. Biol. Chem.* **265**, 7864–7871.
8. Kapust, R.B. and Waugh, D.S. (2000). Controlled intracellular processing of fusion proteins by TEV protease. *Protein Expr Purif.* **19**, 312–318
9. Nallamsetty, S., and Waugh, D.S. (2007) A generic protocol for the expression and purification of recombinant proteins in *Escherichia coli* using a combinatorial His6-maltose binding protein fusion tag. *Nat. Protoc.* **2**, 383–391.
10. Mohanty, A., Simmons, C.R., and Wiener, M.C. (2003) Inhibition of tobacco etch virus protease activity by detergents. *Protein Expr. Purif.* **27**, 109–114.
11. Sheffield, P., Garrard, S., and Derewenda, Z. (1999) Overcoming expression and purification problems of RhoGDI using a family of "parallel" expression vectors. *Protein Expr. Purif.* **15(1)**, 34–39.
12. Kapust, R.B., Tözsér, J., Fox. J.D., Anderson, D.E., Cherry, S., Copeland, T.D., and Waugh, D.S. (2001) Tobacco etch virus protease: mechanism of autolysis and rational design of stable mutants with wild-type catalytic efficiency. *Protein Eng.* **14**, 993–1000.
13. Riggs, P. (2000) Expression and purification of recombinant proteins by fusion to maltose-binding protein. *Mol. Biotechnol.* **15**, 51–63.
14. Nallamsetty, S., Austin, B.P., Penrose, K.J. and Waugh, D.S. (2005) Gateway vectors for the production of combinatorially-tagged His6-MBP fusion proteins in the cytoplasm and periplasm of *Escherichia coli*. *Protein Sci.* **14**, 2964–2971.
15. Smith, D. (2003) Crystal structures of fusion proteins with large-affinity tags. *Protein Sci.* **12**, 1313–1322.
16. Tropea, J.E., Cherry, S., Nallamsetty, S., Bignon, C., and Waugh, D.S. (2007) A generic method for the production of recombinant proteins in *Escherichia coli* using a dual His_6-MBP affinity tag. *Methods Mol. Biol.* **363**, 1–19.
17. Austin, B.P., Nallamsetty, S., and Waugh DS (2009) Hexahistidine-tagged maltose-binding protein as a fusion partner for the production of soluble recombinant proteins in *Escherichia coli*. *Methods Mol. Biol.* **498**, 157–172.
18. Pattenden, L.K., and Thomas, W.G. (2008) Amylose affinity chromatography of maltose-binding protein: purification by both native and novel matrix-assisted dialysis refolding methods. *Methods Mol. Biol.* **421**, 169–189.

Chapter 16

Purification of Proteins from Baculovirus-Infected Insect Cells

Luke O'Shaughnessy and Sean Doyle

Abstract

Expression of recombinant proteins in the baculovirus/insect cell expression system is employed because it enables post-translational protein modification and high yields of recombinant protein. The system is capable of facilitating the functional expression of many proteins – either secreted or intracellularly located within infected insect cells. Strategies for the isolation and extraction of soluble proteins are presented in this chapter and involve selective cell lysis, precipitation and chromatography. Protein insolubility, following recombinant expression in insect cells, can occur. However, using the methods described herein, it is possible to extract and purify insoluble protein using affinity, ion-exchange and gel filtration chromatography. Indeed, protein insolubility often aids protein purification.

Key words: Baculovirus, Insect cell expression system, Post-translational modification, Viral capsid proteins, Ni-NTA affinity chromatography

1. Introduction

Recombinant protein expression in baculovirus-infected insect cells offers a number of advantages over other commonly used expression systems. First, post-translational modifications, such as glycosylation, are possible in the insect cell expression system – unlike the situation in prokaryotic expression systems such as *Escherichia coli* where few cases of post-translational modification of recombinant proteins have been identified (1). Moreover, the likelihood of obtaining functional eukaryotic proteins or enzymes is enhanced using the eukaryotic insect cell expression system and many viral capsid proteins have been successfully expressed in this system (2,3). Recombinant protein expression is also possible in other eukaryotic systems such as yeast (e.g., *Pichia pastoris*) (4) or fungal systems. However, unless recombinant protein secretion

Dermot Walls and Sinéad T. Loughran (eds.), *Protein Chromatography: Methods and Protocols*, Methods in Molecular Biology, vol. 681, DOI 10.1007/978-1-60761-913-0_16, © Springer Science+Business Media, LLC 2011

occurs in these particular systems, then recovery of intracellularly located protein can be challenging and may require deployment of vigorous cell lysis methods such as French Press technology, grinding in liquid N_2 and extensive use of lytic enzymes to degrade yeast or fungal cell wall material (5). Ultimately, the nature of the protein of interest is possibly the key determinant in deciding which expression system to choose. Protein expression is still a largely empirical science with trial and error playing a large part in identifying a specific expression system that yields the required amounts of the protein under study.

The most commonly used baculovirus for recombinant protein expression is *Autographa californica* and is generally termed a recombinant baculovirus as soon as it has been confirmed to contain the gene of interest. Once this recombinant baculovirus is available, it is possible to infect cultured insect cells (e.g., *Spodoptera frugiperda* 9; *Sf9*) to generate recombinant protein as the viral infection cycle proceeds in the *Sf9* insect cells (6). Recombinant protein expression generally commences after 2 days of culture and is optimal between days 3 and 5 (7). Along with the optimal amount of recombinant baculovirus required for successful infection (i.e., the multiplicity of infection; MOI), and the location of the recombinant protein (i.e., intracellular or secreted), the third important parameter that must be established when undertaking recombinant protein expression in insect cells is the optimal post-infection timepoint for recombinant protein expression (8). Armed with these three important pieces of information, the investigator can be sure that subsequent attempts to purify recombinant protein should not be affected by variability associated with the initial expression of the recombinant in infected *Sf9* insect cells.

Most recombinant proteins that are produced in the insect cell expression system, and other systems, for research purposes, are generally engineered to contain either an N- or C-terminal His_6 tag to facilitate post-expression protein identification by immunoblotting and purification by metal chelate (Ni-NTA) affinity chromatography (9–11) (see *also* Chapter 9). Generally, there are only two reasons not to use this or alternative 'tag' systems – either to avoid undesired effects on the solubility or function of the protein of interest or infringement of patents covering the use of His_6 technology should the work be carried out in an industrial environment. Although insect cells can be lysed directly into binding buffers compatible with Ni-NTA resins for affinity purification, initial cell lysis and specific protein extraction can improve overall process efficiency and also lead to higher protein purity.

This chapter provides extensive details on the strategy for, and methods of, assessing protein localisation, solubility, and extraction from insect cells as in our experience these are the key steps that must be optimized to ensure significant recombinant protein recovery. Strategies for chromatographic purification of recombinant proteins from the insect cell expression system are also presented.

2. Materials

2.1. Localisation and Solubility Determination of Recombinant Protein

1. T75 flasks (Sarstedt).
2. 10% (w/v) Trichloroacetic acid (TCA): TCA (1 g) was dissolved in 10 mL deionised water.
3. 5× SDS Solubilisation buffer: Glycerol (8 mL) was added to deionised water (4 mL), containing 1.6 mL of 10% (w/v) SDS and 1 mL of 0.5 M Tris–HCl, (pH 6.8). 2-mercapthoethanol (0.4 mL) was added to the solution along with 0.2 mL of 0.5 % (w/v) bromophenol blue solution.
4. 2 M Tris (unbuffered): Trizma base (48.44 g) was added to distilled water (100 mL) and dissolved on a magnetic stirrer. The volume was adjusted to 200 mL with distilled water. The solution was autoclaved before use and stored at 4°C.
5. Coomassie Brilliant Blue R (Sigma-Aldrich GmbH, Steinheim, Germany).
6. Apparatus and reagents for small- or medium-sized SDS–PAGE; apparatus for western blotting.
7. Antibodies: anti-IgG [anti-His_6] (Roche, GmbH, Germany) and anti-mouse IgG-horseradish peroxidase conjugate (Bio-RAD, Hercules, CA).
8. Diaminobenzidine (DAB, Sigma-Aldrich, UK) or Enhanced Chemiluminescence assay (ECL, Thermo-Scientific, Rockford, IL, USA).

2.2. Ammonium Sulphate Precipitation of Soluble Recombinant Protein from Lysed Insect Cells

1. Lysis buffer (500 mL): 25 mM Tris–HCl, 10 mM NaCl, 5 mM $MgCl_2$, pH 7.5. Store at 4°C.
2. Potter-Elvehjem homogenizer (Sigma-Aldrich, UK).
3. Ammonium sulphate, 100% saturated (Sigma-Aldrich, UK).
4. PBS (Sigma-Aldrich, UK).

2.3. Extraction of Insoluble Recombinant Protein from Insect Cells

1. PBS-EDTA. Phosphate buffered saline containing 1 mM EDTA.
2. Lysis buffer (500 mL): 25 mM Tris–HCl, 10 mM NaCl, 5 mM $MgCl_2$, pH 7.5. Store at 4°C.
3. 10% (w/v) Nonidet P40 (10 mL) (Sigma, UK): Weigh 1 g of Nonidet P-40 and make up to 10 mL with deionised water in a graduated cylinder – use immediately.
4. Leupeptin (2 mg/mL): Dissolve 2 mg leupeptin in 1 mL deionised water. Stable for 1 week at 4°C or up to 6 months at –20°C.
5. Pepstatin (2 mg/mL): Dissolve 2 mg pepstatin in 1 mL methanol. Stable at –20°C for 6 months.

6. Aprotinin: Aprotinin is supplied in solution (2 mg/mL). Stable at 4°C.
7. 100 mM PMSF (17.4 mg/mL methanol). 1 mL. Store at –20°C. PMSF is a neurotoxin, use gloves when handling.
8. 5% (w/v) Sodium deoxycholate (10 mL) (Sigma-Aldrich, UK): Dissolve 0.5 g of sodium deoxycholate and make up to 10 mL with deionised water in a graduated cylinder – use immediately.
9. 5 M Sodium chloride solution (100 mL): Dissolve 29 g sodium chloride in 70 mL deionised water and make up to 100 mL in a graduated cylinder.
10. 1 M $MgSO_4$ (50 mL) (Sigma-Aldrich, UK).
11. 50 mM HEPES buffer, pH 7.5 (100 mL) (Sigma-Aldrich, UK).
12. Preliminary solubilisation buffer (100 mL): 25 mM Tris–HCl, 2 M urea, 1 mM EDTA, pH 8.0.
13. Solubilisation Buffer (100 mL): 25 mM Tris–HCl, 8 M urea (see Note 1), 1 mM EDTA, pH 8.0.
14. Guanidine–HCl Solubilisation Buffer (100 mL): 20 mM Tris–HCl, 6 M guanidine–HCl, 20 mM dithiothreitol (see Note 2), pH 8.0.

2.4. Pre-Chromatographic Sample Preparation

2.4.1. Dialysis

1. Dialysis Tubing (Sigma-Aldrich, UK).
2. Beaker of suitable size.
3. Magnetic stirrer plate.

2.4.2. Gel Filtration

1. PD-10 column, (GE Healthcare, Germany).

2.5. Chromatography

2.5.1. Ni-NTA Affinity Chromatography

2.5.1.1. Purification under Native Conditions

1. Ni-NTA slurry (Qiagen, GmbH, Hilden, Germany).
2. Binding buffer: 50 mM NaH_2PO_4, 300 mM NaCl and 10 mM imidazole, pH 8.0.
3. Wash buffer: 50 mM NaH_2PO_4, 300 mM NaCl and 20 mM imidazole, pH 8.0.
4. Elution buffer: 50 mM NaH_2PO_4, 300 mM NaCl and 250 mM imidazole, pH 8.0.

2.5.1.2. Purification Under Denaturing Conditions

1. Binding buffer A: 100 mM NaH_2PO_4, 10 mM Tris–Cl and 8 M urea, pH 8.0.
2. Wash buffer: 100 mM NaH_2PO_4, 10 mM Tris–Cl and 8 M urea, pH 6.3.

3. Elution buffer A: 100 mM NaH_2PO_4, 10 mM Tris–Cl and 8 M urea, pH 5.7.
4. Elution buffer B: 100 mM NaH_2PO_4, 10 mM Tris–Cl and 8 M urea, pH 4.2.

2.5.2. Ion Exchange Chromatography

1. Cation-exchange resin.
2. Anion-exchange resin.

2.5.2.1. Determination of Binding Conditions for Ion Exchange Chromatography

3. 0.5 M methanesulphonic acid pH 5.5.
4. 0.5 M Tris–HCl pH 7.8.
5. 20 mM methanesulphonic acid pH 5.5.
6. 20 mM Tris–HCl pH 7.8.

2.5.2.2. General Strategy for Ion Exchange Chromatography

1. Typical equilibration buffer for cation exchange: 20 mM methanesulphonic acid pH 5.5.
2. Typical elution buffer for cation exchange: 20 mM methanesulphonic acid, 0.5 M, NaCl, pH 5.5.
3. Typical equilibration buffer for anion exchange: 20 mM Tris-HCL pH 7.8.
4. Typical elution buffer for anion exchange: 20 mM Tris–HCl, 0.5 M NaCl, pH 7.8.

3. Methods

Insect cells can be cultured from small scale (i.e., 5.0×10^5 cells/well in a 24-well plate) to large scale ($2.0–2.5 \times 10^6$ cells/mL; 0.1–5 L) in shaker or spinner flasks, or using Wave Bag® technology (Wave Biotech AG, Switzerland) (12). Preliminary experiments to detect recombinant protein expression, identifying if the protein is secreted or released from infected insect cells or located intracellularly (soluble/insoluble), are generally carried out using small-scale pilot studies [5×10^6 cells/mL in T75 tissue culture flasks; (75 cm^2)]. SDS–PAGE and western blot analysis are usually used to provide the aforementioned information and can also yield information on both the integrity of the recombinant protein and preliminary data on the expression level.

3.1. Localisation and Solubility Determination of Recombinant Protein

1. Prepare T75 flasks containing 9×10^6 insect cells in 10 mL of TC-100 medium. Infect cells with recombinant baculovirus over an MOI range of 0, 1, 2.5, 5, 10, 15 and 20. Allow infection to proceed. After 3 days recover both supernatant and infected cells for analysis by first directly pipetting the supernatant from the T75 flasks followed by centrifugation at $600 \times g$ for 10 min to remove any cell debris or intact cells.

2. Adherent cells should be dislodged from the surface of the T75 flasks by firmly knocking the flasks against a solid surface. Sufficient liquid should be present to allow transfer of insect cells to a clean container.
3. Aliquots of culture supernatant (20 μL) should be transferred to microfuge tubes followed by the addition of 5 μL 5× SDS solubilisation buffer to each tube.
4. Larger volumes of culture supernatant can be precipitated using 1/10 vol of 10% (w/v) TCA, followed by incubation at 4°C for 30 min and centrifugation at 10,000 × *g* for 2 min to collect precipitated protein. Remove supernatants and resolubilise TCA precipitates in 20 μL 1× SDS solubilisation buffer. Add 1–5 μL 2 M Tris (unbuffered) to adjust pH to approximately 8.0 (colour changes from yellow to blue).
5. Analyse all specimens by duplicate SDS–PAGE (Gel 1 and Gel 2). Gel 1 should be stained with Coomassie Brilliant Blue R (Sigma-Aldrich GmbH, Steinheim, Germany) and photographed. Gel 2 should be subjected to electrotransfer and immunological detection (western blotting) of the recombinant protein using monoclonal IgG [anti-His_6] as primary antibody followed by anti-mouse IgG-horseradish peroxidase conjugate. Immune complexes may then be located with either diaminobenzidine (Sigma-Aldrich, UK) or by an ECL assay according to manufacturer's instructions (see Notes 3 and 4) (see Fig. 1). If recombinant protein-specific antisera is available, then this can be used for immunodetection instead of, or as well as, monoclonal IgG [anti-His_6] (see Note 5).
6. Once completed, it will be clear if the recombinant protein is either secreted or located intracellularly in a soluble or insoluble form. Secreted proteins are almost always soluble and this represents the ideal scenario where the recombinant protein is free from intracellular contaminating proteins.
7. If the protein is secreted then it is generally recommended to reduce the volume of the supernatant, and thereby concentrate the recombinant protein, prior to purification or storage, by (i) commercial ultrafiltration units (e.g., Amicon or Centricon devices, Millipore, Billerica, MA, USA), used according to manufacturer's instructions or (ii) selective precipitation using saturated ammonium sulphate.
8. The identification of intracellular protein localization, or solubility, generally requires application of the isolation strategies outlined in either Subheading 3.2 (see Chapter 11) or Subheading 3.3.

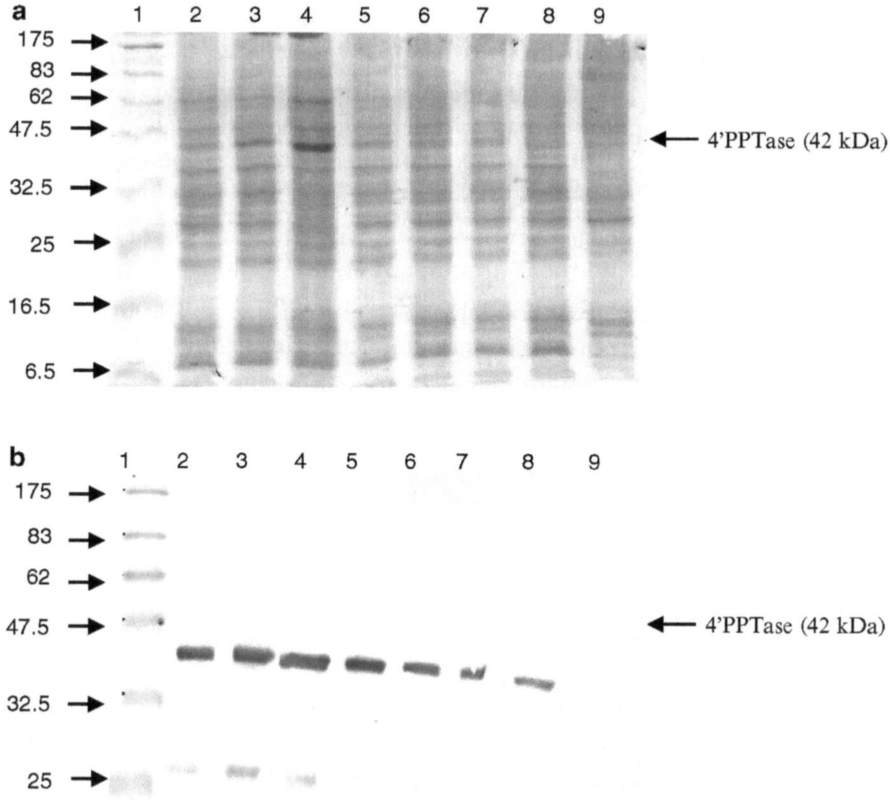

Fig. 1. SDS–PAGE and western blot analysis of *Sf*9 cells following infection, at different MOI with a recombinant baculovirus expressing a 4′-phosphopantetheinyl transferase (4′PPTase; 42 kDa) of fungal origin (16). (**a**). SDS–PAGE. (**b**). Western blot with diaminobenzidine development. Infected cells were loaded at 5×10^5 cells/track. Monoclonal anti-His_6 was used at 1 mg/mL and goat anti-mouse IgG–horseradish peroxidase conjugate at 1/1,000 dilution. Lane 1: Molecular mass marker. Lane 2: MOI_{40}. Lane 3: MOI_{30}. Lane 4: MOI_{20}. Lane 5: MOI_{10}. Lane 6: MOI_5. Lane 7: $MOI_{2.5}$. Lane 8: $MOI_{0.5}$. Lane 9: Cells only.

3.2. Ammonium Sulphate Precipitation of Soluble Recombinant Protein from Lysed Insect Cells

1. Obtain freshly harvested or frozen insect cells, resuspend in minimum volume lysis buffer and pool together. Note the final number of cells (see Note 6).
2. Make up to 5 mL lysis buffer per 10^8 cells. Subject to 2× freeze thaw cycles (−70°C to room temperature) and sonicate if required.
3. Homogenize using a Potter-Elvehjem homogenizer until suspension is homogenous.
4. Centrifuge at $16,000 \times g$ for 10 min.
5. Remove the supernatant to a clean tube. Retain pellet also.
6. Remove 6× 500 μL aliquots of the supernatant to microfuge tubes. Add 50, 100, 200, 400, 600 and 800 μL of 100% (w/v) ammonium sulphate to each tube, respectively.

7. Incubate for 1 h at 4°C, then centrifuge at 10,000 × *g* for 15 min.
8. Remove the supernatants to clean tubes and resuspend pellets in a minimal volume of lysis buffer. Analyse by SDS–PAGE. Identify tubes containing the most recombinant protein (see Notes 7 and 8).
9. Repeat ammonium sulphate precipitation of bulk supernatant using the optimum ammonium sulphate concentration.
10. Incubate for 1 h at 4°C, centrifuge at 10,000 × *g* for 15 min, remove the supernatant to a clean tube.
11. Resuspend pellets in PBS and dialyse (see Subheading 3.4.1) against 2 × 50 vol PBS.
12. Repeat SDS–PAGE analysis.

3.3. Extraction of Insoluble Recombinant Protein from Insect Cells

1. Obtain freshly harvested or frozen insect cells.
2. Carry out the following step if cells have been cultured in the presence of foetal bovine serum, otherwise proceed directly to step 3: Resuspend infected insect cells in PBS (optionally containing 1 mM EDTA) by adding 1.8 mL PBS-EDTA per 10^8 cells (see Note 6). Immediately add protease inhibitors (final concentrations: 0.1 mM PMSF, 2 μg/mL pepstatin and leupeptin, respectively). Centrifuge at 500–600 × *g* for 15 min. Discard the supernatant.
3. Resuspend the cell pellet in the same volume of lysis buffer (1.8 mL per 10^8 cells) and add protease inhibitors (0.1 mM PMSF, 2 μg/mL pepstatin and leupeptin, final concentrations). Homogenize the suspension by pipetting upwards and downwards. Remove a 10 μL aliquot for SDS–PAGE.
4. Add 180 μL of 10% (w/v) Nonidet P-40, per 1.8 mL of lysis buffer, to the remaining cell suspension. Mix by inverting to lyse the cells. Lysis is effectively instantaneous. Centrifuge at 5,000–10,000 × *g* for 30 min.
5. Remove 10 μL of the supernatant for SDS–PAGE and discard the remaining supernatant (if you are sure it does not contain your protein of interest). Wash the pelleted nuclei by resuspending in 1.8 mL PBS-EDTA per 10^8 cells and add protease inhibitors (final concentrations: 0.1 mM PMSF, 2 μg/mL pepstatin and leupeptin, respectively). Centrifuge at 5,000–10,000 × *g* for 10 min.
6. Discard the supernatant and resuspend the pellet in 1.8 mL distilled water per 10^8 cells and immediately add protease inhibitors. Add 100 μL 5% (w/v) sodium deoxycholate per 1.8 mL. Mix by inverting and incubate for 10 min. Note the appearance of particulates and the increase in viscosity

(DNA release) of the lysate. Retain a 10 μL aliquot for SDS–PAGE.

7. The following reagents should then be added sequentially to the lysate (per 10^8 cells)
   (i) 3.1 mL of 50 mM HEPES, pH 7.5
   (ii) 40 μL of 1 M magnesium sulphate solution
   (iii) 770 μL of 5 M sodium chloride solution
   (iv) Protease inhibitors (final concentrations: 0.1 mM PMSF, 2 μg/mL pepstatin and leupeptin, respectively)
   (v) 3.1 mL of 50 mM HEPES, pH 7.5
   (vi) Deoxyribonuclease – 1 RNAase free (10 μg/mL final concentration) (see Note 9).
8. Incubate for 18 h at 4°C and then for 2 h at room temperature. Centrifuge at 5,000–10,000 × *g* for 15 min.
9. Discard the supernatant and resuspend the pellet in 8 mL Preliminary Solubilisation Buffer per 10^8 cells. Centrifuge at 10,000 × *g* for 15 min.
10. Resuspend the pellet in 2 mL Solubilisation Buffer per 10^8 cells. If the protein of interest solubilises at this point then proceed to pre-chromatographic sample preparation. Otherwise, centrifuge at 10,000 × *g* for 15 min.
11. Discard the supernatant and resuspend the pellet in 2 mL guanidine–HCl Solubilisation Buffer per 10^8 cells and incubate at room temperature for up to 1 h. Protein solubilisation is indicated by disappearance of gelatinous pellet. (see Note 10).
12. Perform sequential dialysis against 50 volumes of 6 and 3 M urea (buffered with 50 mM Tris–HCl, pH 8.0) to remove guanidine–HCl. The extracted protein should be stored at –20 to –70°C or ideally utilized immediately.

3.4. Pre-Chromatographic Sample Preparation

Prior to either affinity chromatography or ion-exchange chromatography (IEX), it is imperative that the concentrated culture supernatant, crude cell lysate or semi-pure protein extract is present under buffer conditions, which will promote binding of the recombinant protein to the chromatographic matrix so that it can be efficiently displaced by salt gradient or pulse (generally used for affinity chromatography) elution. This usually means that a buffer exchange step, generally by dialysis (see Note 11) into the chromatographic binding buffer, is carried out prior to chromatography. Dialysis is used for both the removal of low molecular mass contaminants and buffer exchange of protein solutions. The optimal protein concentration range for samples is 0.5–5 mg/mL.

3.4.1. Dialysis

1. Cut the dialysis membrane to the required length and soak in deionised water for at least 5 min prior to use.
2. Clip or tie a double knot in one end of the membrane and dispense the sample, of known volume, into the dialysis bag using a plastic pipette.
3. Remove excess air from the bag and leave at least 25% extra space for volume expansion. Clip or tie a double knot at the open end to seal the bag.
4. Commence dialysis against at least 50 volumes of relevant pre-chilled buffer at 4°C, twice, for at least 3 h on each occasion (i.e., if sample volume is 5 mL then buffer volume should be at least 250 mL). Ensure that the dialysate is gently stirring throughout the procedure. Maximum dialysis time will depend on sample stability and generally should not exceed 48 h.
5. Upon completion of dialysis, lightly dry the outside of the dialysis membrane with tissue, cut the end of the dialysis bag and carefully remove the sample using a plastic pipette.
6. Measure sample volume post-dialysis and store appropriately prior to further use.

3.4.2. Gel Filtration

1. Commercial gel filtration columns (e.g., GE Healthcare PD-10 columns) are available for low volume (up to 2.5 mL) buffer exchange or salt removal and can be used according to manufacturer's instructions.

3.5. Chromatography

The chromatographic strategy used for recombinant protein purification is generally precisely tuned to obtaining purification of the specific protein. Consequently, rather than describing the purification strategy for an individual protein, we will describe the key issues that must be addressed to ensure successful protein purification.

3.5.1. Ni-NTA Affinity Chromatography

Ni-NTA Purification Systems are designed for the purification of His_6-tagged recombinant proteins. Ni-NTA affinity chromatography uses nitrilotriacetic acid (NTA), tetradentate chelating ligand which is highly cross-linked to an agarose matrix. NTA binds four of the six ligand binding of the nickel ion, leaving two sites free to interact with the His_6-tagged protein (10). The His_6 region binds to Ni^{2+}, which is immobilized on the Ni-NTA resin. After unbound proteins are washed away, the target protein is recovered by elution with imidazole (under native conditions) or with low pH urea-containing buffer (under denaturing conditions). The principle and applications of this technology, also known as immobilized metal affinity chromatography (IMAC), are presented in Chapters 9 and 17, respectively.

3.5.1.1. Purification Under Native conditions

1. Add 1 mL of the 50% (v/v) Ni-NTA slurry (Qiagen, GmbH, Hilden, Germany) to 4 mL cell lysate supernatant and mix gently by shaking (200 rpm on a rotary shaker) at 4°C for 60 min (see Note 12 and 13).
2. Load the lysate–Ni-NTA mixture into a column. Allow the lysate–Ni-NTA resin to settle and collect the column flow through. Save flow through for SDS–PAGE analysis.
3. Wash the gel or resin with four column volumes of wash buffer; collect the fractions for SDS–PAGE analysis.
4. Elute the protein with four column volumes of elution buffer (see Notes 14 and 15). Collect 0.5 mL fractions for SDS–PAGE analysis.

3.5.1.2. Purification Under Denaturing Conditions

1. Urea (up to 8 M) can be added to Ni-NTA buffers to facilitate purification of denatured proteins. Repeat steps 1–3 (from subheading "Purification Under Native Conditions" above) using the denaturing buffers (see Note 5).
2. Elution of the target protein can be achieved by the use of low pH elution buffers (e.g. elution buffer A pH 5.7 and/or elution buffer B pH 4.2; in the absence or presence of 300 mM imidazole). This is dependent on the solubility of the target protein.

3.5.2. Ion Exchange Chromatography

Ion exchange (IEX) chromatography uses either cation- or anion-exchangers to effect protein fractionation and most commercially available ion exchangers can bind in the order of 20–30 mg total protein per mL of resin or gel. Cation-exchange resins are negatively charged resins and will bind positively charged proteins (using buffer conditions at pH values < protein isoelectric point [pI]). Conversely, anion exchangers are positively charged resins and bind negatively charged proteins (using buffer conditions at pH values > protein pI).

The starting pH should be at least 1 pH unit above the isoelectric point for anion-exchanger chromatography or at least 1 pH unit below the isoelectric point for cation-exchange chromatography to facilitate adequate binding (13,14). Proteins tend to dissociate from IEX resins about 0.5 pH units from their isoelectric points at an ionic strength of 0.1 M (15). There are a variety of groups, which have been chosen for use in ion exchangers (13): cation exchangers include carboxymethyl (CM), sulphopropyl (SP), methyl sulphonate (S), while anion exchange resins include diethylaminoethyl (DEAE), quaternary aminoethyl (QAE) and quaternary ammonium (Q). Occasionally, one will find the term 'strong' or 'weak' associated with a particular ion-exchanger, which actually defines the pH range over which the resin maintains charge and is *not* an indication of the strength of protein binding to the ion exchanger.

Use of IEX chromatography for recombinant protein purification generally requires either (i) a knowledge of the approximate pI of the protein of interest or (ii) a pre-screening step whereby the propensity of the protein of interest to bind to IEX resins (0.5 mL each) over a particular pH range is determined.

3.5.2.1. Determination of Binding Conditions for IEX Chromatography

1. Set up a series of ten 15 mL tubes.
2. Add 1 mL of 50% (w/v) IEX slurry to each tube: five cation- and five anion exchangers, respectively.
3. Equilibrate the gel in each tube to a different pH by washing ten times with 10 mL of 0.5 M equilibration buffer (e.g., 0.5 M methanesulphonic acid pH 5.5 for cation exchanger or 0.5 M Tris–HCl pH 7.8 for anion exchanger. Use a range of pH 5–9 for anion and pH 4–8 for cation exchangers, with 0.5–1.0 pH unit intervals between tubes and buffer within 0.5 pH unit of p*K*a.
4. Equilibrate the resin in each tube at a lower ionic strength (0.01–0.02 M) by washing five times with 10 mL of buffer of the same pH but lower ionic strength (see Note 17).
5. Add identical volumes of sample (each not containing greater than 5 mg total protein and at the appropriate pH) to each tube.
6. Mix the contents of the tubes for 5–10 min.
7. Allow the gel to settle and assay the supernatant for the protein of interest. Absence of the target protein, or its activity, in the supernatant indicates that optimum binding conditions have been achieved.

3.5.2.2. General Strategy for IEX Chromatography

1. Equilibrate the IEX resin in 10–15 column volumes of binding or equilibration buffer (generally low ionic strength).
2. Apply the protein solution (salt-free as far as practicable) to the IEX column. The protein solution should contain identical buffer constituents to that used for column equilibration.
3. Wash through unbound protein using equilibration buffer. Measure the A280 nm of the column effluent and when this reaches zero you can be confident all unbound protein has been washed through the column.
4. Once all unbound protein has been removed, specifically elute the bound protein fraction using either a salt gradient (e.g. 0–0.5 M NaCl over 7–10 column volumes) or pulse elution with equilibration buffer containing 0.5 M NaCl.
5. Individual fractions can be collected during elution as screened for the protein of interest by enzyme activity, immunological or alternative strategies.

4. Notes

1. Weigh out and add urea to a minimum volume of deionised water for preparation of buffers containing high concentrations of urea. Then add additional reagents (e.g. stock buffers, EDTA, etc.)
2. Prepare 1 M dithiothreitol (DTT) stock (154 mg/mL) and store at −20°C. Add to guanidine–HCl Solubilisation Buffer immediately before use (20 mM final).
3. If enhanced chemiluminescent (ECL) detection, as opposed to diaminobenzidine, is required for western blot analysis then possibly too little useful protein may be present.
4. Coomassie detection of proteins in insect cells is optimal at 10^5 cells/track for SDS–PAGE. Breakdown/sub-fragments observed on western blots indicate protein degradation or protein truncation. To avoid protein degradation conduct procedures in a cold room and use ice buckets for storage of extraction buffers.
5. Make sure to use the correct anti-species-enzyme conjugate if using non-murine IgG for recombinant protein detection.
6. All reagents should be kept at 1–5°C unless otherwise stated. To facilitate this, conduct procedures in a cold room and use ice buckets. Keep intervals between steps as short as possible. All volumes given are based on 10^8 cells. Volumes should be adjusted according to the cell number available.
7. Precipitate sample overnight if the protein concentration is very low, this may increase precipitate concentration.
8. The presence of high amounts of $(NH_4)_2SO_4$ can interfere with SDS–PAGE. However, we have not observed excessive interference at the concentrations used here. In addition, initial selective precipitation of contaminating proteins at low concentrations of $(NH_4)_2SO_4$ can aid purification of the protein of interest if it is found to precipitate at higher $(NH_4)_2SO_4$ concentrations.
9. Commercial sources of DNase can contain low levels of serine protease activity and should always be pre-treated with 0.1 mM PMSF prior to use.
10. If protein solubilisation cannot be achieved using guanidine–HCl and DTT, then insoluble protein after the Solubilisation Buffer treatment (Subheading 3.3, step 10) can be solubilised using 50 mM Tris–HCl pH 8.0 containing 0.1–0.5% (w/v) SDS in the presence of 20 mM DTT. Such protein is unsuitable for IEX chromatography but may be further purified by gel filtration chromatography.

11. If initial chromatographic purification is by gel filtration chromatography, then a buffer exchange step is generally not required. IEX or gel filtration chromatography can be carried out in the presence of 8 M urea.
12. The amount of lysate required depends on the expression level of the 6× His-tagged protein and the expression system used. The binding capacity of Ni-NTA resins is protein-dependent and normally lies between 5 and 10 mg/mL.
13. The 10–20 mM imidazole in the lysis buffer reduces non-specific binding. If the tagged protein does not bind under these conditions, the amount of imidazole should be reduced to 1–5 mM or the incubation time extended (11).
14. Elution conditions from Ni-NTA columns must be optimized for each protein (10). One way to optimize conditions is to use an imidazole gradient for elution, rather than a single imidazole concentration; however, this can only be performed using an automated chromatography system.
15. Imidazole exhibits 280 nm absorbtion. Use elution buffer as the blank when monitoring absorbance.
16. Re-use of Ni-NTA affinity columns should only be performed with identical samples to prevent cross contamination.
17. The pH to be used in the experiment should allow the protein to be bound, but should be as close to the point of release as possible. If too low (or high) a pH is chosen, elution may become more difficult and high salt concentrations may have to be used.

Acknowledgments

Luke O'Shaughnessy is supported by a Strategic Research Cluster Grant (BioNanoInteract; 07/SRC/B1155) from Science Foundation Ireland.

References

1. Miller, L. K. (1988) Baculoviruses for foreign gene expression in insect cells. *Biotechnology* **10,** 457–465.
2. Michel, P. O., Makela, A. R., Korhonen, E., Toivola, J., Hedman, L., Soderlund-Venermo, M., Hedman, K. & Oker-Blom, C. (2008) Purification and analysis of polyhistidine-tagged human parvovirus B19 VP1 and VP2 expressed in insect cells. *J Virol Methods* **152,** 1–5.
3. Viscidi, R. P. & Clayman, B. (2006) Serological cross reactivity between polyomavirus capsids. *Adv Exp Med Biol.* **577,** 73–84.
4. Tolner, B., Smith, L., Begent, R. H. & Chester, K. A. (2006) Production of recombinant protein in Pichia pastoris by fermentation. *Nat Protoc.* **1,** 1006–1021.
5. Reeves, E. P., Reiber, K., Neville, C., Scheibner, O., Kavanagh, K. & Doyle, S. (2006) A nonribosomal peptide synthetase (pes1) confers

protection against oxidative stress in *Aspergillus fumigatus*. *FEBS J.*, **273,** 3038–3053.

6. King, L. A., Hitchman, R. & Possee, R. D. (2007) Recombinant baculovirus isolation. *Methods Mol Biol.* **388,** 77–94.
7. Harrison, R. L. & Jarvis, D. L. (2007) Transforming lepidopteran insect cells for continuous recombinant protein expression. *Methods Mol Biol.* **388,** 299–316.
8. Janakiraman, V., Forrest, W. F., Chow, B. & Seshagiri, S. (2006) A rapid method for estimation of baculovirus titer based on viable cell size. *J Virol Methods* **132,** 48–58.
9. Gupta, G., Kim, J., Yang, L., Sturley, S. L. & Danziger, R. S. (1997) Expression and purification of soluble, active heterodimeric guanylyl cyclase from baculovirus. *Protein Expr Purif.* **10,** 325–330.
10. Crowe, J., Masone, B. S. & Ribbe, J. (1996) One-step purification of recombinant proteins with the 6xHis tag and Ni-NTA resin. *Methods Mol Biol.* **58,** 491–510.
11. Khan, F., He, M. & Taussig, M. J. (2006) Double-hexahistidine tag with high-affinity binding for protein immobilization, purifica tion, and detection on ni-nitrilotriacetic acid surfaces. *Anal Chem.,* **78,** 3072–3079.
12. Kadwell, S. H. & Hardwicke, P. I. (2007) Production of baculovirus-expressed recombinant proteins in wave bioreactors. *Methods Mol Biol.* **388,** 247–266.
13. Himmelhoch, S. R (1971) Chromatography of proteins on ion-exchange adsorbents. *Meth. Enzymol.* **22,** 273–286.
14. Lampson G. P., Tytell, A. A. (1965) A simple method for estimating isoelectric points. *Anal Biochem.* **11,** 374–377.
15. Kelley, B. D., Switzer, M., Bastek, P., Kramarczyk, J. F., Molnar, K., Yu, T. & Coffman, J. (2008) High-throughput screening of chromatographic separations: IV. Ion-exchange. *Biotechnol Bioeng.* **100,** 950–963.
16. Neville, C., Murphy, A., Kavanagh, K. & Doyle, S. (2005) A 4'- phosphopantetheinyl transferase mediates non-ribosomal peptide synthetase activation in *Aspergillus fumigatus*. *Chembiochem* 6, 679–685.

Chapter 17

Purification of Poly-Histidine-Tagged Proteins

Sinéad T. Loughran and Dermot Walls

Abstract

His-tagging is the most widespread and versatile strategy used to purify recombinant proteins for biochemical and structural studies. Recombinant DNA methods are first used to engineer the addition of a short tract of poly-histidine tag (His-tag) to the N terminus or C terminus of a target protein. The His-tag is then exploited to enable purification of the "tagged" protein by immobilised metal affinity chromatography (IMAC). Here, we describe efficient procedures for the isolation of highly purified His-tagged target proteins from an *Escherichia coli* host using IMAC.

Key words: His-Tag, Recombinant protein expression, IMAC, Protein purification, Poly-histidine, 6×Histidine, Affinity chromatography, Ni-NTA

1. Introduction

Affinity tags are highly efficient tools for purifying recombinant proteins from crude extracts (see Chapter 9). The use of genetically engineered affinity tags for improved protein purification has increased greatly in recent years and affinity tags have become indispensable tools for structural and functional proteomics initiatives. The most commonly employed method utilises immobilised metal affinity chromatography (IMAC) to purify recombinant proteins containing a short affinity-tag consisting of poly-histidine (poly His) residues (1). IMAC is based on the interaction between a transition metal ion (in this case the Nickel ion, Ni^{2+}) immobilised on a matrix and specific amino acid side chains (usually 6×Histidine residues). Hochuli et al. developed a tetradentate chelating adsorbent, nitrilotriacetic acid (NTA), for metal-chelate affinity chromatography (2). NTA occupies four of the six ligand binding sites in the coordination sphere of the

Dermot Walls and Sinéad T. Loughran (eds.), *Protein Chromatography: Methods and Protocols*, Methods in Molecular Biology, vol. 681, DOI 10.1007/978-1-60761-913-0_17, © Springer Science+Business Media, LLC 2011

nickel ion, leaving two sites free to interact with the 6×His-tag. Histidine is the amino acid that exhibits the strongest interaction with immobilised metal-ion matrices, as electron donor groups on the histidine imidazole ring readily form coordination bonds with the immobilised transition metal (3).

High-level expression of 6×His-tagged proteins in *Escherichia coli* is achieved in this chapter using vectors from the pQE range (QIAgen) (see Table 1) which contain a regulatable promoter-operator element consisting of the T5 promoter and two *lac* operator sequences that increase *lac* repressor binding and ensure efficient repression of the powerful T5 promoter (recognised by the *E. coli* RNA polymerase). The *lac* operon promoter system is the most widely used bacterial expression system and encodes proteins that facilitate the uptake and metabolism of β-galactosides and is subject to both negative regulation (by binding of the *lac* repressor protein to the *lac* operator and preventing transcription) and positive regulation (by binding of an activator to the *lac* promoter and stimulating transcription). IPTG (an allolactose analogue; isopropyl-1-thio-β-D-galactopyranoside) is a gratuitous inducer of the *lac* operon, insofar as it competes strongly with the *lac* operator for binding to the *lac* repressor, yet itself is not metabolised in the process and can, thus, be used to induce expression. Recombinant constructs derived from the pQE vectors can be generated such that the encoded 6×His tag is located at the N terminus or the C terminus of the protein of interest (see Table 1).

Table 1
Expression systems for His-tagged recombinant protein expression

Plasmid	Source	Description
pQE60/pQE30	QIAgen (http://www.qiagen.com)	High-copy number expression vectors for expression of C- or N-terminal 6×His-tagged fusion proteins, respectively
pGSLink	(4)	Expression vector for expression of N- or C-terminal 6×His-tagged fusion proteins linked to protein of interest via a flexible peptide linker (4)
pTrcHis A, B, and C	Invitrogen (http://www.invitrogen.com)	Expression vectors for C- or N-terminal 6×His-tagged fusion proteins
pET vector series 14–16 and 19–52	Novagen (http://www.merck-chemicals.com)	pET vectors permit cloning, detection, and purification of target proteins

The pQE30 vector allows for the incorporation of a N-terminal 6×His tag, while the pQE60 vector facilitates C-terminal tagging with 6×His. The C-terminal placement of the 6×His tag ensures that only full-length proteins are subsequently purified.

Histidine residues are infrequent amino acids in globular proteins, amounting to approximately 2% of the amino acid content. At pH 8.0, the tag is small (0.84 kDa), uncharged, and therefore does not generally affect secretion, compartmentalisation, or folding of the fusion protein within the cell. The popularity of His-tag IMAC is due in part to its high affinity with Ni-NTA which is a relatively inexpensive matrix capable of withstanding multiple regeneration cycles under stringent conditions, but also due to the ease of controlled release using mild (e.g. non-denaturing) conditions. As the tertiary structure of the His-tag is not important for purification, an insoluble His-tagged recombinant protein can be purified by IMAC under denaturing conditions and subsequently refolded (for a review see ref. 5). The 6×His affinity-tag is poorly immunogenic; therefore, it is usually not necessary to remove the tag for the purposes of antibody generation. In addition, in most cases, the 6×His tag does not interfere with the structure or function of the purified protein as demonstrated for a wide variety of proteins, including enzymes, transcription factors, and vaccines. An additional advantage of this system is that anti-His antibodies can be used for the detection of tagged recombinant proteins during expression and purification steps.

Generally, most recombinant proteins can be cloned and expressed at high levels in *E. coli* (see Note 1). However, many polypeptide gene products expressed in this host accumulate as insoluble aggregates that lack functional activity. Furthermore, problems with toxicity towards the host cell, protein instability, improper processing or post-translational modification, and inefficient translation may also be experienced. The use of *E. coli* strain M15 [pREP4] (see Table 2 (7)) is advantageous for high-level recombinant protein expression using pQE vectors as the pREP4 plasmid constitutively expresses the Lac repressor protein. *Escherichia coli* strains that harbour the *lacI*^{q} mutation [including RosettaBlue™ (Novagen) and XL 10-Gold™ (Stratagene) (see Table 2)] also produce enough Lac repressor to efficiently repress the T5 promoter. The formation of stable disulphide bonds is a requirement for proper folding and activity of some recombinant proteins. Without disulphide bonds, these proteins may be degraded or accumulate as inclusion bodies. Bacterial strains with glutathione reductase (*gor*) and thioredoxin reductase (*trxB*) mutations [e.g. Origami™ strains; Novagen (see Table 2)] greatly enhance the formation of disulphide bonds in the *E. coli* cytoplasm (8). Codon usage can be an issue when expressing recombinant genes in a heterologous cell context. Codon use preferences reflect the amounts of corresponding cellular tRNA

Table 2
***Escherichia coli* host strains for recombinant protein expression**

Bacterial strain	Description	Genotype
E. coli M15 [pREP4] (QIAgen)	General expression host; cells contain pREP4 plasmid encoding *lac* repressor in trans, ensuring tightly regulated expression	*Nal*^{S}, *Str*^{S}, *Rif*^{S}, *Thi*$^-$, *Lac*$^-$, *Ara*$^+$, *Gal*$^+$, *Mtl*$^-$, *F*$^-$, *RecA*$^+$, *Uvr*$^+$, *Lon*$^+$ pREP4 (*lacI*^{q}) (Kan^R)
E. coli RosettaBlue™ (Novagen)	Expression host; provides six rare codon tRNAs	*end*A1, *hsd*R17(r_{K12}^-, m_{K12}^+), *sup*E44, *thi-1*, *rec*A1, *gyr*A96, *rel*A1, *lac*[F′ *pro*A$^+$B$^+$ *lacI*^{q}ZΔM15::Tn10(tet^R)] pRARE(argU, argW, ileX, glyT, leuW, proL) (Cm^R)
E. coli Origami™ (DE3) pLysS (Novagen)	Expression host; two mutations in cytoplasmic disulphide reduction pathway enhance disulphide bond formation in *E. coli* cytoplasm	Δ*ara–leu*7697 Δ*lac*X74 Δ*phoA PvuII phoR araD139 ahpC galE galK rpsL F′ [lac*$^+$ *lacI*^{q} *pro]* (DE3) *gor522::*Tn*10 trxB* pLysS (Cam^R, Kan^R, Str^R, Tet^R)
E. coli BL21(DE3) Δ*cyoA*Δ*yfbG*Δ*adhP* (6)	BL21(DE3) mutant knock-out strain deficient in three prevalent host proteins found in a strategic fraction of an elution profile Ni-IMAC	F – *ompT hsdS*$_B$(r_B– m_B–) *gal dcm* (DE3) Δ*cyoA*Δ*yfbG*Δ*adhP*
E. coli XL10-Gold™ (Stratagene)	General expression host, high efficiency transformation	Tet^R, Δ(μχρA)183 Δ(mcrCB-hsdSMR-mrr)173, *end*A1, *sup*E44, *thi-1*, *rec*A1, *gyr*A96, *rel*A1, *lac* Hte[F′ *pro*AB *lacI*^{q}ZΔM15 Tn10(tet^R) Amy Cam^R]

levels suggesting that recombinant genes containing rare codons may be subject to slower translation due to non-saturating amounts of corresponding tRNAs in the host cell (9, 10). If the coding sequence contains codons that are infrequently used by *E. coli*, then the protein may not be expressed due to this translational limitation (see Note 2). Finally, in another recent development, proteomics has been used to design a host cell tailored for highly efficient protein purification in isolation from co-purifying contaminants (6) (see Table 2 and Note 3). In practice, it is usually worthwhile to test a matrix of several different vector/host combinations to obtain the best possible yield of protein in its desired form.

In this chapter, we outline an example of the expression in *E. coli* (RosettaBlue™ host strain) of a His-tagged recombinant protein, and describe procedures for His-tagged fusion protein purification using IMAC. Here, we express recombinant human Bfl-1, a member of the intensively researched *bcl-2* gene family that is associated with regulating apoptosis in eukaryotic cells (11).

2. Materials

Use Milli-Q-purified water or equivalent for the preparation of all buffers. Reagents may be acquired from Sigma-Aldrich unless otherwise stated. Use high-quality chemicals. Solutions should be filtered through 0.45 or 0.22 μm filters.

2.1. High-Level Expression of Recombinant His-Tagged Proteins

1. Glycerol stock of transformed *E. coli* (see Table 2) cells expressing the His-tagged protein of interest in a suitable vector (see Table 1).
2. Luria–Bertani (LB) medium per litre: 10 g tryptone (Oxoid), 5 g yeast (Oxoid), 5 g NaCl. Adjust to pH 7.2 with NaOH. Autoclave and store at 4°C.
3. Ampicillin stock solution: 100 mg/mL in deionised water, filter sterilise, and store at −20°C. Use at 100 μg/mL within 1 month.
4. Chloramphenicol stock solution: 34 mg/mL in ethanol, store at −20°C. Use at 34 μg/mL (to maintain pRARE plasmid in RosettaBlue™ host strain).
5. Tetracycline stock solution: 5 mg/mL in ethanol, store at −20°C. Use at 12.5 μg/mL (to maintain pRARE plasmid in RosettaBlue™ host strain).
6. Orbital shaker (e.g. Gallenkamp Model H400).
7. Spectrophotometer.
8. Cuvettes (Sarstedt).
9. IPTG, 100 mM stock: dissolve 2.38 g IPTG in 100 mL deionised water. Filter sterilise and store at −20°C.
10. Glucose: 20% (w/v) D-glucose solution in deionised H_2O. Autoclave and store sterile solution at room temperature. Add glucose to LB agar with antibiotics to a final concentration of 0.5–1%.
11. Centrifuge (e.g. Eppendorf 5415D).
12. Sonicator (e.g. Vibra Cell).
13. Lysis Buffer A (1 L): 100 mM NaH_2PO_4; 13.8 g $NaH_2PO_4 \cdot H_2O$ (MW 137.99 g/mol), 10 mM Tris–HCl; 1.2 g Tris base

(MW 121.1 g/mol), 6 M guanidine hydrochloride; 573 g. Adjust pH to 8.0 using NaOH.

14. Lysis Buffer B (1 L): 100 mM NaH_2PO_4; 13.8 g $NaH_2PO_4 \cdot H_2O$ (MW 137.99 g/mol), 10 mM Tris–HCl; 1.2 g Tris base (MW 121.1 g/mol), 8 M urea; 480.5 g (MW 60.06 g/mol). Adjust pH to 8.0 using NaOH.

2.1.1. Protein Analysis

1. SDS loading buffer (2×): 100 mM Tris–HCl (pH 7.6), 4% (w/v) SDS, 20% (w/v) glycerol, 10% (v/v) 2-mercaptoethanol, 0.2% bromophenol blue. Store at room temperature.

2.1.2. SDS-Polyacrylamide Gel Electrophoresis of Proteins

1. ATTO protein gel electrophoresis system (ATTO).

2.1.2.1. Preparation of SDS-Polyacrylamide Gels

1. Rainbow full range molecular weight marker (GE Healthcare).
2. Acrylagel, Bis-acrylagel (National Diagnostics): acrylagel is toxic and a known carcinogen. Bis-acrylagel is an irritant. Consult the corresponding material safety data sheet (MSDS) before use.
3. 1.5 M Tris–HCl (pH 8.8).
4. 1 M Tris–HCl (pH 6.8).
5. Deionised water (dH_2O).
6. 10% (v/v) SDS (dissolved in dH_2O).
7. 10% (v/v) Ammonium persulphate (APS) (dissolved in dH_2O): APS is a strong and harmful oxidising agent. Consult the corresponding MSDS before use.
8. TEMED (*N,N,N′,N′*-tetramethylethylenediamine).
9. 10% (v/v) Resolving gels and 5% (v/v) stacking polyacrylamide gels (see Table 3).
10. 5× Tris–glycine running buffer: 15.1 g Tris base, 95.4 g glycine (pH 8.3), 50 mL 10% (w/v) SDS. Make up to 1 L with dH_2O and store at room temperature.
11. 1× Tris–glycine running buffer: 200 mL 5× Tris–glycine running buffer, 800 mL dH_2O.

2.1.3. Coomassie Blue Staining

1. Destain: 450 mL methanol, 450 mL dH_2O, 100 mL glacial acetic acid. Store at room temperature.
2. Coomassie blue stain: 0.25 g Coomassie Brilliant Blue R250, 100 mL destain. Store at room temperature.

2.1.4. Western Blotting

2.1.4.1. Transfer of Protein to Nitrocellulose Filters

1. Bio-Rad Trans-Blot® SD semi-dry electrophoretic transfer cell.
2. Transfer buffer: 750 mL dH_2O, 2.9 g glycine, 5.8 g Tris base, 3.7 mL 10% (w/v) SDS, 200 mL methanol. Adjust volume to 1 L with dH_2O and store at 4°C.

Table 3
SDS-polyacrylamide gel composition

Component	10% Resolving gel (mL)	5% Stacking gel (mL)
Acrylagel	3.33	0.42
Bis-acrylagel	1.35	0.168
1.5 M Tris-HCL (pH 8.8)	2.5	0
1 M Tris-HCL (pH 6.8)	0	0.312
dH_2O	2.61	1.5475
10% (v/v) SDS	0.10	0.025
10% (v/v) APS	0.10	0.025
TEMED	0.01	0.0025
Total	10	2.5

2.1.4.2. Staining of Proteins Immobilised on Nitrocellulose Filters

1. 1× Tris-buffered saline (TBS): 6.1 g Tris base, 8.8 g NaCl, 800 mL, dH_2O. Adjust the pH to 7.5 with HCl and the volume to 1 L. Store at room temperature.
2. TBS-T: 1 L 1× TBS, 1 mL Tween 20. Store at room temperature.
3. Blocking buffer: 3 g bovine serum albumin, 100 mL TBS-T. Prepare immediately before use.
4. Nitrocellulose blotting membrane (Labkem).
5. 3-MM filter paper (Whatman).
6. Scalpel blade (Swann-Morton).
7. Ponceau S.

2.1.4.3. Immunological Probing

1. Mouse anti-His-HRP; a monoclonal antibody that reacts with poly-histidine residues, and is conjugated to horseradish peroxidase (see Note 4). Dilute to 1/1,000 in blocking buffer ("Immunological Probing" in Subheading 3.1.4) immediately before use.
2. Substrate [e.g. 5-bromo-4-chloro-3-indolyl phosphate/Nitro Blue Tetrazolium (BCIP/NBT) or 3,3′,5,5′-tetramethylbenzidine (TMB)]. These substrates are light sensitive, protect from light.
3. Stop solution: alkaline phosphatise; 3 M NaOH; Peroxidase; 3 M HCl or 3 M H_2SO_4.
4. Gel documentation system [e.g. ChemiBIS 1.6-Imaging System (Berthold Technologies)].

2.2. Determination of Protein Solubility

1. Materials from Subheading 2.1.
2. Lysozyme 100 mg/mL: dissolve 10 mg lysozyme in 1 mL dH_2O.
3. Lysis buffer (1 L): 50 mM NaH_2PO_4; 6.90 g $NaH_2PO_4{\cdot}H_2O$ (MW 137.99 g/mol), 300 mM NaCl; 17.54 g NaCl (MW 58.44 g/mol), 10 mM imidazole; 0.68 g imidazole (MW 68.08 g/mol). Adjust pH to 8.0 using NaOH.

2.3. IMAC Purification of His-Tagged Recombinant Proteins

2.3.1. Denaturing and Native Solutions

2.3.1.1. Denaturing Solutions

Caution: the buffers listed below contain sodium hydroxide which is an irritant. Lysis Buffer A contains guanidine hydrochloride which is harmful and an irritant. Consult the corresponding MSDS before use:

1. Lysis Buffer A (1 L): 100 mM NaH_2PO_4; 13.8 g $NaH_2PO_4{\cdot}H_2O$ (MW 137.99 g/mol), 10 mM Tris–HCl; 1.2 g Tris base (MW 121.1 g/mol), 6 M guanidine hydrochloride; 573 g. Adjust pH to 8.0 using NaOH.
2. Lysis Buffer B (1 L): 100 mM NaH_2PO_4; 13.8 g $NaH_2PO_4{\cdot}H_2O$ (MW 137.99 g/mol), 10 mM Tris–HCl; 1.2 g Tris base (MW 121.1 g/mol), 8 M urea; 480.5 g (MW 60.06 g/mol). Adjust pH to 8.0 using NaOH.
3. Wash Buffer C (1 L): 100 mM NaH_2PO_4; 13.8 g $NaH_2PO_4{\cdot}H_2O$ (MW 137.99 g/mol), 10 mM Tris–HCl; 1.2 g Tris base (MW 121.1 g/mol), 8 M urea; 480.5 g (MW 60.06 g/mol). Adjust pH to 6.3 using HCl.
4. Elution Buffer D (1 L): 100 mM NaH_2PO_4; 13.8 g $NaH_2PO_4{\cdot}H_2O$ (MW 137.99 g/mol), 10 mM Tris–HCl; 1.2 g Tris base (MW 121.1 g/mol), 8 M urea; 480.5 g (MW 60.06 g/mol). Adjust pH to 5.9 using HCl.
5. Elution Buffer E (1 L): 100 mM NaH_2PO_4; 13.8 g $NaH_2PO_4{\cdot}H_2O$ (MW 137.99 g/mol), 10 mM Tris–HCl; 1.2 g Tris base (MW 121.1 g/mol), 8 M urea; 480.5 g (MW 60.06 g/mol). Adjust pH to 4.5 using HCl.

2.3.1.2. Native Solutions

Caution: the buffers listed below contain sodium hydroxide and imidazole which are irritants. Consult the corresponding MSDS before use:

1. Lysis buffer (1 L): 50 mM NaH_2PO_4; 6.90 g $NaH_2PO_4{\cdot}H_2O$ (MW 137.99 g/mol), 300 mM NaCl; 17.54 g NaCl (MW 58.44 g/mol), 10 mM imidazole; 0.68 g imidazole (MW 68.08 g/mol). Adjust pH to 8.0 using NaOH. If the tagged protein does not bind under these conditions, the amount of imidazole should be reduced to 1–5 mM.
2. Native wash buffer (1 L): 50 mM NaH_2PO_4; 6.90 g $NaH_2PO_4{\cdot}H_2O$ (MW 137.99 g/mol), 300 mM NaCl; 17.54 g NaCl (MW 58.44 g/mol), 20 mM imidazole; 1.36 g

imidazole (MW 68.08 g/mol). Adjust pH to 8.0 using NaOH.

3. Native elution buffer (1 L): 50 mM NaH_2PO_4; 6.90 g $NaH_2PO_4 \cdot H_2O$ (MW 137.99 g/mol), 300 mM NaCl; 17.54 g NaCl (MW 58.44 g/mol), 250 mM imidazole; 17.00 g imidazole (MW 68.08 g/mol). Adjust pH to 8.0 using NaOH.
4. Small bench rotary shaker (e.g. IKA, KS130).

2.3.2. Batch Purification

Ni-NTA agarose is composed of Ni-NTA coupled to Sepharose® CL-6B and the manufacturer's reported binding capacity is 5–10 mg of 6×His-tagged protein per millilitre of resin. Caution: nickel-NTA is harmful, a sensitizer, and flammable. Consult the corresponding MSDS before use:

1. Ni-NTA resin (QIAgen).
2. Buffers from Subheading 2.3.

2.3.3. Column Purification

1. Ni-NTA spin columns (QIAgen).
2. Buffers from Subheading 2.3.

2.4. Growth for Preparative Purification

1. Materials as described in Subheadings 2.1 and 2.3.

2.5. Buffer Exchange and Concentration of Purified Protein

2. Amicon Ultrafilter (UFC8 00508) Millipore.
3. Phosphate-buffered saline (PBS).

3. Methods

The process strategy for the purification of poly-histidine-tagged proteins from the initial design stage to the final purification steps in outlined in Fig. 1. Initially, an expression screening procedure should be performed under denaturing conditions (Subheading 3.1), which will lead to the isolation of any tagged protein, independent of its location within the cell. Following expression screening, protein solubility can be determined (Subheading 3.2) which will guide the decision to purify under native or denaturing conditions (Subheading 3.3).

3.1. High-Level Expression of Recombinant His-Tagged Proteins

Preliminary analysis of expression levels, cellular localization, and solubility of the target protein should be performed before purification or activity measurements using the methods described in this section. It is useful at this point to predict the physico-chemical properties of the fusion protein (see Note 5) using bioinformatics

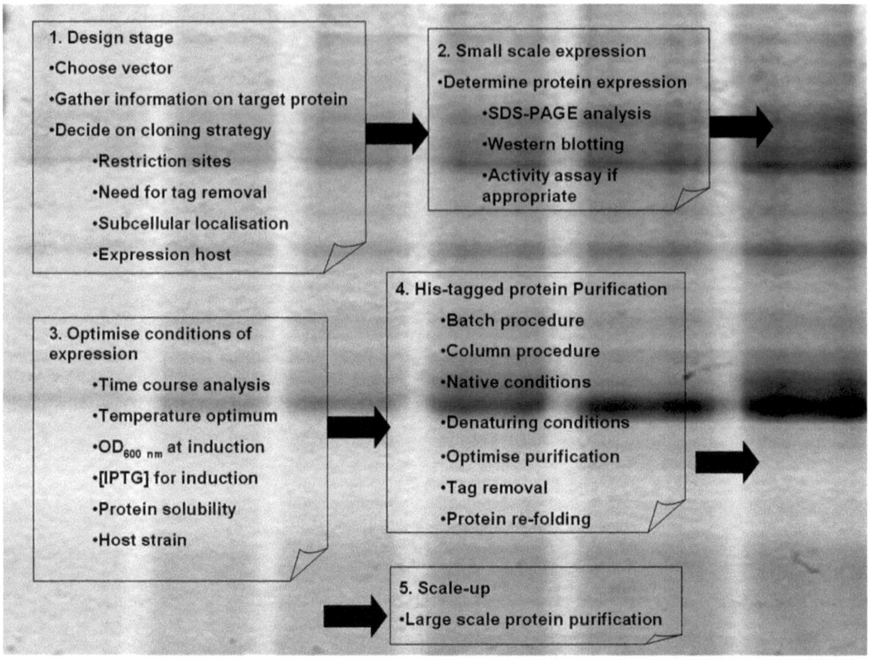

Fig. 1. Strategy for the purification of poly-histidine-tagged proteins. The flow chart shows the various stages in the overall process leading to the purification of a His-tagged recombinant protein.

tools such as the ProtParam tool (see Chapter 1: Table 2 for more detail). Here, high-level expression of recombinant proteins is achieved by induction with IPTG:

1. Inoculate 1.5 mL of LB broth supplemented with appropriate antibiotics (pQE vector series – ampicillin at 100 μg/mL; RosettaBlue™ host – chloramphenicol at 34 μg/mL and tetracycline at 12.5 μg/mL) with single colonies of transformants and grow overnight at 37°C in an orbital shaker set at 200 rpm (see Notes 6 and 7).
2. Use 500 μL of these cultures to inoculate 10 mL (see Notes 6 and 8) of supplemented LB broth, and grow as before for 3–4 h until logarithmic phase [optical density at 600 nm wavelength (OD_{600nm}) of 0.5] is reached (see Note 9). At the same time, set up a control culture which will serve as an uninduced control in step 3.
3. Induce the cultures to express fusion proteins by the addition of IPTG at a final concentration of 1.0 mM followed by growth for a further 5 h as above (see Notes 10 and 11). At the same time, set up an uninduced control culture.
4. Collect cells by centrifugation at 15,000 × *g* for 5 min. Discard the supernatant and lyse the cell pellets in 400 μL of lysis buffer (Buffer B) (see Notes 12 and 13). Sonicate on ice for 6 × 10 s with 10 s pauses at 200 W (see Notes 14 and 15).

5. Remove cell debris by centrifugation at 15,000 ×*g* for 15 min and analyse the supernatant (cleared cell lysate) by SDS-PAGE (see Subheading 3.1.2) to confirm protein expression at the predicted protein molecular weight.

To optimise the expression of a given recombinant protein, a time-course analysis of the level of expression can be performed by harvesting cells from expression cultures at time zero and at time intervals of 1 h for 5 h and at a final timepoint of 12 h following induction (see Notes 7, 10, and 16; see Fig. 2). The level of recombinant protein expression can be optimised by setting up small-scale expression cultures as above with variations of the following relevant growth parameters: IPTG concentration for induction (varied over the range 0.0025–1.5 mM), optical density measured at 600 nm at the time of induction (varied over the range 0.3–1.0 absorbance units), and growth after induction (varied over the range 25–37°C).

If proteolysis is occurring, take steps to avoid degradation as outlined in Notes 12, 18–20 (see also Chapter 4). Perform western blotting (as described in Subheading 3.1.4) to verify His-fusion protein expression. Immunoreactivity with the anti-His HRP-conjugated monoclonal antibody will be evident upon chromogenic development following the addition of TMB substrate to the immunoprobed nitrocellulose membrane. The solubility of recombinant proteins can also determined as described in Subheading 3.2. Finally, if no recombinant protein expression is evident, check that the coding sequence is ligated into the correct reading frame by sequencing. Also, variation of growth conditions as above or a change of expression host may permit recombinant protein expression.

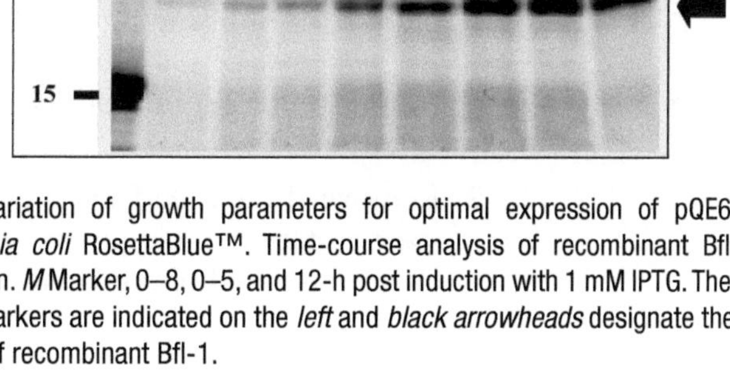

Fig. 2. Variation of growth parameters for optimal expression of pQE60-Bfl-1 in *Escherichia coli* RosettaBlue™. Time-course analysis of recombinant Bfl-1 protein expression. *M* Marker, 0–8, 0–5, and 12-h post induction with 1 mM IPTG. The molecular weight markers are indicated on the *left* and *black arrowheads* designate the predicted location of recombinant Bfl-1.

3.1.1. Protein Analysis

Prepare total cellular proteins for analysis by SDS-polyacrylamide gel electrophoresis (SDS-PAGE) and by western blotting:

1. To the supernatant prepared in Subheading 3.1, add an equal volume of 2× SDS gel loading buffer.
2. Immediately heat the sample for 5 min at 85°C to denature the proteins. Clarify the lysate by centrifugation at 12,000 × *g* for 10 min at room temperature and store at –20°C until SDS-PAGE analysis.

3.1.2. SDS-Polyacrylamide Gel Electrophoresis of Proteins

Perform SDS-PAGE using 10% (v/v) resolving gels and 5% (v/v) stacking polyacrylamide gels prepared as described in "Preparation of SDS-Polyacrylamide Gels" in Subheading 2.1.2:

3.1.2.1. Preparation of SDS-Polyacrylamide Gels

1. Wash glass plates with detergent, then rinse with tap water, then with dH_2O, and finally wipe in one direction with tissue soaked in 100% (v/v) ethanol.
2. Place the gasket around the ridged plate; assemble the plates; and secure with clamps.
3. Pour the resolving gel to within 2 cm of the top of the larger plate and overlay with 100% (v/v) ethanol.
4. When set, remove the ethanol and pour the stacking gel. Insert a clean comb that has been wiped in 100% (v/v) ethanol and allow the gel to polymerise for at least 20 min.
5. Fill the electrophoresis tank with 1× Tris–glycine running buffer to a level of about 5 cm deep.
6. After polymerisation, remove the gaskets and clamps and lower the pre-poured gels into the buffer at an angle to exclude air bubbles from the gel-buffer interface. Fill the tank completely with 1× running buffer and remove the comb from the gel. Fix the gel plates firmly in place with the pressure plates. Fill the chamber formed by the inner plates (notched plate facing inwards) with 1× running buffer.
7. Load the samples and attach the electrodes. Perform electrophoresis at a constant current of 30 mA per gel until the blue dye front has reached the bottom of the gel. When complete, separate the plates and place the gel in either transfer buffer prior to western blotting (Subheading 3.1.4) or in Coomassie blue stain (Subheading 3.1.3).

3.1.3. Coomassie Blue Staining

1. Immerse the gel in Coomassie stain for 30 min with gentle agitation.
2. Remove the gel and immerse in destain solution with constant agitation. Change this solution four or five times at 1-h interval until all background staining has been removed from the gel.

3. Capture an image of the gel in black and white using a UV trans-illuminator switched to white light only and using a white tray.

3.1.4. Western Blotting

During western blotting, electrophoretically separated proteins are transferred from the polyacrylamide gel to a solid support, usually a nitrocellulose membrane, and probed with antibodies that react specifically with antigenic epitopes present on the target protein that is attached to the solid support. In this chapter, the bound antibody is detected by a secondary immunological reagent, conjugated to either the alkaline phosphatase or horseradish peroxidase enzyme for detection:

3.1.4.1. Transfer of Protein to Nitrocellulose Filters

1. Following gel electrophoresis, equilibrate the gel(s) by immersion in transfer buffer for at least 15 min (see Note 21).
2. Cut the nitrocellulose membrane according to the dimensions of the gel, along with six pieces of 3-MM filter paper required for the gel/membrane sandwich. Immerse nitrocellulose and filter paper in transfer buffer for 15 min.
3. Perform protein transfer as follows (the procedure described is for a Bio-Rad Trans-Blot® SD semi-dry electrophoretic transfer cell): place a pre-soaked sheet of filter paper onto the platinum anode. Use a pipette to roll over the surface of the filter paper to exclude all air bubbles. Repeat this step with two more sheets of filter paper. Then, place the pre-wetted blotting membrane on top of the filter paper and roll all bubbles out. Place the equilibrated gel carefully on top of the nitrocellulose membrane, aligning the gel on the centre of the membrane. Roll out any air bubbles. Place another three sheets of pre-wetted filter on top of the gel, with care taken to remove air bubbles. Place the cathode on top of the stack and replace the safety cover on the transfer unit. Transfer proteins for 22 min at 15 V.

3.1.4.2. Staining of Proteins Immobilised on Nitrocellulose Filters

Ponceau S staining can be used to determine whether uniform transfer of proteins to the nitrocellulose membrane has taken place. Transferred proteins are detected as red bands on a white background. This staining technique is reversible thus allowing further analysis of blotted proteins by immunological probing. Ponceau S is a negative stain, which binds to positively charged amino acid groups of proteins. It also binds non-covalently to non-polar regions of proteins:

1. Following electrophoretic transfer, immerse the nitrocellulose membrane in 20 mL of Ponceau S solution and stain for 5 min with constant agitation.
2. Once proteins are visible, verify that transfer has occurred evenly across the membrane. Wash the membrane in several

changes of dH_2O until all the stain has been washed away. The membrane can then be used for immunological probing.

3.1.4.3. Immunological Probing

His-tagged proteins can be readily detected by Western blotting using anti-His antibodies. However, cross-reactivity of the primary antibody with endogenous histidines may be a considerable drawback in mammalian and insect expression systems (12):

1. Following Ponceau S staining, incubate the membrane in blocking buffer for 2 h at room temperature with constant agitation.
2. Remove the blocking buffer and incubate the membrane with the anti-His-HRP antibody (diluted in blocking buffer) at 4°C overnight with constant agitation.
3. Following overnight incubation, wash the membrane three times in TBS-T for 15 min (see Note 22).
4. Place the membrane in a clean container and cover with the appropriate substrate; BCIP/NBT for alkaline-phosphatase-conjugated antibodies or TMB for horseradish peroxidase-conjugated antibodies. When incubated with alkaline phosphatase or hydrogen peroxidase, bound enzyme catalyses the production of a coloured product that is easily observable.
5. Rinse the membrane in distilled water or add stop solution to stop the reaction. Capture the resultant image using a camera/scanner or gel documentation system.

3.2. Determination of Protein Solubility

Many polypeptide gene products expressed in *E. coli* accumulate as insoluble aggregates (see Note 23). Eukaryotic proteins expressed intracellularly in *E. coli* are frequently sequestered into insoluble inclusion bodies. Since the interaction between Ni-NTA and the 6×His tag of the recombinant protein does not depend on the latter's tertiary structure, proteins can be purified either under native or denaturing conditions.

In order to determine if the protein is soluble in the cytoplasm and therefore purifiable under native conditions, the soluble and insoluble fractions of expression lysates are first examined by SDS-PAGE analysis (see Fig. 3). If the protein of interest resolves in the soluble fraction, purification under native conditions is possible (proceed to "Under Native Conditions" in Subheading 3.3.1 for batch purification and "Under Native Conditions" in Subheading 3.3.2 for column purification). Alternatively, if the protein of interest resolves in the insoluble fraction, the protein can be purified under denaturing conditions (proceed to "Under Denaturing Conditions" in Subheadings 3.3.1 and "Under Denaturing Conditions" in 3.3.2 for batch and column purification procedures, respectively). The Bfl-1-His protein

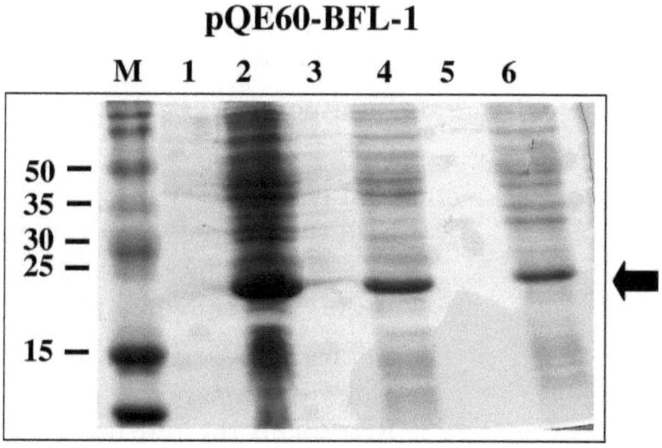

Fig. 3. Determination of protein solubility. Expression cultures harvested at 5-h post induction were lysed in native lysis buffer followed by incubation with lysozyme and sonication. Soluble and insoluble extracts were prepared and analysed by SDS-PAGE. *Lanes*: *M* Marker (GE Healthcare), 1 and 2: unrelated experiment; 3 and 5: blank; 4: Extract A, soluble extract; 6: Extract B, insoluble extract. The molecular weight markers are indicated on the *left* and *black arrowheads* designate the predicted location of recombinant Bfl-1.

(see Fig. 3) was resolved in the soluble fraction as can be seen by comparing the soluble and insoluble extracts in lanes 4 and 6, respectively, indicating that the protein was therefore purifiable under native conditions.

If the tagged protein is insoluble, the expression conditions can be modified in an attempt to enhance solubility. A reduction in growth temperature following induction may lead to an increase in soluble protein. A change of host strain or a reduction in the expression level as induced by IPTG (reduce from 1 to 0.005 mM) may permit higher levels of protein expression before inclusion body formation:

1. Inoculate 1.5 mL of LB broth supplemented with appropriate antibiotics (pQE vector series – ampicillin at 100 μg/mL; RosettaBlue™ host – chloramphenicol at 34 μg/mL and tetracycline at 12.5 μg/mL) with single colonies of transformants and grow overnight at 37°C in an orbital shaker set at 200 rpm (see Notes 6 and 7).
2. Use 500 μL of these cultures to inoculate 10 mL (see Notes 6 and 8) of LB broth with antibiotics and grow overnight at 37°C in an orbital shaker set at 200 rpm.
3. Use 2.5 mL of these cultures to inoculate 50 mL of LB broth with antibiotics and grow as before for 3–4 h until logarithmic phase [OD_{600nm} of 0.5] is reached (see Note 9).
4. Induce the cultures to express fusion proteins by the addition of IPTG at a final concentration of 1.0 mM followed by growth for a further 5 h as above (see Notes 10 and 11).

5. Harvest the expression cultures by centrifugation at 4,000 × *g* for 20 min.
6. Resuspend pelleted cells in 5 mL of native lysis buffer (see Notes 24 and 25).
7. Add lysozyme (1 mg/mL) (see Notes 26 and 17) and incubate the sample for 30 min at room temperature (see Note 27) followed by sonication on ice for 6 × 10 s with 10 s pauses at 200 W (see Notes 14 and 15).
8. Centrifuge the lysate at 10,000 × *g* at 4°C for 20 min and aspirate the supernatant and store on ice (soluble extract). Resuspend the remaining pellet in 5 mL of lysis buffer and store on ice (insoluble matter).
9. Examine the soluble and insoluble extracts for recombinant protein expression by SDS-PAGE analysis (Subheading 3.1.2).

3.3. IMAC Purification of His-Tagged Recombinant Proteins

The purification of His-tagged recombinant proteins can be performed using batch and column procedures under native and denaturing conditions. Native protein purification involves the use of buffers that preserve the native, three-dimensional structure and surface charge characteristics of a selected soluble protein during harvest from an expression host. However, if the tagged protein of interest is insoluble, purification under denaturing conditions may be necessary. Denaturants, such as 8 M urea or 6 M guanidine, can be used to enhance protein solubility (see Note 28). Incompatible reagents to be avoided during IMAC purification are listed in Table 4 (see Note 29). Finally, if protein activity/native structure is important (e.g. for biochemical or structural studies), protein refolding following purification under denaturing conditions can be performed (Subheading 3.4).

During the process of purification, collect all eluates and fractions for analysis by SDS-PAGE and Coomassie blue staining. Protein purification steps can also be monitored by UV absorbance at 280 nm (see Chapter 13 for details on UV protein absorption assay using the Nanodrop® ND-1000).

3.3.1. Batch Purification

3.3.1.1. Under Denaturing Conditions

1. Equilibrate 1 mL of 50% Ni-NTA resin by adding of 10 mL of Buffer B.
2. Centrifuge the resin-buffer mixture at 1,200 × *g* for 1 min and discard the supernatant.
3. Add 4 mL of cleared lysate (from 100 mL culture volumes scaled up by a factor of 10 from Subheading 3.1) (see Notes 29 and 30) to the equilibrated resin and mix gently for 60 min at room temperature with rotation at 200 rpm on a rotary shaker.
4. Centrifuge the lysate–resin mixture at 1,200 × *g* for 1 min and remove the supernatant.
5. Add 4 mL wash buffer (Buffer C) (see Note 31) to the resin, mix thoroughly, and centrifuge at 1,200 × *g* for 1 min.

Table 4
Reagent compatibility with IMAC resin

Reagent	Compatibility	Acceptable concentration
CHAPS	Yes	1% (with caution)
DTE (dithioerythritol)	No	–
DTT (dithiothreitol)	No	–
EDTA	No	–
EGTA	No	–
Ethanol[a]	Yes	30%
Ethylene glycol	Yes	30%
Glycerol	Yes	20%
Guanidine[b]	Yes	6 M
HEPES[e]	Yes	100 mM
Imidazole[c]	Yes	200 mM at pH 7.0–8.0, for elution
KCl	Yes	500 mM
MES	Yes	20 mM
MOPS[e]	Yes	100 mM
NaCl	Yes	1 mM
NP-40	Yes	1%
SDS	Yes	1% with caution
Tris[d,e]	Yes	100 mM
Triton-X 100	Yes	<2%
Tween	Yes	2%
Urea	Yes	8 M
β-Mercaptoethanol[b]	Yes	20 mM (with caution)

[a]Ethanol may precipitate proteins, causing low yields and column clogging
[b]Do not store resin in buffers containing these reagents. Use resin immediately after equilibrating with buffers containing these reagents
[c]For IMAC column binding, imidazole cannot be used at concentrations higher than 5–10 mM due to competition for binding to the immobilized metal ions
[d]Tris coordinates weakly with metal ions, causing a decrease in capacity
[e]Sodium phosphate or phosphate–citrate buffer is recommended

6. Repeat the wash step twice more.
7. Elute by adding four volumes of 500 μL of elution buffer (Buffer D/E) (see Note 32) to the resin and centrifuge at 1,200 × *g* for 1 min. Collect the supernatant/eluate.

3.3.1.2. Under Native Conditions

1. Prepare expression cultures as in Subheading 3.1 but using 100 mL culture volumes (scaled up by a factor of 10 from Subheading 3.1).
2. Lyse the cells in native lysis buffer.
3. Add lysozyme (1 mg/mL) (see Notes 26 and 17) and incubate the sample for 30 min at room temperature (see Note 27) followed by sonication on ice for 6 × 10 s with 10 s pauses at 200 W (see Notes 14 and 15).
4. Remove cell debris by centrifugation at 15,000 × *g* for 15 min.
5. Equilibrate 1 mL of 50% Ni-NTA resin by adding 10 mL of native lysis buffer, centrifuge at 1,200 × *g* for 1 min, and remove the supernatant.
6. Add 4 mL of cleared lysate to the equilibrated resin and mix gently for 60 min at room temperature with rotation at 200 rpm on a rotary shaker.
7. Centrifuge the lysate–resin mixture at 1,200 × *g* for 1 min and remove the supernatant.
8. Add 4 mL of native wash buffer to the resin (see Note 24 and Table 4), mix thoroughly, and centrifuge at 1,200 × *g* for 1 min.
9. Repeat the wash step twice more.
10. Elute by adding four volumes of 500 μL of native elution buffer to the resin and centrifuge at 1,200 × *g* for 1 min. Collect the supernatant/eluate.

3.3.2. Column Purification

3.3.2.1. Under Denaturing Conditions

1. Equilibrate a Ni-NTA spin column using 600 μL Buffer B. Centrifuge for 5 min at 700 × *g*.
2. Load cell lysates from Subheading 3.1 onto pre-equilibrated Ni-NTA IMAC minicolumns. Centrifuge for 5 min at 700 × *g*. Collect the flow through.
3. Perform two wash steps by adding 600 μL wash buffer (Buffer C) (see Note 31) followed by centrifugation at 700 × *g* for 5 min. Collect wash fractions.
4. Add 2× 200 μL elution buffer to the column (Buffer E; pH 4.5). Centrifuge for 5 min at 700 × *g*. Collect the eluates.

3.3.2.2. Under Native Conditions

1. Equilibrate a Ni-NTA spin column using 600 μL native lysis buffer. Centrifuge for 5 min at 700 × *g*.
2. Add cell lysates (prepared as described in "Under Native Conditions" in Subheading 3.3.1) onto pre-equilibrated Ni-NTA IMAC minicolumns. Centrifuge for 5 min at 700 × *g*. Collect the flow through.

3. Perform two wash steps by adding 600 μL native wash buffer (see Note 31) followed by centrifugation at 700 × *g* for 5 min. Collect wash fractions.
4. Add 2× 200 μL native elution buffer to the column. Centrifuge for 5 min at 700 × *g*. Collect the eluates.

3.4. Downstream Processes: His-Tag Removal by Proteolytic Cleavage and Protein Refolding

Purification may include enzymatic cleavage to remove part or all of the fusion tag, using endopeptidases including enterokinase, factor Xa, thrombin, or HRV 3C proteases. The 6×His tag can be easily and efficiently removed from His-tagged proteins expressed in the pQE30-Xa vector using Factor Xa Protease. For detailed discussion and protocols regarding tag removal see Chapter 19. Detailed protocols for protein precipitation and differential solubilisation following purification are described in Chapter 11.

3.5. Large-Scale Growth for Preparative Purification of Recombinant Protein

Once optimal purification conditions have been determined empirically using small-scale expression and purification experiments, growth for preparative-scale purification can be performed. The procedure below is outlined for native purification and can be modified for large-scale purification under denaturing conditions according to the steps outlined in "Under Denaturing Conditions" in Subheading 3.3.2. Monitor purification as before by collecting fractions for analysis by SDS-PAGE (see Fig. 4):

1. Inoculate 1.5 mL of LB broth (see Note 1) supplemented with antibiotics (pQE vector series – ampicillin at 100 μg/mL; RosettaBlue™ host – chloramphenicol at 34 μg/mL and tetracycline at 12.5 μg/mL) with single colonies of transformants and grow overnight at 37°C in an orbital shaker set at 200 rpm (see Notes 6 and 7).

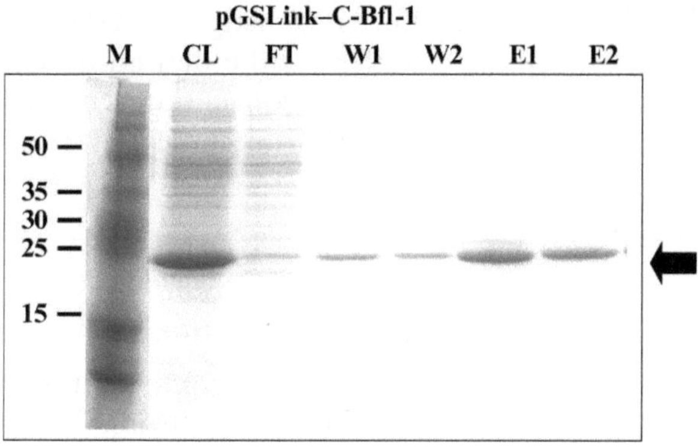

Fig. 4. Optimisation of purification of His-Linker-Bfl-1 under native conditions. Recombinant protein purification of His-Linker-Bfl-1 protein under native conditions using Ni-NTA resin. *M* Marker (Amersham), *CL* Cleared cell lysate, *FT* Flow through, *W1* and *W2* Wash with native wash buffer, *E1–2* Elution with native elution buffer. The molecular weight markers are indicated on the *left* and *black arrowheads* designate the location of recombinant Bfl-1.

2. Use 500 μL of these cultures to inoculate 20 mL (see Notes 6–8) of supplemented LB broth, and incubate overnight at 37°C in an orbital shaker set at 200 rpm.
3. Inoculate LB broth (1 L) 1:50 with these cultures (see Note 8) and grow as above for 3–4 h until logarithmic phase [OD_{600nm} of 0.8] is reached (see Note 9).
4. Induce the expression of tagged protein by adding IPTG at a final concentration of 0.1 mM followed by growth for a further 5 h at 37°C in an orbital shaker set at 200 rpm.
5. Collect cells by centrifugation at 4,000 × *g* for 20 min.
6. Discard the supernatant and lyse the cell pellets in native lysis buffer (5 mL/g wet weight), followed by incubation in lysozyme [1 mg/mL for 30 min at room temperature (see Notes 25–27)] and sonication on ice for six 10-s pulses with 10 s breaks between pulses at an amplitude: 40/output, 200 W (see Notes 14 and 15).
7. Remove cell debris by centrifugation at 15,000 × *g* for 25 min at 4°C.
8. Apply the supernatant to 2 mL Ni-NTA matrix that has been packed into a disposable column and pre-equilibrated with native lysis buffer.
9. Collect the flow through and reapply to the resin a further two times.
10. Perform two washes with wash buffer.
11. Elute purified proteins twice using 5 mL of elution buffer.

3.6. Buffer Exchange and Concentration of Purified Protein

1. Pool the eluates from preparative purification and apply to an Amicon ultrafilter.
2. Buffer exchange the 10 mL eluate to PBS (or the appropriate buffer for the desired downstream application) and simultaneously concentrate (by reducing volume from 10 mL to 500 μL three times) by centrifugation at 1,252 × *g* and 4°C.

4. Notes

1. While most recombinant proteins can be cloned and expressed at high levels in *E. coli*, there are exceptions. Some His-tagged proteins show weak binding to the metal chelating resin which can be due to concealment of the His-tag, and may be alleviated by switching its position to the other terminus of the protein (13) or by introducing a linker to separate the His-tag from the partner protein (4). In addition, these difficulties can be overcome by increasing the length of the His-tag to eight or ten histidines (14, 15).

2. *Escherichia coli* RosettaBlue™ (see Table 2) and *E. coli* Rosetta™ (Novagen) strains have been engineered to express the tRNAs for rarely expressed codons from a chloramphenicol-resistant plasmid, pRARE. The use of *E. coli* RosettaBlue™ or *E. coli* Rosetta™ strains as expression hosts facilitates the expression of proteins that would otherwise be limited by codon bias in *E. coli*.
3. IMAC is very sensitive to the presence of metal chelators (16). In *E. coli* expression systems, the cell lysate contains many unspecific weak chelators such as dicarboxylic acids from the citric acid cycle. Under stress conditions, *E. coli* can also produce highly specific metal chelators known as metallophores (17). A recent report linked the failure to purify low-abundance His-tagged proteins from *E. coli* lysates to metal-ion leakage from purification columns (18). In that study, the authors used His-tagged GFP (His_6-GFP) to examine the effect of *E. coli* lysate on the protein binding capacity of IMAC columns and concluded that low molecular weight components of the lysate, associated with the periplasm, severely reduced the binding capacity of the column. By removing the periplasmic material before cell lysis, the authors observed a tenfold increase in the yield of His_6-GFP when it was diluted with *E. coli* lysate before purification to simulate a low-abundance protein.
4. Sodium azide can be added to each antibody solution to a final concentration of 0.02% (w/v) as a preservative thus permitting reuse of the antibody. Do not use sodium azide as a preservative when diluting HRP-conjugated antibodies as it is an established peroxidase inhibitor. Also note that the phosphate in PBS can inhibit the alkaline phosphatase reporter enzyme.
5. Predict codon usage, molecular weight, molar extinction coefficient using bioinformatics tools (see Chapter 1: Table 2).
6. In order to ensure good aeration, add medium up to only 20% of the total flask volume. For 1.5 mL culture volumes, use 17 × 100 mm sterile snap-cap tube (VWR International); for 10 mL culture volumes, use 125 mL Erlenmeyer flasks; for 20 mL culture volumes, use 250 mL Erlenmeyer flasks.
7. If cultures are grown overnight, 0.5–1% glucose may be added to the media in order to reduce target protein expression prior to induction (19).
8. The staging procedure will minimise the shock-induced lag phase in growth caused by the transfer of a small inoculum to a larger volume of fresh medium and diffusion of vitamins, minerals, and cofactors from the cells (20).

9. Monitor the OD_{600} during growth by removing aliquots aseptically.
10. When directing fusion proteins to the periplasmic space, leakage of the protein to the medium might be enhanced by prolonged inductions (16 h or more).
11. The Overnight Express™ Autoinduction System (Novagen) is designed for high-level protein expression with IPTG-inducible bacterial expression systems (21) without the need to monitor cell growth. The method is based on media components that are metabolised differentially to promote growth to high density and automatically induce protein expression from *lac* promoters.
12. Protease inhibitors can be added at this point as an option (see Chapter 4). Serine protease inhibitors should be avoided if the target protein is to be treated with Thrombin, Factor Xa, or Recombinant Enterokinase. Cysteine protease inhibitors should be avoided if the target protein is to be treated with HRV 3C. Although purification may remove active inhibitors, dialysis or gel filtration is recommended before cleavage.
13. Lysing cells in Buffer B allows solubilisation of most proteins and inclusion bodies, and facilitates their direct analysis by SDS-PAGE. To solubilise very hydrophobic receptor or membrane proteins, Buffer A, containing guanidine hydrochloride (Gu-HCl), can be used, sometimes in combination with detergents.
14. Sonication conditions may vary with the equipment. Alternatively, if a sonicator is not available, pass the sample through a 27-gauge needle several times to reduce the viscosity or use a French press to lyse cells (perform two passes at 20,000 psi using a chilled pressure cell).
15. Excessive sonication can destroy protein functionality.
16. When analyzing the time course of expression, begin with a 100-mL culture in a flask, and take 10 mL samples at each timepoint after induction.
17. DNase I (5 μg/mL) and RNase A (10.5 μg/mL) can be added at this point (followed by 10–15 min incubation on ice) to reduce the viscosity of the lysate, allowing for more efficient removal of cellular debris.
18. In order to reduce proteolytic activity, maintain protein stability, and improve yield, perform all manipulations at 4–8°C. A reducing agent, such as 10 mM β-ME, or a protease inhibitor, such as PMSF, in the equilibration/wash buffer, may improve the structural stability of fragile proteins during sample preparation (see Table 4 for reagent compatibility).

19. In addition, 1 mM EDTA can be added to the equilibration/wash buffer (pH 7.0) to inhibit metalloproteases during the extraction. However, EDTA must be removed before applying the sample to the resin. A gel filtration column (such as PD-10, GE Healthcare) equilibrated with the equilibration buffer can be used for this. In some cases, the host cell produces low molecular weight chelators that can also be removed using gel filtration.
20. Use protease-deficient *E. coli*; strains B834, BL21, BLR, Origami™ B, Rosetta™, Rosetta 2, Rosetta-gami™ B, and Tuner™ are deficient in the Lon protease and lack the OmpT outer membrane protease that can degrade proteins during purification (22). For further detailed procedures on avoiding proteolysis see Chapter 4.
21. Equilibration facilitates the removal of electrophoresis salts and detergents. If not removed, salts will increase the conductivity of the transfer and the amount of heat generated during transfer.
22. If the primary antibody is not HRP- or AP-conjugated, repeat steps 2 and 3 with an appropriate conjugated secondary antibody, incubating for 2 h at room temperature. Proceed to step 4.
23. Depending on the intended application, preferential localization of the recombinant protein to inclusion bodies, secretion to the medium, or the periplasmic space can be advantageous for rapid purification by relatively simple procedures. Vectors are available which can provide a signal sequence for translocation into the periplasmic space (pelB, DsbA, DsbC; Novagen).
24. The salt content of this buffer may be increased up to 0.5 M as some proteins may exhibit greater solubility when the cells are lysed in a buffer containing salt. Other proteins, such as those associated with membranes, may partition into the soluble fraction if a zwitterionic detergent (e.g. 10 mM CHAPS) is added to the lysis buffer.
25. An additional option is to freeze the pellet completely at −20 or −70°C, then thaw completely. The freeze/thaw step ruptures the outer membrane allowing lysozyme to access the cell wall.
26. If the bacterial strain contains a plasmid encoding lysozyme (e.g. pLysS or pLysE; Novagen), then additional lysozyme treatment is not necessary.
27. Incubations at room temperature results in elevated proteolytic activities. Alternatively, to avoid proteolysis incubate at 4°C.
28. Samples containing 6 M guanidine must be dialyzed overnight against buffer containing 8 M urea before loading on an SDS-PAGE gel.

29. A desalting column can be used to adjust the pH of the lysate, to change buffer, to remove low molecular weight contaminants, and/or to concentrate sample before applying to the column.

30. If the cell lysate is still not clear after centrifugation, filter using a 1 μm filter. Membrane filters that give the least amount of non-specific binding of proteins are composed of cellulose acetate or PVDF.

31. Wash steps can be monitored for efficient removal of unbound material from the column by analyzing protein content of wash fractions using UV absorbance at 280 nm (see Chapter 13 for detailed protocols). Elution can begin when all unbound material has been washed through. This will improve the purity of the eluted target substance.

32. Elution Buffer D and/or E may be required for elution. Monomers generally elute in Buffer D, while multimers, aggregates, and proteins with two 6×His tags will generally elute in Buffer E.

References

1. Porath, J., Carlsson, J., Olsson, I., and Belfrage, G. (1975) Metal Chelate Affinity Chromatography, a New Approach to Protein Fractionation. *Nature.* **258**, 598–599.
2. Hochuli, E., Dobeli, H., and Schacher, A. (1987) New Metal Chelate Adsorbent Selective for Proteins and Peptides Containing Neighbouring Histidine Residues. *J. Chromatogr.* **411**, 177–184.
3. Terpe, K. (2003) Overview of Tag Protein Fusions: From Molecular and Biochemical Fundamentals to Commercial Systems. *Appl. Microbiol. Biotechnol.* **60**, 523–533.
4. Loughran, S. T., Loughran, N. B., Ryan, B. J., D'Souza, B. N., and Walls, D. (2006) Modified His-Tag Fusion Vector for Enhanced Protein Purification by Immobilized Metal Affinity Chromatography. *Anal. Biochem.* **355**, 148–150.
5. Li, M., Su, Z. G., and Janson, J. C. (2004) In Vitro Protein Refolding by Chromatographic Procedures. *Protein Expr. Purif.* **33**, 1–10.
6. Liu, Z., Bartlow, P., Varakala, R., Beitle, R., Koepsel, R., and Ataai, M. M. (2009) Use of Proteomics for Design of a Tailored Host Cell for Highly Efficient Protein Purification. *J. Chromatogr. A.* **1216**, 2433–2438.
7. Villarejo, M. R. and Zabin, I. (1974) Beta-Galactosidase from Termination and Deletion Mutant Strains. *J. Bacteriol.* **120**, 466–474.
8. Prinz, W. A., Aslund, F., Holmgren, A., and Beckwith, J. (1997) The Role of the Thioredoxin and Glutaredoxin Pathways in Reducing Protein Disulfide Bonds in the *Escherichia coli* Cytoplasm. *J. Biol. Chem.* **272**, 15661–15667.
9. Grosjean, H. and Fiers, W. (1982) Preferential Codon Usage in Prokaryotic Genes: The Optimal Codon-Anticodon Interaction Energy and the Selective Codon Usage in Efficiently Expressed Genes. *Gene.* **18**, 199–209.
10. Sorensen, M. A., Kurland, C. G., and Pedersen, S. (1989) Codon Usage Determines Translation Rate in *Escherichia coli. J. Mol. Biol.* **207**, 365–377.
11. Ghiotto, F., Fais, F., and Bruno, S. (2010) BH3-Only Proteins: The Death-Puppeteer's Wires. *Cytometry A.* **77**, 11–21.
12. Kimple, M. E. and Sondek, J. (2004) Overview of Affinity Tags for Protein Purification. *Curr. Protoc. Protein Sci. Chapter 9*, Unit 9.9.
13. Xu, C. G., Fan, X. J., Fu, Y. J., and Liang, A. H. (2008) Effect of Location of the His-Tag on the Production of Soluble and Functional Buthus Martensii Karsch Insect Toxin. *Protein Expr. Purif.* **59**, 103–109.
14. Grisshammer, R., White, J. F., Trinh, L. B., and Shiloach, J. (2005) Large-Scale Expression and Purification of a G-Protein-Coupled Receptor for Structure Determination – an

Overview. *J. Struct. Funct. Genomics.* **6**, 159–163.

15. Yeliseev, A. A., Wong, K. K., Soubias, O., and Gawrisch, K. (2005) Expression of Human Peripheral Cannabinoid Receptor for Structural Studies. *Protein Sci.* **14**, 2638–2653.
16. Chaga, G. S. (2001) Twenty-Five Years of Immobilized Metal Ion Affinity Chromatography: Past, Present and Future. *J. Biochem. Biophys. Methods.* **49**, 313–334.
17. Manjasetty, B. A., Turnbull, A. P., Panjikar, S., Bussow, K., and Chance, M. R. (2008) Automated Technologies and Novel Techniques to Accelerate Protein Crystallography for Structural Genomics. *Proteomics.* **8**, 612–625.
18. Magnusdottir, A., Johansson, I., Dahlgren, L. G., Nordlund, P., and Berglund, H. (2009) Enabling IMAC Purification of Low Abundance Recombinant Proteins from *E. coli* Lysates. *Nat. Methods.* **6**, 477–478.
19. Grossman, T. H., Kawasaki, E. S., Punreddy, S. R., and Osburne, M. S. (1998) Spontaneous cAMP-Dependent Derepression of Gene Expression in Stationary Phase Plays a Role in Recombinant Expression Instability. *Gene.* **209**, 95–103.
20. Hunt, G. R. and Stieber, R. W., (1986) Inoculum Development, in Manual of Industrial Microbiology and Biotechnology. Demain, A. L. and Solomon, N. A., Eds. ASM, Washington, DC, 32–40.
21. Studier, F. W. (2005) Protein Production by Auto-Induction in High Density Shaking Cultures. *Protein Expr. Purif.* **41**, 207–234.
22. Grodberg, J. and Dunn, J. J. (1988) OmpT Encodes the *Escherichia coli* Outer Membrane Protease that Cleaves T7 RNA Polymerase During Purification. *J. Bacteriol.* **170**, 1245–1253.

Chapter 18

Immobilized Metal Affinity Chromatography/Reversed-Phase Enrichment of Phosphopeptides and Analysis by CID/ETD Tandem Mass Spectrometry

Rosana Navajas, Alberto Paradela, and Juan Pablo Albar

Abstract

Major difficulties in phosphoprotein analysis relate to the presence of a huge number of nonphosphorylated proteins and to the wide concentration dynamic range among them. In order to overcome the analysis complexity, specific clean-up and highly efficient enrichment procedures are mandatory prior to the chromatographic separation and identification by tandem mass spectrometry. In this chapter, a procedure based on immobilized metal affinity chromatography (IMAC)/reversed-phase phosphopeptide purification and analysis by nanoHPLC-ESI-MS/MS with ion trap is described in detail. CID (collision-induced dissociation) and ETD (electron-transfer dissociation) fragmentation techniques are used in combination to specifically determine phosphorylation sites inside the peptide sequences, through the analysis of MS/MS spectra.

Key words: Phosphoproteomics, LC-MS/MS, Electron-transfer dissociation, Immobilized metal affinity chromatography, CID/ETD MS/MS spectra, Phosphopeptide purification

1. Introduction

Protein phosphorylation is a reversible post-translational modification mediated by specific protein kinases and phosphatases that is involved in the regulation of many critical cellular processes (1). The most frequent protein phosphorylation sites are those in which the phosphate group is attached to the side chain of serine and threonine residues and less frequently to tyrosine residues (2). In addition, phosphorylation may also occur on arginine, lysine, and histidine as well as in glutamic and aspartic acid residues, although these are extremely rare. The identification of phosphoproteins is particularly complex due to their low abundance: the ratio of the phosphorylated form to its nonphosphorylated

Dermot Walls and Sinéad T. Loughran (eds.), *Protein Chromatography: Methods and Protocols*, Methods in Molecular Biology, vol. 681, DOI 10.1007/978-1-60761-913-0_18, © Springer Science+Business Media, LLC 2011

counterpart is often much less than one, and therefore only a small subset of the proteome is composed of phosphoproteins. This ratio is even smaller considering the large number of proteolytic nonphosphopeptides generated after digestion with a specific protease in the bottom-up strategy, which is the most common proteomics approach used for phosphoprotein analysis.

In order to overcome the aforementioned issues, different enrichment and purification strategies have been developed over the last few years in order to selectively isolate phosphopeptides prior to their separation and identification by high-performance liquid chromatography (HPLC) coupled to mass spectrometry.

Among these strategies, immobilized metal affinity chromatography (IMAC) is one of the most frequently used techniques (3–5). In IMAC, positively charged metal ions (Fe^{3+}, Ga^{3+}, Cu^{2+}) bound to a stationary phase through different linkers (nitriloacetic acid, iminodiacetic acid, tris(carboxymethyl) ethylenediamine), form chelation complexes by electrostatic interactions with negatively charged phosphate groups that are retained on the stationary phase. Therefore, phosphorylated peptides can be isolated because of the phosphate group's affinity toward the metal ion. However, an important drawback of this technique is that nonphosphopeptides containing acidic residues (such as aspartic and glutamic acid) can also be retained due to their affinity for the metal. This undesired co-purification process can be minimized by adjusting the pH of the loading solution to assure that the phosphate group is dissociated while the carboxyl groups of glutamic/aspartic acid residues remain protonated, and hence unable to bind the metal ion. Therefore, in order to enhance the efficiency of the enrichment process and to avoid the isolation of nonspecific binding peptides, optimization of different experimental parameters should be carried out, such as pH, ionic strength, and solvent composition of the different loading – washing – eluting solutions.

The current procedure concatenates two in-house packed microcolumns, the IMAC microcolumn and the Oligo R3 reversed-phase one. Oligo R3 medium is a polymeric packing designed for desalting hydrophilic peptides such as phosphopeptides. The use of both microcolumns provides selective purification and sample clean-up prior to nanoHPLC-ESI-MS/MS analysis (6).

Tandem mass spectrometry is the most powerful technique for the identification and characterization of proteins as well as for the analysis of complex peptide mixtures generated after proteolytic digestion (7). Peptide sequences and post-translational modifications, such as phosphorylation, are determined through the analysis of tandem MS/MS spectra by using either CID (collision-induced dissociation) and/or ETD (electron-transfer dissociation) fragmentation modes.

In CID fragmentation, the collision of peptides with inert gas molecules (helium or argon) primarily induces dissociation of the peptide amide bonds, between the carbonyl and the amine groups. Depending on whether the end where the charge is retained is either at the N- or at the C-terminus, the ion is known as either a b-type or a y-type ion, respectively. Although CID is the most commonly used fragmentation technique in bottom-up proteomic approaches, phosphoester bonds are very labile in these conditions, resulting in an extremely efficient loss of the phosphate group from the peptide ion. Thus, tandem MS/MS spectra are in most occasions clearly incomplete and insufficient to elucidate the sequence of the phosphopeptide and to specifically determine the site of phosphorylation. In order to overcome this problem, complementary fragmentation techniques have been recently developed, such as MS/MS ETD-based mode (8, 9).

In ETD fragmentation, radical anions are generated by a reactant (fluoranthene) by negative chemical ionization, followed by electron-transfer reactions from reactant radical anions to positive peptide ions that produce peptide cation radicals. As these radicals are not stable, dissociation of the bonds between amine and α-carbon groups occurs, while preserving the putative post-translational modifications attached to peptides. Two main types of peptide fragment ions are predominantly generated: if the charge is retained on the N-terminal end it is known as a c-type ion, whereas if the charge is retained at the C-terminus the ion is known as a z-type ion. Despite the fact that in ETD spectra the signal-to-noise ratios obtained are usually lower than those observed when CID fragmentation is used, the number of fragment ions and the coverage of the spectrum mass range is greater than those with CID. As both CID and ETD fragmentation modes are complementary and compatible, the analysis alternating both techniques is a good approach for the elucidation of phosphopeptide sequences (10).

Using different search engines such as Mascot (www.matrixscience.com; Matrix Science, London, UK), experimental tandem mass spectra are compared and matched with those theoretical ones available in databases. Candidate peptide sequences are obtained and scored according to the quality of the match, and based on the set of peptide assignments reported, proteins are also scored and assigned by the search engine.

2. Materials

2.1. Chloroform/Methanol Precipitation

1. Chloroform, ultrapure grade.
2. Methanol, ultrapure grade.
3. Deionized H_2O, from a Milli-Q water purification system or equivalent.

2.2. Sample Digestion

1. Digestion solution: 8 M urea, 25 mM NH_4HCO_3, pH 7.8. Store in aliquots at −20°C.
2. Reduction solution: 0.25 M dithiothreitol (DDT), 25 mM NH_4HCO_3. Store in aliquots at −20°C.
3. Alkylation reagent: 0.5 M iodoacetamide, 25 mM NH_4HCO_3 (see Note 1).
4. Stock solution: 100 mM NH_4HCO_3. Store in aliquots at −20°C and dilute to 25 mM NH_4HCO_3.
5. Sequencing-grade modified porcine trypsin (Promega, Madison, WI).

2.3. Phosphopeptide Purification

All the solutions for the phosphopeptide purification should be prepared fresh prior to use to optimize the efficiency of the process.

1. Phos-Select gel (IMAC-Fe^{3+} type) (Sigma).
2. Oligo R3 reversed-phase packing medium (Applied Biosystems). Prepare 50 Oligo R3: 50 acetonitrile (v/v) slurry.
3. 35 μm pore size filter (MoBiTec).
4. Two in-house 200 μL pipette tips: one for the IMAC enrichment and the other for the Oligo R3 reversed-phase material.
5. 1 mL syringe with its corresponding adaptor.
6. Phos-Select loading/washing solution: acetonitrile/H_2O/trifluoroacetic acid (50:50:0.3).
7. Transfer solution: phosphoric acid (1%).
8. Reversed-phase washing solution: acetonitrile/H_2O/trifluoroacetic acid (5:95:0.1).
9. Acidic elution solution: acetonitrile/H_2O/trifluoroacetic acid (50:50:0.1).
10. Basic elution solution: 90 mM NH_4OH/acetonitrile (70:30).
11. Formic acid, ultrapure grade.

2.4. Analysis by Tandem Mass Spectrometry, Using a NanoHPLC-ESI-MS/MS Ion Trap System

1. Ultimate 3000 nanoHPLC system (Dionex).
2. Analytical column: C18 PepMap 100, 3 μm particle size, 100 Å, 15 cm × 75 μm I.D (Dionex).
3. Trapping column: C18 PepMap, 5 μm, 100 Å, 5 mm × 300 μm I.D (Dionex).
4. Mobile phase (loading pump): 98:2:0.05 (H_2O/acetonitrile/trifluoro acetic acid).
5. Mobile phase (micro pump):
   - A. – 100:0.1 (H_2O/formic acid)
   - B. – 20:80:0.1 (H_2O/acetonitrile/formic acid)

6. HCT Ultra Ion Trap Mass Spectrometer (Bruker Daltonics).
7. DataAnalysis 3.4 software (Bruker Daltonics).
8. MS-Biotools 3.1 software (Bruker Daltonics).
9. Licensed version (v. 2.2.04) of the Mascot search engine (MatrixScience).

3. Methods

A summarizing general scheme for the analysis of phosphorylated proteins is shown in Fig. 1, and the procedure taken from which is described in detail below.

3.1. Sample Preparation

Before proceeding to the purification process, it is highly recommended to eliminate all the contaminants present in the sample that could interfere with the analysis, such as salts, detergents, nucleic acids, lipids, etc. For this purpose, the following chloroform/methanol-based protein precipitation protocol may be used:

1. Dissolve the sample in 100 μL of aqueous buffer. If necessary, use a larger volume and divide it into 100 μL aliquots after resuspending the sample. It is recommended to perform the entire process in 1.5 mL polypropylene tubes.
2. Add 4 volumes of methanol and vortex well.
3. Add 1 volume of chloroform and vortex.
4. Add 3 volumes of Milli-Q water and vortex well.
5. Centrifuge for 2 min at 15,000 × *g* at room temperature. Proteins should be found at the interface.
6. Remove carefully the aqueous top layer and discard it.
7. Add 4 volumes of methanol and vortex well.
8. Centrifuge for 2 min at 15,000 × *g*. Proteins should be seen as a precipitate.
9. Remove carefully as much liquid as possible, without disturbing the precipitate.
10. Dry the sample using a speed-vacuum system for 10–15 min.
11. Store the sample at −20°C until needed.

3.2. In-Solution Sample Digestion

1. Dissolve the sample in an appropriate volume of 8 M urea, 25 mM NH_4HCO_3 solution (pH 7.8) and vortex well to make sure that the solubilization is complete.
2. Reduce disulfide bridges by adding reduction solution to a final DTT concentration of 10 mM.
3. Incubate the protein mixture at 37°C for 1 h (see Note 2).

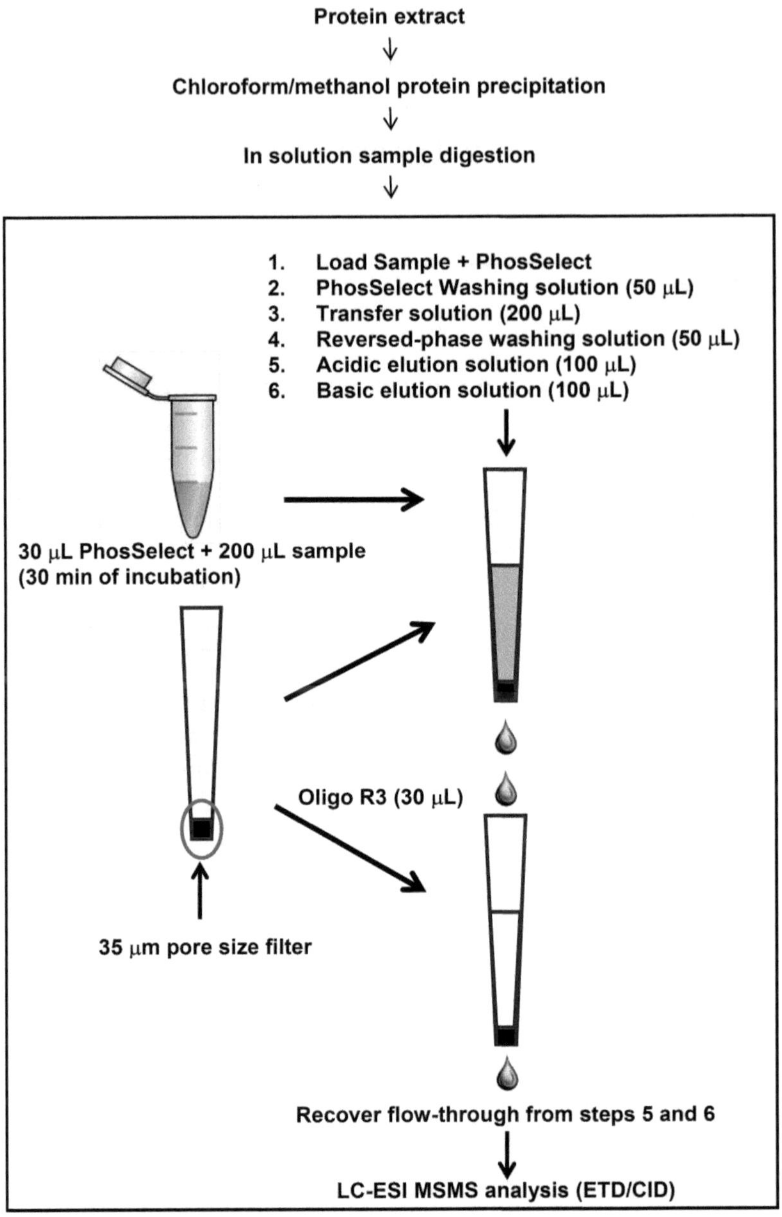

Fig. 1. General scheme for the phosphoprotein procedure according to the bottom-up strategy followed, including sample preparation, in-solution digestion, phosphopeptide selective purification and sample clean-up, and analysis by nanoHPLC-ESI-MS/MS.

4. Add 0.5 M iodoacetamide solution to a final concentration of 50 mM, to alkylate by carbamidomethylation the cysteine residues previously reduced, thus preventing the reversible reaction with the formation of the disulfide bridges (see Note 1).
5. Incubate the sample in darkness at room temperature, for 45 min.

6. Dilute the sample fourfold with 25 mM NH_4HCO_3 solution to get urea at a final concentration of 2 M. Add trypsin in a 1:20 protease-to-protein ratio (see Note 3).
7. Incubate overnight at 37°C.
8. Dry the sample using a speed-vacuum system.

3.3. Phosphopeptide Purification

The current procedure concatenates two in-house packed micro-columns, the IMAC pipette tip and the Oligo R3 reversed-phase one, which allows selective enrichment, concentration, and desalting of phosphopeptides. All the solutions used in the different steps for conditioning, washing, and elution are first loaded on the IMAC pipette tip and passed over both microcolumns. Gentle air pressure from an attached 1 mL syringe is used to push the different solutions through both columns. The syringe adaptor is based on a 1 mL syringe that is attached to the tip by using the upper part of a 200 μL pipette tip. Always keep the slurry wet. Drying results in decreased chromatographic efficiency.

Procedure: Prepare two 200 μL pipette tip microcolumns, one for the IMAC enrichment and the other for the reversed-phase material.

1. Cut a 0.8 cm long portion from the end of a 200 μL pipette tip with a surgical blade (see Note 4).
2. Pack a MoBiTec filter (35 μm pore size) into the 200 μL tip. The function of this filter is to serve as a frit to retain the Phos-Select gel or the Oligo R3 reversed-phase resin.
3. Packing the Oligo R3 medium: resuspend the Oligo R3 resin in acetonitrile. Load a 30 μL aliquot of this slurry into the tip, wash with 30 μL of acetonitrile, and apply gentle air pressure with a plastic syringe.
4. Place a 30 μL aliquot of the Phos-Select gel suspension in a 0.5 mL polypropylene tube, add 200 μL of loading solution (50% acetonitrile:50% H_2O:0.3% trifluoroacetic acid), and vortex well.
5. Centrifuge for 1 min at 1,200 × *g*, aspirate the supernatant off the gel, and discard it.
6. Dissolve the sample in 200 μL of loading solution (50% acetonitrile:50% H_2O:0.3% trifluoroacetic acid) or dilute it 20-fold in the loading solution (see Note 5).
7. Load the sample onto the equilibrated Phos-Select gel and vortex well.
8. Incubate at room temperature while gently mixing for 30 min, to improve phosphopeptide binding to the Phos-Select stationary phase.
9. Load the resulting slurry (sample and Phos-Select gel) into the 200 μL pipette tip, previously packed with the 35 μm filter (see Note 6).

10. Wash the tips with 50 μL of loading solution. Discard flow-through from the Oligo R3 microcolumn.
11. Wash the tips with 200 μL of transfer solution (1% phosphoric acid) (see Note 7). Discard the flow-through from the Oligo R3 tip.
12. Add 50 μL of reversed-phase washing solution (5% acetonitrile:95% H_2O:0.1% trifluoroacetic acid) (see Note 8). Discard the flow-through from the Oligo R3 microcolumn.
13. Elute with 100 μL of acidic elution solution (50% acetonitrile:50% H_2O:0.1% trifluoroacetic acid). Recover the flow-through from the Oligo R3 microcolumn.
14. Elute with 100 μL of basic elution solution (70% 90 mM NH_4OH:30% acetonitrile), recover the flow-through from the Oligo R3 tip and acidify the sample with 2 μL of formic acid (see Note 9).

3.4. Analysis by Tandem Mass Spectrometry, Using a NanoHPLC-ESI-MS/MS Ion Trap System

Here, a bottom-up strategy for the phosphopeptide analysis based on a previous separation of the enriched phosphopeptides by nanoHPLC using a reversed-phase capillary column (see Note 10) coupled to a 3D ion trap mass spectrometer is followed. To obtain tandem MS/MS spectra amenable for phosphopeptide structural characterization, two different and complementary fragmentation techniques are alternatively used in the acquisition method: collision-induced dissociation (CID) and electron-transfer dissociation (ETD).

Procedure:

3.4.1. Sample Reconstitution

Reconstitute the sample in 5 μL of 98% H_2O:2% acetonitrile:0.1% formic acid, vortex briefly, sonicate it for 2 min and inject it onto the chromatographic system.

3.4.2. Chromatographic Conditions

1. Concentrate and desalt the sample onto the C18 trapping column using 98% H_2O:2% acetonitrile:0.05% trifluoroacetic acid at a flow-rate of 30 μL/min for 1 min. The trapping column is switched online with the analytical column through a 10-port valve (see Note 11).
2. Elute the sample from the trapping column at 300 nL/min under the following gradient elution conditions:

   4% of mobile phase B for 5 min, a linear increase to 40% B in 60 min, followed by a linear increase to 95% B in 1 min that is maintained for 7 min, and finally return to the initial conditions and maintain those for 10 min (see Note 12).
3. UV wavelengths are simultaneously monitored at 214 and 280 nm.

3.4.3. Mass Spectrometer Operating Conditions

1. The effluent from the chromatographic system is connected to distal coated emitter tips (new objective) via a nanospray ion source operating in positive ion mode, with capillary voltage set at 1,400 V.
2. MS scan range: *m/z* 300–1,500.
3. Data-dependent acquisition of the two most abundant ions in the survey scan in MS mode, which are isolated and analyzed by alternating CID and ETD fragmentation (see Note 13).
4. Ion charge control is set to 300,000 and maximum accumulation time set to 100 ms (see Note 14).
5. Dynamic exclusion option is set to 1 min (see Note 15).

3.4.4. MS/MS Database Searching

1. MS/MS raw data were processed using Data Analysis 3.4.
2. CID and ETD MS/MS data were searched via MS-Biotools 3.1 against NCBI nonredundant protein sequence databases (http://www.ncbi.nlm.nih.gov/), using a licensed version (v.2.2.04) of the Mascot search engine (Matrix Science), considering the following parameters:
   (a) Carbamidomethylation of cysteines as a fixed modification.
   (b) Oxidized methionine and phosphorylation (serine/threonine or tyrosine) as a variable modification.

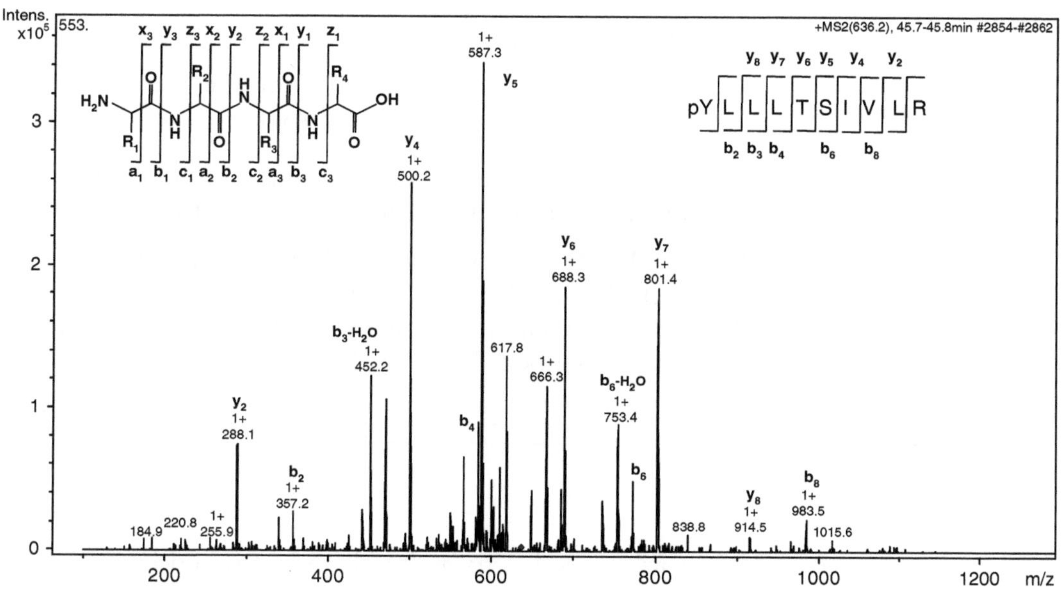

Fig. 2. Ion trap CID MS/MS spectrum of the doubly charged ion (*m/z* 636.2) corresponding to the tryptic phosphopeptide pYLLLTSIVLR from the human transmembrane protein 168 (Q9H0V1). The peptide sequence is shown together with the corresponding identified b-/y-type ions. pY indicates phosphorylated tyrosine residue. A peptide fragmentation scheme is also depicted.

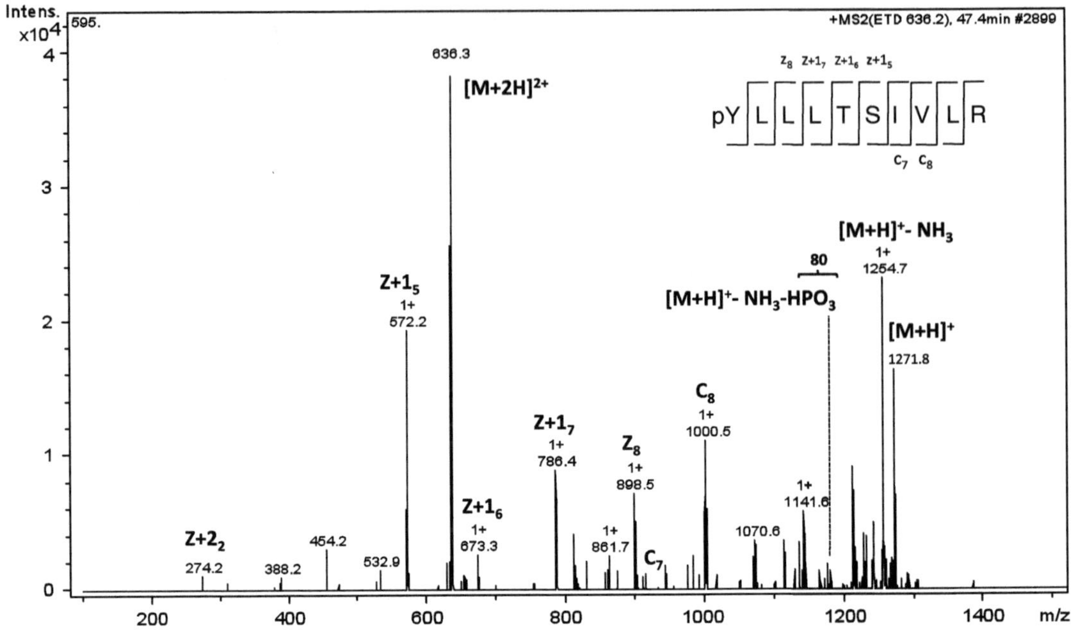

Fig. 3. Ion trap ETD MS/MS spectrum of the doubly charged ion (*m/z* 636.2) corresponding to the tryptic phosphopeptide pYLLLTSIVLR from the human transmembrane protein 168 (Q9HOV1). The peptide sequence is shown together with the corresponding identified c-/z-type ions. pY indicates a phosphorylated tyrosine residue. Fragment ions resulting from charge stripping of the precursor ion are labeled as $[M+2H]^{2+}$ and $[M+H]^{+}$, corresponding to the doubly and singly charged versions of the *m/z* 636.2 ion.

(c) Two trypsin missed cleavages.

(d) A mass tolerance of ±0.8 Da for the precursor mass and ±0.6 Da for the fragment masses.

3. MS/MS spectra with scores above the threshold defined by Mascot were manually validated (see Note 16).

CID and ETD MS/MS spectra are shown for the phosphopeptide pYLLLTSIVLR from the human transmembrane protein 168 (Q9HOV1), identified an alternating CID/ETD experiment performed as described above (see Figs. 2 and 3).

4. Notes

1. Iodoacetamide has to be freshly prepared and kept in darkness, in order to avoid the oxidation to the corresponding carboxylic acid.
2. Avoid higher temperatures in the presence of urea for protein reduction, to prevent the carbamylation of proteins.
3. The initial urea concentration of 8 M has to be diluted fourfold as 2 M is the highest urea concentration compatible

with trypsin activity. Check the pH of the solution so that it should be around 8 for optimal activity of the trypsin.

4. Eppendorf tips have been proven not to nonspecifically adsorb peptides.
5. The pH of the loading solution should be lower than 3.65 (the pKa of the β-COOH of aspartic acid) for two main reasons: (1) in order to have phosphoric acid dissociated as a phosphate group that is going to be used as a metal binding ligand in the IMAC technique and (2) so as not to competitively bind acidic residues (aspartic and glutamic acid) that would be dissociated as aspartate and glutamate residues and could also be retained on the IMAC gel.

   The percentage of trifluoroacetic acid used (0.3%) is the optimum for the protonation of carboxyl groups and at the same time dissociates the phosphoric acid to allow the interaction during IMAC. The 50% of acetonitrile is effective to avoid interferences from hydrophobic interactions between nonphosphopeptides and the IMAC gel.
6. Sample washing and elution is performed by loading the liquid on the top of the gel and by applying gentle air pressure to generate a low flow through the column.
7. The 1% phosphoric acid solution transfers the phosphopeptides from the IMAC gel to the reversed-phase.
8. Undesirable interferences are washed away, whereas phosphopeptides are bound to the reversed-phase.
9. In order to improve the efficiency of recovery of the phosphorylated peptides from the IMAC gel, an additional basic elution step is included, taking advantage in the extended pH range stability of the R3 polymeric reversed-phase (pH range: 1–14). It is recommendable to acidify the sample to prevent the phosphopeptides from undergoing hydrolysis at basic pH values.
10. Reversed-phase liquid chromatography is the most widely used separation technique due to its excellent compatibility with electrospray interfaced tandem mass spectrometry.
11. Short desalting and concentration times are recommended so as to prevent the hydrophilic phosphopeptides (mostly those with multiple phosphorylation sites) from not being retained and hence lost from the C18-trapping column.
12. Phosphopeptides are eluted from the analytical column with the linear gradient from 4 to 40% B over 60 min. The gradient from 40 to 95% B is used to wash the column and to eliminate retained and undesirable compounds.
13. MS/MS on an ion trap is performed "in time," which means that a serial process takes place for collection, isolation of precursor ions, fragmentation and detection of product ions.

14. This device is used to control the ion population inside the trap and to prevent space-charge problems.
15. The dynamic exclusion prevents the same *m/z* ion from being isolated for 1 min after its fragmentation.
16. Mascot scores over 40/50 usually correspond to correct identifications.

References

1. Graves, J. D., Krebs, E. G. (1999) Protein phosphorylation and signal transduction. *Pharmacol. Ther.* **82**, 111–121.
2. Kersten, B., Agrawal, G. K., Iwahashi, H., Rakwal, R. (2006) Plant phosphoproteomics: a long road ahead. *Proteomics* **6** (20), 5517–5528.
3. Ishihama, Y., Wei, F. Y., Aoshima, K., Sato, T., Kuromitsu, J., Oda, Y. (2007) Enhancement of the efficiency of phosphoproteomic identification by removing phosphates after phosphopeptide enrichment. *J. Proteome Res.* **6** (3), 1139–1144.
4. Kobuku, M., Ishihama, Y., Sato, T., Nagasu, T., Oda, Y. (2005) Specificity of immobilized metal affinity-based IMAC/C18 tip enrichment of phosphopeptides for protein phosphorylation analysis. *Anal. Chem.* 77 (16), 5144–5154.
5. Carrascal, M., Ovelleiro, D., Casas, V., Gay, M., Abian, J. (2008) Phosphorylation analysis of primary human T lymphocytes using sequential IMAC and titanium oxide enrichment. *J. Proteome Res.* 7 (12), 5167–5176.
6. Thingholm, T. E., Jensen, O. N., Robinson, P. J., Larsen, M. R. (2008) SIMAC (sequential elution from IMAC), a phosphoproteomics strategy for the rapid separation of monophosphorylated from multiply phosphorylated peptides. *Mol. Cell. Proteomics* 7 (4), 661–671.
7. Aebersold, R., Mann, M. (2003) Mass spectrometry-based proteomics. *Nature* **422** (6928), 198–207.
8. Mikesh, L. M., Ueberheide, B., Chi, A., Coon, J. J., Syka, J. E. P., Shabanowitz, J., Hunt, D. F. (2006) The utility of ETD mass spectrometry in proteomic analysis. *Biochim. Biophys. Acta* **1764** (12), 1811–1822.
9. Molina, H., Matthiesen, R., Kandasamy, K., Pandey, A. (2008) Comprehensive comparison of collision induced dissociation and electron transfer dissociation. *Anal. Chem.* **80** (13), 4825–4835.
10. Molina, H., Horn, D. M., Tang, N., Mathivanan, S., Pandey, A. (2007) Global Proteomic profiling of phosphopeptides using electron transfer dissociation tandem mass spectrometry. *Proc. Natl. Acad. Sci. U. S. A.* **104** (7), 2199–2204.

Chapter 19

Tag Removal by Site-Specific Cleavage of Recombinant Fusion Proteins

Adam Charlton and Michael Zachariou

Abstract

Where an affinity tag has served its purpose it may become desirable to remove it from the protein of interest. This chapter describes the removal of such fusion partners from the intended protein product by cleavage with site-specific endoproteases. Methods to achieve proteolytic cleavage of the fusion proteins are provided, along with techniques for optimising the yield of authentic product.

Key words: Fusion protein, Affinity tag, Site-specific proteolysis, Protease, Proteolytic cleavage

1. Introduction

The use and benefits of affinity tags are the subject of this chapter, although when the tag has served its purpose it is often desirable to remove it to obtain homogeneous protein product of native size and sequence. The use of site-specific endoproteases to facilitate this removal is an approach that has gained considerable favour in recent times. There are many reasons for this widespread adoption, but foremost amongst these are that site-specific proteases recognize long, uncommon amino acid sequences that are highly unlikely to be found within the protein of interest. Also, as proteases are themselves quite labile proteins, sensitive to extremes of temperature or chemical environment, proteolytic cleavage systems tend to function in mild conditions that may enhance protein product stability. Finally, many site-specific proteases act after their recognition sequence, rather than within it. This therefore provides the opportunity to generate the exact sequence for the target protein, as no contribution to catalysis needs to be made by any element of the target protein itself.

Dermot Walls and Sinéad T. Loughran (eds.), *Protein Chromatography: Methods and Protocols*, Methods in Molecular Biology, vol. 681, DOI 10.1007/978-1-60761-913-0_19, © Springer Science+Business Media, LLC 2011

A very limited number of all proteases display suitable site specificity for a sufficiently long amino acid sequence to be useful for fusion protein cleavage. These proteases are frequently isolated as proprotein activation enzymes, where evolutionary pressure has led toward site specificity. This is the case with many of the proteases covered in this chapter, which represent those that are both readily commercially available and have a long history of application in fusion protein cleavage.

1.1. Recombinant Fusion Proteins

The fusion protein strategy is a popular approach to the expression of recombinant proteins in bacteria. The fusion of the protein of interest to another, unrelated protein, or fusion partner, can improve yields of the target protein. The fusion partner can provide protection against proteolysis, enable in vivo folding of the target protein, or facilitate recovery by acting as affinity tags (1, 2).

The protease substrate numbering convention of Schechter and Berger (3) will be used for this chapter, where the amino acids of the substrate (the fusion protein) N-terminal to the site of cleavage are designated P and those of C-terminal are P′. The residues are numbered with increasing distance from the scissile bond (Fig. 1). The fusion partner may be incorporated at the N- or C-terminal end of the target protein, but for the purposes of this chapter, N-terminal fusions will be specifically covered. As all the specific proteases detailed cleave on the carbonyl side of the P1 residue, less or no non-native sequence elements are retained from these fusions. The methods are valid for C-terminal fusions, but the recognition sequences will remain attached as a C-terminus extension of the protein product. Figure 1 depicts an N-terminal fusion protein.

In the design of a fusion protein strategy, the selection of the protease to effect the final cleavage may be as important as the selection of the fusion partner itself. Where available, sequence, and structural information can guide this decision, as can the final application of the target protein. When a protease has been selected, the recognition sequence for that protease must be inserted between the fusion partner and the target protein as a linker peptide, as shown in Fig. 1.

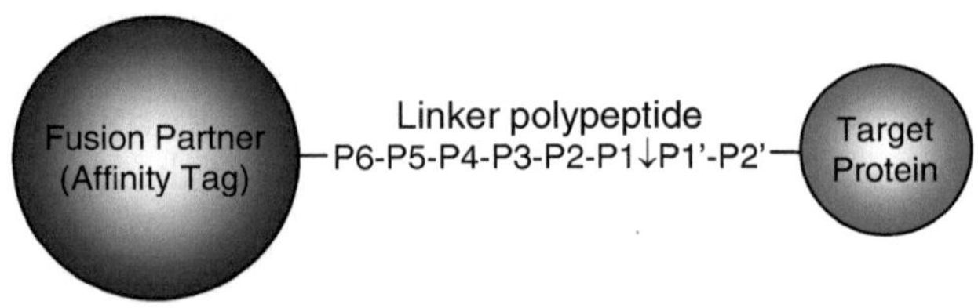

Fig. 1. A schematic representation of an N-terminal fusion protein.

1.2. Site-Specific Proteases: Enterokinase

Enterokinase (EC 3.4.21.9) is a mammalian gastric serine protease. The in vivo function of this enzyme is the activation of trypsin by cleavage of the trypsinogen zymogen to its active form. The cleavage site for this enzyme with its natural substrate is C-terminal to the recognition sequence pentapeptide (aspartate)$_4$–lysine (4). As enterokinase cuts C-terminal to its recognition sequence, without requiring the interaction of residues on the other side of the scissile bond, it is capable of generating a native N-terminus for the protein product. The high charge density of the recognition sequence will increase the likelihood of solvent exposure at the site, maximising protease accessibility and also serving to improve the overall solubility of the fusion protein (5).

The unique nature of the cleavage motif should preclude its occurrence within a protein product; however, enterokinase largely recognises the charge density of its recognition sequence rather than the precise amino acid sequence. Cleavage by enterokinase is possible down to sequences as short as Asp-Asp-Lys (4) and activity is permitted with substitution of the motif residues with their charge equivalents (6). Therefore, similar apparent charge densities in the target protein may also be susceptible to enterokinase cleavage.

Enterokinase is available as a recombinant enzyme, in many cases as only the catalytic subunit of the holoenzyme. It must be noted that not all vendors offer the recombinant protein, so care must be taken in obtaining the enzyme if this is important.

1.3. Site-Specific Proteases: Factor Xa

Factor Xa (EC 3.4.21.6) is an enzyme of the mammalian blood clotting cascade. Upon its own activation, this enzyme in turn activates the next enzyme in the cascade by cleavage of prothrombin, liberating active thrombin. Factor Xa is highly specific for cleavage following the tetrapeptide sequence isoleucine–(glutamate/aspartate)–glycine–arginine, allowing for the generation of an authentic N-terminus for the protein product (7).

Factor Xa is not currently produced recombinantly, and therefore must be isolated from mammalian plasma (usually bovine). This should be considered when selecting Factor Xa for a fusion protein system, depending on the intended final use of the target protein product.

1.4. Site-Specific Proteases: Thrombin

Thrombin (EC 3.4.21.5) is another enzyme of the mammalian blood-clotting cascade, acting downstream of Factor Xa its function in vivo is the cleavage of fibrinogen to generate fibrin (8). Unlike the other specific proteases in this chapter, thrombin does not have a long-defined specificity sequence, with the only absolute requirement for cleavage being that it occurs after an arginine, especially where the arginine residue is preceded by a glycine or proline at P2 and followed by a glycine at P1′ (9). Although lacking a long recognition sequence, thrombin cleavage can be further targeted by inclusion of hydrophobic residues in the P4 and P3 positions (9). Thrombin cleavage is also improved with non-acidic P1′ and P2′ residues, but these will be determined by the target protein's sequence and may not usually be available for substitution.

Thrombin distinctly prefers cleavage within a P-R↓G sequence, so much so that it should be considered to cleave within this recognition sequence, and as such a protein released from a fusion by this protease will have a residual N-terminal glycine. Thrombin is therefore unlikely to produce the target protein with fully authentic sequence, except in cases where the first residue of the protein is glycine. There are examples of thrombin cleavage prior to residues other than glycine, but these are uncommon (10).

Thrombin possesses high intrinsic activity, so can function at relatively low enzyme concentrations and is tolerant of a wider range of buffer conditions than other mammalian proteases. Like factor Xa, thrombin is not commercially available as a recombinant product, so consideration of the purpose for the target protein must be made before designing a fusion protein regime around this protease.

1.5. Site-Specific Proteases: Genenase I

Genenase I is unique amongst the selected proteases, as it represents both the only example of a bacterial enzyme and of a protease with engineered specificity. The parent enzyme for this rationally designed protease is subtilisin BPN′ from the bacterium *Bacillus subtilis* (11). Genenase I was developed by mutation of a necessary active-site histidine residue (changed to alanine), resulting in a non-functional enzyme. The functionality of the protease can be restored if the side chain of the Histidine residue is supplied by the substrate at the P2 or P1′ position, this mechanism is known as substrate-assisted catalysis (11, 12).

Cleaving C-terminal to its ideal recognition sequence, genenase I is capable of producing the correct N-terminus for the product. As this sequence is not based around a charged amino acid, as is the case with many of the other proteases, genenase I offers a quite different cleavage mechanism. It is tolerant of somewhat harsher conditions than its mammalian counterparts.

Owing to the requirement for substrate-assisted catalysis, the overall activity of this enzyme is considerably lower than other, fully self-functional proteases. This often translates to a requirement for higher enzyme:substrate ratios. As a licenced product, genenase I is only available from one manufacturer and may impose a cost limitation to future scale-up of a cleavage system.

1.6. Site-Specific Proteases: Viral Cysteine Proteases

To obtain novel site-specific proteases, attention has turned to the enzymes of RNA viruses. Upon infection, the genomes of these viruses are translated as one large polyprotein (13). The proteases act to specifically cleave the polyprotein into its individual structural and functional components. A major feature that distinguishes this group of proteases is that they employ a cysteine residue at the core of their catalytic mechanism, as opposed to the serine of the mammalian and bacterial proteases. The overall fold of these viral enzymes is very similar to that of the serine proteases, in some cases the active-site cysteine can be substituted with serine to achieve an active enzyme, albeit with significantly diminished activity (14).

Many viral proteases are highly specific for very long recognition sequences, but the two that have made the greatest impact in fusion protein cleavage are the proteases of tobacco etch virus (TEV) and human rhinovirus (HRV). The recognition sequence for these enzymes spans at least seven and eight residues, respectively, with little divergence from the wild-type sequence of the natural polyprotein junctions possible. The minimum cleavage site for TEV protease is of the form E-X-X-Y-X-Q↓(G/S), with a consensus sequence of E-N-L-Y-F-Q↓(G/S) (15, 16). The site for HRV follows a similar general theme, with a consensus sequence of L-E-V-L-F-Q↓G-P (17). As can be seen from these sequences, the viral proteases cleave within their recognition sequences and will hence leave a non-natural mono- or dipeptide extension on the N-terminus of the target protein. TEV protease is somewhat more flexible in its P1′ requirements, with peptide studies suggesting that it may tolerate glycine, serine, alanine, or methionine at P1′ (18). Although for initial proof of concept cleavage trials, it would be advisable to maintain the wild-type glycine or serine.

High purity recombinant preparations of TEV and HRV proteases are available for fusion protein cleavage. Many manufacturers' implementations of these enzymes also bear an affinity tag to facilitate later removal of the protease from the protein preparation.

2. Materials

2.1. Selection of the Appropriate Protease

Endoprotease(s) are commercially available from a range of sources.

2.2. Cleavage of Recombinant Fusion Proteins with Serine Proteases

1. Cleavage buffer: 50 mM Tris–HCl (see Note 1), 50 mM NaCl, 2 mM $CaCl_2$ (see Note 2), pH 8.
2. Microfuge tubes.
3. Pipettes and tips for accurate liquid dispensation in the 10, 100, and 1,000 μL ranges.
4. Ice.
5. Heating block.

6. Reducing sodium dodecyl sulphate–polyacrylamide gel electrophoresis (SDS–PAGE) loading buffer: 50 mM Tris–HCl, 2% SDS, 10% glycerol, bromophenol blue, 0.002%, pH 6.8. Alternatively, as supplied for proprietary PAGE systems.
7. Dialysis equipment (if required), for example, tubing or centrifugal concentrator.
8. 1 M HCl.

2.3. Cleavage of Recombinant Fusion Proteins with Cysteine Proteases

1. Cleavage buffer: 50 mM Tris–HCl (see Note 1), 150 mM NaCl, 1 mM ethylenediamine tetraacetic acid (EDTA), 1 mM dithiothreitol (DTT) (see Note 3), pH 7.5.
2. Microfuge tubes.
3. Pipettes and tips for accurate liquid dispensation in the 10, 100, and 1,000 μL ranges.
4. Ice.
5. Two times reducing SDS–PAGE loading buffer: 50 mM Tris–HCl, 2% SDS, 10% Glycerol. 0.02% Bromophenol blue, pH 6.8, or as supplied for proprietary PAGE systems.
6. Dialysis equipment (if required), e.g. tubing or centrifugal concentrator.
7. 1 M HCl.

3. Methods

3.1. Selection of the Appropriate Protease

1. Based on the background information and the data in Table 1, select a protease appropriate for the fusion protein of interest.
2. Examine the target protein amino acid sequence for complete or partial occurrences of the recognition sequence for the intended protease. Where that sequence, or the two or three residues around the cleavage site, exists in the target protein product, avoid the use of that protease.
3. Use recombinant DNA methods to construct the corresponding expression vector, in which the coding sequence for the endopeptidase cleavage site is located between the coding sequences for the fusion partner and the desired protein product. Verify the DNA sequence of the corresponding portion of the expression vector so as to ensure the correct insertion of the protease recognition sequence.
4. Obtain the selected protease from the appropriate commercial source. Always use the highest purity protease preparations to avoid non-specific cleavage of the target protein by contaminating proteases.

Table 1
Properties of specific proteases for fusion protein cleavage

Protease	Protease type	Cleavage site	Unlikely to cleave before	Suppliers	Notes
Enterokinase	Serine	D-D-D-D-K↓	P	I, R, M, N, S	
Factor Xa	Serine	I-E-G-R↓	P, R	S, M, P, R, Q, N, G	
Genenase I	Serine	P-G-A-A-H-Y↓	P, I, D*, E*	N	*4
Thrombin	Serine	(G/P)-R↓G	n/a	M, G, S, R	
Tobacco etch virus protease	Cysteine	E-N-L-Y-F-Q↓(G/S)	n/a	I, U	
Rhinovirus 3C proteinase	Cysteine	L-E-V-L-F-Q↓G-P	n/a	M, G	

M Merck Biosciences, *S* Sigma Aldrich, *I* Invitrogen, *P* Pierce, *R* Roche Diagnostics, *Q* Qiagen, *N* New England Biolabs, *G* GE Healthcare (Amersham Biosciences), *U* US Biological

5. Go to the Subheading for the protease type, serine (Subheading 3.2) or cysteine (Subheading 3.3), of the selected protease system.

3.2. Cleavage of Recombinant Fusion Proteins with Serine Proteases

1. If the fusion protein sample contains urea or guanidine (see Note 5), salts >250 mM (see Note 6), imidazole >50 mM, ionic detergents >0.01% (see Note 7), reducing agents or known protease inhibitors (see Note 8), then dialyse into cleavage buffer.
2. Concentrate or dilute the fusion protein preparation to approximately 0.5 mg/mL (see Note 9).
3. Dilute the protease preparation to 0.05 U/μL (or 0.05 μg/μL) in cleavage buffer (see Note 10). Keep protease preparations and stock on ice until needed.
4. Set up a pilot cleavage by mixing 100 μL of fusion protein (50 μg at 0.5 μg/mL, from step 2) with 10 μL of protease dilution (from step 3). Prepare a negative control reaction by adding 2 μL of cleavage buffer to 20 μL of fusion protein preparation. Incubate these reactions at approximately 21°C (see Note 11). If a positive cleavage control was supplied, prepare this reaction according to manufacturer's directions.
5. Remove 22 μL samples of the cleavage reaction at 1, 2, 4, 8, and 24 h. Terminate the reaction by adding 22 μL of 2× reducing SDS–PAGE loading buffer (see Notes 12 and 13). Terminate the negative control at 24 h. Store at −20°C until all of the samples are ready to run on SDS–PAGE (see Note 14).
6. Analyse the samples and the negative control by SDS–PAGE.

7. If there is significant degradation of the target protein (see Note 15) go to step 8. If there is incomplete cleavage (see Note 16), or no cleavage apparent where a positive control was successful, go to step 10. If the cleavage was successful, go to step 12.
8. Incubation with a lower amount of protease may help to minimise (see Note 17) internal cleavage of the target protein. Dilute the protease preparation to 0.005 and 0.0005 U/μL (or 5 and 0.5 ng/μL). To 2 × 20 μL of fusion protein from step 2, add 2 μL each of these protease dilutions. Incubate at approximately 21°C (see Note 11) for 1 h (see Note 18). Terminate the reaction (see Note 13) and analyse as before. If these reactions yield sufficient correctly cleaved target protein, go to step 12.
9. If overdegradation is still observed, reduce the concentration of protease further and repeat the reaction. Further improvement in the yield of correct protein product may be possible by altering the structural properties of the target protein, see step 11.
10. Increasing the concentration of protease may enable cleavage. Dilute the protease stock to 0.25 and 0.5 U/μL (or μg/μL). Add 4 μL of each protease dilution to 40 μL of fusion protein from step 2. To another 40 μL of fusion protein, add 4 μL of neat protease stock. Incubate at approximately 21°C (see Note 11). Remove 22 μL aliquots at 4 and 24 h. Terminate the reactions (see Note 13) and analyse by SDS–PAGE. If these reactions yield sufficient correctly cleaved target protein, go to step 12. If these protease concentrations remain unable to produce adequate levels of correctly cleaved material, or if significant degradation of the target protein is observed (see Note 15), go to step 11.
11. Alter reaction conditions (see Note 19).
    (a) Select one factor at one concentration/level from Table 2 to alter and prepare fusion protein at 0.5 mg/mL in this variant cleavage buffer by dialysis into the new system or by adjustment of the original cleavage buffer to include the new factor.
    (b) Dilute the protease preparation to 0.05 U/μL (or 0.05 μg/μL) in the variant cleavage buffer. Keep protease preparations and stock on ice until needed.
    (c) To 20 μL of the new fusion protein preparation (step 11a) add 2 μL of the new protease dilution (step 11b) and incubate at the desired temperature for either 1 h where reduction of internal cleavage is desired or 24 h where improvement of incomplete cleavage is the intended outcome. Terminate the reaction at the appropriate time and analyse.

Table 2
Conditions that can alter protease specificity that are compatible with serine proteases

pH	Non-ionic detergent (%, v/v)	Ionic detergent (%, w/v)	Chaotrope (M)	NaCl (mM)	Temperature (°C)
6.5	0.1	0.01	0.5	100	4
7.0	0.5	0.05	1	200	16
7.5	1	0.1	2	300	21–25
8.0	1.5	0.5	3	400	37
8.5	2		4	500	
9.0					
9.5					
Note 19a, b	Note 19c, d	Note 19c, e	Note 19c, f	Note 19g	Note 19h

(d) If the degree of correct cleavage is increased, but not sufficiently, further improvement may be possible by altering the selected factor up or down, and repeating steps 11a–c. If further improvement within one factor class is not possible, hold this first factor constant at the level that gave the best result and introduce a second variant factor, repeating steps 11a–c with both factors.

(e) See Note 20 for other avenues to achieve successful cleavage.

12. Scale up the successful reaction conditions tenfold to provide a working preparation of cleaved protein. Although individual reaction conditions and incubation times will vary depending on those determined in steps 4–11, a generic reaction protocol would be as follows: Mix 1 mL of fusion protein preparation (from step 2) with 100 μL of protease dilution (from steps 3, 8, 9, or 10). Incubate at the required temperature (from step 4 or 11) for 1 h (if step 8 was followed) or 4–24 h (if steps 10–11 were followed). Terminate the reaction by the addition of 50 μL of 1 M HCl or addition of appropriate protease inhibitors (Table 3).
13. For notes on product purification and reaction cleanup, see Note 22.

3.3. Cleavage of Recombinant Fusion Proteins with Cysteine Proteases

1. If the fusion protein sample contains urea or guanidine (see Note 5), ionic detergents >0.01% (see Note 6), Zn^{2+} >5 mM (see Note 23) or known protease inhibitors (see Note 8), then dialyse into cleavage buffer.
2. Concentrate or dilute the fusion protein preparation to approximately 0.5 mg/mL (see Note 9).

Table 3
Common protease inhibitors

Inhibitor	Protease class	Molecular weight (Da)	Effective concentration	Notes
Aprotinin	S	6,500	10–250 μg/mL	
Leupeptin hemisulphate	S/C	475.6	1–100 μM	21
Phenylmethylsulfonyl fluoride (PMSF)	S	174.2	0.1–1 mM	
Iodoacetic acid	C	207.9	1–10 mM	
Pefabloc® SC (AEBSF)	S	239.7	0.1–2 mM	
Pepstatin A	A	685.9	0.5–1 μg/mL	
Bestatin	M (E)	344.8	1–150 μM	
EDTA	M	372.3	1–10 mM	
E-64	C	357.4	1–10 μM	

S serine, *C* cysteine, *M* metalloprotease, *(E)* exoprotease, *A* aspartic

3. Dilute the protease preparation to 0.05 U/μL (or 0.05 μg/μL) in cleavage buffer (see Note 10). Keep protease preparations and stock on ice until needed.
4. Set up a pilot cleavage by mixing 100 μL of fusion protein (50 μg at 0.5 μg/μL, from step 2) with 10 μL of protease dilution (from step 3). Prepare a negative control reaction by adding 2 μL of cleavage buffer to 20 μL of fusion protein preparation. Incubate these reactions at 4°C (see Note 24). If a positive cleavage control was supplied, prepare this reaction according to the manufacturer's directions.
5. Terminate the reactions after 24 h by adding 22 μL of 2× reducing SDS–PAGE loading buffer (see Notes 12 and 13). Store at -20°C until the samples are ready to run on SDS–PAGE (see Note 14).
6. Analyse the time point samples and the control(s) on SDS–PAGE.
7. If there is significant degradation of the target protein (see Note 25), go to step 8. If there is incomplete cleavage (see Note 16) or no cleavage apparent where a positive control was successful, go to step 10. If the cleavage was successful, go to step 12.
8. Carefully analyse the negative (no protease) control (see Note 26), if degradation is observed in this reaction consider expression in a host protease deficient bacterial strain such as *Escherichia coli* BL21(DE3). The inclusion of protease inhibitors

that do not affect cysteine proteases may also be beneficial, see Table 3. Return to step 4 with inhibitor inclusions or new host strain. Where the degradation is observed to be attributable to the viral protease, continue to step 9.

9. Incubation with a lower amount of protease may help to minimise internal cleavage of the target protein (see Note 17). Dilute the protease preparation to 0.005 and 0.0005 U/μL (or 5 and 0.5 ng/μL). To 2×20 μL of fusion protein from step 2, add 2 μL each of these protease dilutions. Incubate at 4°C for 24 h. Terminate the reaction (see Note 13) and analyse by SDS–PAGE. If these reactions yield sufficient correctly cleaved target protein, go to step 12. Otherwise continue to step 11.
10. Increasing the concentration of protease may enable cleavage. Dilute the protease preparation to 0.25 and 0.5 U/μL (or μg/μL). Add 4 μL of each protease dilution to 40 μL of fusion protein from step 2. To another 40 μL of fusion protein, add 4 μL of neat protease stock. Incubate the reactions at 4°C for 24 h. Terminate the reactions (see Note 13) and analyse by SDS–PAGE. If these reactions yield sufficient correctly cleaved target protein, go to step 12. If these protease concentrations remain unable to produce adequate levels of correctly cleaved material, then go to step 11.
11. Alter reaction conditions (see Note 19): Select one factor at one concentration/level from Table 4 to alter and prepare fusion protein at 0.5 mg/mL in this variant cleavage buffer

Table 4
Conditions that can alter protease specificity that are compatible with cysteine proteases

pH	Non-ionic detergent (%, v/v)	NaCl (mM)	Temperature (°C)
6.5	0.1	200	4
7.0	0.5	300	16
7.5	1	400	21–25
8.0	1.5	500	34
8.5	2	800	
9.0		1,000	
9.5			
Note 19a, b	Note 19c, d	Notes 19g and 27	Note 19h

by dialysis into the new system or by adjustment of the original cleavage buffer to include the new factor.

(a) Dilute the protease preparation to 0.05 U/μL (or 0.05 μg/μL) in the variant cleavage buffer. Keep protease preparations and stock on ice until needed.

(b) To 20 μL of the new fusion protein preparation (step 11a), add 2 μL of the new protease dilution (step 11b) and incubate at the desired temperature for 24 h. Terminate the reaction and analyse.

(c) Increase or decrease the factor iteratively by repeating steps 11a–c until successful cleavage is obtained.

(d) See Note 20 for other avenues to achieve successful cleavage.

12. Scale up the successful reaction conditions tenfold to provide a working preparation of cleaved protein. Although individual reaction conditions and incubation times will vary depending on those determined in steps 4–11, a generic reaction protocol would be as follows: Mix 1 mL of fusion protein preparation (from step 2) with 100 μL of protease dilution (from steps 3, 9, 10, or 11). Incubate at the required temperature (from step 4 or 11) for 24 h. Terminate the reaction by the addition of 50 μL of 1 M HCl or addition of appropriate protease inhibitors (Table 3).

13. For notes on product purification and reaction cleanup, see Note 22.

4. Notes

1. Other buffers for this pH are acceptable, such as *N*-(2-hydroxyethyl) piperazine-*N′*-(2-ethanesulfonic acid) (HEPES).

2. The action of these proteases is enhanced by inclusion of low levels of NaCl and trace $CaCl_2$ (19).

3. The catalytic mechanism of cysteine proteases relies on the active-site cysteine thiol nucleophile. It is therefore vital to the activity of these enzymes that this thiol be preserved, with the state of this functional group ensured by maintaining a reducing environment. If this concentration of DTT causes reduction of labile disulphide bonds in the target protein [determined by incubation of the protein in cleavage buffer followed by analysis by a technique such as reversed phase high performance liquid chromatography (rp-HPLC)], then a milder redox pair such as 3 mM reduced glutathione + 0.3 mM oxidised glutathione might be more appropriate.

4. Cleavage prior to aspartate or glutamate can be improved more than tenfold by cleavage in 2 M KCl (manufacturer's recommendation).
5. Chaotropes such as urea or guanidine–HCl are known to severely inhibit cleavage by many proteases. The activity of these enzymes falls off sharply in the presence of any chaotrope, often with undetectable activity in concentrations above 2 M urea/1.5 M guanidine–HCl. Aside from decreased protease activity, the presence of chaotropes can alter the specificity profile of the enzyme, potentially giving rise to cleavage at unintended sites. It is therefore recommended that chaotropes be avoided in the pilot cleavage experiments to avoid their unexpected interference.
6. Enterokinase and factor Xa are inhibited by concentrations of salts (such as NaCl) over 250 mM, as such it is recommended that the total concentration of all salts not exceed this level in initial experiments. Imidazole is known to inhibit these enzymes at concentrations over 50 mM. Although thrombin is generally more salt and imidazole tolerant, with successful cleavage reported in 500 mM NaCl and 500 mM imidazole (20), it is again advised that the total concentration be kept below the stated thresholds if possible. Viral proteases are far more salt tolerant than the serine proteases with activity reported in 800 mM NaCl (18).
7. Sodium dodecyl sulphate (SDS), an anionic detergent, can inhibit cleavage at concentrations as low as 0.001%, but in practice the effect of less than 0.01% should be negligible. Although less information exists for enzyme inhibition by other charged detergents, it is likely that they too cause a very similar loss of activity and as such, their presence in pilot cleavage experiments is not recommended.
8. Protease inhibitors may have been added at the cell lysis stage of protein purification.
9. Substrate concentration can have an effect on the rate of enzyme reactions. Keep fusion protein concentrations as consistent as possible in pilot cleavage experiments. Concentration of the fusion protein preparation can be performed simultaneously with step 1.
10. 1% Concentration of protease (relative to fusion protein) is the goal. Use 1 U of enzyme where the supplier defines a unit as having the ability to cleave >90% of 100 μg. Some manufacturers may use a different unit definition, in these circumstances adjust the volume of protease added accordingly. For example, if a particular manufacturer's protease preparation defines one unit as having the ability to cleave 50 μg of control protein, then double the volume of protease added.

Where both the mass (e.g. mg/mL) and the activity (Units) of the protease preparation are supplied, use the activity measure to determine the amount of protease to use.

11. Room temperature is acceptable if constant and within 20–25°C.
12. A reducing SDS–PAGE will show if the protease has cleaved protein product internally. An adventitiously cut protein product may appear intact on a non-reducing gel if held together by disulphide bonds.
13. The constituents of SDS–PAGE loading buffer, particularly the high concentration of SDS, will very effectively terminate all protease activity. If not using SDS–PAGE analysis, the reaction may be terminated by acidification, e.g. add 3–5 μL 1 M HCl or by the addition of a protease inhibitors against the added enzyme, as listed in Table 3.
14. Select a SDS–PAGE system that will allow separation in the range between the size of the full-size fusion protein and the successfully cleaved target protein. Bear in mind that there may be smaller fragments present if the protein has been overdegraded.
15. Degradation of the protein product is indicated by a decreased abundance of material with the correct mass and the appearance of smaller products that were not present in the initial preparation or in the negative control sample. These effects will usually become more pronounced over the time course. In some cases, the fusion partner may be visible by SDS–PAGE, ensure its presence is not mistaken for an internal cleavage fragment. If degradation is observed in the protease negative control, there may be contamination of the fusion protein sample by other proteases. Consider further purification.
16. In many cases, incomplete cleavage is preferable to overdegradation as intact fusion protein is more readily separated from the correct protein product than that protein will be from internal cleavage fragments, see Note 22.
17. Where internal cleavage of the protein has occurred, it is unlikely to be completely avoided. If the presence of these breakdown fragments or the associated yield losses cannot be tolerated consider using another protease system.
18. It is assumed that the protein was overdegraded at the 1 h point in the initial time course. In most cases, a lower protease concentration will not change the cleavage profile (the products that are generated) substantially, but will instead increase the time taken to achieve the same profile. Performing the reaction at a lower protease concentration can be thought of as somewhat analogous to expanding the time taken to create the reaction products. Thus it is possible to collect the

reaction products at time points that would have been impractical to capture at the initial reaction ratio, such as those that formed in the first few minutes or seconds of the reaction.

19. Table 2 (Subheading 3.2 serine proteases) or Table 4 (see Subheading 3.3, cysteine proteases) suggest a range of potential reaction condition variations in which the specificity of the protease may be sufficiently altered to enable hydrolysis at the intended site. The concentration/level value ranges provided are intended as a guide only, with any amount within those ranges acceptable as circumstances may dictate. However, deviation outside the upper and lower limits specified is unlikely to meet with a successful cleavage reaction outcome. The tertiary structure of the protein can either inhibit protease action at the intended site by sterically hindering accessibility or promote incorrect internal cleavage by exposing labile surface motifs. The non-exhaustive lists in Table 2 or 4 suggest conditions that will mildly alter the protein structure without denaturing the protease or protein product. Modification of multiple factors in concert may be required for optimal outcomes. If the degree of correct cleavage is increased by a factor, but not sufficiently so at any concentration/level, further cleavage improvements may be made by holding this first factor constant at the level that gave the best result and introduce a second variant factor and repeating the optimisation experiments.
    (a) Whilst not significantly altering the structure of the protein, the pH at which the reaction is performed may be particularly useful for reducing non-specific cleavage within the protein. As can be seen in Table 1, many of the proteases recognise charge amino acid groupings, therefore altering the pH of the buffer can move toward or away from the pK_a of the side chains of ionisable amino acids. This can alter local charge environments and can be sufficient to mask the secondary sites and prevent cleavage. Similarly, varying the pH can cause localized charge modifications in the protease active site that can shift the specificity of the enzyme enough to discourage secondary cleavage.
    (b) Table 5 lists some common buffers that will be effective at the stated pH points. 50 mM solutions of each will provide sufficient buffering.
    (c) Inclusion of chaotropes or detergents will relax the structure of the protein. These agents allow the normally buried hydrophobic residues of the protein to become more solvent exposed by disrupting hydrogen bonding and hydrophobic interactions. This can perturb the original structure of the protein, providing greater exposure of

Table 5
Suitable buffers at given pH ranges

6.5	7.0	7.5	8.0	8.5	9.0	9.5
Citrate						
MES						
	MOPS	MOPS				
		Tris–HCl	Tris–HCl	Tris–HCl		
		HEPES	HEPES			
		Tricine	Tricine	Tricine		
				Borate	Borate	Borate
					CHES	CHES

the expected target cleavage site, potentially shifting the equilibrium of the cleavage reaction away from the secondary site and toward the primary. Although Note 5 cautions against the use of chaotropes, successful cleavage is indeed possible under these conditions, with successful cleavage reported by Enterokinase in 2 M urea (21) and genenase I in 2.5 M urea (22, 23). However, the activity of the proteases will most likely be significantly decreased, requiring a higher concentration of enzyme. The inclusion of chaotropes will most likely require a concurrent re-examination of the amount of enzyme used, as in step 10.

(d) Examples of common non-ionic detergents are Tween-20 and Triton X-100.

(e) An example of a common ionic detergent is SDS. Ionic detergents should be used sparingly, as they are powerful protein denaturants.

(f) The most commonly used chaotropes are urea and guanidine–HCl. The concentrations in Tables 2 and 4 are based on urea; if guanidine–HCl is used instead, decrease these values by 25%.

(g) The inclusion of NaCl can relax protein structure by reducing the stabilizing effect of salt bridges. The inclusion of NaCl alone is unlikely to alter the initial cleavage profile, but can synergistically act with the other suggested factors to improve the overall specificity of the protease.

(h) Aside from directly contributing to the rate of the protease reaction in a manner much similar to alteration in the enzyme:substrate ratio, the temperature of the incubation can also have an effect on protein structure. As decreased temperatures weaken hydrophobic interactions and strengthen hydrogen bonds and vice versa, there exists the potential to alter the cleavage profile of the system by simply altering the incubation temperature (author's personal observations).

20. If successful cleavage is still not obtained but the use of the selected protease is still desired, consider the insertion of a tetra- to hexa-peptide spacer sequence N-terminal to the protease recognition sequence. The inclusion of a flexible spacer peptide sequence can allow greater access to the intended cleavage site by minimising steric inhibition by the fusion partner. The steric inhibition effect can be particularly prevalent when dealing with small, largely unstructured peptide fusions that are able to fold back onto the protein structure, occluding the cleavage site (author's personal observations). For serine proteases, sequences such as S_3G (24), SG_4A (25), and SG_5 (26) have been used successfully for this purpose. As viral proteases tolerate little deviation from the wild-type recognition sequence, an upstream spacer derived from their wild-type polyprotein sequences may be more useful than an artificial polypeptide at reducing steric interference. In the case of TEV, such a sequence is DYDIPTT (27) and for HRV a similar candidate is KMQITDS (28). Return to Subheading 3.2, step 1 or Subheading 3.3, step 1 with the new fusion construct.

21. Leupeptin may also inhibit viral cysteine proteases at concentrations over 100 μM (29).

22. The full-length fusion protein and the separated affinity tag will both bind to the affinity column under the same conditions employed to generate the fusion protein initially. The correctly cleaved protein, now lacking an affinity tag, will not be bound by the column and will thus flow through. It should be noted that internal cleavage fragments of the product (if generated) will not be separated by this technique. If an internally cut protein is held together by disulphide bonds (see Note 12), it may be successfully separated from intact protein by ion-exchange chromatography due to the extra surface charges provided by the hydrolysis sites. Where the internal cleavage fragments are not held together, size-exclusion chromatography may provide separation.

23. Zinc ions are quite potent inhibitors of cysteine protease activity, with concentrations as low as 5 mM resulting in significant loss of activity. This inactivation is thought to occur due to the formation of a complex between the zinc ion and three amino acids in the active-site pocket, including the catalytic cysteine (21).

24. Although not the optimal temperature for these enzymes, it has been shown, at least in the case of TEV protease, that incubation at 4°C results in only a threefold reduction in overall activity compared with room temperature (20°C) (30). The benefit to product stability at low temperature is, in most cases, well worth a slightly longer incubation time.
25. Internal cleavage by viral cysteine proteases is highly unlikely, with no reported cleavage at sites other than the minimum penta- or hexa-peptide recognition sequences in fusion proteins.
26. Degradation may be due to the action of bacterial host proteases that have co-purified with the fusion protein.

References

1. Marston, F.A. (1986) The purification of eukaryotic polypeptides synthesized in *Escherichia coli*. *Biochem. J.* **240,** 1–12.
2. Nilsson, J., Stahl, S., Lundeberg, J., Uhlen, M. and Nygren, P.A. (1997) Affinity fusion strategies for detection, purification, and immobilization of recombinant proteins. *Protein Expr. Purif.* **11,** 1–16.
3. Schechter, I. and Berger, A. (1967) On the size of the active site in proteases. I. Papain. *Biochem. Biophys. Res. Commun.* **27,** 157–162.
4. Maroux, S., Baratti, J. and Desnuelle, P. (1971) Purification and specificity of porcine enterokinase. *J. Biol. Chem.* **246,** 5031–5039.
5. Prickett, K.S., Amberg, D.C. and Hopp, T.P. (1989) A calcium-dependent antibody for identification and purification of recombinant proteins. *Biotechniques.* **7,** 580–587.
6. Light, A. and Janska, H. (1989) Enterokinase (enteropeptidase): comparative aspects. *Trends Biochem. Sci.* **14,** 110–112.
7. Nagai, K. and Thøgersen, H.C. (1984) Generation of beta-globin by sequence-specific proteolysis of a hybrid protein produced in *Escherichia coli*. *Nature* **309,** 810–812.
8. Blomback, B., Blomback, M., Hessel, B. and Iwanaga, S. (1967) Structure of N-terminal fragments of fibrinogen and specificity of thrombin. *Nature* **215,** 1445–1448.
9. Chang, J.-Y. (1985) Thrombin specificity. *Eur. J. Biochem.* **151,** 217–224.
10. Forsberg, G., Baastrup, B., Rondahl, H., Holmgren, E., Pohl, G., Hartmanis, M. and Lake, M. (1992) An evaluation of different enzymatic cleavage methods for recombinant fusion proteins, applied of des(1-3)insulin-like growth factor I. *J. Prot. Chem.* **11,** 201–211.
11. Carter, P. and Wells, J.A. (1987) Engineering enzyme specificity by "substrate-assisted catalysis". *Science* **237,** 394–399.
12. Carter, P., Nilsson, B., Burnier, J.P., Burdick, D. and Wells, J.A. (1989) Engineering subtilisin BPN' for site-specific proteolysis. *Proteins* **6,** 240–248.
13. Allison, R., Johnston, R.E. and Dougherty, W.G. (1986) The nucleotide sequence of the coding region of tobacco etch virus genomic RNA: evidence for the synthesis of a single polyprotein. *Virology* **154,** 9–20.
14. Lawson, M.A. and Semler, B.L. (1991) Poliovirus thiol proteinase 3C can utilize a serine nucleophile within the putative catalytic triad. *Proc. Natl. Acad. Sci. U.S.A.* **88,** 9919–9923.
15. Carrington, J.C. and Dougherty, W.G. (1988) A viral cleavage site cassette: identification of amino acid sequences required for tobacco etch virus polyprotein processing. *Proc. Natl. Acad. Sci. U.S.A.* **85,** 3391–3395.
16. Dougherty, W.G. and Parks, T.D. (1989) Molecular genetic and biochemical evidence for the involvement of the heptapeptide cleavage sequence in determining the reaction profile at two tobacco etch virus cleavage sites in cell-free assays. *Virology.* **172,** 145–155.
17. Cordingley, M.G., Callahan, P.L., Sardana, V.V., Garsky, V.M. and Colonno, R.J. (1990) Substrate requirements of human rhinovirus 3C protease for peptide cleavage *in vitro*. *J. Biol. Chem.* **265,** 9062–9065.
18. Kapust, R.B., Tozer, J., Copeland, T.D. and Waugh, D.S. (2002) The P1' specificity of tobacco etch virus protease. *Biochem. Biophys. Res. Comm.* **294,** 949–955.

19. Baratti, J., Maroux, S. and Louvard, D. (1973) Effect of ionic strength and calcium ions on the activation of trypsinogen by enterokinase. *Biochim. Biophys. Acta.* **321,** 632–638.
20. Forstner, M., Peters-Libeu, C., Contreras-Forrest, E., Newhouse, Y., Knapp, M., Rupp, B. and Weisgraber, K.H. (1999) Carboxyl-terminal domain of human apolipoprotein E: expression, purification, and crystallization. *Prot. Expr. Purif.* **17,** 267–272.
21. Zhang, H., Yuan, Q., Zhu, Y. and Ma, R. (2005) Expression and preparation of recombinant hepcidin in *Escherichia coli. Prot. Expr. Purif.* **41,** 409–416.
22. Lien, S., Milner, S.J., Graham, D.L., Wallace, J.C. and Francis, G.L. (2001) Linkers for improved cleavage of fusion proteins with an engineered α-lytic protease. *Biotechnol. Bioeng.* **74,** 335–343.
23. Francis, G.L., Aplin, S.E., Milner, S.J., McNeil, K.A., Ballard, F.J. and Wallace, J.C. (1993) Insulin-like growth factor (IGF)-II binding to IGF-binding proteins and IGF receptors is modified by deletion of the N-terminal hexapeptide or substitution of arginine for glutamate-6 in IGF-II. *Biochem. J.* **293,** 713–719.
24. Holowachuk, E.W. and Ruhoff, M.S. (1995) Biologically active recombinant rat granulocyte macrophage colony-stimulating factor produced in *Escherichia coli. Prot. Expr. Purif.* **6,** 588–596.
25. Polyak, S.W., Forsberg, G., Forbes, B.E., McNeil, K.A., Aplin, S.E. and Wallace, J.C. (1998) Introduction of spacer peptides N-terminal to a cleavage recognition motif in recombinant fusion proteins can improve site-specific cleavage. *Prot. Eng.* **10,** 615–619.
26. Hakes, D.J. and Dixon, J.E. (1992) New vectors for high level expression of recombinant proteins in bacteria. *Anal. Biochem.* **202,** 293–298.
27. Allison, R.F., Sorenson, J.C., Kelly, M.E., Armstrong, F.B. and Dougherty, W.G. (1985) Sequence determination of the capsid protein gene and flanking regions of the tobacco etch virus: evidence for synthesis and processing of a polyprotein in potyvirus genome expression. *Proc. Natl. Acad. Sci. U.S.A.* **82,** 3969–3972.
28. Stanway, G., Hughes, P.J., Mountford, R.C., Minor, D.P. and Almond, J.W. (1984) The complete nucleotide sequence of a common cold virus: human rhinovirus 14. *Nucl. Acids. Res.* **12,** 7859–7875.
29. Dougherty, W.G., Parks, T.D., Cary, S.M., Bazan, J.F. and Fletterick, R.J. (1989) Characterization of the catalytic residues of the tobacco etch virus 49-kDa proteinase. *Virology.* **172,** 302–310.
30. Nallamsetty, S., Kapust, R.B., Tozser, J., Cherry, S., Tropea, J.E., Copeland, T.D. and Waugh, D.S. (2004) Efficient site-specific processing of fusion proteins by tobacco vein mottling virus protease in vivo and in vitro. *Prot. Expr. Purif.* **38,** 108–115.

Chapter 20

Purification of Antibodies Using Affinity Chromatography

Elaine Darcy, Paul Leonard, Jenny Fitzgerald, Martin Danaher, and Richard O' Kennedy

Abstract

Affinity chromatography permits the isolation of a target analyte from a complex mixture and can be utilised to purify proteins, carbohydrates, drugs, haptens, or any analyte of interest once an affinity pair is available. It involves the exploitation of specific interactions between a binding affinity pair, such as those between an antibody and its associated antigen, or between any ligand and its associated binding receptor/protein. With the discovery of protein A in 1970, and, subsequently proteins G and L, immuno-affinity chromatography has grown in popularity and is now the standard methodology for the purification of antibodies which may be implemented for a selection of different applications such as immunodiagnostics. This chapter is designed to inform the researcher about the basic techniques involved in the affinity chromatography-based purification of monoclonal, polyclonal, and recombinant antibodies. Examples are provided for the use of proteins A and G. In addition, tables are provided that allow the reader to select the most appropriate protein for use in the isolation of their antibody.

Key words: Protein A, Protein G, Protein L, Affinity chromatography, Monoclonal, Polyclonal, Recombinant antibodies

1. Introduction

Affinity chromatography was first introduced by Cuatracasas, Antinson, and Wilchek in 1968 (1). Using the basic concept that every bio-molecule usually recognises another natural or artificial molecule, they developed a technique that is used extensively in the majority of scientific fields such as biotechnology and molecular biology (2). In general terms, affinity chromatography is a separation technique in which a molecule with a selective binding affinity is covalently coupled to a support matrix. This permits the ligand to bind to, and extract its complementary target from a mixture in solution or suspension. Affinity chromatography uses

Dermot Walls and Sinéad T. Loughran (eds.), *Protein Chromatography: Methods and Protocols*, Methods in Molecular Biology, vol. 681, DOI 10.1007/978-1-60761-913-0_20, © Springer Science+Business Media, LLC 2011

specific and reversible biological interactions between a ligand and a complementary target molecule (3). It is frequently used in the down-stream processing and isolation of antibodies and is advantageous as it is "user-friendly," rapid, and a selective technique for capturing target proteins (4).

Protein A, from the bacterial strain *Staphylcoccus aureus*, is routinely used for immunoglobulin G (IgG) purification, and the observed interaction between these entities was one of the first native interactions upon which its methodology was based (5). Alternatively, *Streptococcal* protein G, which differs in its binding properties through the target of $F(ab')_2$ and Fc regions, can be used. Furthermore, an additional protein from *Peptostreptococcus magnus* (protein L) which targets variable domains of immunoglobulin light chains without any detrimental effect on the ability of the antibody to interact with its cognate antigen is also applicable. The versatility of this method can be further improved by use in conjunction with sepharose which allows for the purification of mouse IgG, IgM, IgA, human IgG, antibody binding fragment (Fab) including mouse/human chimeric recombinant antibodies (6). Herein, we provide a brief overview of how each of these proteins can be implemented as purification-associated ligands.

1.1. Protein A

Protein A, a binding receptor isolated from the cell wall of the prokaryote *S. aureus*, has a high affinity for the Fc region of different immunoglobulin iso-types. The first example of this purification was demonstrated by Kronvall and Frommel in 1970 (7) and later by Sjoquist in 1973 (8). Hjelm and co-workers (9) observed that protein A reacts strongly with the Fc part of an immunoglobulin and weakly with the Fab region. The non-immune IgG or Fc regions of an IgG coupled to sepharose could be used as an immunosorbent for its isolation. Furthermore, they established that this protein A also functioned to remove IgG from preparations of other Ig classes and, thus, facilitates the separation of IgA, IgM, IgD, and IgE which do not react with protein A. Further studies showed that protein A is restricted to binding only IgG subclasses IgG_1, IgG_2, and IgG_4, with a very low reactivity with IgG_3, which makes up approximately 8% of the total IgG (10).

1.2. Protein G

In 1984, protein G, a bacterial cell wall protein isolated from human G148 *Streptococcal* strain was shown to exhibit a more diverse binding range than protein A (10). It demonstrated binding to intact IgG as well as antibody fragments such as $F(ab')_2$ and Fc regions. Protein G also binds to immunoglobulins of most species including rat and goat and recognises most classes and subclasses (see Tables 1 and 2 for a complete list of binding properties) (11, 12). While protein G also has a high affinity for the protein albumin which may cause problems with contamination,

Table 1
Binding affinities of proteins A, G, and L to murine and human classes and subclasses of immunoglobulins

Species	Immunoglobulin class/ subclasses	Protein A	Protein G	Protein L
Murine	IgG_1	+	++	++
	IgG_{2a}	++	++	++
	IgG_{2b}	++	++	++
	IgG_3	+	++	++
	IgM	++	+	-----
	IgA	++	++	+
Human	IgG_1	+++	+++	++
	IgG_2	+++	+++	++
	IgG_3	----	+++	++
	IgG_4	+++	-----	++
	IgA	+	-----	++
	IgM	+	-----	++
	IgE	+	-----	++

Key: +++ strong binding, ++ moderate binding, + weak binding, ---- no binding

Table 2
The binding of immunoglobulins for different animal species to proteins A, G, and L

Species	Immunoglobulin class	Protein A	Protein G	Protein L
Chicken egg	IgY	----	----	+++
Cow	IgG	----	+	----
Goat	IgG	----	+++	----
	IgM	----	----	----
Horse	IgG	+++	+++	----
Rabbit	IgG	+++	+++	++
	IgM	----	----	++
Rat	IgG	+	++	+++
	IgM	----	----	+++
Sheep	IgG	+++	+++	----

Key: +++ strong binding, ++ moderate binding, + weak binding, ---- no binding

this problem was circumvented by the production of recombinant forms of this binding ligand which lack the albumin binding site. Hence, this makes protein G extremely useful as an affinity chromatography ligand (13, 14).

1.3. Protein L

Protein L, a protein identified on the surface of *P. magnus* by Myhre and Erntell (1988), (15) is also widely used in antibody purification. This elongated protein was found to have a molecular weight of 75 kDa and binds to the framework region of the variable light chain domains without Fc site interference. In 1993, De Chateau and collaborators (16) analysed a panel of antibodies of the IgG, IgM, IgA, and IgE classes, and strong protein L-binding activity was identified with antibodies carrying kappa (κ) light chains. In addition to human immunoglobulins, a strong protein L-binding activity was detected in antibodies of primates, rats, mice, and rabbits. It was later shown that protein L is suitable for the purification of Fab, scFv (short chain variable region), $F(ab')_2$, and antibody derivatives, and that it binds with high affinity to a large number of immunoglobulins with κ1, κ3, and κ4 light chains, but not to κ2 and λ subgroups. Hence, it recognises 50% of human and more than 75% of murine immunoglobulins (17). Akerstrom and Bjorck, (18) compared proteins A, G, and L showing that all were fibrous proteins with no inter- or intra-chain disulphide bonds. Subsequent sequencing of the three showed that there was no homology, and also verified that the equilibrium constants for the binding reactions between protein L and IgM, IgA, and IgG are equal. However, the strength of the binding to the kappa chains alone was approximately one-seventh to that of the whole immunoglobulin. This suggested that the interactions of heavy and light chains in the whole immunoglobulin molecule gave a particular conformation to the light chains which is much more favourable to protein L binding. Thus, protein L binds to a wider range of immunoglobulin classes that contain a kappa light chain than either protein A or protein G, and it is a useful tool in affinity chromatography and for antibody immobilisation.

1.4. Protein A/G

Protein A/G is a gene fusion product secreted from non-pathogenic forms of *Escherichia coli* which combines the IgG-binding domains of proteins A and G (19). This genetically engineered protein has four Fc-binding domains from protein A and two Fc-binding domains from protein G with a mass of 50 kDa. It binds to all subclasses of human IgG (20, 21), and is particularly useful for purifying polyclonal and monoclonal immunoglobulin subclasses where the class/subclass is unknown. This recombinant fusion protein also facilitates the purification of human IgM and IgA but it binds to a lesser extent to IgD. In relation to mouse immunoglobulins, binding to all subclasses of IgG but not to IgM, IgA, or serum albumin occurs (19), and therefore, its use for the purification and detection of mouse monoclonal IgGs without interference from IgM, IgA, and serum albumin is feasible. It was also shown by Eliasson and co-workers (20) that mouse monoclonal antibodies have a stronger binding affinity to protein A/G than protein A or protein G.

1.5. Salt Precipitation of Immunoglobulins

Crude antibody preparations are often subjected to salt precipitation as a primary purification step before affinity chromatography. This removes a large number of contaminants and often helps to concentrate the antibody in a smaller volume with minimum reduction in yield. Salt precipitation is based on the fact that at a high salt concentration, the natural tendency of proteins not to aggregate is overcome as the surface charges are neutralised, which ultimately results in the formation of large complexes. These complexes are easy to separate by mild centrifugation. This technique provides an easy way of enriching for particular proteins in a mixture, since each protein will begin to aggregate at a characteristic salt concentration. Salts, such as ammonium or sodium sulphate, are commonly used and permit the precipitation of Igs from all mammals (22) and can be applied to serum, plasma, ascites, fluid, and cultured hybridoma supernatant (23). Alternative precipitation methodology involves the use of caprylic acid. This technique is routinely used for mammalian IgG and, again, can be employed for the precipitation of IgG from similar sources. Caprylic acid-based precipitation works by precipitating non-IgG proteins, and the concentration of caprylic acid required to precipitate the protein is species dependant (24). However, Mohanty and co-workers (25) carried out an experiment to determine whether ammonium sulphate or ammonium sulphate followed by caprylic acid was a more effective means of precipitation, and determined that the former is more suited to the extraction of IgG. Other reported methods of precipitation involve using poly ethylene glycol (PEG) as a preliminary purification step for monoclonal antibodies from cell culture supernatant (26) and as an effective means for the precipitation of IgY from egg yolk (27).

Precipitated IgG is extremely stable and is capable of withstanding long-term storage at 4°C. Ammonium sulphate is the most adaptable and commonly used salt for precipitation and using 40% ammonium sulphate yields preparations of approximately 40% purity. Sodium sulphate may yield a purer preparation for some species such as human and monkey (23).

2. Materials

2.1. Purification of Polyclonal Antibodies

2.1.1. Ammonium Sulphate Precipitation

All reagents were sourced from Sigma-Aldrich unless otherwise stated.

1. Ammonium sulphate 99.0% purity.
2. Phosphate-buffered saline (PBS): 136 mM NaCl, 99.5% purity, 2.6 mM KCl, 99.0% purity, 5.37 mM Na_2HPO_4, 99.0% purity, 1.76 mM KH_2PO_4, 98.0% purity. All components are dissolved in 1 L of deionised water and adjusted to pH 7.4

with concentrated hydrochloric acid (HCl) ACS reagent 37%. This solution is stored at room temperature.

3. Phosphate buffer saline-Tween (PBST): Tween surfactant 20 is added to PBS to a final concentration of 0.05% (v/v), pH 7.4. This solution is stored at room temperature.
4. Dialysis tubing: Snake skin tubing (Pierce) (see Note 1).

2.1.2. Protein A Purification of Polyclonal Antibodies

1. Protein A fast flow sepharose in 20% (v/v) ethanol (5 mL).
2. Equilibration buffer: 0.5 M NaCl, 0.6 mM KCl, 5.37 mM Na_2PO_4, 1.76 mM KH_2PO_4. Adjusted to pH 7.4 with HCl.
3. Wash buffer: 0.3 M NaCl, 0.6 mM KCl, 5.37 mM Na_2PO_4, 1.76 mM KH_2PO_4 with 0.05% (v/v) Tween surfactant 20. This solution is adjusted to pH 7.4 with HCl.
4. Elution buffer: 0.1 M glycine–HCl made up in 100 mL of deionised water. This solution is adjusted to pH 2.2 with HCl.
5. Neutralisation buffer: 1.5 M Tris–HCl in 100 mL. This solution is adjusted to pH 8.7 with NaOH.
6. Viva-spin columns (AGB).
7. Sodium azide 20% (w/v).

2.2. Purification of Monoclonal Antibodies

2.2.1. Iso-Typing of Monoclonal Antibodies by ELISA

1. Nunc ELISA maxisorb plates (Thermo Scientific Inc.).
2. Mouse monoclonal iso-typing kit (Pierce).
3. Phosphate-buffered saline (PBS): as described in Subheading 2.1.1.
4. Phosphate-buffered saline with added Tween 20 (PBST): as described in Subheading 2.1.1.
5. Blocking solution: 5% (w/v) dried milk in PBS.
6. Primary and secondary antibody diluents: 0.5% (w/v) milk powder in PBST.
7. Phosphate citrate buffer with sodium perborate. One tablet is dissolved in 100 mL of deionised water which gives a solution 0.05 M phosphate citrate buffer, pH 5.0, with 0.03% (w/v) sodium perborate.
8. Substrate: 3,3′,5,5′ tetramethyl benzidine (TMB) for use with a horseradish peroxidase-labelled antibody. One tablet is dissolved in 10 mL of phosphate citrate buffer with sodium perborate and covered due to photosensitivity.
9. 4-Nitrophenyl phosphate disodium salt tablets (pNPP) for use with an alkaline phosphatase-labelled secondary antibody. One tablet of Tris–HCl (provided) is dissolved in 20 mL of deionised water and the pNPP tablet is subsequently added and allowed to dissolve.

2.2.2. Lateral Flow-Based Assay for Iso-Typing of Monoclonal Antibodies

1. Pierce rapid iso-typing kit plus kappa and lambda-mouse (Pierce).

2.2.3. Protein G Purification of Monoclonal Antibodies

1. Protein G fast flow sepharose in 20% (v/v) ethanol (5 mL).
2. Equilibration buffer, wash buffer, elution buffer, and neutralisation buffer, as described in Subheading 2.1.2.

3. Methods

This section outlines the salt precipitation, iso-typing, and purification of monoclonal and polyclonal antibodies using proteins A and G. Protein L is not outlined but the same method applies.

3.1. Purification of Polyclonal Antibodies

Protein A matrices are effective for affinity purification of IgG from serum and other substances such as ascites fluids of many species including pig, dog, and cat. However, this protein is especially suited for polyclonal antibody purification from rabbits and is routinely used after salt precipitation.

3.1.1. Saturated Ammonium Sulphate Precipitation

1. Prepare saturated ammonium sulphate (SAS) by dissolving 500 g of ammonium sulphate in 500 mL of distilled water at 50°C. Store this solution overnight at 4°C and adjust to pH 7.2 with HCl, this is 100% (w/v) SAS.
2. Add 10 mL of rabbit serum (which is stored on ice) to 10 mL of cold SAS drop-wise, continuous stirring is required at 85 rpm.
3. Stir this mixture for 30 min at room temperature at 85 rpm.
4. Centrifuge the resulting precipitate at 4,500 ×*g* for 20 min at room temperature.
5. Discard the supernatant carefully and wash the precipitated pellet twice, by resuspending the pellet in 10 mL of 45% (w/v) SAS, centrifuging for 20 min at 4,500 ×*g* at room temperature and carefully discarding the supernatant each time.
6. Reconstitute the resulting pellet in 5 mL of PBS by pipetting and dialyse overnight in 5 L of PBS at 4°C (see Note 2).

3.1.2. Protein A Chromatography

1. Prepare a protein A column by pouring 2 mL of Protein A fast flow sepharose in 20% (v/v) ethanol into a clean column (see Note 3). Allow to settle, and drain any excess ethanol off. Equilibrate the column with 50 mL of 0.3 M NaCl PBS, pH 7.4.
2. Apply 2 mL of the dialysed ammonium sulphate rabbit concentrated immunoglobulin to the column. Adjust the flow rate

to 1 mL/min or allow the dialysed fractions to flow-through by gravity. Collect this flow-through and reapply to the column two more times.

3. Wash the column with 20 mL of PBS, pH 7.4, containing 0.5 M of NaCl and 0.05% (v/v) Tween.
4. Elute the affinity captured antibody with 0.1 M glycine–HCl, pH 2.2, and collect the eluate in 500 μL to 1 mL fractions in 1.5 mL Eppendorf tubes containing 50–100 μL (one-tenth volume of glycine) of 1.5 M Tris–HCl, pH 8.7.
5. Determine the protein concentration of the samples by UV absorbance (280 nm) and then pool the positive samples.
6. Add the pooled fractions to a viva-spin column and centrifuge at 4°C at 5,500 ×*g* until the volume reaches 500 μL.
7. Resuspend the filter cake gently in approximately 5 mL of PBS, pH 7.4, containing 0.5 M NaCl; centrifuge again at 4°C at 5,500 ×*g* until the final volume is 500 μL. This results in a concentrated and buffer exchanged antibody. Alternatively, dialyse overnight in PBS pH 7.4, containing 0.5 M NaCl at 4°C.
8. Add a preservative agent such as sodium azide (extreme care should be taken when handling this product as it is highly toxic) to a final concentration of 0.02% (w/v). Aliquot these samples and store at −20°C.
9. Verify the purification by loading samples of both flow-through and eluted protein onto a sodium dodecyl sulphate polyacrylamide gel electrophoresis (SDS-PAGE) gel (Fig. 1). Further characterisation can be achieved by Western blotting (28) as this type of analysis can confirm the presence of a desired protein in a complex mixture of proteins and provides the user with information relating to the size of the protein of interest.

3.2. Purification of Monoclonal Antibodies

3.2.1. Iso-Typing of Monoclonal Antibodies

Prior to purification of monoclonal antibodies, knowledge of its class and isoforms permits the selection of the most suitable affinity ligand to be applied. To determine monoclonal antibody class, subclass, and light chain type/class/family, two methods have been found to be extremely useful. The first of these involves an ELISA-based assay and the second is an iso-typing lateral flow strip kit. Both of these methods are described below.

3.2.2. ELISA-Based Iso-Typing Assay

1. Collect supernatant from the hybridoma cells by continuously culturing cells in Dulbecco's Modified Eagles Medium (DMEM)/Roswell Park Memorial Institute media (RPMI 1640). Store at 4°C with the addition of sodium azide (0.02% [w/v]).
2. Coat a Nunc maxisorb plate with the antigen of choice (see Note 4). Incubate overnight at 4°C or for 1 h at 37°C.

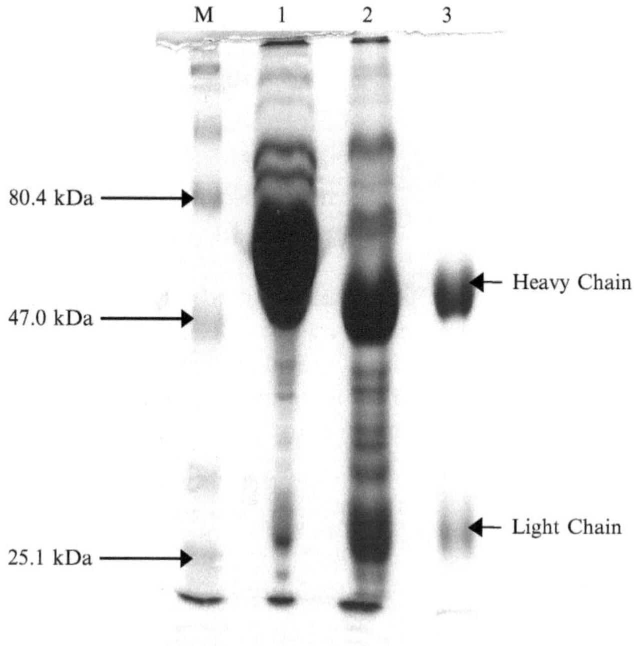

Fig. 1. Denaturing SDS-PAGE gel of anti-cephalexin polyclonal antibody following purification. The serum, SAS, and protein A-purified samples of the anti-cephalexin polyclonal antibody were loaded from left to right, respectively. The gel was stained with Coomasie blue. *Lane M*, represents Blue Ranger pre-stained standard molecular weight marker, obtained from Pierce; *Lane 1* is the crude serum diluted 1:20 in PBS; *Lane 2* is the SAS (saturated ammonium sulphate precipitate) diluted 1:2 in PBS; *Lane 3* is the neat protein A affinity-purified pooled fractions. Heavy chain and light chain domains are highlighted at approximately 50 and 25 kDa, respectively.

3. Displace the unbound antigen from the plate after incubation and add 200 μL of blocking agent, and incubate at 37°C for 1 h.
4. Displace the blocking agent from the plate and wash once with PBS containing 0.15 M NaCl, pH 7.4. Dilute the culture supernatant (primary antibody) 1/100 in PBST containing 0.5% (w/v) of blocking agent. Incubate for 1 h at 37°C.
5. Displace the unbound primary antibody from the plate and wash three times with PBST containing 0.15 M NaCl, pH 7.4, and 0.05% (v/v) Tween, and three times with PBS containing 0.15 M NaCl, pH 7.4. Enzyme-labelled secondary antibodies to IgG_1, IgG_{2a}, IgG_{2b}, IgG_3, IgM, IgA, kappa (κ), and lambda (λ) chains can be purchased from most major suppliers such as Pierce, Sigma-Aldrich, and BD Biosciences. Use the dilution recommended by the supplier, using the same diluents as in the previous step. Incubate for 1 h at 37°C.

6. Displace the unbound secondary antibody from the plate and wash three times with PBST. Add 100 μL of enzyme substrate appropriate to the enzyme label and read at absorbance 405 nm for alkaline phosphatase or 450 nm for horseradish peroxidase (see Note 6).
7. A high absorbance reading is indicative of the antibody isotype.

3.2.3. Iso-Strip Lateral Flow Assay Kit

The Pierce Rapid Iso-typing Kit (Pierce Biotechnology) is a lateral flow assay (LFA) that takes approximately 5 min to perform and determines the class and subclass to which the monoclonal antibodies belong. This method can be used with tissue culture supernatant or mouse ascitic fluid.

1. Dilute the sample 1:100 (see Notes 7 and 8) by adding 5 μL of supernatant to 495 μL of sample diluent provided.
2. Open the pouch containing the cassettes and add 150 μL of the diluted sample to each of the wells of the three cassettes.
3. A successful test will result in a red band at the "C" or control location and another band at one of the three iso-types location on each cassette as illustrated in Fig. 2.

3.2.4. Ammonium Sulphate Precipitation of Monoclonal Antibodies for Protein G Purification

As described in Subheading 1.5, polyclonal and monoclonal antibodies are usually treated with either a pre-concentration step or a precipitation step before purification. The method below describes the SAS precipitation of monoclonal antibodies.

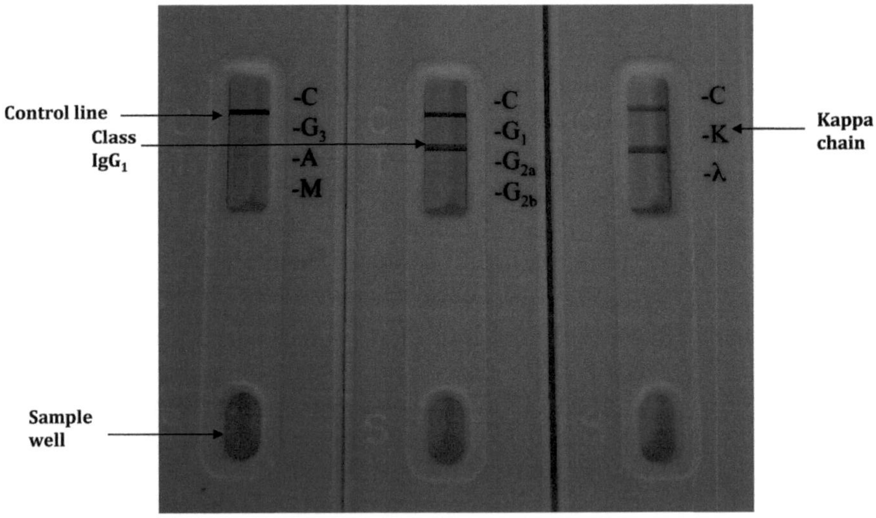

Fig. 2. The "Pierce Rapid Iso-typing Kit plus Kappa and Lambda-mouse" for iso-typing anti-thiabendazole (TBZ) monoclonal antibody. Cassette 1 shows that the anti-TBZ monoclonal antibody is not of the IgG_3, IgM, or IgA isoform. Cassette 2 shows the anti-TBZ monoclonal antibody to be of the subclass IgG_1. Cassette 3 shows anti-TBZ monoclonal antibody is kappa chain-specific.

1. Add ammonium sulphate (99% purity) to the culture supernatant to a final concentration of 45% (w/v).
2. Stir the mixture at 85 rpm at room temperature for 30 min.
3. Centrifuge the resulting precipitate at 4,500 × *g* for 20 min at 4°C.
4. Resuspend the pellet with one-fifth of the original volume of PBS containing 0.15 M NaCl, pH 7.4.
5. Add ammonium sulphate to a final concentration of 40% (w/v), and stir at 85 rpm until fully precipitated.
6. Centrifuge this mixture at 4,500 × *g* at 4°C for 20 min.
7. Resuspend this pelleted precipitate with half of the original volume of PBS containing 0.15 M NaCl, pH 7.4.
8. Dialyse the precipitate overnight with 5 L of PBS containing 0.15 M NaCl, pH 7.4 (see Notes 1 and 2).

3.2.5. Protein G Chromatography

Monoclonal antibodies are important reagents used in biomedical research, diagnostics, and therapeutics. Protein G is effective for affinity purification of IgG from serum and other fluids of many species. It is especially suited for purification of monoclonal antibodies from immune hosts and the broadest spectrum of species and IgG subclasses from human, goat, and sheep samples.

1. Add 2 mL of Protein G fast flow sepharose in 20% (v/v) ethanol into a clean column (see Note 3). Allow this to settle, and drain any excess ethanol off without letting the column run dry. Equilibrate the column with approximately 20 mL of sterile-filtered PBS containing 0.5 M NaCl, pH 7.4.
2. Apply 2 mL of the dialysed ammonium sulphate precipitated immunoglobulin to the column. Adjust the flow rate to 1 mL/min or alternatively allow the dialysed fractions to flow-through by gravity. Collect the flow-through and apply this to the column two more times.
3. Pass 25 mL of wash buffer PBS containing 0.3 M NaCl, pH 7.4 through the column. Store the eluent for further analysis.
4. Elute the affinity captured antibody with 0.1 M glycine–HCl buffer (pH 2.5) and collect 500 μL fractions. Neutralise immediately with 150 μL of 2 M Tris–HCl, pH 8.7 (see Note 5).
5. Determine the protein concentration of the samples by UV absorbance (280 nm) and then pool the samples containing the most protein. Concentrate and buffer exchange the antibody by the use of a viva-spin column as described in Subheading 3.1.2.
6. Verify the purification by loading a sample of both flow-through and eluted protein onto a SDS-PAGE gel.

3.2.6. Protein L Affinity Chromatography

Protein L binds to representatives of all classes of Ig, including IgG, IgM, IgA, IgE, and IgD. Single chain variable fragments (scFv) and Fab fragments also bind to Protein L. Despite this wide-ranging binding capability with respect to Ig classes, protein L is not a universal immunoglobulin-binding protein. Binding of protein L to immunoglobulins is restricted to those containing kappa light chains (i.e. κ chain of the VL domain) (6). In humans and mice, κ light chains predominate. The remaining immunoglobulins have λ light chains. Furthermore, protein L is effective in binding only certain subtypes of κ light chains. For example, it binds human VκI, VκIII, and VκIV subtypes but does not bind the VκII subtype. Binding of mouse immunoglobulins is restricted to those having VκI light chains (6).

Given these specific requirements for effective binding, the main application for protein L is purification of monoclonal antibodies from ascites or culture supernatant that are known to have the kappa light chain. Protein L is extremely useful for purification of VLk-containing monoclonal antibodies from culture supernatant because it does not bind bovine immunoglobulins which are present in the media serum supplement. Furthermore, protein L does not interfere with the antigen-binding site of the antibody, making it useful for immunoprecipitation assays, even when using IgM (29).

4. Notes

1. Take care while working with tubing as the use of any sharp object, such as micro pipette tips, can puncture the tubing resulting in a loss of sample. Wear gloves at all times to prevent cross contamination. Tie the tubing at least three times on each end to prevent leaking.
2. For improved dialysis change the PBS buffer four to five times (once every hour) before overnight dialysis.
3. Empty disposable PD-10 columns which have a capacity of 13 mL of chromatographic material and have a diameter of 1.5 cm can be purchased from GE Healthcare.
4. Antigen coating concentrations may be pre-determined by performing a checkerboard ELISA with varying concentrations of antigen (1 μg/mL–10 ng/mL) with varying antibody dilutions. Alternatively, a coating concentration of 1 μg/mL in PBS will suffice for this experiment. Antigen are frequently diluted in buffers with a high pH 9.6, carbonate/bicarbonate, a neutral buffer pH 7.4, PBS, or a low acetate buffer, pH 4.0 (30).

5. This solution serves to rapidly neutralise the highly acidic environment of the elution buffer, thereby, preventing denaturation of the eluted IgG fraction.
6. For horse radish peroxidase (HRP)-labelled antibodies, when TMB is used the absorbance can be determined at 600 nm. Cover with tinfoil as the substrate is light-sensitive and will need to be made up fresh before use. Alternatively leave this for 15 min and stop the reaction with the addition of 50 μL of 10% (v/v) HCl and measure absorbance at 450 nm.
7. Mouse ascites are diluted 1:8,000. If the sample concentration is known, dilute to 10 ng/mL.
8. For supernatant containing greater than 1 μg/mL of monoclonal antibody, dilute the sample 1:10 by adding 50 μL of supernatant to 450 μL of sample diluent and vortex to mix.

Acknowledgements

This material is based upon works supported by the Science Foundation Ireland under Grant No. 05/CE3/B754 and the Food Institutional Research Measure (FIRM) Department of Agriculture, under Grant No. 05/RND/TN355.

References

1. Cuatrecasas P, Wilchek M, Anfinsen CB. (1968) Selective enzyme purification by affinity chromatography. *Proc. Natl. Acad. Sci. U.S.A.* **61**, 636–43.
2. Wilchek M, Miron T. (1999) Thirty years of affinity chromatography. *React. Funct. Polym.* **41**, 263–8.
3. Van Emon J, Gerlach C, Bowman K. (1998) Bioseparation and bioanalytical techniques in environmental monitoring. *J. Chromatogr. B.* **715**, 211–28.
4. Hober S, Nord K, Linhult M. (2007) Protein A chromatography for antibody purification. *J. Chromatogr. B.* **848**, 40–7.
5. Duhamel RC, Schur PH, Brendel K, Meezan E. (1979) pH gradient elution of human IgG1, IgG2 and IgG4 from protein A-sepharose. *J. Immunol. Methods.* **31**, 211–7.
6. Nilson H, Logdberg L, Kastern W, Bjorck L, Akerstrom B. (1993) Purification of antibodies using protein L-binding framework structures in the light chain variable domain. *J. Immunol. Methods.* **164**, 33–40.
7. Kronvall G, Frommel D. (1970) Definition of staphylococcal protein A reactivity for human immunoglobulin G fragments. *Immunochemistry.* 7, 124–7.
8. Sjoquist J. (1973) Structure and immunology of protein A. *Contrib. Microbiol.* **1**, 83–92.
9. Hjelm H, Hjelm K, Sjoquist J. (1972) Protein A from *Staphylococcus aureus.* Its isolation by affinity chromatography and its use as an immunosorbent for isolation of immunoglobulins. *FEBS Lett.* **28**, 73–6.
10. Bjorck L. (1984) Purification and some properties of streptococcal protein G, a novel IgG-binding reagent. *J. Immunol.* **133**, 969–74.
11. Derrick JP, Wigley DB. (1994) The third IgG-binding domain from Streptococcal protein G: an analysis by X-ray crystallography of the structure alone and in a complex with Fab. *J. Mol. Biol.* **243**, 906–18.
12. Stone G, Sjobring U, Bjork L, Sjoquist J, Barber C, Nardella F. (1989) The Fc binding site for streptococcal protein G is in the Cg2–Cg3 interface region of IgG and is related to the sites that bind staphylococcal

protein A and human rheumatoid factors. *J. Immunol.* **143**, 565–70.

13. Ey PL, Prowse SJ, Jenkin CR. (1978) Isolation of pure IgG1, IgG2a and IgG2b immunoglobulins from mouse serum using protein A-sepharose. *Immunochemistry.* **15**, 429–36.
14. Lindmark R, Thorén-Tolling K, Sjöquist J. (1983) Binding of immunoglobulins to protein A and immunoglobulin levels in mammalian sera. *J. Immunol. Methods.* **62**, 1–13.
15. Myhre EB, Erntell M. (1985) A non-immune interaction between the light chain of human immunoglobulin and a surface component of a *Peptococcus magnus* strain. *Mol. Immunol.* **22**, 879–85.
16. Chateau M, Nilson B, Erntell M, et al. (1993) On the interaction between protein L and immunoglobulins of various mammalian species. *Scand. J. Immunol.* **37**, 399–405.
17. Stura EA, Graille M, Housden NG, Gore MG. (2002) Protein L mutants for the crystallization of antibody fragments. *Biol. Crystal.* **58**, 1744–8.
18. Akerstrom B, Bjorck L. (1989) Protein L: an immunoglobulin light chain-binding bacterial protein. Characterization of binding and physicochemical properties. *J. Biol. Chem.* **264**, 19740–6.
19. Thermo scientific, pierce protein research. (2009) Protein A/G agarose or ultralink resins http://www.piercenet.com/Objects/View.cfm?typr=ProductFamily&ID=01010321. (Accessed 5/2/09)
20. Eliasson M. (1989) IgG-binding characteristics of staphylococcal protein A, streptococcal protein G, and a chimeric protein AG. *J. Immunol.* **142**, 575–81.
21. Hage DS, Ruhn PF. (2006) Introduction to affinity chromatography, Chapter 1, pp. 3–11. Hage DS (ed) Handbook of Affinity Chromatography. CRC press, Taylor and Francis group, Boca Raton, FL.
22. Baines MG, Thorpe R. (1992) Purification of immunoglobulin G (IgG). *Methods Mol. Biol.* **10**, 79–104.
23. Page M, Thorpe R. (2002) Purification of IgG using caprylic acid, Chapter 138, pp 985. Walker JM (ed) The Protein Protocols Handbook. Humana Pr Inc., Totowa, NJ.
24. Steinbuch M, Audran R. (1969) The isolation of IgG from mammalian sera with the aid of caprylic acid. *Arch. Biochem. Biophys.* **134**, 279–84.
25. Mohanty JG, Elazhary Y. (1989) Purification of IgG from serum with caprylic acid and ammonium sulphate precipitation is not superior to ammonium sulphate precipitation alone. *Comp. Immunol. Microbiol. Infect. Dis.* **12**, 153–60.
26. Brooks DA, Bradford TM, Hopwood JJ. (1992) An improved method for the purification of IgG monoclonal antibodies from culture supernatants. *J. Immunol. Methods.* **155**, 129–32.
27. Gassmann M, Thommes P, Weiser T, Hubscher U. (1990) Efficient production of chicken egg yolk antibodies against a conserved mammalian protein. *FASEB J.* **4**, 2528–32.
28. Ma H, Shieh KJ. (2006) Western blotting method. *Am. J. Sci.* **2**, 23–7.
29. Bjorck L. (1988) Protein L. A novel bacterial cell wall protein with affinity for Ig L chains. *J. Immunol.* **140**, 1194–7.
30. Crowther JR. (2001) The ELISA Guidebook: Theory and Practice, 2nd edition. Humana press, Totowa, NJ.

Chapter 21

Optimized Generation of High-Affinity, High-Specificity Single-Chain Fv Antibodies from Multiantigen Immunized Chickens

William J.J. Finlay, Laird Bloom, and Orla Cunningham

Abstract

High-affinity, highly specific binding proteins are a key class of molecules used in the development of new affinity chromatography methods. Traditionally, antibody-based methods have relied on the use of whole immunoglobulins purified from immune animal sera, from egg yolks, or from murine monoclonal hybridoma supernatants. To accelerate and refine the reagent antibody generation process, we have developed optimized methods that allow the rapid assembly of scFv libraries from chickens immunized with pools of immunogens. These methods allow the simplified generation of a single moderately sized library of single-chain Fv (scFv) and the subsequent isolation of diverse, high-affinity, and high-specificity monoclonals for each individual immunogen, via phage display. Using these methods, antibodies can be derived that exhibit the desired selectivity, such as complete specificity or cross-reactivity to multiple orthologues of the same protein.

Key words: Chicken, scFv, Recombinant, Affinity, Phage display, Chromatography, Epitope, Monoclonal

1. Introduction

The close phylogenetic relationship between humans and rodents can frequently result in tolerance to immunization if the protein immunogen is highly conserved across mammals. This hinders the generation of high-quality antibodies and thereby limits the experimental options of a researcher who would like to have an immunochemical reagent for purifying or tracking a novel protein. We have sought to identify and improve antibody generation processes that bypass immune tolerance issues, can be performed simply, reliably, and affordably by a single operator, and lead to

Dermot Walls and Sinéad T. Loughran (eds.), *Protein Chromatography: Methods and Protocols*, Methods in Molecular Biology, vol. 681, DOI 10.1007/978-1-60761-913-0_21, © Springer Science+Business Media, LLC 2011

the isolation of fully specific antibodies that are easily characterized and economical to express. These optimized methods combine the selective power of phage display with the high-affinity immunoglobulin response generated when avians are immunized with mammalian proteins (1, 2). The antibody fragments (scFv) generated in these protocols are expressed and characterized via *Escherichia coli*, rendering their production cost-effective. As the chicken immunoglobulin system is based on frameworks that express well in *E. coli* and the molecules are selected via phage display, the final clones often exhibit high expression in simple bacterial culture (3).

The protocols outlined here have proven to be highly successful in our hands, simplifying the generation of high-affinity scFv antibodies to many proteins. Multiprotein immunization schemes have proven to be very efficient in the recent literature, with five or more proteins being addressed in a single project (4–6). The phagemid vectors required to perform these experiments are now freely available from the leaders of a number of academic labs, and we recommend the use of vector systems such as pCOMB3X (7), which combine the elements required for both phage display selections and the subsequent expression of soluble scFv for screening and purification. Alternative vectors such as pAK100 (8) can be simply substituted by making the requisite changes in polymerase chain reaction (PCR) primers and antibiotic selection conditions.

2. Materials

2.1. Immunizations

1. Purified proteins of interest.
2. Complete and incomplete Freund's adjuvants.
3. PBS.

2.2. cDNA Generation

1. TRIzol® Plus RNA Purification Kit (Invitrogen).
2. SuperScript III First-Strand Synthesis System for RT-PCR (Invitrogen).
3. Solvents: Bichloropropane (BCP), 100 and 70% ethanol (30% nuclease-free water).
4. Nuclease-free plastics: 1.5- and 2-mL microcentrifuge tubes and pipette tips.
5. Nuclease-free water.
6. RNAse-free 3 M sodium acetate, pH 5.2.

2.3. Phage Library Generation

1. Restriction enzymes (*Sfi* I, New England Biolabs).
2. Pellet paint (Novagen).

3. Gel extraction kit (QiaQuick kit, Qiagen).
4. T4 DNA ligase (NEB).
5. Phusion polymerase (Finnzymes).
6. Midiprep kit (Plasmid Midi kit, Qiagen).
7. Electroporator (Gene Pulser, Bio-Rad) with 2-mm gap cuvettes.
8. Electrocompetent *E. coli* TG1 (Stratagene).
9. *E. coli* SCS110 (Stratagene).
10. Media: SOC medium (Invitrogen), solid media as below.
11. 100 and 70% ethanol.
12. 50× Tris-Acetate-EDTA electrophoresis buffer (TAE): 242 g Tris base, 57.1 mL glacial acetic acid, 100 mL 0.5 M NaOH–EDTA (pH 8.0). Adjust to 1 L with dH_2O and store at room temperature.

2.4. Library Rescue and Selections

1. Liquid media (all constituents from Sigma-Aldrich): 2xTY medium. Plus 2% glucose = 2xTY-G, plus 2% glucose, 100 μg/mL carbenicillin = 2xTY-GC, plus 5 μg/mL tetracycline = 2xTY-T, plus 100 μg/mL carbenicillin, 50 μg/mL kanamycin = 2xTY-CK.
2. Solid media (all constituents – Sigma-Aldrich): (2% glucose in all petri dishes): 2xTY-TG, 2xTY-CG/Tet, 2xTYCK plates (2% glucose/50 μg/mL carbenicillin plates/25 μg/mL Kan), 2xTY-CG/Tet bioassay trays (large 22-cm petri dishes). Prepare 2xTY medium with 15 g/L agar as described above, cooling to below 50°C before adding antibiotics and glucose and pouring into plates.
3. Solutions (all constituents – Sigma-Aldrich): 20% PEG-8000/2.5 M NaCl solution (autoclaved), 3% M-PBS (3% dried milk powder in PBS, make fresh daily), TEA (triethylamine), 50 mM sodium carbonate buffer (pH 9.5), PBS and PBS/0.1% Tween-20, 1 M Tris–HCl pH 7.5, 10% bleach, 50% glycerol (v/v with sterile water).
4. Nunc Maxisorp Immunotubes (Star immunotubes).
5. Disposable sterile microbiological spreaders (VWR).
6. *E. coli* ER2738 (New England Biolabs).
7. Helper phage M13K07 (New England Biolabs).
8. 85-mL Oak Ridge centrifuge tubes.

2.5. Output Screening

1. 2xTY-GC.
2. 1 M IPTG (Sigma-Aldrich).
3. Periprep buffer (50 mM Tris–HCl, pH 8.0, 1 mM EDTA, pH 8.0, 20% sucrose).
4. ELISA plates (Nunc Maxisorp).

5. Deepwell plates and breathable seals (Greiner).
6. HRP-labeled secondary antibody appropriate for epitope tags on vector.
7. TMB ELISA substrate (Pierce).
8. 0.18 M phosphoric acid (Sigma-Aldrich).
9. PBS/0.1% Tween (Sigma-Aldrich).

3. Methods

In chicken immunizations, we have included a mix of up to five different proteins at any one time and other groups have used as many as eight proteins successfully (6). The proteins are mixed in a single adjuvant preparation and each animal immunized receives all proteins simultaneously. The 1-mL total inoculum volume may be spread out over two to four sites if necessary. We routinely use two to four animals per library generation project. The basic steps involved in the workflow of library generation are outlined in Fig. 1.

3.1. Chicken Immunization Regime

Immunization regime: Day 1 – 0.5 mL per animal of mixed immunogens in PBS (50 μg of each protein), emulsified in 0.5 mL Complete Freund's adjuvant, to be delivered subcutaneously. Day 21 – as on day 1, but using Incomplete Freund's adjuvant and 50 μg of each protein in PBS. Days 42 and 63 – as on day 21, but 25 μg of each protein. A response to antigen can be established simply by performing a titration ELISA of sera (which may be extracted at the time of tissue harvest) against each of the proteins of interest, plus an unrelated negative control protein. We routinely observe serum titers >100,000 against each antigen used to immunize, with <1,000 titer against control protein.

3.2. cDNA Preparation from Spleen and Bone Marrow

Before harvesting tissues, seek the advice of a veterinarian or animal handling technician who is experienced in avian anatomy (see Note 1).

1. Seven days after the final immunization, harvest femur bone marrow and spleen and transfer to separate 50-mL tubes, each containing 10-mL TRIzol® reagent.
2. Homogenize the samples at room temperature in TRIzol® using a homogenizer (e.g., Tissumizer, Tekmar) for 1 min. Incubate for 5 min at room temperature. Samples may be frozen at −80°C at this stage.
3. Add 10 mL more Trizol to each tube, then remove particulates before RNA extraction by centrifuging all samples at 2,500 × *g* for 10 min. Carefully remove the clarified supernatant (avoid carry-over of particulates if possible) and transfer 2 × 1.5 mL

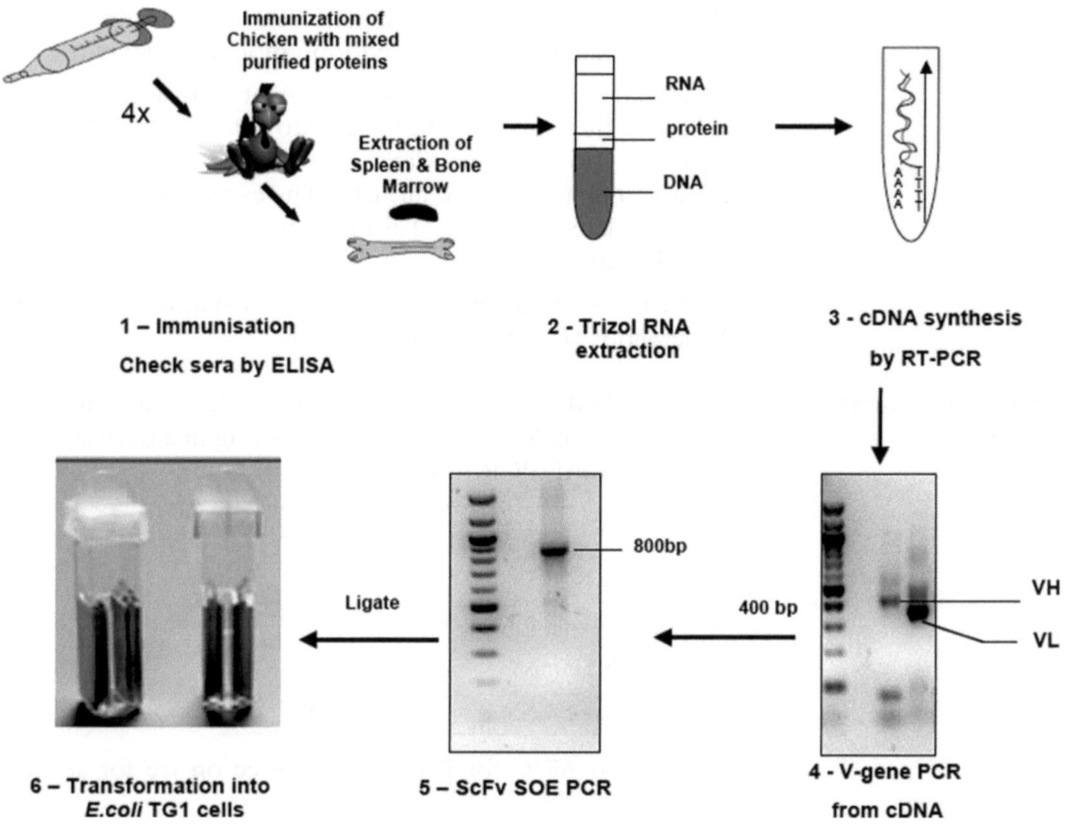

Fig. 1. A stepwise schematic illustrating the simple process of chicken scFv library assembly: (1) Each chicken is immunized four times with mixed purified proteins of interest and the spleen, femur bone marrow, and serum sample extracted. (2) Total RNA is prepared from the tissues via Trizol extraction. (3) cDNA is synthesized by RT-PCR. (4) Only four PCR primers are required to amplify the total V_H and V_L populations from the cDNA. (5) The scFv library is created by randomly recombining the purified V_H and V_L products in a Splice Overlap Extension PCR. (6) The *Sfi* I digested scFv PCR product and phagemid vector are ligated together and transformed into *E. coli* TG1.

each of spleen and bone marrow extract to fresh nuclease-free tubes for RNA extraction.

4. Add 0.15-mL BCP per 1.5-mL TRIzol® reagent, cap the tubes securely, and shake vigorously by hand for 15 s. Incubate at room temperature for 2–3 min.
5. Centrifuge at 12,000 × *g*, 15 min, 4°C in a microcentrifuge.
6. Remove 750 μL of the colorless upper aqueous phase to a fresh 2-mL tube, add an equal volume of 70% ethanol, and mix well by shaking vigorously.
7. Perform column purification of the total RNA from each 1.5-mL sample using the TRIzol Plus kit according to manufacturer's instructions.
8. After elution from the purification column, combine the bone marrow and spleen samples for each chicken. Remove a 5-μL

aliquot of pooled RNA and dilute 1:200 in RNAse-free water. (Store remaining RNA on ice for no more than 30 min.)

9. Determine the OD of diluted RNA at 260 and 280 nm. Calculate the ratio of OD 260/280 to determine RNA purity (typically 1.6–1.9) and determine the RNA concentration (OD 260 nm of 1.0 = 40 ng/μL). The expected yield of total RNA isolated from a 1.5-mL sample is approximately 250–350 μg.
10. Immediately perform first-strand cDNA synthesis using the SuperScript III Kit as outlined below.

3.3. First-Strand cDNA Synthesis

1. Perform 100 μL first-strand synthesis reactions using the SuperScript III Kit. Combine the following in a nuclease-free 0.2- or 0.5-mL PCR tube:

Component	Volume (μL)
25 μg total RNA	*n*
50 ng/μL random hexamers	5
10 mM dNTP mix	5
DEPC-treated water	To 50

2. Incubate at 65°C for 5 min, then place on ice for at least 1 min.
3. Prepare the following cDNA synthesis mix:

Component	Volume per 100 μL reaction (μL)
10× RT buffer	10
25 mM $MgCl_2$	20
0.1 M DTT	10
RNaseOUT (40 U/μL)	5
SuperScript III RT (200 U/μL)	5

4. Add 50 μL synthesis mix per reaction (final volume 100 μL) and incubate for 10 min at 25°C, then 50 min at 50°C.
5. Terminate the reactions at 85°C for 5 min. Chill on ice.
6. Add 5 μL of RNase H to each tube, mix thoroughly and incubate for 20 min at 37°C.
7. Store finished cDNA synthesis reactions at −20°C.

3.4. PCR Assembly of Chicken scFv Libraries

To capture the V-gene sequences from each animal's immune repertoire, we recommend the use of the previously described primer set (7) outlined below (see Note 2). The *Sfi* I cloning sites (underlined) can be replaced by restriction sites appropriate for any phagemid vector. This process should be performed for each animal, to create a sublibrary that pairs the appropriate heavy and light chain repertoires.

V_H amplification primers:

CSCVHo-FL
5′-GGTCAGTCCTCTAGATCTTCCGGCGGTGGTGG
CAGCTCCGGTGGTGGCGGTTCCGCCCTG
ACGTTGGACGAG-3′

CSCG-B 5′-CTGGCCGGCCTGGCCACTAGTGGAGGAG
ACGATGACTTCGGTCC-3′

V_L amplification primers:

CSCVK 5′-GTGGCCCAGGCGGCCCTGACTCAGCCGT
CCTCGGTGTC-3′

CKJo-B 5′-GGAAGATCTAGAGGACTGACCTAGGACGG
TCAGG-3′

SOE-PCR primers:

CSC-F5′-GAGGAGGAGGAGGAGGAGGTGGCCCAGGC
GGCCCTGACTCAG-3′

CSC-B 5′-GAGGAGGAGGAGGAGGAGGAGCTGGCCG
GCCTGGCCACTAGTGGAGG-3′

1. Test PCR amplifications for heavy and light chain V-regions: Perform one 50 μL PCR for each cDNA with each primer set (for Phusion polymerase: add 1 μL of cDNA, 5 μL 5× HF buffer, 1 μL 10 mM dNTP mix, 1 μL each primer at 50 pmoles/μL, water to 50 μL total). Cycling parameters: 98°C – 2 min, 30 cycles of 98°C – 15 s, 50°C – 15 s, 72°C – 30 s, and final extension 72°C – 10 min.
2. Analyze 5 μL of each reaction on a 2% TAE (1×) agarose gel. The expected product sizes are: V_H ~ 450 bp and V_L ~ 350 bp.
3. If the test PCRs were successful, perform an additional ten PCRs for each primer set, then pool each set of reactions and run on a 2% TAE (1×) agarose gel.
4. Excise the desired PCR product from the gel and extract the DNA using the Qiagen gel extraction kit, using 400-mg agarose per tube. Elute in 50 μL H_2O per tube.
5. Quantify the purified DNA by measuring the OD of each sample at 260 nm (OD 1.0 = 50 μg/mL). Also measure the OD at 280 nm and calculate the OD 260/280 ratio (>1.6 is acceptable). Expect >2 μg of DNA for each purified PCR product.
6. Set up a test Splice Overlap Extension PCR to create the final scFv library construct: One reaction per sublibrary as in step 2, but adding 50 ng per fragment of V_H and V_L DNA from step 5 (instead of cDNA) and using the primers CSC-F and CSC-B.

7. Analyze 5 μL of each PCR sample by electrophoresis on a 1.5% TAE (1×) agarose gel. A successful SOE-PCR should produce a dominant band of ~850–900 bp, with little or none of the two original ~400-bp fragments being evident.
8. If steps 6 and 7 have been successful, perform an additional 20 PCRs as in step 6 and then concentrate the DNA by extraction using the Qiagen PCR cleanup kit according to manufacturer's instructions (use four columns in total). Elute in 50 μL H_2O per tube.
9. Finally, purify the PCR products using gel purification as in step 5 and quantify as in step 6. Expect >10 μg of purified scFv DNA per sublibrary.

3.5. Cloning the Chicken scFv-Phagemid Library

Separate sublibrary purified scFv products may be combined at this stage to make a single final library or kept separate to ensure equal representation per animal (our preferred approach).

1. Prepare a Qiagen "Midiprep" (according to manufacturer's instructions), from a single colony of *E. coli* SCS110 (see Note 3) harboring your phagemid vector.
2. Perform restriction digests on 40 μg of the phagemid vector and 10 μg of the library DNA sample. Reaction conditions: for each digest add 20 μL 10× NEB Buffer 2, 2 μL 100× BSA (supplied with the enzyme), 10 μL (200 U) of *Sfi* I enzyme and water to 200 μL. Mix thoroughly and incubate at 50°C for 8 h.
3. Gel-purify (as above) the digested scFv DNA on a 1.5% TAE (1×) agarose gel and the digested vector on a 0.8% gel. Take great care to run the vector for long enough that the double-cut vector band (e.g., 3.3 kb) is clearly separated from the single-cut vector band >1.0 kb above. Also purify the stuffer DNA sequence (e.g., 1.5 kb) released from the vector. Quantify the purified DNA fragments as above. Store the vector DNA in 2 μg aliquots and scFv DNA in 1 μg aliquots, at −80°C.
4. Test ligations: In each of three PCR tubes combine 200 ng *Sfi* I digested plasmid, plus one of– (a) 100 ng scFv insert (library test), (b) 200 ng stuffer (ligation efficiency test), or (c) plasmid alone (background ligation test). To each tube also add 2 μL 10× NEB T4 ligase buffer (fresh as possible for high ATP content), 2 μL NEB T4 ligase (4×10^5 U/mL), and nuclease-free water to 20 μL. Incubate at 16°C for 12 h, then 80°C for 30 min to denature the ligase.
5. Transform 1 μL of each ligation reaction from step 4 into 50 μL of electrocompetent *E. coli* TG1 (e.g., 2.5 kV, 25 μF, 200 Ω in a Bio-Rad "Gene Pulser").

6. Rescue the cells by washing the cuvette three times with 1 mL room temperature SOC medium, combining into a final volume of 3 mL in a sterile 50-mL tube. Incubate at 37°C for 1 h, with 200 rpm shaking.
7. Take a 100-μL aliquot from this transformed culture and add to 900 μL of LB broth. Mix this 10^{-1} dilution thoroughly and perform five further 1:10 dilutions to 10^{-6}. Plate 100-μL aliquots of each dilution on 2xYT-CG agar and incubate at 37°C overnight.
8. Count the colonies on dilution plates and calculate the total number of transformants created using the following calculation: no. of transformants = no. of colonies/100 (μL) × dilution factor × 3,000 (original culture volume in μL). Expect similar numbers of transformants for reactions A and B (~10^6 total), but approximately 100-fold fewer transformants for reaction C (vector self-ligation). If this is the case, proceed to library ligations.
9. Set up each library ligation reaction as follows: In a PCR tube combine 2 μg *Sfi* I digested plasmid, 1 μg scFv insert, 20 μL 10× NEB T4 ligase buffer (fresh as possible for high ATP content), and 20 μL NEB T4 ligase (4×10^5 U/mL). Add nuclease-free water to 200 μL and mix thoroughly. Incubate as in step 4. Perform two ligations per sublibrary.
10. Add 20 μL 3 M NaAc (pH 5.4), 2 μL "Pellet Paint," and 440 μL 100% ethanol. Mix thoroughly and place at −20°C overnight to precipitate the DNA.
11. Centrifuge the precipitated ligations at 15,000 × *g*/4°C/15 min and decant the supernatant. Wash the pellet by adding 1 mL 70% ethanol and centrifuging again for 5 min. Repeat this wash process three times (see Note 4).
12. Resuspend the pellet in 30-μL nuclease-free dH_2O and heat to 50°C for 30 min. The pellet should go fully into solution, turning the sample pink (see Note 5).
13. Begin thawing *E. coli* TG1 cells (300 μL of cells per ligation) and chilling 2-mm gap electroporation cuvettes on ice. For maximum efficiency, do not leave thawed cells on ice for >20 min. Briefly chill the resuspended ligation mix on ice (5 min).
14. Add 300 μL of cells to the 30 μL DNA and mix. Transfer all 330 μL to a cuvette (on ice) and transform immediately as above.
15. Wash the cuvette five times with 1 mL room temperature SOC medium, combining the 5 mL total in a sterile 50-mL tube. Incubate for 1 h at 37°C/200 rpm shaking.
16. Take a 100-μL aliquot from this transformed culture and add to 900 μL of LB broth (10^{-1} dilution, carried on to

step 17). Centrifuge the remaining 5-mL culture at 4,000×*g*/4°C/15 min, then resuspend the cell pellet in 1-mL 2xYT medium and plate onto a 22-cm bioassay tray containing 2xYT-CG agar (see Note 6). Incubate at 30°C overnight.

17. Mix the 10^{-1} dilution and perform five further 1:10 dilutions to 10^{-6}. Plate 100-μL aliquots of each dilution on 2xYT-CG agar and incubate at 37°C overnight.
18. Count the colonies on dilution plates and calculate the total number of transformants created (library size) using the following calculation: no. of transformants = no. of colonies/100 (μL) × dilution factor × 5,000 (original culture volume in μL). Therefore, for example, 100 colonies on the 10^{-5} plate = 5×10^{8} total transformants. Expected number of transformants per electroporation is >1×10^{8}.
19. If the total transformant number is satisfactory (>1×10^{8}), archive the library: resuspend the cells from the 22-cm dish by scraping into 20-mL 2xYT medium. Homogenize the cell suspension by vortexing, centrifuge as in step 11, resuspend in 5-mL 2xYT/20% V/V glycerol, and freeze in 0.5-mL aliquots at –80°C.

3.6. Protocol for the Rescue of scFv-Phagemid Libraries

The effective rescue of the full diversity of clones created in protocol 3.5 is critical to the optimal function of the library. This protocol ensures the full rescue of library diversity.

1. Dilute 10 μL of library glycerol stock (from Subheading 3.5, step 19) into 1,990-μL freezing medium and measure the OD at 600 nm. Calculate the number of bacteria per mL in the stock: an OD of 1.0 corresponds to ~3×10^{8} bacteria/mL for TG1 cells. Calculate the amount of broth culture needed to accept an inoculum of cells totaling 10× the original calculated total transformant number (not cell stock concentration!) at an OD < 0.1 (3×10^{7} cells/mL).

   EXAMPLE: total transformants = 8×10^{8} clones. Therefore, the required number of cells for inoculation of this stock is 8×10^{9} cells. These cells must be diluted to a starting concentration of ~1.5×10^{7} cells/mL to reach an OD of ~0.05. Therefore, 8×10^{9} cells/1.5×10^{7} cells/mL = 533 mL culture. If the cell stock concentration from step 1 = 1.6×10^{10}/mL, then dilute 0.5 mL of cell stock into 533 mL of medium.
2. Add the inoculum into sterile 2-L flasks containing 2xTY-CG broth, with a maximum of approximately 500 mL per flask. Grow with shaking at 37°C to reach an OD of 0.4–0.6 at 600 nm. Check the OD every 20 min to avoid overgrowth.
3. Add helper phage M13KO7 at a ratio of 20:1 phage to bacteria. An OD of 0.5 = 1.5×10^{8} bacteria/mL. Therefore, add 3.0×10^{9} phage/mL.

4. Incubate the flasks for 30 min at 30°C without shaking, then continue for a further 90 min while shaking at 150 rpm. After 120 min, take an aliquot of 100 μL per flask and serially dilute to 10^{-9}. Plate 100-μL volumes of 10^{-6} to 10^{-9} dilutions on both 2xTYAG and 2xTYAKG plates and incubate overnight at 37°C to provide a titer of viable cells (Amp resistant) vs. helper phage infected viable cells (Amp + Kan resistant). Expect the total number of infected cells per flask (by Amp/Kan resistance) to be larger than the theoretical library diversity (see Note 7).
5. Centrifuge cultures at 4,000×*g*/15 min/4°C. Dispose of supernatant (add bleach to 20% to kill helper phage; see Note 8) and resuspend bacterial pellets in 2xTY-AK (using the same volume of media used to start the culture, e.g., 533 mL). Incubate overnight at 30°C/250 rpm.
6. The next day, centrifuge cultures at 4,000×*g*/15 min/4°C. Dispose of the pellets and add one fifth of a volume of 5× PEG/NaCl solution to the culture supernatant. Place on ice for 1 h.
7. Centrifuge at 15,000×*g*/15 min/4°C to pellet crude phage. Resuspend pellets in 40-mL PBS per initial culture, then centrifuge at 9,000×*g*/15 min/4°C to remove bacterial debris. Transfer the supernatant to a new centrifuge tube and add 10 mL of 5× PEG/NaCl solution. Place on ice for 30 min.
8. Centrifuge at 15,000×*g*/15 min/4°C. Resuspend phage pellets in 2 mL PBS per initial culture.
9. Proceed immediately to selections (Subheading 3.7 below) or add glycerol to 15% V/V and freeze at –80°C in aliquots of 250 μL.

3.7. Phage Selection Protocol

Each selection round is carried out over a period of 3 days. For round 1, the elements of "day 1," outlined below, can be carried out on the same day as the rescue of the unselected library begins, so that the fresh library phage may be used in selections immediately. As the library will likely contain antibodies specific to all proteins used during the immunizations, each protein should be selected against separately. A single operator can comfortably perform three or more selection campaigns simultaneously. If species cross-reactive antibodies are required, perform two branches of selection with one concentrating on the protein used to immunize and another substituting orthologues of the protein at each selection round (see Fig. 2a). It is routine to perform four rounds of selection in total, but using these protocols, diverse hits are usually identified by round 2.

3.7.1. Day 1

1. Inoculate 5 mL of 2xTY-T broth with *E. coli* ER2738 from a single colony (from a fresh streak on 2xTY-TG plates; see Note 9) and incubate overnight (250 rpm at 37°C).

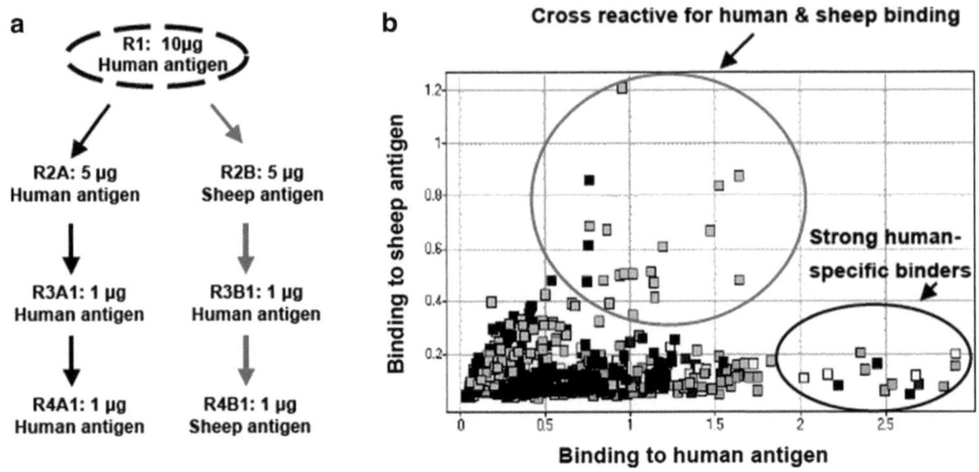

Fig. 2. An example of typical selection campaign conditions used to successfully isolate human/sheep cross-reactive antibodies from a multiprotein immunized library. (**a**) Example screening data from the same selection campaign, showing the ELISA signal (OD 450 nm on *both axes*) for selected clones against the human and sheep orthologs of one of five different proteins used during immunization. (**b**) Cross-reactive binding signals (*circled in gray*) derive predominantly from selection branch B, while those human-specific signals (*circled in black*) derive predominantly from branch A.

2. Rescue phage (for round 2 selections and beyond).
3. Coat immunotubes with target antigen. Add 10–50 µg protein per tube in 1 mL 50 mM sodium carbonate buffer (pH 9.6) in round 1, 5–20 µg or less in subsequent rounds (see Fig. 2a, Note 10). Fully seal the tubes with two layers of parafilm and place on a rotary mixer at 20 rpm/4°C overnight.

3.7.2. Day 2

1. Phage preparation (round 2 and beyond): Split the culture (see "day 3") into two 85-mL Oak Ridge tubes and spin at 15,000×*g*/15 min/4°C. Transfer supernatant to a fresh tube. To 24 mL of supernatant, add 6 mL of 5× PEG/NaCl and mix well. Chill on ice for 30 min. Centrifuge at 15,000×*g*/4°C/15 min. Carefully pour off supernatant into 10% bleach. Briefly spin again (<1 min at 15,000×*g*/4°C) to bring down the last of the supernatant (see Note 11).
2. Carefully pipette off the last of the PEG supernatant and then resuspend the pellet in 1.0-mL PBS per tube (from 24 mL initial supernatant). Transfer both 1.0 mL samples to two 1.5-mL microcentrifuge tubes, then centrifuge immediately for 5 min at 15,000×*g*/4°C to pellet bacterial debris.
3. Remove supernatants to fresh tubes and add 0.25-mL PEG/NaCl. Mix and leave on ice for 15 min. Centrifuge again at 15,000×*g*/4°C for 15 min, aspirate off, and discard the supernatant into 10% bleach.

4. Remove all PEG supernatant, then resuspend each of the pellets in 0.5-mL blocking solution (3% M-PBS). Combine the two tubes of prepared phage to make 1.0 mL.
5. Inoculate 0.5 mL of an overnight culture of ER2738 into 50 mL 2xTY-T in a 250-mL flask. Expect about 2.5 h to reach A600 = 0.45–0.6. Start checking at about 1.5 h, and every 15 min thereafter. Prepare 10-mL culture per selection pool, plus 10 mL for taking OD, titering input, etc. Do not let cells overgrow (see Note 12). While the phage from step 3 are blocking, wash the antigen-coated immunotubes five times with PBS and block by filling with 3% M-PBS. Allow to block for 1 h at room temperature.
6. Target binding: Decant the block from each immunotube and add the 1.0 mL of blocked phage preparation (*remember to keep approximately 50 μL for input phage counts!*). Fully seal the tubes with two layers of parafilm and place on a rotary mixer at 20 rpm. In round 1, mix for 1 h at room temperature, then sitting static for 1 h. In subsequent rounds, halve these times (30 and 30 min).
7. Decant the unbound phage from the tubes into 10% bleach. Wash the tubes by adding 5-mL PBS-T and discarding all washes in 10% bleach. Be sure to use a single pipette and a single tube of wash buffer for each selection if you are doing more than one selection at once (see Note 13). Typical washing regime: 10× washes PBS-T, plus 5 PBS in rounds 1 and 2, 10× washes PBS-T/10× washes PBS at later rounds.
8. Elute phage: Add 1 mL of freshly made 100 mM TEA. Incubate at RT/20 rpm for no more than 10 min. (Longer exposure will damage phage.) Transfer eluted phage to a 50-mL Falcon tube containing 500 μL of sterile 1 M Tris–HCl pH 7.5 to neutralize the TEA.
9. Reinfect eluted phage: To each eluted phage sample, add 10 mL of *E. coli* ER2738 culture (at OD 600 nm ~ 0.4–0.6) from step 5. Save 100-μL uninfected *E. coli* cells each (200 μL total) for control plate and input infections. Mix thoroughly by hand (swirling) and incubate for 30 min at 37°C without shaking.
10. Following infection, remove 100 μL of cells for titering output (see below).
11. Recovery of infected cells: Centrifuge the 50-mL tubes of infected cells at 2,000 × *g*/4°C for 15 min. Carefully discard the supernatant and resuspend cells in 1 mL 2xTY medium. Spread the cells onto a bioassay tray containing 2xTYCG agar and incubate overnight at 30°C.
12. Plate 100 μL of the uninfected ER2738 culture (saved in step 9) on 2xTYCG as a negative control.

13. Titer numbers of phage before and after immunotube selection. Prepare serial tenfold dilutions of the cultures infected with pre- and postselection phage and plate 100 μL of 10^0, 10^1, 10^2, 10^3, 10^4, 10^5, and 10^6-fold dilutions of *E. coli* cells on 2xTYCG plates. Incubate plates overnight at 30°C.
14. Titer calculations: cfu/mL = no. of colonies on plate × dilution factor plated × original sample volume in μL (×10^3 for input titer and ×10^4 for the output titer).

3.7.3. Day 3: Recovery of Cells and Phage Rescue for the Next Selection Round

1. Count colonies on input and output titration plates and record the values. (Wrap these plates in parafilm and save at 4°C for testing individual clones later.)
2. Add 20 mL 2xTY to each bioassay plate. Resuspend cells by scraping bacterial lawn thoroughly with a sterile plate spreader. Transfer resuspended cells to a 50-mL tube and spin down at 2,000 × *g*/4°C for 15 min.
3. Resuspend in 6 mL 2xTY/20% glycerol and freeze in 1-mL aliquots at −80°C.
4. Use 50 μL of the cell suspension from step 3 above (or 50 μL of a thawed glycerol stock) to inoculate 50 mL 2xTYAG/Tet (5 μg/mL tetracycline) in a 250-mL flask. Adjust to OD 0.1 and grow at 37°C (250 rpm) to an OD 600 nm of 0.3–0.5.
5. Superinfect cells with M13K07 helper phage (~4×10^{11} cfu = 20 cfu:cell) and incubate at 37°C (static) for 30 min, then shake at 37°C (150 rpm) for 60 min.
6. After the full 90 min, take an aliquot of 100 μL per flask and serially dilute to 10^{-9}. Plate 100 μL volumes of 10^{-6} to 10^{-9} dilutions on both 2xTYAG and 2xTYAKG plates and incubate overnight at 37°C to provide a titer of viable cells vs. helper-infected viable cells. Expect the number of infected cells to be at least tenfold greater than the diversity of the inoculum (i.e., the number of total output cfu counted from the previous round).
7. Centrifuge cells at 2,000 × *g* for 10 min, resuspend the pellet in 50 mL 2xTYAK, and transfer to a 250-mL culture flask. Incubate at 30°C/300 rpm overnight.
8. Set up an overnight culture of ER2738 cells in 5 mL 2xTYT for selection the following day.
9. After overnight growth, perform two rounds of PEG precipitation as described for day 1 and proceed with the next round of panning.

3.8. Screening of Output Clones

Output clones can be selected for screening from 10 cm^2 dishes prepared for selection output titrations (as in step 3 above). If cross-reactive antibodies are sought, each clone may be simultaneously screened on separate ELISA plates for binding to each of the target proteins selected against (see Fig. 2b).

1. Plate the outputs of the second and subsequent rounds of selection onto 10-cm 2xTYCG plates so that single colonies can easily be picked. It is usually convenient to save the output titer plates from each selection (step 3 above) for this purpose. Using sterile toothpicks or sterile 10-μL filtered tips, pick colonies into 160 μL of 2YT-GC in each well of a 96-well sterile plate (see Notes 14 and 15).
2. Seal the plates using breathable seals and incubate overnight with shaking at 37°C and 80% humidity (see Note 16). The following morning, add 60 μL of 50% glycerol to each well and store at −80°C (these are now "master plates" for any further analyses), or replicate immediately into deepwell plates for secondary cultures to perform binding assays.
3. Inoculate ~0.5 μL/well of thawed glycerol stocks or fresh overnight cultures of picked colonies into 96-well deepwell plates containing 900 μL of 2YT with 0.1% glucose. Cover with breathable seals and grow at 37°C with shaking at 80% humidity.
4. When the OD 600 reaches 0.6–0.8 (about 3–4 h), add 100 μL of 0.2 mM IPTG to each well for a final concentration of 0.02 mM. Reduce the temperature of the shaking incubator to 28°C and incubate overnight.
5. Coat Nunc Maxisorp ELISA plates with 100-μL antigen per well overnight at 4°C. Most antigens coat more efficiently in 50 mM sodium carbonate buffer, pH 9.6 compared to PBS, and generally a coating concentration of 1 μg/mL is sufficient to achieve saturation of wells. However, this is not always the case, and if a control antibody is commercially available, these conditions should be optimized prior to scFv testing. Coat negative control plates with 2% BSA to allow determination of antigen-specific binding.
6. After overnight induction, pellet cells by centrifugation at 1500 × *g* for 10 min, decant supernatants, and add 150 μL of ice-cold periprep buffer per well. Vortex briefly to resuspend.
7. To release the periplasmic fraction by osmotic shock, add 150 μL/well of a 1:5 dilution of periprep buffer in H_2O. Incubate at 4°C for 30 min. Centrifuge plates at 3,500 × *g* for 15 min at 4°C.
8. Transfer 110 μL of the resulting supernatant from each well to a low-binding 96-well plate containing an equal volume of 2× blocking solution (6% milk, 2% BSA in PBS). Block for at least 1 h at room temperature.

9. In the meantime, wash coated ELISA plates 3× with 200 μL of PBS-T (PBS containing 0.1% Tween-20) and add 200 μL of 1× blocking solution to each well. Block for at least 1 h at room temperature.
10. Remove blocking solution from the ELISA plates and add 100 μL of blocked periprep to both the antigen- and BSA-coated plates. Add 100 μL of blocking solution to negative control wells and 100 μL of a 1 μg/mL solution of control antibody prepared in blocking solution to positive control wells. Incubate for 1 h at room temperature, then wash plates 5× with PBS-T.
11. Add 100 μL of relevant secondary antibody (e.g., an HRP-labeled antibody that recognizes the tagged scFv such as anti-HA or anti-c-Myc depending on the phagemid vector) prepared in blocking solution to each well and incubate at room temperature for 1 h. Wash plates 5× with PBS-T.
12. Add 75 μL of TMB to each well. Once the blue color develops, stop the reaction by adding an equal volume of 0.18 M phosphoric acid.
13. Read absorbance at 450 nm using a microtiter plate reader.
14. For all clones exhibiting positive signal (e.g., >threefold the values for negative control wells), pick from the master plates generated to subculture, bank, sequence, and express as needed for further characterization.

4. Notes

1. It is important to know the differences between avian and mammalian anatomy before beginning. For example, the avian spleen does not look like the mammalian version; rather, it looks like a large purple "kalamata" olive.
2. Some of these PCR primers are long. For maximal function, all primers should be purified by the manufacturer, either by SDS–PAGE (particularly CSCVHo-FL) or by HPLC, to remove incomplete synthesis products.
3. *E. coli* SCS 110 is our phagemid preparation strain of choice as it is negative for both Dam and Dcm DNA methylation, rendering all potential restriction digests optimal.
4. Adequate cleanup at this stage is critical for removing all traces of salt and obtaining high transformation efficiency.
5. Care is needed at the library resuspension and electroporation stages to ensure high recovery of ligated clones. Full solubilization of the precipitated plasmid is critical for obtaining high levels of transformation.

6. Note that TG1 cells are tetracycline-sensitive. If you are using the tetracycline-resistant strain ER2738 for phage selections (not available commercially as electrocompetent cells, but preferred for selections), remember that your plates and liquid media for handling the initial library in TG1 must be *free of tetracycline.*
7. When superinfecting with helper phage, it is important to determine the fraction of infected cells vs. viable cells. Even if helper phage infection is poor, the small fraction of infected cells can grow and produce a high phage titer (but a very low-diversity population, which will not be evident unless you include this control).
8. Remember that the M13 bacteriophage is extremely tough. Careful handling of all phage-containing materials is important for avoiding cross-contamination of phage libraries. The use of a dedicated set of pipettors and filtered tips is critical to reduce cross-contamination. Liquids containing phage (discarded culture supernatants, selection supernatants, washes) should be treated with 10% bleach overnight before disposal. Reusable culture vessels and centrifuge tubes should be treated with 20% bleach overnight after use, followed by standard washing and baking at 110°C for 4 h. Work surfaces and pipettes should be decontaminated with 10% bleach and UV treatment after use.
9. Cells for phage infection should be kept under selection to maintain expression of the F pilus, which M13 phage use to infect *E. coli.* ER2738 has an F'episome with a tetracycline-selectable marker, and a streak of this strain onto 2xYT-TG can be prepared and stored for up to 2 weeks.
10. Surprisingly, coating concentrations that are apparently above the saturating concentration for saturating a mAb signal can still make a significant difference for phage selections. 10–20 μg/mL is a reasonable concentration at which to start for most antigens, although many selections will work with 1–5 μg/mL, and some may require as high as 50–100 μg/mL.
11. Failure to remove all supernatant during PEG precipitation can lead to later loss of the phage pellet.
12. It is better to grow multiple cultures and let OD go up to about 0.6 than it is to keep cells on ice for an extended period (>30 min).
13. When washing tubes from multiple phage selections, be sure to avoid using the same pipette for multiple selections. Also, avoid the use of squeeze bottles for washing tubes, since this can lead to the spread of phage in aerosols. It is often convenient to do all of the washes for one selection and leave the final wash in the tube, and then move on to all of the washes for the next selection, and so on. Do not allow the tube to dry.

14. In order to get a handle on sequence diversity and binding function within each selection round, we tend to pick 96 colonies from each round. However, this can be reduced to 48 or 24 colonies/round depending on individual throughput capabilities.
15. We often leave the last four wells (i.e., E12, F12, G12, H12) of a 96-well plate empty so that we can add duplicate positive and duplicate negative controls to each plate during the assay stage. Bacteria containing positive and negative control phagemids can also be inoculated into these wells when colonies are picked, and this provides an internal control for scFv expression.
16. Appropriate speed for 96-well plates depends on the shaker. A shaker platform with a 3-mm orbit, designed for 96-well plates, will keep cells well suspended at 600 rpm. Shakers designed for larger flasks (at 25 mm orbit) will require slower speed to avoid splashing from well to well.

Acknowledgments

The authors would like to acknowledge the work of Benedicte Autin, Cliona O'Sullivan, Ciara Mahon, Matthew Lambert, Emma Cummins, Alfredo Sheehan, Jason Wade, Brian Fennell, and Angela Widom for their assistance in establishing these methods.

References

1. Leonard, P., Safsten, P., Hearty, S., McDonnell, B., Finlay, W. & O'Kennedy, R. (2007) High throughput ranking of recombinant avian scFv antibody fragments from crude lysates using the Biacore A100. *J Immunol Methods* **323**, 172–9.
2. Nishibori, N., Shimamoto, T., Nakamura, N., Shimokawa, M., Horiuchi, H., Furusawa, S. & Matsuda, H. (2004) Expression vectors for chicken-human chimeric antibodies. *Biologicals* **32**, 213–8.
3. Finlay, W. J., Shaw, I., Reilly, J. P. & Kane, M. (2006) Generation of high-affinity chicken single-chain Fv antibody fragments for measurement of the Pseudonitzschia pungens toxin domoic acid. *Appl Environ Microbiol* **72**, 3343–9.
4. Finlay, W. J., deVore, N. C., Dobrovolskaia, E. N., Gam, A., Goodyear, C. S. & Slater, J. E. (2005) Exploiting the avian immunoglobulin system to simplify the generation of recombinant antibodies to allergenic proteins. *Clin Exp Allergy* **35**, 1040–8.
5. Chiliza, T. E., Van Wyngaardt, W. & Du Plessis, D. H. (2008) Single-chain antibody fragments from a display library derived from chickens immunized with a mixture of parasite and viral antigens. *Hybridoma (Larchmt)* **27**, 413–21.
6. Hof, D., Hoeke, M. O. & Raats, J. M. (2008) Multiple-antigen immunization of chickens facilitates the generation of recombinant antibodies to autoantigens. *Clin Exp Immunol* **151**, 367–77.
7. Andris-Widhopf, J., Rader, C., Steinberger, P., Fuller, R. & Barbas, C. F. III (2000) Methods for the generation of chicken monoclonal antibody fragments by phage display. *J Immunol Methods* **242**, 159–81.

8. Krebber, A., Bornhauser, S., Burmester, J., Honegger, A., Willuda, J., Bosshard, H. R. & Pluckthun, A. (1997) Reliable cloning of functional antibody variable domains from hybridomas and spleen cell repertoires employing a reengineered phage display system. *J Immunol Methods* **201**, 35–55.

Chapter 22

Measuring Protein–Protein Interactions Using Biacore

Paul Leonard, Stephen Hearty, and Richard O'Kennedy

Abstract

The use of optical biosensors for studying macromolecular interactions is gaining increasing popularity. In one study, 1,179 papers that involved the application of biosensor data were identified for the year 2007 alone (Rich and Myszka, J Mol Recognit 21:355–400, 2008), the sheer volume and variety of which present a daunting task for the burgeoning biosensor user to accumulate and decipher. This chapter is designed to provide the reader with the tools necessary to prepare, design, and efficiently execute a kinetic experiment on Biacore. It is written to guide the Biacore user through basic theory, system maintenance, and assay set-up while also offering some practical tips that we find useful for Biacore-based studies. Many kinetic-based screening assays require rigorous sample preparation and purification prior to analysis. To highlight these procedures, this protocol describes the kinetic characterisation of single chain Fv (scFv) antibody fragments from crude bacterial lysates using an antibody affinity capture approach. Even though we specifically describe the capture of HA-tagged scFv antibody fragments to an anti-HA tag monoclonal antibody-immobilised surface prior to kinetic analysis, the same methodologies are universally applicable and can be used for practically any affinity pair and most Biacore systems.

Key words: Antibody, Surface plasmon resonance, Biacore, Kinetics, Biosensor analysis, Protein–protein interactions.

1. Introduction

The past decade has seen radical advances in both protein discovery and antibody production methodology, creating an unprecedented need for rapid methods of studying protein–protein interactions on a molecular scale. During this period, surface plasmon resonance (SPR) biosensors, such as Biacore systems (2), have co-evolved as standard tools for protein characterisation and binding interaction studies. The potential of this maturing biosensor market is evident from the expanding number of companies providing SPR-based sensors (3), from the quantity and variety of published material using commercial biosensors

Dermot Walls and Sinéad T. Loughran (eds.), *Protein Chromatography: Methods and Protocols*, Methods in Molecular Biology, vol. 681, DOI 10.1007/978-1-60761-913-0_22, © Springer Science+Business Media, LLC 2011

(reviewed annually by Rich and Myszka (1, 3–11)) and as a result of the quality of Biacore data generated from recent comparative studies with methods such as KinExA (12) and solution-based calorimetry measurements (13). While Biacore-based biosensors have been used for the analysis of mammalian (14–16) and bacterial cells (17–20), viruses (21), proteins (12, 22–26), toxins (27, 28), and haptens (13, 29–33), antibody characterisation is by far the most common application (1, 3, 5–11).

1.1. Theory

Surface plasmon resonance is an optical phenomenon which occurs as a result of total internal reflection of light at a metal film–liquid interface. Under conditions of total internal reflection, incident light excites plasmons in the gold film; the angle at which this occurs (known as the SPR angle) is very sensitive to refractive index changes of the solution. Biacore biosensors measure the shift in SPR angle due to mass changes at the surface of the chip and, therefore, enable the detection and measurement of protein–protein interactions in real-time, without the use of labels. The most basic description of a biomolecular interaction between a soluble monovalent analyte (A) and immobilised monovalent ligand (B) can be interpreted as

$$[\mathrm{A}]+[\mathrm{B}] \leftrightarrows [\mathrm{AB}]$$

where one molecule [A] (usually of known concentration and injected over the surface) reversibly binds to another [B] (usually immobilised on the sensor surface) (34). The strength of the interaction can be described by

$$\frac{[\mathrm{A}]\times[\mathrm{B}]}{[\mathrm{AB}]} = K_{\mathrm{D}} \quad (1)$$

The rate of formation of complex [AB] is described by the differential equation

$$\frac{\mathrm{d}[\mathrm{AB}]}{\mathrm{d}t} = k_{\mathrm{a}} \times [\mathrm{A}][\mathrm{B}] - k_{\mathrm{d}} \times [\mathrm{AB}] \quad (2)$$

where the ratio of the association rate constant k_{a} ($\mathrm{M}^{-1}\mathrm{s}^{-1}$) over the ratio of the dissociation rate constant k_{d} (s^{-1}) describes the affinity constant *KA*. At equilibrium, the change in the concentration of the complex [AB] is zero, and therefore, the equilibrium dissociation constant (or affinity) K_{D} can be derived from Eq. 2 such that

$$0 = k_{\mathrm{a}} \times [\mathrm{A}][\mathrm{B}] - k_{\mathrm{d}} \times [\mathrm{AB}]$$

$$k_{\mathrm{d}} \times [\mathrm{AB}] = k_{\mathrm{a}} \times [\mathrm{A}][\mathrm{B}]$$

$$\frac{k_{\mathrm{d}}}{k_{\mathrm{a}}} = \frac{[\mathrm{A}][\mathrm{B}]}{[\mathrm{AB}]} = (\text{from Eq.1 above}) K_{\mathrm{D}}$$

2. Materials

The materials listed in this section may be obtained from Biacore, GE Healthcare, Uppsala, Sweden unless otherwise stated.

2.1. Instrument Cleaning and Surface Preparation

1. Biacore CM5 and maintenance sensor chips.
2. BIAdesorb solution 1 consisting of 0.5% (w/v) SDS.
3. BIAdesorb solution 2 consisting of 50 mM glycine–NaOH, pH 9.5.
4. HBS running buffer (10 mM HEPES, 150 mM NaCl, 3 mM EDTA, and 0.05% (v/v) surfactant P20).
5. Ultrapure H_2O.
6. 100 mM HCl and 50 mM NaOH.

2.2. Ligand Preconcentration

1. HBS running buffer.
2. 10 mM Sodium acetate pH 4–5 with 10% (v/v) acetic acid.
3. Anti-HA tag monoclonal antibody (Affinity BioReagents).
4. 20 mM NaOH.

2.3. Immobilisation

1. HBS running buffer.
2. 0.4 M EDC (*N*-ethyl-*N*′-(dimethylaminopropyl)carbodiimide).
3. 0.1 M NHS (*N*-hydroxysuccinimide).
4. Anti-HA tag monoclonal antibody diluted in sodium acetate buffer at the appropriate pH.
5. 1 M Ethanolamine–HCl, pH 8.5.
6. 20 mM NaOH.
7. *Optional:* Control protein for immobilisation onto the reference flow cell.

2.4. Surface Regeneration

1. HBS running buffer.
2. Full range of regeneration buffers.
3. Anti-HA tag monoclonal antibody.

2.5. Kinetic Studies and Data Analysis

1. HBS running buffer.
2. Bacterial lysate containing expressed scFv antibody fragment.
3. Purified monomeric analyte of choice.
4. 120 mg/mL CM Dextran in HBS running buffer.
5. 120 mg/mL bovine serum albumin (BSA) in HBS running buffer.
6. Optimised regeneration buffer.

3. Methods

Although a panoply of sensorgrams, assay formats, and detailed protocols can be found described in the literature from a variety of instruments (35–39), the analysis of antibody/antigen complexes using Biacore technology usually requires the immobilisation of one interactant (in relatively pure form) on the sensor chip surface followed by the injection of the second interactant, the concentration of which must be known for the calculation of the association constant, k_a. Affinity-based immobilisation (affinity capture) (23–25) is an alternative format that facilitates surface regeneration and the directed orientation of the antibody (40). Essentially, the affinity-based immobilisation protocol we describe involves the non-reversible immobilisation (amine coupling) of a high-affinity ligand (anti-HA tag monoclonal antibody) to reversibly anchor recombinant HA-tagged scFv antibody fragments to the surface prior to analyte injection. Replacement (regeneration) of the scFv antibody fragment after each binding cycle avoids losses in antibody binding activity, assuming scFv antibody capture is reproducible (see Subheading 3.4 on surface regeneration).

Regardless of experimental design, the quality of one's biosensor data is always directly dependent on the quality of the reagents used (41). While capture assays can be used to rank antibodies from crude matrices (25), the analyte should be well characterised (exact concentration known), of high purity (>95%), of good integrity (functional with no aggregates), and for 1:1 interaction studies be monovalent. The Biacore user should be fastidious in instrument preparation (discussed in Subheading 3.1), spend time choosing the right assay set-up for their interaction pair (see Note 1), try to eliminate artefacts such as non-specific binding (see Note 2), evaluate the impact of wash (see Note 3) and regeneration steps (see Note 4), and always include sample replicates (see Note 5).

3.1. Instrument Cleaning and Surface Preparation

The method described in this chapter was performed on a Biacore 3000 instrument, but irrespective of the instrument used, proper and regular system maintenance is essential when aiming to generate high-quality data. In addition to the recommended "desorb" and "sanitize" maintenance protocols listed in the Biacore 3000 handbook, we recommend performing a "super-desorb" (13, 26, 42) followed by preconditioning of a fresh CM5 sensor chip.

1. Undock the instrument sensor chip and dock a maintenance chip (see Note 6).
2. Prime the instrument five times with 0.5% (w/v) SDS (BIAdesorb solution 1).

3. Replace the 0.5% (w/v) SDS solution with ultrapure water and prime the system once.
4. Prime the instrument five times with 50 mM glycine–NaOH, pH 9.5 (BIAdesorb solution 2).
5. Replace the 50 mM glycine–NaOH, pH 9.5 solution with ultrapure water and prime the system another five times.
6. Undock the instrument maintenance chip and dock a fresh CM5 sensor chip and prime three times with running buffer.
7. Set the instrument flow rate to 100 μL/min and precondition the chip with two consecutive injections each of 100 mM HCl, 50 mM NaOH, and 0.5% (w/v) SDS for 10 s per injection.

3.2. Ligand Preconcentration

Low ionic strength buffers allow the electrostatic adsorption of positively charged proteins (proteins in low ionic buffer at a pH below the proteins' isolectric point [pI]) to the negatively charged dextran surface, maximising "*preconcentration*" of protein to the sensor surface for subsequent immobilisation and thus increasing the amount of ligand at the surface that can be efficiently covalently coupled on suitably activated chip surfaces. Electrostatic binding of protonated amine groups on the biological component to negatively charged carboxyl groups on the chip surface is facilitated by adjusting the pH below the pI of the protein. Therefore, protein solutions should be prepared in low ionic strength buffers such as 10 mM sodium acetate at a range of different pHs and these solutions passed over an underivatised chip surface, with the degree of electrostatic binding monitored. The highest pH (see Note 7) at which preconcentration of protein onto the underivatised surface is observed to be sufficient should be chosen as the pH for immobilisation. Depending on user preference, the methods described below can be performed using the manual operation function, the application wizards, or the method control function.

1. Prepare 25 μg/mL anti-HA tag monoclonal antibody (see Note 8) in 10 mM sodium acetate buffer at pH increments from pH 4 to 5.
2. Inject each sample over flow cell 2 of a preconditioned CM5 sensor chip (see Subheading 3.1) for 30 s at a flow rate of 10 μL/min (see Note 9).
3. *Optional:* If the response does not return to baseline after ligand preconcentration (some ligands can have a tendency to non-specifically stick to the dextran surface), inject 20 mM NaOH for 15 s after each preconcentration.
4. Evaluate the magnitude and slope of the preconcentration response (see Fig. 1) and determine if the desired ligand density can be achieved. Use the mildest pH conditions that will enable the achievement of the desired immobilisation level.

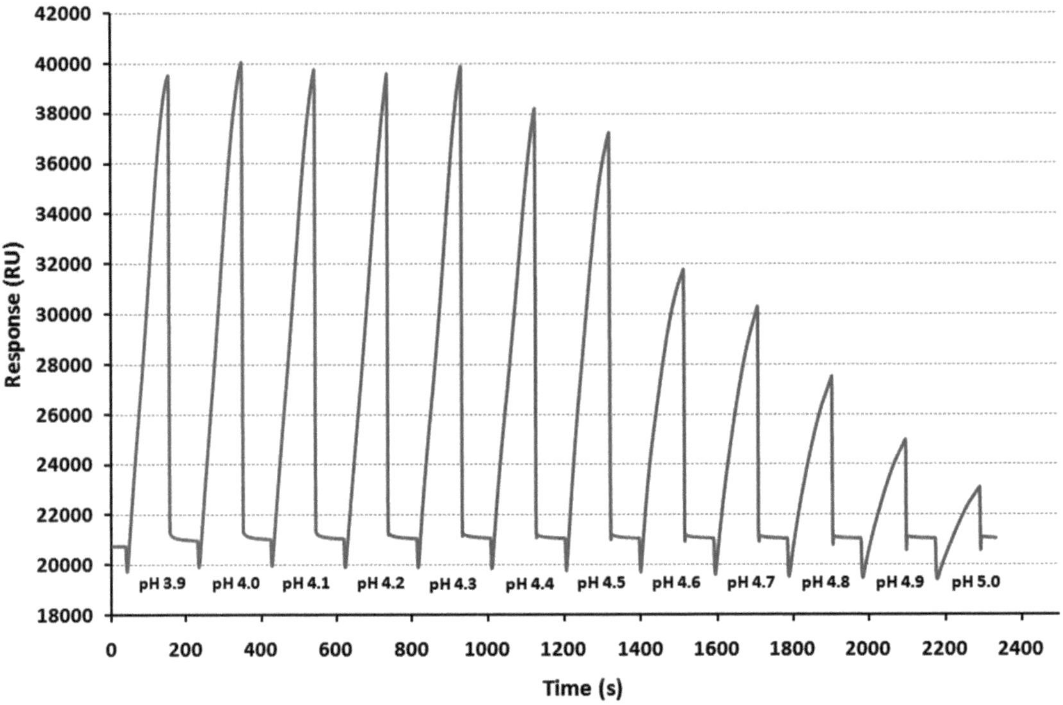

Fig. 1. Typical preconcentration profile of a monoclonal capture antibody electrostatically interacting with the CM-dextran sensor chip surface. Solutions of 50 μg/mL of antibody in 10 mM sodium acetate buffer at various pH increments (indicated on the sensorgram shown here) were passed over an untreated CM-dextran surface at 10 μL/min for a period of 2 min. The low ionic strength of the acetate buffer favours the electrostatic attraction between the negatively charged dextran layer and the positively charged protein (i.e. below its isoelectric point). The degree of preconcentration was measured from the response prior to the end of each sample injection with the ionic strength of the HBS running buffer (containing 150 mM NaCl) sufficient for the removal of electrostatically attracted antibody from the surface. The pH used for the high density immobilisation (>10,000 RU) of the monoclonal antibody onto the CM-dextran chip surface was pH 4.5 (see Note 7).

3.3. Immobilisation

Immobilisation of ligand to a CM5 sensor chip surface involves activation of the surface, followed by ligand preconcentration/immobilisation and finally surface deactivation. Even though there are a wide variety of immobilisation chemistries available for the coupling of ligands to the CM dextran surface, depending on the functional groups available for coupling, the most commonly employed strategy is the use of EDC (*N*-ethyl-*N'*-(dimethylaminopropyl) carbodiimide) and NHS (*N*-hydroxysuccinimide) chemistry. Using the carbodiimide EDC, the CM dextran carboxyl groups can be transformed into active ester functional groups in the presence of NHS. The surface NHS esters can then react with suitable amino groups in the protein and deactivated or "capped" by the injection of 1 M ethanolamine–HCl (pH 8.5), which also serves to remove non-covalently bound protein by reducing the electrostatic attraction between the protein and the CM-dextran (carboxymethyl dextran sodium salt).

1. Prepare 100 μL of 0.4 M EDC, 100 μL of 0.1 M NHS, 200 μL of 25 μg/mL anti-HA tag monoclonal antibody (capture antibody), and 100 μL of 1 M ethanolamine–HCl.
2. Mix equal volumes of 0.4 M EDC and 0.1 M NHS and inject 70 μL over flow cell 2 (same flow cell as preconcentration) at 10 μL/min flow rate (see Note 10).
3. Inject 150 μL of capture antibody over the activated surface at 10 μL/min (see Note 11).
4. Inject 70 μL of 1 M ethanolamine–HCl at a flow rate of 10 μL/min to block residual activated carboxyl moities on the surface.
5. Post-condition the surface with five 15 s injections of 20 mM NaOH.
6. *Optional:* Prepare a reference surface on flow cell 1 by repeating steps 1–5 above replacing the capture antibody in step 3 with 25 μg/mL of suitable control protein of choice (see Note 12).

3.4. Surface Regeneration

The efficient and reproducible regeneration of antibody-binding surfaces is of major importance for reusable sensor formats such as required for direct "real-time" biosensing technologies and is often difficult to achieve (40). For accurate kinetic analysis, surface regeneration while maintaining ligand integrity is usually of paramount importance (see Note 13). Surface regeneration studies usually involve the identification of a suitable regeneration buffer (regeneration scouting) and the performance evaluation of the surface (surface performance test). The aim of the regeneration studies is to identify a suitable buffer to remove all the bound analyte without compromising functionality of the surface ligand. In order not to destroy the ligand surface, it is always recommended to start buffer scouting with mild buffers working up to more harsh buffers if regeneration is not readily achieved. Once a suitable buffer (or possibly a number of buffers) is identified, its effect on ligand functionality over time is evaluated by repeated binding studies to determine its suitability for kinetic studies (i.e. does it maintain a fully active surface over the study life time?). For multiple analyte screening, capture kinetic experiments simplify regeneration studies as conditions for only one binding interaction (the capture ligand binding interaction) are required compared to direct binding studies where the vast range of analyte affinities may require different buffer conditions to achieve regeneration.

3.4.1. Regeneration Scouting

1. Prepare a panel of regeneration buffers to be tested (see Note 14).
2. Inject 60 μL of a high concentration of HA-tagged scFv antibody (see Note 15) over the anti-HA tag monoclonal

antibody immobilised surface and reference surface at 30 μL/min. Monitor both the baseline and binding levels.

3. Starting from the mildest buffer, inject 15 μL of the regeneration buffer at 30 μL/min.
4. Repeat steps 2 and 3 five times and then change to the next mildest buffer.
5. Follow the trend in scFv-capture response (relative response measured from the baseline before injection of the scFv) and the absolute baseline level (see Fig. 2).

3.4.2. Surface Performance Test

1. Prepare 1 mL regeneration buffer identified in Subheading 3.4.1 and 2 mL scFv antibody lysate.
2. Inject 30 μL scFv antibody over flow cells 1 and 2 (reference and capture surfaces) at 30 μL/min. Monitor both the baseline and binding levels.
3. Inject 15 μL regeneration buffer over flow cells 1 and 2 at 30 μL/min.
4. Repeat the binding/regeneration cycle for 30–40 cycles or as required for the number of cycles to be used in the final assay.

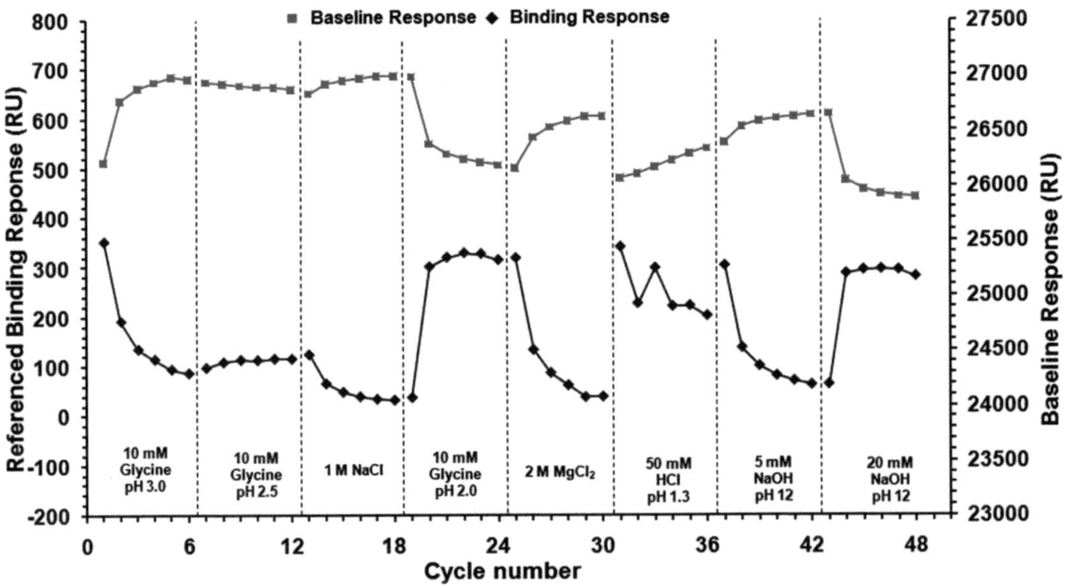

Fig. 2. Anti-HA tag antibody surface performance study using repeated scFv lysate binding cycles regenerated with mild to harsh buffer conditions (indicated on the graph). Six binding cycles for each buffer were performed and the baseline (*squares*) and binding response (*diamonds*) levels recorded. A decrease in binding response with an increase in baseline response was seen with 10 mM glycine, pH 3.0, 1 M NaCl, 2 M $MgCl_2$, 50 mM HCl, pH 1.3, and 5 mM NaOH, pH 12, indicating these buffers were too mild. While both the baseline and binding response levels for 10 mM glycine, pH 2.5, were stable, only 10 mM glycine, pH 2.0, and 20 mM NaOH, pH 12, restored both the baseline and binding responses back to original levels in cycle 1. The latter two buffers were subsequently evaluated as described in Subheading 3.4.2 and 20 mM NaOH chosen as the regeneration buffer for kinetic analysis.

5. Evaluate the ligand stability and baseline level over time. If the baseline increases (with a subsequent decrease in binding level), a harsher regeneration solution may be needed. If the binding capacity decreases significantly while the baseline remains constant or even decreases, the ligand stability is being compromised resulting in a lower analyte binding capacity and a different regeneration solution should be used. Regeneration studies, albeit tedious, are essential for the optimum assay performance and, therefore, time invested in identifying the correct conditions is time well spent.

3.5. Kinetic Studies and Data Analysis

The design of a kinetic experiment must include sufficient experimental data for the reliable calculation of accurate rate constants. Experimental parameters such as (a) temperature, (b) immobilisation level, (c) injection time and flow rate, (d) dissociation time and (e) analyte concentration range should be evaluated prior to running the kinetic experiment (35). Experimental design should also include blank injections for elimination of systematic artefacts, repeat analyte injections, and start up cycles to condition the flow system and sensor surface (35, 41).

1. Prepare doubling dilutions of analyte from 100 to 1.56 nM (see Note 16) in running buffer and include a buffer sample for double referencing to remove instrument artefacts.
2. Dilute bacterial lysate containing expressed scFv antibody 1/10 in running buffer containing 12 mg/mL BSA and 12 mg/mL CM dextran.
3. Perform three to five "*start up*" cycles to confirm that the surface performance is reproducible and to condition the flow system and surface.
4. Inject a sufficient concentration of bacterial lysate, containing expressed scFv, to give an analyte R_{max} of approximately 50 RU (response units) (see Note 17). For bacterial lysates expressing avian scFv antibodies, we use a 2–3 min injection at 20 μL/min.
5. Inject 90 μL of 100 nM analyte at 30 μL/min and allow dissociation in buffer for 10 min.
6. Inject 15 μL of optimised regeneration solution at 30 μL/min.
7. Repeat steps 4–6 above three to five times and evaluate the binding responses which should be reproducible.
8. Perform the kinetic analysis on the range of analyte concentrations listed in step 1, running at least one analyte concentration in duplicate or triplicate and running a buffer injection for double referencing.

9. Inject 60 μL of bacterial cell lysate containing scFv (prepared in step 2) at 20 μL/min over flow cells 1 and 2 (over the reference and active surface).
10. For each analysis cycle, inject 90 μL of analyte (inject analyte concentrations randomly including replicates and buffer injections) at 30 μL/min and allow dissociation in buffer for 15–20 min.
11. Inject 15 μL of optimised regeneration solution at 30 μL/min.
12. *Optional:* After regeneration, perform an IFC (integrated micro fluidics cartridge) wash procedure followed by short buffer injection to clean the injection needle, tubing, and flow path (see Note 3).
13. Repeat steps 8–12 above until all analyte concentrations have been analysed.
14. Open the above kinetic analysis experimental result file in BIAevaluation and prepare the raw data for analysis.
15. Overlay the reference-subtracted data and adjust the baseline prior to analyte injection to zero and align the overlaid sensorgrams with time on the *x*-axis.
16. Remove the regeneration data, capture data, and any other unnecessary data from the sensorgram.
17. Subtract the zero analyte response from the analyte responses ("*double referencing*") to remove systematic deviations.
18. Select the interaction model that best suits your data and start the fitting procedure. The curve fitting procedure is an iterative mathematical process whereby the evaluation software, using default starting values for the parameters to be determined, calculates theoretical binding curves and compares them with experimental data by least square fitting of residuals (see Fig. 3).

4. Notes

1. When designing a binding assay on Biacore, one of the interactants is generally captured or immobilised onto the surface (the ligand) and the other interactant passed over the surface (the analyte). Choosing which interactant from an affinity pair to immobilise will depend on its stability (effect of immobilisation and regeneration on its function), its non-specific binding (NSB) tendency (proteins with a high pI tend to show NSB to the dextran surface and are better used as the ligand), its valency (if possible always use a monovalent

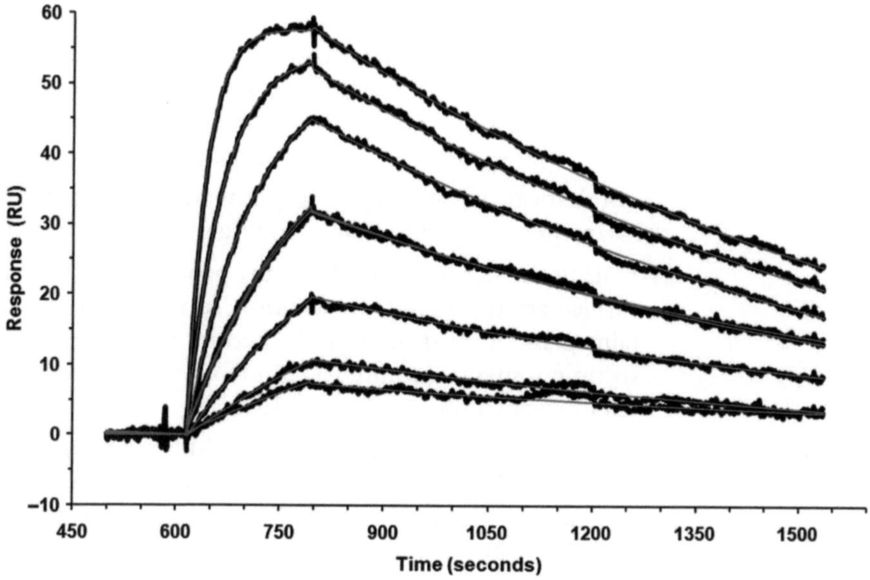

Fig. 3. Example of fitted kinetic data from the analysis of a captured HA tagged avian scFv antibody fragment binding to a human disease protein biomarker (unpublished work using the method described in Subheading 3.5). The fitted curves (fitted with a 1:1 with drifting baseline model) are shown as *solid thin grey lines* overlaid on the raw analyte responses (*thick black lines*) at 100, 50, 25, 12.5, 6.25, 3.125, and 1.56 nM (the 12.5 nM analyte concentration was analysed in duplicate). The values of the calculated kinetic parameters were 3.72×10^5 $M^{-1}s^{-1}$ (k_a), 1.07×10^{-3} s^{-1} (k_d), and 2.88×10^{-9} M (K_D).

molecule as the analyte), and its size (larger molecules are preferred as the analyte even though, with good experimental design, the latest Biacore instruments are capable of measuring binding responses from analytes as small as 100 Da [see http://www.biacore.com]).

2. Prior to assay set-up, evaluation of NSB is required as many biological materials can often exhibit a tendency to non-specifically bind to the sensor surface. This can be done by simply injecting each reagent separately over a blank surface at the highest concentration to be used in the assay. If NSB is high, consider changing the experimental buffers (e.g. adding 12 mg/mL CM dextran and 12 mg/mL BSA in the sample buffer can help reduce NSB) or changing the sensor chip chemistry (e.g. use another sensor chip type or consider deactivating with ethylenediamine instead of ethanolamine after amine coupling).

3. Kinetic analysis, especially direct binding studies (i.e. not following a capture step) can be complicated by small amounts of regeneration buffer carried over between binding cycles which can contaminate the sample plug and affect the next binding reaction (41). Performing a wash step and buffer injection between cycles can help eliminate this.

4. Maintaining ligand function while fully removing bound analyte is essential for reproducible kinetic experiments, and considerable effort spent scouting the right conditions will pay dividends in the quality of the kinetic data obtained.
5. Replicate analyses (duplicate or triplicate) and analyte randomisation are necessary to show the repeatability of the assay, the regeneration step, and ligand activity over time.
6. During the desorb procedure, any proteins on the sensor chip surface are destroyed. In addition, a maintenance chip (contains glass only) should be docked to avoid the possibility of stripping small amounts of gold off the sensor chip which could stick to the instrument IFC.
7. The immobilisation pH to use should be the highest pH that gives suitable immobilisation levels. The buffer pH to be used will depend on the immobilisation level required (low pH buffer may be needed to provide adequate preconcentration of sample for high capacity surfaces used for concentration analysis studies, while lower levels and, thus, milder pH can be used for kinetic surfaces), the sample stability (some proteins are not stable at low pH), and sample concentration (some samples may be in limited supply and maximum preconcentration conditions will be needed).
8. To avoid possible NSB and assay interference, the ligand immobilised should be at least 95% pure if possible. Many commercial antibodies come with additional stabilisers which can contaminate and reduce immobilisation efficiency. If these are present, the antibody should be purified by affinity chromatography (see Chapter 3).
9. The protein concentration to use for preconcentration studies will depend on ligand availability and desired immobilisation level. Generally, proteins are diluted to 1–50 μg/mL in appropriate buffer.
10. Approximately 40% of the carboxyl groups on a CM5 chip surface are activated after a 7 min activation injection. If high-density surfaces are required, this injection time can be increased.
11. The injection times and immobilisation level needed for capture experiments will vary with assay requirements and ligand properties. If the concentration of the scFv antibody to be studied is known *a priori*, the desired anti-HA capture antibody immobilisation levels could be theoretically calculated (see Note 16). In this experiment, we aimed to characterise hundreds of antibody clones differing over 1,000-fold in antibody expression levels and, therefore, a high capacity surface (~10,000 RU) was chosen.

12. On Biacore, the reference surface may be (a) an unmodified surface, (b) an activated-deactivated surface, or (c) a surface immobilised with a non-interacting "dummy" ligand. A prerequisite for choosing which reference surface to use is the determination of the level of NSB to the surface. This should be performed by injecting the analyte at the highest concentration over the reference surface while evaluating the binding levels, if any. If non-specific binding is not an issue, try treating the reference surface with the same immobilisation levels and conditions used for the active test surface as this creates a similar environment between the flow cells.
13. Of increasing popularity is the sequential analysis of binding interactions with reduced regeneration cycles ("*single cycle kinetics*") for kinetics (43) and concentration analysis (44) or accurately choosing a small number of analyte concentrations that are confidently measured (34). These approaches can be very useful where regeneration scouting is difficult due to ligand instability.
14. We usually prepare and filter a range of solutions such as 10 mM glycine buffers, pH 1.5–3.0, 10–100 mM phosphoric acid, HCl, and NaOH and high ionic strength buffers such as 1 M NaCl and 2 M $MgCl_2$.
15. Inject a concentration of HA-tagged scFv antibody equal to or greater than the highest concentration used in the study. This could be a stock lysate of a high expression clone.
16. As a general rule, multiple analyte concentrations ranging from 0.1 to 10 times K_D should be used for kinetic studies with at least one (or preferably more) concentration performed in duplicate. From our experience, the affinity of antibody clones stringently selected (with no *in vitro* affinity maturation steps incorporated) from immune antibody libraries generally range from ~50 to 0.5 nM. Therefore, choosing an analyte concentration range that can be "confidently measured" (34) for *all* clones in a high-throughput screening run will depend on the screening criteria set by the user (i.e. if detailed kinetic analysis is needed, a large range of concentrations will have to be analysed and those concentrations that can be confidently measured used for curve fitting and statistical analysis. However, if the experimental aim is to kinetically rank a high number of clones, a lower analyte concentration range can be used and selected clones reanalysed over an analyte range more suited to its affinity).
17. Information on the optimal ligand immobilisation level relative to binding stoichiometry [binding stoichiometry = (analyte R_{max}/ligand density)×(ligand MW/analyte MW)], where MW is molecular weight, required to achieve the most appropriate maximum analyte binding response

(R_{max}) for kinetic analysis (i.e. <100 RU). Combining low R_{max} levels with high analyte sample flow rates is essential to counteract mass-transfer limitations. If the experimental stoichiometric R_{max} is less than expected, this can indicate that the ligand has lost some activity, a commonly seen artefact with random immobilisation coupling chemistries (37). When multi-valent interaction partners (e.g. bivalent IgG) are to be studied, they should preferentially be used as the surface-ligand in kinetic experiments. If the cognate analyte is truly monomeric then the interaction can be reliably studied using a 1:1 model. For the analysis of a low number of samples, the experimental scFv capture level required to achieve the desired R_{max} can be determined by the iterative injection of diluted samples over the capture surface. However, for high-throughput screening of hundreds of antibodies from crude bacterial lysates, with greater than 1,000-fold differences in expression level, this is not practical, and from our experience, a 2–3 min injection of a 1/10 dilution of scFv containing lysate injected over a high capacity anti-HA monoclonal antibody surface is sufficient for most avian scFv antibody clones.

References

1. Rich, R. L., and Myszka, D. G. (2008) Survey of the year 2007 commercial optical biosensor literature. *J. Mol. Recognit.* **21**, 355–400.
2. Jason-Moller, L., Murphy, M., and Bruno, J. (2006) Overview of Biacore systems and their applications. *Curr. Protoc. Protein Sci. Chapter* **19**, Unit 19 13.
3. Rich, R. L., and Myszka, D. G. (2007) Survey of the year 2006 commercial optical biosensor literature. *J. Mol. Recognit.* **20**, 300–66.
4. Myszka, D. G. (1999) Survey of the 1998 optical biosensor literature. *J. Mol. Recognit.* **12**, 390–408.
5. Rich, R. L., and Myszka, D. G. (2000) Survey of the 1999 surface plasmon resonance biosensor literature. *J. Mol. Recognit.* **13**, 388–407.
6. Rich, R. L., and Myszka, D. G. (2001) Survey of the year 2000 commercial optical biosensor literature. *J. Mol. Recognit.* **14**, 273–94.
7. Rich, R. L., and Myszka, D. G. (2002) Survey of the year 2001 commercial optical biosensor literature. *J. Mol. Recognit.* **15**, 352–76.
8. Rich, R. L., and Myszka, D. G. (2003) A survey of the year 2002 commercial optical biosensor literature. *J. Mol. Recognit.* **16**, 351–82.
9. Rich, R. L., and Myszka, D. G. (2005) Survey of the year 2003 commercial optical biosensor literature. *J. Mol. Recognit.* **18**, 1–39.
10. Rich, R. L., and Myszka, D. G. (2006) Survey of the year 2005 commercial optical biosensor literature. *J. Mol. Recognit.* **19**, 478–534.
11. Rich, R. L., and Myszka, D. G. (2005) Survey of the year 2004 commercial optical biosensor literature. *J. Mol. Recognit.* **18**, 431–78.
12. Drake, A. W., Myszka, D. G., and Klakamp, S. L. (2004) Characterizing high-affinity antigen/antibody complexes by kinetic- and equilibrium-based methods. *Anal. Biochem.* **328**, 35–43.
13. Papalia, G. A., Leavitt, S., Bynum, M. A., Katsamba, P. S., Wilton, R., Qiu, H., Steukers, M., Wang, S., Bindu, L., Phogat, S., Giannetti, A. M., Ryan, T. E., Pudlak, V. A., Matusiewicz, K., Michelson, K. M., Nowakowski, A., Pham-Baginski, A., Brooks, J., Tieman, B. C., Bruce, B. D., Vaughn, M., Baksh, M., Cho, Y. H., Wit, M. D., Smets, A., Vandersmissen, J., Michiels, L., and Myszka, D. G. (2006) Comparative analysis of 10 small molecules binding to carbonic anhydrase II by different investigators using Biacore technology. *Anal. Biochem.* **359**, 94–105.
14. Skottrup, P., Hearty, S., Frokiaer, H., Leonard, P., Hejgaard, J., O'Kennedy, R., Nicolaisen, M., and Justesen, A. F. (2007) Detection of fungal spores using a generic surface plasmon resonance immunoassay. *Biosens. Bioelectron.* **22**, 2724–9.

15. Quinn, J. G., O'Kennedy, R., Smyth, M., Moulds, J., and Frame, T. (1997) Detection of blood group antigens utilising immobilised antibodies and surface plasmon resonance. *J. Immunol. Methods.* **206**, 87–96.
16. Quinn, J. G., O'Neill, S., Doyle, A., McAtamney, C., Diamond, D., MacCraith, B. D., and O'Kennedy, R. (2000) Development and application of surface plasmon resonance-based biosensors for the detection of cell–ligand interactions. *Anal. Biochem.* **281**, 135–43.
17. Hearty, S., Leonard, P., Quinn, J., and O'Kennedy, R. (2006) Production, characterisation and potential application of a novel monoclonal antibody for rapid identification of virulent *Listeria monocytogenes*. *J. Microbiol. Methods.* **66**, 294–312.
18. Leonard, P., Hearty, S., Quinn, J., and O'Kennedy, R. (2004) A generic approach for the detection of whole *Listeria monocytogenes* cells in contaminated samples using surface plasmon resonance. *Biosens. Bioelectron.* **19**, 1331–5.
19. Leonard, P., Hearty, S., Wyatt, G., Quinn, J., and O'Kennedy, R. (2005) Development of a surface plasmon resonance-based immunoassay for *Listeria monocytogenes*. *J. Food Prot.* **68**, 728–35.
20. Reeves, E. P., Ali, T., Leonard, P., Hearty, S., O'Kennedy, R., May, F. E., Westley, B. R., Josenhans, C., Rust, M., Suerbaum, S., Smith, A., Drumm, B., and Clyne, M. (2008) *Helicobacter pylori* Lipopolysaccharide interacts with TFF1 in a pH-dependent manner. *Gastroenterology* **135**(6), 2043–2054.e2.
21. Gomes, P., Giralt, E., and Andreu, D. (1999) Surface plasmon resonance screening of synthetic peptides mimicking the immunodominant region of C-S8c1 foot-and-mouth disease virus. *Vaccine.* **18**, 362–70.
22. Papalia, G. A., Baer, M., Luehrsen, K., Nordin, H., Flynn, P., and Myszka, D. G. (2006) High-resolution characterization of antibody fragment/antigen interactions using Biacore T100. *Anal. Biochem.* **359**, 112–9.
23. Safsten, P., Klakamp, S. L., Drake, A. W., Karlsson, R., and Myszka, D. G. (2006) Screening antibody–antigen interactions in parallel using Biacore A100. *Anal. Biochem.* **353**, 181–90.
24. Canziani, G. A., Klakamp, S., and Myszka, D. G. (2004) Kinetic screening of antibodies from crude hybridoma samples using Biacore. *Anal. Biochem.* **325**, 301–7.
25. Leonard, P., Safsten, P., Hearty, S., McDonnell, B., Finlay, W., and O'Kennedy, R. (2007) High throughput ranking of recombinant avian scFv antibody fragments from crude lysates using the Biacore A100. *J. Immunol. Methods.* **323**, 172–9.
26. Katsamba, P. S., Navratilova, I., Calderon-Cacia, M., Fan, L., Thornton, K., Zhu, M., Bos, T. V., Forte, C., Friend, D., Laird-Offringa, I., Tavares, G., Whatley, J., Shi, E., Widom, A., Lindquist, K. C., Klakamp, S., Drake, A., Bohmann, D., Roell, M., Rose, L., Dorocke, J., Roth, B., Luginbuhl, B., and Myszka, D. G. (2006) Kinetic analysis of a high-affinity antibody/antigen interaction performed by multiple Biacore users. *Anal. Biochem.* **352**, 208–21.
27. Rasooly, A. (2001) Surface plasmon resonance analysis of staphylococcal enterotoxin B in food. *J. Food Prot.* **64**, 37–43.
28. Daly, S. J., Keating, G. J., Dillon, P. P., Manning, B. M., O'Kennedy, R., Lee, H. A., and Morgan, M. R. (2000) Development of surface plasmon resonance-based immunoassay for aflatoxin B(1). *J. Agric. Food Chem.* **48**, 5097–104.
29. Lacy, A., Dunne, L., Fitzpatrick, B., Daly, S., Keating, G., Baxter, A., Hearty, S., and O'Kennedy, R. (2006) Rapid analysis of coumarins using surface plasmon resonance. *J. AOAC Int.* **89**, 884–92.
30. Fitzpatrick, B., and O'Kennedy, R. (2004) The development and application of a surface plasmon resonance-based inhibition immunoassay for the determination of warfarin in plasma ultrafiltrate. *J. Immunol. Methods.* **291**, 11–25.
31. Dillon, P. P., Manning, B. M., Daly, S. J., Killard, A. J., and O'Kennedy, R. (2003) Production of a recombinant anti-morphine-3-glucuronide single-chain variable fragment (scFv) antibody for the development of a "real-time" biosensor-based immunoassay. *J. Immunol. Methods.* **276**, 151–61.
32. Brennan, J., Dillon, P., and O'Kennedy, R. (2003) Production, purification and characterisation of genetically derived scFv and bifunctional antibody fragments capable of detecting illicit drug residues. *J. Chromatogr. B Analyt. Technol. Biomed. Life Sci.* **786**, 327–42.
33. Dillon, P. P., Daly, S. J., Manning, B. M., and O'Kennedy, R. (2003) Immunoassay for the determination of morphine-3-glucuronide using a surface plasmon resonance-based biosensor. *Biosens. Bioelectron.* **18**, 217–27.
34. Onell, A., and Andersson, K. (2005) Kinetic determinations of molecular interactions using Biacore – minimum data requirements for efficient experimental design. *J. Mol. Recognit.* **18**, 307–17.
35. Karlsson, R., and Larsson, A. (2004) Affinity measurement using surface plasmon resonance. *Methods Mol. Biol.* **248**, 389–415.

36. Murphy, M., Jason-Moller, L., and Bruno, J. (2006) Using Biacore to measure the binding kinetics of an antibody-antigen interaction. *Curr. Protoc. Protein Sci. Chapter* **19**, Unit 19 14.
37. Morton, T. A., and Myszka, D. G. (1998) Kinetic analysis of macromolecular interactions using surface plasmon resonance biosensors. *Methods Enzymol.* **295**, 268–94.
38. Myszka, D. G. (2000) Kinetic, equilibrium, and thermodynamic analysis of macromolecular interactions with BIACORE. *Methods Enzymol.* **323**, 325–40.
39. Gomes, P., and Andreu, D. (2002) Direct kinetic assay of interactions between small peptides and immobilized antibodies using a surface plasmon resonance biosensor. *J. Immunol. Methods.* **259**, 217–30.
40. Quinn, J., Patel, P., Fitzpatrick, B., Manning, B., Dillon, P., Daly, S., Okennedy, R., Alcocer, M., Lee, H., Morgan, M., and Lang, K. (1999) The use of regenerable, affinity ligand-based surfaces for immunosensor applications. *Biosens. Bioelectron.* **14**, 587–95.
41. Myszka, D. G. (1999) Improving biosensor analysis. *J. Mol. Recognit.* **12**, 279–84.
42. Navratilova, I., Papalia, G. A., Rich, R. L., Bedinger, D., Brophy, S., Condon, B., Deng, T., Emerick, A. W., Guan, H. W., Hayden, T., Heutmekers, T., Hoorelbeke, B., McCroskey, M. C., Murphy, M. M., Nakagawa, T., Parmeggiani, F., Qin, X., Rebe, S., Tomasevic, N., Tsang, T., Waddell, M. B., Zhang, F. F., Leavitt, S., and Myszka, D. G. (2007) Thermodynamic benchmark study using Biacore technology. *Anal. Biochem.* **364**, 67–77.
43. Karlsson, R., Katsamba, P. S., Nordin, H., Pol, E., and Myszka, D. G. (2006) Analyzing a kinetic titration series using affinity biosensors. *Anal. Biochem.* **349**, 136–47.
44. Dillon, P. P., Killard, A. J., Daly, S. J., Leonard, P., and O'Kennedy, R. (2005) Novel assay format permitting the prolonged use of regeneration-based sensor chip technology. *J. Immunol. Methods.* **296**, 77–82.

Chapter 23

Ultra-Performance Liquid Chromatography–Mass Spectrometry of Proteins

Timothy R. Croley

Abstract

The commercialization of ultra-performance liquid chromatography (UPLC) has allowed more researchers to take advantage of the benefits of this work. Many researchers are exploring this technique to reduce analytical throughput and to increase resolution. The majority of this work has focused on small molecule analysis; however, this technique can provide the same advantages for the analysis of proteins. Traditionally, protein chromatography has suffered from a number of issues such as carryover, peak splitting, peak broadening, and poor peak shape. Because UPLC utilizes a smaller particle at a higher pressure and flow rate, many of these issues are remedied. When used in combination with mass spectrometry, UPLC becomes a powerful tool for protein identification, characterization, and quantitation. In this work we show how UPLC/MS can be used to separate and identify intact proteins.

Key words: UPLC, Intact proteins, Mass spectrometry, Protein identification, Chromatography

1. Introduction

Electrospray ionization was first reported by Fenn et al. (1) in 1989. This work was the first instance of the analysis of large biomolecules from a liquid phase directly into the gas phase. Subsequently, electrospray ionization allowed the combination of high performance liquid chromatography with mass spectrometry (HPLC/MS). HPLC/MS has been shown to be a powerful tool for the characterization of proteins (i.e., proteomics), and the analysis of proteins has grown rapidly in a number of disciplines that were once dominated strictly by biologists and biochemists (e.g., chemistry and physics). For example, two major areas of research in proteomics using mass spectrometry are the MudPIT (2) and top-down (3) approaches. MudPIT, which stands for multidimensional protein identification technology, uses a 2D

Dermot Walls and Sinéad T. Loughran (eds.), *Protein Chromatography: Methods and Protocols,* Methods in Molecular Biology, vol. 681, DOI 10.1007/978-1-60761-913-0_23, © Springer Science+Business Media, LLC 2011

separation to discriminate peptides and proteins for mass spectral identification and typically comparing the results to a database. In top-down proteomics, the intact protein is analyzed by HPLC/MS or electrospray/MS and then the proteins of interest are further dissociated for identification. The application described herein is based on the latter approach, which requires chromatography of the intact protein for characterization and differentiation of pathogens. This report describes the essential parameters for obtaining optimum UPLC/MS for intact proteins.

Reversed-phase chromatography of proteins has been historically difficult where carryover, peak splitting, and broad or misshapen peaks are common. These resolution problems result from two major factors. First, the size of the proteins, compared to peptides or other small molecules, causes them to have slow diffusion times in the pores of the chromatographic column, and, second, the protein interaction with both the mobile and stationary phases can affect the confirmation of the protein within the column. The traditional solution to poor resolution was to increase the run time, which negatively affects the analytical throughput and does not solve other issues. One solution is the use of smaller particles to increase chromatographic resolution and to decrease run time based on the lower plate height minimum in Van Deemter equation. Mathematically, chromatographic resolution (R_s) is defined by the equation

$$R_S = \frac{\sqrt{N}}{4}\left(\frac{\alpha-1}{\alpha}\right)\left(\frac{k}{k+1}\right)$$

where N is separation efficiency, α is selectivity, and k is the retention factor. The separation efficiency is described by

$$N = \frac{L}{H} = \frac{L}{hd_p}$$

where L is the column length, H is the plate height, h is the reduced plate height, and d_p is the diameter of the particle. Since N is proportional to $\sqrt{N}$, R_s is inversely proportional to d_p, and chromatographic resolution can be defined by:

$$R_S \propto N \propto \frac{1}{d_p}$$

This relationship shows the importance of particle size and the role it plays in the resolution in chromatography. When the advantages of small molecules are combined with higher pressures in UPLC, the technique becomes a powerful analytical tool.

Jorgenson et al. introduced ultra-high pressure liquid chromatography for intact proteins (4). This work demonstrated that reducing the column particle size below 2-μm increased resolution while decreasing elution time. Later this same group has shown that the increased pressure also increases the recovery of proteins (5). However, in this work the particle used was nonporous which has a reduced surface area compared to porous particles, which subsequently reduces the loading capacity and limits the application. The loading capacity is especially important when you consider that a 1 μm particle with 10 nm pores has 22× more loading capacity than a 1-μm nonporous particle (6).

The work in our group is dedicated to using UPLC/MS to characterize intact proteins. The speed of UPLC allows the analysis to be completed at times commensurate with MALDI-TOF (7) and we are able to utilize the retention time data for further characterization. In addition, we have been able to observe more proteins at higher masses using this technique which provides more opportunities to identify unique proteins. Briefly, bacterial cells are grown in culture and harvested. These cells are chemically lysed and an extract from the lysis process is analyzed using full-scan UPLC/MS. The resulting protein profile from the bacteria allows differentiation at the strain and substrain level to compliment current biological assays.

2. Materials

2.1. Protein Standards

All materials can be sourced from Sigma-Aldrich and MP Biomedicals unless otherwise indicated. Individual solutions of each protein are prepared in 20 mM NH_4OAc at a concentration of 1 mg/mL. These ten proteins are combined and diluted to 1.5 μM in 50:50 MeOH:H_2O (0.1% v/v formic acid). The working solution is stored at −20°C.

1. Bovine serum albumin (BSA).
2. Horse heart myoglobin.
3. Cytochrome *c*.
4. Chymotrypsinogen A.
5. Ovalbumin.
6. Trypsin inhibitor (soybean).
7. Insulin (bovine pancreas).
8. Ribonuclease A (RNase A, bovine pancreas).
9. α-Lactalbumin (bovine milk).
10. Lysozyme (HEWL).
11. NaCsI (Sodium iodide and cesium iodide solution) (Waters, Milford, MA).

2.2. Solvents

All solvents should be HPLC grade. Organic-free >18 MΩ-cm water should be used for all chromatography. For bacteria preparation sterile water is autoclaved and purified. All solutions are sonicated for approximately 5 min and filtered through a 0.2 μm PTFE filter.

1. 2-Propanol.
2. Acetonitrile (ACN).
3. Methanol.
4. Formic acid.
5. Trifluoroacetic acid (TFA).

2.3. Cell Culture and Lysis

1. Trypticase soy agar plates (5% sheep's blood).
2. Sterile water.
3. Turbidity meter.
4. 1.5 mL Lobind Eppendorf tubes.
5. Centrifuge capable of 6,000 ×*g*.
6. Acetonitrile (ACN).
7. Trifluoracetic acid (TFA).
8. Vortex mixer.
9. 0.22 μm, 4 mm PVDF (polyvinylidene difluoride) low-binding GV filter.

2.4. Chromatography

1. X-bridge C18 BEH 300 Å 150 mm × 2.1 mm, 1.7 μm (Waters, Milford, MA).
2. Acquity UPLC (Waters).

2.5. Mass Spectrometry

1. QTOF-Premier (Waters).

2.6. Data Analysis

1. ProTrawler 6 software (BioAnalyte, Portland, ME).
2. MS Manager software (Advanced Chemistry Development Laboratories, Toronto, ON).

3. Methods

In our work, we are concerned with a wide range of proteins from cell lysates varying in weight from 1 to 120 kDa. To that end, we use a range of protein standards to check instrument performance. Depending on the application, the standards are chosen to bracket the range of proteins of interest. Herein, we describe an examination of general parameters (e.g., temperature, organic modifier, and particle size) for UPLC to obtain sufficient separation for our need. For comparison purposes, HPLC data will be presented that may also be applicable.

3.1. Cell Culture and Lysis

1. *E. coli* O157:H7 cells are grown on trypticase soy agar plates (5% sheep's blood) at 37°C for 24 h.
2. Remove the cells from the plate and place in 1 mL of water until the optical density reaches 1.0 using a turbidity meter.
3. Transfer 500 μL of the suspension into a 1.5 mL Lobind Eppendorf tube (Westbury, NY). Add 500 μL of sterile water to the suspension and centrifuge at room temperature for 5 min at 6,000 × *g*. Repeat twice, discard the supernatants each time.
4. Suspend the pellet in 150 μL of 50:50 H_2O:ACN (0.1% v/v TFA) and vortex mix briefly (approximately 5 s).
5. Centrifuge this solution for 4 min at 4,100 × *g* at room temperature.
6. Filter the supernatant through a 0.22 μm, 4 mm PVDF (polyvinylidene difluoride) low-binding GV filter (Millipore, Billerica, MA) prior to analysis.

3.2. Chromatography

1. Follow the manufacturer's recommendations to maintain column lifetime.
2. The temperature of the column should not exceed 70°C and the backpressure should not exceed 10 kpsi (see Note 1).
3. Perform a gradient elution using A = H_2O (1% formic acid) and B = ACN (1% formic acid), increasing the %B from 0 to 55% over 60 min at 75 μL/min.

3.3. Mass Spectrometry

1. This work was performed on a Waters QTOF-Premier (see Note 2).
2. The mass spectrometric conditions used here are a source temperature of 115°C with a desolvation gas flow of 900 L/h at 500°C.
3. Monitor ions over a mass-to-charge (m/z) range of 620–2,450 Da (or full-scan of this instrument) and in reflectron (V) mode using an acquisition rate of two spectra/sec.
4. Optimize the instrument for both sensitivity and resolution using BSA.
5. Use NaCsI for mass calibration (see Note 3).

3.4. Data Analysis

1. Use ProTrawler 6 software is to perform automated charge state deconvolution across a selected range of the chromatogram. The software generates a text file that contains the

neutral protein mass and other data such as retention time and intensity, which are useful in differentiation of strains.

2. MS Manager software (Advanced Chemistry Development Laboratories, Toronto, ON) is utilized to display the data and remove common masses; however, the removal of common masses can be performed with ProTrawler (see Note 4).

3.5. Chromatographic Optimization

1. An HPLC/MS chromatogram of the standard protein mixture utilizing common LC conditions – ACN at 5% and ramped to 55% over 60 min (see Fig. 1).
2. The column used is a C18 (150 mm × 2.1 mm × 3.5 μm) column with a 300-Å pore size to accommodate the larger molecules. As can be seen in the figure, the resolution between four of the proteins was not adequate.
3. Increased column temperatures provide two benefits: the viscosity of the mobile phase is decreased allowing for higher speeds due to the reduced backpressure and the analyte diffusivity is increased which increases the mass transfer and column efficiency (9). These advantages also hold true for intact proteins (10, 11). Recent work from our group demonstrated that at a constant flow rate increased temperature improved the separation of some proteins; however, this improvement is not universal (7).

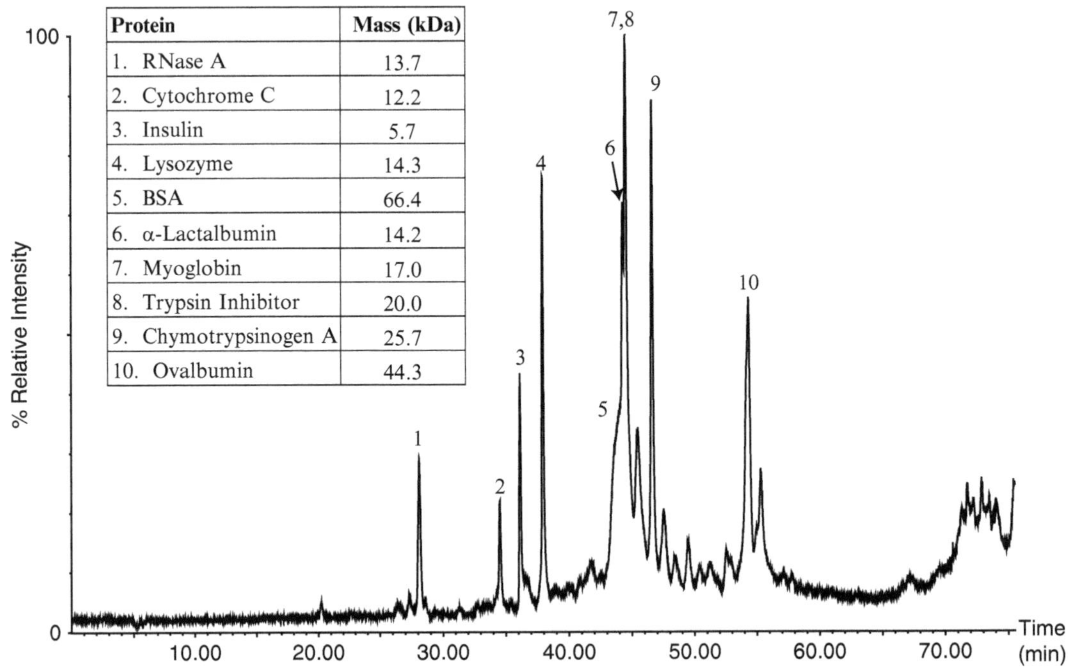

Protein	Mass (kDa)
1. RNase A	13.7
2. Cytochrome C	12.2
3. Insulin	5.7
4. Lysozyme	14.3
5. BSA	66.4
6. α-Lactalbumin	14.2
7. Myoglobin	17.0
8. Trypsin Inhibitor	20.0
9. Chymotrypsinogen A	25.7
10. Ovalbumin	44.3

Fig. 1. HPLC/MS chromatogram of ten protein standards (*inset*) at 1.5 μM using common conditions. The column had 3.5 μm particles at 35°C. A gradient elution was performed using A = H_2O (1% formic acid) and B = ACN (1% formic acid), where the %B was increased from 0 to 55% over 60 min at 75 μL/min. Reprinted from (7) Copyright 2008, with permission from Elsevier.

For example, when the temperature increases from 50 to 65°C co-elution of insulin and lysozyme are observed.

4. Most applications utilize acetonitrile as the organic modifier. This solvent works very well for intact protein separations; however, alternative modifiers such as alcohols (e.g., methanol) should also be considered. In our work we use isopropyl alcohol (IPA) for separation of proteins. IPA provides a number of advantages for intact protein separation. Hydrophobic proteins tend to be more soluble in IPA (12), the elution power is stronger than ACN (7), and the electrospray response with IPA is improved (13, 14). The HPLC/MS analysis comparing ACN (top) to IPA (bottom) displays that the elution time for all proteins is reduced significantly and the benefits of this change when applied to UPLC are clear (see Fig. 2).

3.6. UPLC/MS

1. With UPLC smaller particles are utilized at higher pressures to increase resolution while decreasing carryover and analytical throughput. While the most profound effects of smaller particles are seen at lower flow rates, some of these benefits are still observed at higher flow rates and pressures. These benefits are best observed where the protein mixture is comparatively analyzed by HPLC using a 3.5 μm particle, UPLC

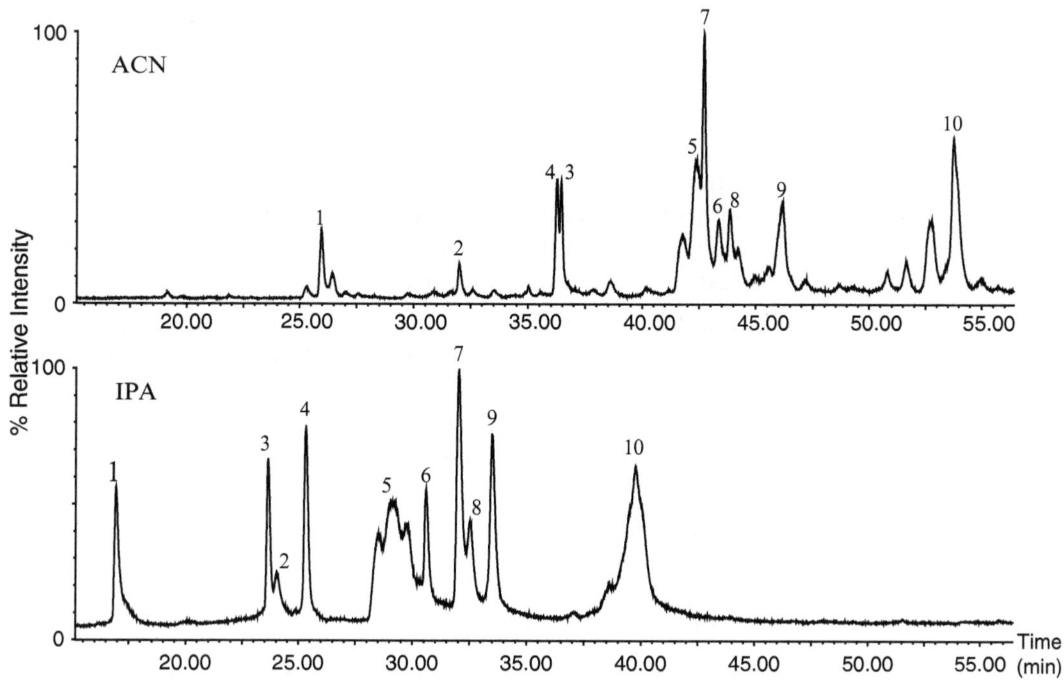

Fig. 2. Under the same experimental conditions, the performance of ACN (*top*) is compared to IPA (*bottom*) for the ten protein standards. As can be seen, the change to IPA reduced the chromatographic time by 15 min. Reprinted from (7) Copyright 2008, with permission from Elsevier.

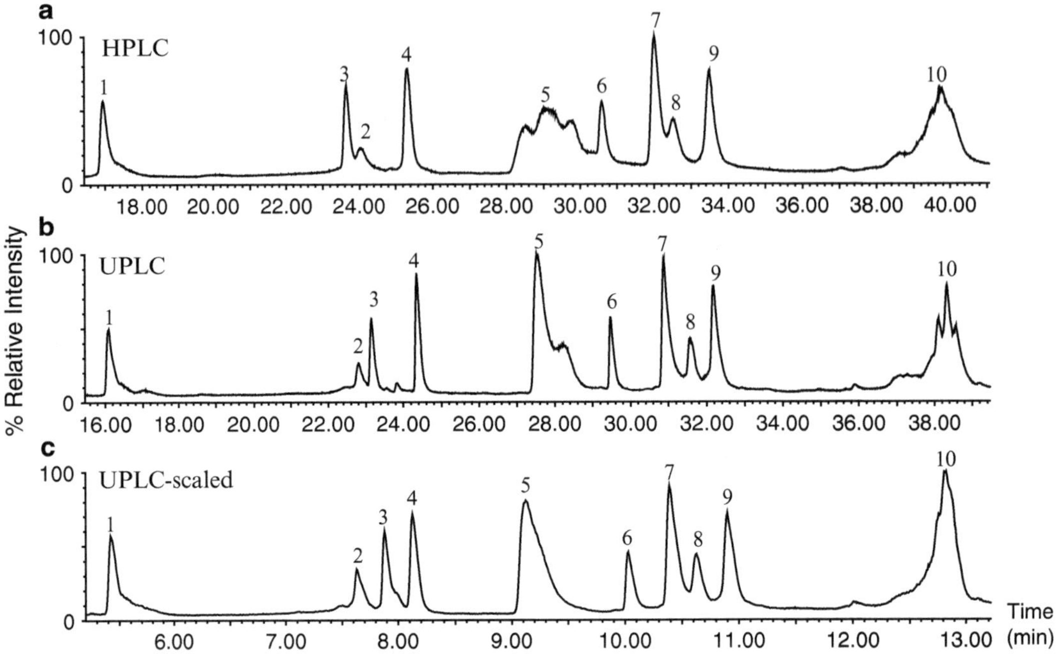

Fig. 3. HPLC/MS compared to UPLC/MS. (**a**) Demonstrates the advantage of IPA. (**b**) Shows the increased resolution achieved by using a smaller particle (1.7 vs. 3.5 μm in (**a**)). (**c**) is the optimized method employing the smaller particle and scaling the gradient from 60 min at 75 μL/min to 20 min at 225 μL/min. A threefold increase in throughput is achieved with no discernable loss in resolution. Reprinted from (7) Copyright 2008, with permission from Elsevier.

using a 1.7 μm particle, and then ultimately when compared to UPLC using a scaled down method (see Fig. 3).

2. In the particle size comparison, the gradient conditions were kept consistent between the two methods. A significant improvement in the separation of no. 2 and no. 3 (cytochrome *c* and insulin, respectively) can be observed as well as increased resolution for no. 5 (BSA) and no. 10 (ovalbumin) just by using the smaller particle.
3. For UPLC/MS analysis, the gradient is reduced from 60 to 20 min and the flow rate increased from 75 to 225 μL/min. Here the separation between no. 2–4 does not suffer greatly and the resolution between proteins no. 5 and no. 10 has decreased; however, there is a threefold reduction in time (see Note 5).

3.7. Application to Cell Lysates

1. When this method is applied to cell lysates the advantages of UPLC vs. HPLC are difficult to discern; however, the overall performance of UPLC is consistent with what was observed with HPLC (see Fig. 4).
2. The mass spectrometric data after processing with ProTrawler (see Note 6) demonstrates equivalent data quality between HPLC and UPLC (see Fig. 4b). Faster chromatography with UPLC allows faster data processing and quicker identification of unique proteins in a lysate.

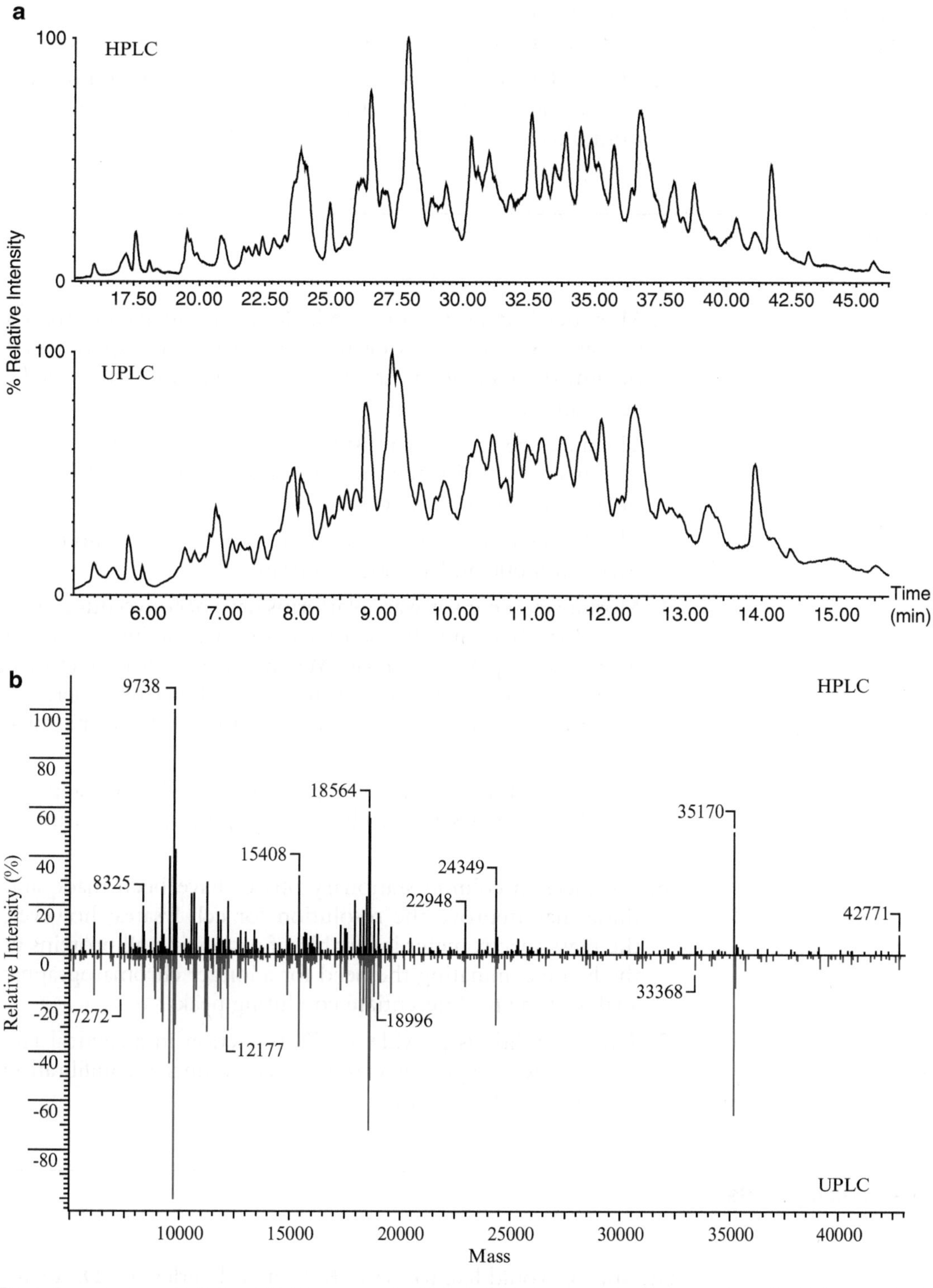

Fig. 4. A comparison of HPLC vs. UPLC for an *E. coli* cell lysate (**a**) and the resulting mass spectrum (**b**). While it is difficult to observe any distinct improvements from the UPLC analysis, examination of the mass spectrum shows that the quality of the mass spectral data remained the same. This allows both a reduction in analytical throughput and in data processing time. Reprinted from (7) Copyright 2008, with permission from Elsevier.

3.8. Optimized Conditions

1. The final optimized conditions for the separation of these ten proteins utilize a column with 1.7 μm particle size at 65°C.
2. The gradient consists of A = H_2O (1% formic acid) and B = IPA (1% formic acid) at 225 μL/min, where the %B increases from 0 to 55% over 20 min (see Note 7).

4. Notes

1. More detailed instructions can be found at the manufacturer's website (8). When reviewing these data and experimental design, the parameters should be optimized for the specific application.
2. The mass spectrometer must be capable of scanning at speeds commensurate with the increased resolution of UPLC (i.e., obtain enough scans across the more narrow UPLC peaks).
3. The target resolution for the ion at *m/z* 1,071 should be between 8,000 and 11,000 (<5 ppm).
4. Most instrument software platforms have deconvolution software for obtaining the neutral mass of proteins generated from electrospray ionization. When dealing with intact proteins, this software should be sufficient. Alternately, there are a number of web-based applications that will also perform this task.
5. ACN provides better throughput; however, for our application IPA provides better chromatographic resolution under UPLC conditions.
6. Advances in column stationary phases have been made and these may improve the resolution for cell lysates; however, the mass spectrometer is capable of processing all proteins in the lysate eliminating the need for a longer chromatographic analysis time to differentiate co-eluting peaks.
7. These conditions provide a 33% reduction in analytical run time, reduce the protein carryover, and reduce reequilibration time after each analysis.

Acknowledgments

The author would like to thank Robert A. Everley, Ph.D. for his contributions to this work. A portion of this work was funded by the Centers for Disease Control and Prevention (CDC) grant # U90/CCU317014.

References

1. Fenn, J.B., Mann, M., Meng, C.K., Wong, S.F., Whitehouse, C.M. (1989) Electrospray ionization for mass spectrometry of large biomolecules. *Science*. **246**, 64–71.
2. Washburn, M.P., Wolters, D., Yates, J.R. III. (2001) Large-scale analysis of the yeast proteome by multidimensional protein identification technology. *Nat. Biotechnol.* **19**, 242–247.
3. Kelleher, N.L., Lin, H.Y., Valaskovic, G.A., Aaserud, D.J., Fridriksson, E.K., McLafferty, F.W. (1999) Top down versus bottom up protein characterization by tandem high-resolution mass spectrometry. *J. Am. Chem. Soc.* **121**, 806–812.
4. MacNair, J.E., Opiteck, G.J., Jorgenson, J.W., Moseley, M.A. III. (1997) Rapid separation and characterization of protein and peptide mixtures using 1.5 µm diameter non-porous silica in packed capillary liquid chromatography/mass spectrometry. *Rapid Commun. Mass Spectrom.* **11**, 1279–1285.
5. Eschelbach, J.W., Jorgenson, J.W. (2006) Improved protein recovery in reversed-phase liquid chromatography by the use of ultrahigh pressures. *Anal. Chem.* **78**, 1697–1706.
6. Mellors, J.S., Jorgenson, J.W. (2004) Use of 1.5 µm porous ethyl-bridged hybrid particles as a stationary-phase support for reversed-phase ultrahigh-pressure liquid chromatography. *Anal. Chem.* **76**, 5441–5450.
7. Everley, R.A., Croley, T.R. (2008) Ultra-performance liquid chromatography/mass spectrometry for intact proteins. *J. Chromatogr. A.* **1192**, 239–247.
8. Waters Corporation (2008) ACQUITY UPLC BEH Columns Care and Use Manual. (Accessed April 28, 2009 at http://www.waters.com/webassets/cms/support/docs/715001371.pdf)
9. Greibrokk, T., Andersen, T. (2003) High-temperature liquid chromatography. *J. Chromatogr. A.* **1000**, 743–755.
10. Wang, Y., Balgley, B.M., Rudnick, P.A., Lee, C.S. (2005) Effects of chromatography conditions on intact protein separations for top–down proteomics. *J. Chromatogr. A.* **1073**, 35–41.
11. Yang, X., Ma, L., Carr, P.W. (2005) High temperature fast chromatography of proteins using a silica-based stationary phase with greatly enhanced low pH stability. *J. Chromatogr. A.* **1079**, 213–220.
12. Schlüter, H. Reversed-Phase Chromatography. In Kastner, M. ed. *Protein Liquid Chromatography*. 1st ed. Amsterdam; New York: Elsevier, 2000: 147–234.
13. Rehder, D.S., Dillon, T.M., Pipes, G.D., Bondarenko, P.V. (2006) Reversed-phase liquid chromatography/mass spectrometry analysis of reduced monoclonal antibodies in pharmaceutics. *J. Chromatogr. A.* **1102**, 164–175.
14. Dillon, T.M., Bondarenko, P.V., Rehder, D.S., Pipes, G.D., Kleeman, G.R., Ricci, M.S. (2006) Optimization of a reversed-phase high-performance liquid chromatography/mass spectrometry method for characterizing recombinant antibody heterogeneity and stability. *J. Chromatogr. A.* **1120**, 112–120.

Chapter 24

Hydrophobic Interaction Chromatography

Philip M. Cummins and Brendan F. O'Connor

Abstract

Most proteins and large polypeptides have hydrophobic regions at their surface. These hydrophobic "patches" are due to the presence of the side chains of hydrophobic or nonpolar amino acids such as phenylalanine, tryptophan, alanine, and methionine. These surface hydrophobic regions are interspersed between more hydrophilic and polar regions, and the number, size, and distribution of them are a specific characteristic of each protein. Hydrophobic interaction chromatography (HIC) is a commonly used technique that exploits these hydrophobic features of proteins as a basis for their separation even in complex biological mixtures (Queiroz et al., J Biotechnol 87:143–159, 2001; Eisenberg and McLachlan, Nature 319:199–203, 1986). In general, the conditions under which HIC is used are relatively mild and "protein friendly" resulting in good biological recoveries. Hydrophobic binding is relatively strong and is maintained even in the presence of chaotropic agents, organic solvents, and detergents. For these reasons, this technique has a widespread use for the purification of proteins and large polypeptides.

Key words: Nonpolar, Hydrophobic amino acids, Phenyl-Sepharose, Butyl-Sepharose, Octyl-Sepharose, Chaotropic agents, Glycerol gradients, Detergents, Biological activity

1. Introduction

In hydrophobic interaction chromatography (HIC), the separation of proteins primarily occurs by the reversible binding of the proteins' hydrophobic surface regions to an interacting nonionic group such as a hydrocarbon tail (octyl-, butyl-, or hexyl-) or an aromatic ring (phenyl-) immobilized to an inert stationary microporous matrix such as cross-linked agarose or sepharose (see Fig. 1).

It is vital to the technique that this support matrix does not carry any ionic charge and it is usually rendered inert by alkaline hydrolysis under reduced conditions. This treatment effectively desulphates the cross-linked agarose/sepharose matrix. Next, an uncharged "spacer arm" (usually a tricarbon hydrocarbon chain)

Dermot Walls and Sinéad T. Loughran (eds.), *Protein Chromatography: Methods and Protocols*, Methods in Molecular Biology, vol. 681, DOI 10.1007/978-1-60761-913-0_24, © Springer Science+Business Media, LLC 2011

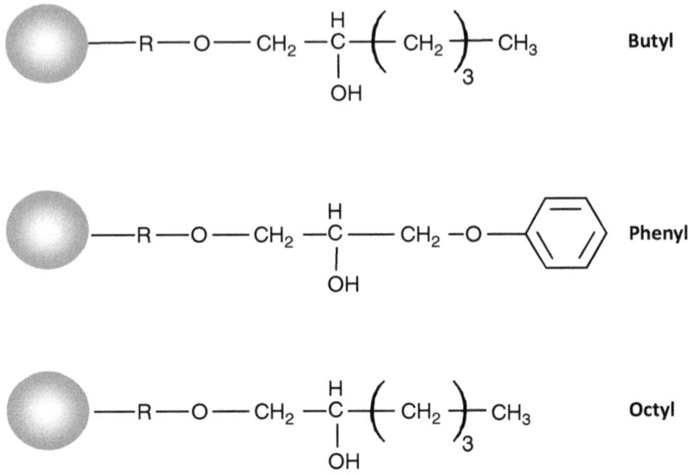

Fig. 1. Diagrammatic representation of butyl-, phenyl-, and octyl-chemical groups covalently linked to sepharose matrix (indicated as *grey spheres*).

is attached to the stable and inert matrix and the hydrophobic groups (e.g. octyl-, butyl-, or phenyl- see Fig. 1) introduced by reaction with the corresponding glycidyl ether. The hydrophobic groups are now attached to the matrix via uncharged, chemically stable ether linkages. This results in a truly "hydrophobic gel" with no ionic properties. These gels are unaffected by chaotropic agents such as urea and guanidine hydrochloride, most organic solvents such as methanol and isopropanol, and detergents such as Triton X-100, sodium dodecyl sulphate (SDS), and sodium deoxycholate. Detergents are routinely used to release and solubilise membrane-bound hydrophobic proteins but then must be kept in the running buffers of any chromatographic columns running to prevent protein aggregation and precipitation.

The ionic properties of both the protein and the running buffer also have an important effect on hydrophobic interaction and binding. A protein with no overall net charge (i.e. where the pH of the running buffer is the same as the p*I* value of the protein) will have maximum hydrophobic properties, whereas a charged protein will be potentially repulsed from the hydrophobic gel to a greater or lesser extent depending on the magnitude of its charge. For example, when the pH is increased, most proteins become negatively charged and therefore become more hydrophilic even under mildly alkaline conditions. The adsorption step often requires the presence of salting-out ions such as sodium chloride, potassium chloride, or ammonium sulphate. These "salting-out" ions decrease the availability of water molecules in solution, thus increasing surface tension and enhancing hydrophobic interaction (1, 2). High salt concentrations may also cause a partial denaturation and unfolding of the proteins with the resultant partially unfolded polypeptide chain(s) often exposing internal hydrophobic regions which will, in turn, enhance hydrophobic interaction and binding. In contrast, low salt concentrations or the use of chaotropic ions will actively reduce or prevent nonionic or hydrophobic interactions. HIC has been also used to separate nucleic acids. This exploits the more hydrophobic nature of single-stranded nucleic acids when compared with double-stranded plasmid DNA (3, 4).

Regarding proteins, HIC is ideal for use after an initial salt precipitation step where the high ionic strength of the sample will greatly enhance hydrophobic adsorption. Also, if the protein of interest is eluted with a positive glycerol and a negative salt gradient, it can then be subjected to ion-exchange chromatography with no need to change either the buffer or pH. In addition, eluting a protein with/into glycerol has been known to stabilize many proteins. Finally, the diversity of potential eluting conditions from the hydrophobic matrices allows for the resolution of even the most complex protein mixtures.

2. Materials

1. Phenyl-Sepharose CL-4B (Pharmacia).
2. Octyl-Sepharose CL-4B (Pharmacia).
3. Loading buffer: 100 mM potassium phosphate buffer pH 7.2.
4. Eluting buffer: 100 mM potassium phosphate buffer pH 7.2 with 2 M ammonium sulphate.
5. Eluting buffer: 100 mM potassium phosphate buffer pH 7.2 with 2 M KCl.
6. Eluting buffer: 100 mM potassium phosphate buffer pH 7.2 with 40% glycerol.
7. Detergent: 1% (w/v) Triton X-100.

3. Methods

All stages should preferably be performed at 4°C:

1. Filter, degas, and dialyze the protein sample into potassium phosphate buffer (see Notes 1 and 2).
2. Pour the appropriate hydrophobic gel (see Note 3) into the column of appropriate dimensions (see Note 4).
3. Equilibrate the chromatography column with four to five column volumes of potassium phosphate buffer with either ammonium sulphate or KCl salts (see Note 5).

4. Load the protein sample onto the top of the column (see Notes 6 and 7).
5. Wash the column with three to four column volumes of potassium phosphate buffer with either ammonium sulphate or KCl salts (see Note 8).
6. Elute the bound protein with a negative gradient of 10–20 column volumes of potassium phosphate buffer (with no salt) and a positive gradient of glycerol in potassium phosphate buffer (see Note 9). Temperature also has an effect elution from hydrophobic gels (see Note 10).
7. Collect 1 ml fractions with a flow-rate in the range of 0.1–0.2 ml/min (see Note 11).
8. Monitor the protein at 280 nm using a standard spectrophotometer (see Note 12).
9. Analyze the fractions for the biological activity of the desired protein by SDS-PAGE.
10. Regenerate the hydrophobic resin by removing tightly bound hydrophobic protein or detergents by washing with ten column volumes of potassium phosphate buffer with either KCl or ammonium sulphate. Ethanol (20–30%) may also be used in this buffer to remove tightly bound detergents.
11. Phenyl- or Octyl-Sepharose gels may be efficiently reused for up to 10–12 separation runs (see Note 13).
12. Store the hydrophobic gels in a potassium phosphate buffer with ethanol and sodium azide (anti-microbial agent) at 4°C (see Note 14).

Figure 2 shows the purification of a prolyl oligopeptidase from the soluble fraction of bovine brain using Phenyl-Sepharose CL-4B (5).

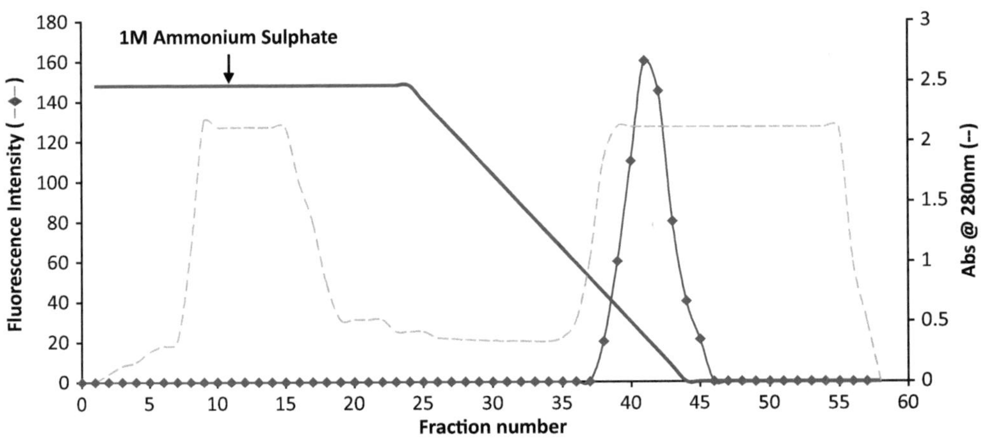

Fig. 2. Purification of a prolyl oligopeptidase from bovine brain using Phenyl-Sepharose (7).

4. Notes

1. All buffers should be filtered and degassed as this prevents the build-up of gases such as CO2 in the column if a pumping system is employed to regulate flow-rate.
2. The starting pH should ideally be between pH 6.5 and 7.5.
3. The choice of hydrophobic gel/ligand should be determined to a large extent by the degree of hydrophobicity of the protein to be purified and/or analyzed. Phenyl-Sepharose is often classified as a weak "hydrophobic gel" and should therefore be used to bind strongly hydrophobic proteins such as membrane-bound enzymes and receptors. This is because of the nature of hydrophobic binding, which is a much stronger type of interaction than ionic binding. Therefore, if a strongly hydrophobic protein was bound to a strongly hydrophobic gel, denaturing conditions would probably have to be employed to effect release/elution. It is of interest to note that pie–pie interactions have been noted to occasionally occur between the aromatic phenyl rings of the gel and the aromatic rings of the side chains of certain amino acids of proteins, and consequently protein binding to Phenyl-Sepharose may not always be purely through hydrophobic interaction. In the case of less hydrophobic proteins not binding/adsorbing to the Phenyl-Sepharose gel, a more hydrophobic matrix, such as Octyl-Sepharose may be used. Octyl-Sepharose is often classified as a strong "hydrophobic gel" and is routinely used to bind weakly hydrophobic proteins. These proteins may not bind until the ionic strength of the running buffer is increased to just below that required for precipitation (of the protein of interest), and the buffer pH adjusted to near the p*I* of the protein of interest, all of which will enhance nonionic or hydrophobic adsorption. The binding capacity of both Phenyl-Sepharose and Octyl-Sepharose is reported by Pharmacia to be around 20 mg of protein per millilitre of gel. This compares very favourably with the binding capacities of most ion-exchange matrices.
4. The optimal column dimensions should favour larger cross-sectional areas, i.e. larger radius with smaller column heights. This allows for a better "concentration" of the protein sample during the binding phase.
5. Ammonium sulphate salt is a stronger salt than KCl and may be used for less hydrophobic proteins as it will cause a greater degree of protein unfolding thus promoting more hydrophobic binding.

6. Protein concentration should be ideally in the 1–5 mg/ml concentration range so as not to exceed the binding capacity of the hydrophobic gel.
7. Sample volumes may be larger than the column volume as this is a binding chromatography and thus effectively "concentrates" the sample.
8. This wash should remove any unbound or loosely bound protein. A low concentration of detergent (0.1% w/v Triton X-100) may be used in this wash if using a membrane-released sample which will have a complex mix of proteins.
9. Negative salt gradients are routinely used to elute proteins from hydrophobic interaction columns. These conditions tend to reduce overall hydrophobic conditions and thus favour elution of the bound protein. Gradients of chaotropic ions, such as urea and guanidine HCl, are rarely used to elute proteins from hydrophobic gels as they often denature the protein with a resultant loss of biological activity. Positive gradients of polarity-reducing agents such as glycerol or the water-miscible solvents such as methanol or ethanol are routinely used to elute bound proteins from hydrophobic gels. A combination of a positive gradient of a polarity-reducing substance such as glycerol with a negative gradient of a "strongly ionic" salt such as ammonium sulphate is one of the commonly used methods for the elution of bound proteins during HIC. However, the water-miscible solvents are rarely used for elution, as they may cause denaturation of proteins with subsequent loss of biological activity. Detergents (especially the milder nonionic detergents such as Triton X-100), while not affecting the hydrophobic ligand/gel itself, do reduce the degree of hydrophobic binding and may therefore also be used for the elution of bound material. The concentrations of these polarity-reducing agents typically used are: glycerol (0–40%), methanol (0–30%), ethanol (0–30%), and ethylene glycol (0–80%). All are routinely used to elute bound proteins from hydrophobic gels.
10. Temperature also has an effect on hydrophobic interaction. Lower temperatures reduce hydrophobic binding between two nonionic groups. This is because the free energy of the interaction becomes more positive with decreasing temperature (6), thus making the interaction less likely to occur spontaneously. However, this reduction in the strength of hydrophobic interaction at lower temperatures is not routinely used as a means for effecting elution of bound material. There is up to a 20–30% reduction in the strength of hydrophobic binding when the temperature is reduced from 25 to 4°C. The strength of hydrophobic interaction increases with temperature up to about 60°C at which point the additional

stability provided by electrostatic and Van der Waal's forces disappears. Above this temperature, the strength of hydrophobic binding falls off rapidly.

11. The flow-rate has to be slow enough to allow for hydrophobic binding to occur efficiently as it is a thermodynamically "slow" reaction.
12. If glycerol is used in the elution buffer then other analytical methods for monitoring protein must be employed as glycerol interferes directly with absorbance at 280 nm.
13. After a certain number of separation and regeneration runs, the binding capacity of the gels becomes reduced especially if detergents are used in the chromatography.
14. Most of the hydrophobic matrices are completely reusable, and may be regenerated by firstly removing any tightly bound protein with 6 M urea and then subjecting them to repeated washing with starting buffer (with salt). Other cleaning methods involve washing with strong bases such as 1 M sodium hydroxide. Detergents may be removed by sequential ethanol, butan-1-ol, ethanol, and distilled water washes. For long-term storage, the hydrophobic gels may be suspended in 20% (v/v) ethanol and 0.05% sodium azide and stored at 4°C.

Acknowledgements

The authors would like to thank Dr. Oonagh Dowling for her work on the purification of the prolyl oligopeptidase. This work was supported by the Health Research Board (HRB) and Science Foundation Ireland (SFI).

References

1. Porath, J. (1986) Salt-promoted adsorption: recent developments. *J Chromatogr* **376**, 331–341.
2. Porath, J. (1990) Salt-promoted adsorption chromatography. *J Chromatogr* **510**, 47–48.
3. Diogo, M.M., Queiroz, J.A., Monteiro, G.A. and Prazeres, D.M.F. (1999) Separation and analysis of plasmid denatured forms using hydrophobic interaction chromatography. *Anal Biochem* **275**, 122–124.
4. Diogo, M.M., Queiroz, J.A., Monteiro, G.A., Martins, S.A.M., Ferreira, G.N.M. and Prazeres, D.M.F. (2000) Purification of a cystic fibrosis plasmid vector for gene therapy using hydrophobic interaction chromatography. *Biotechnol Bioeng* **68**, 576–583.
5. Dowling, O. (1998) The purification and characterization of a prolyl oligopeptidase from the cytosolic fraction of bovine brain. *Ph.D. Thesis*, Dublin City University, Ireland.
6. Ochoa, J.L. (1978) Hydrophobic interaction chromatography. *Biochimie* **60**, 1–15.

Chapter 25

Fast Protein Liquid Chromatography

Ashkan Madadlou, Siobhan O'Sullivan, and David Sheehan

Abstract

Fast protein liquid chromatography (FPLC) is a form of high-performance chromatography that takes advantage of high resolution made possible by small-diameter stationary phases. It was originally developed for proteins and features high loading capacity, biocompatible aqueous buffer systems, fast flow rates, and availability of stationary phases in most common chromatography modes (e.g., ion exchange, gel filtration, reversed phase, and affinity). The system makes reproducible separation possible by incorporating a high level of automation including autosamplers, gradient program control, and peak collection. In addition to proteins, the method is applicable to other kinds of biological samples including oligonucleotides and plasmids. The most common type of FPLC experiment is anion exchange of proteins. This chapter describes such an experiment carried out using an ÄKTA FPLC explorer system (Amersham Pharmacia Biotech, Sweden).

Key words: FPLC, Protein, Purification, Chromatography, Ion exchange

1. Introduction

High-resolution chromatography critically depends on the availability of column packings of small average particle size (1, 2). This gives a minimum of peak broadening on the column owing to the direct relationship between the theoretical plate height parameter, H, and particle size (lowest values of H give highest resolution). High-performance liquid chromatography (HPLC) procedures exploit column packings with average diameters as small as 5–40 μm. However, these require high-pressure systems (up to 400 bar) often with organic solvents and are generally limited to rather low sample loadings (3). To provide a more biocompatible high-resolution separation of biopolymers, including (although not exclusive to) proteins, Pharmacia LKB (Uppsala, Sweden) developed fast protein liquid chromatography (FPLC)

Dermot Walls and Sinéad T. Loughran (eds.), *Protein Chromatography: Methods and Protocols*, Methods in Molecular Biology, vol. 681, DOI 10.1007/978-1-60761-913-0_25, © Springer Science+Business Media, LLC 2011

in 1982 (4). ÄKTA FPLC (the most recent system produced by what is now Amersham Pharmacia Biotech, Sweden) is a fully automated liquid chromatography system designed for research-scale protein purification. In FPLC, the cost per test can be nearly 30 times cheaper when compared with HPLC. Also, the cost of the FPLC column is approximately ten times less than that of a corresponding HPLC column (5). Accelerated purification of RNA polymerase and separation of RNA polymerase II from RNA polymerase III with good yield by ÄKTA FPLC have been reported (6).

FPLC provides the full range of chromatography modes, such as ion exchange (7–9), chromatofocusing (8), gel filtration (10), hydrophobic interaction (9), and reverse phase (11), based on particles with average diameter sizes in a similar range as those used for HPLC separations. These columns can accommodate much higher protein loadings than conventional HPLC, however, and use a wide range of aqueous, biocompatible buffer systems. Although thousands of reports of the use of FPLC have appeared in the literature in the past, two of the most popular FPLC modes are anion exchange and gel filtration. Figure 1 demonstrates the most commonly used configuration for ion-exchange FPLC.

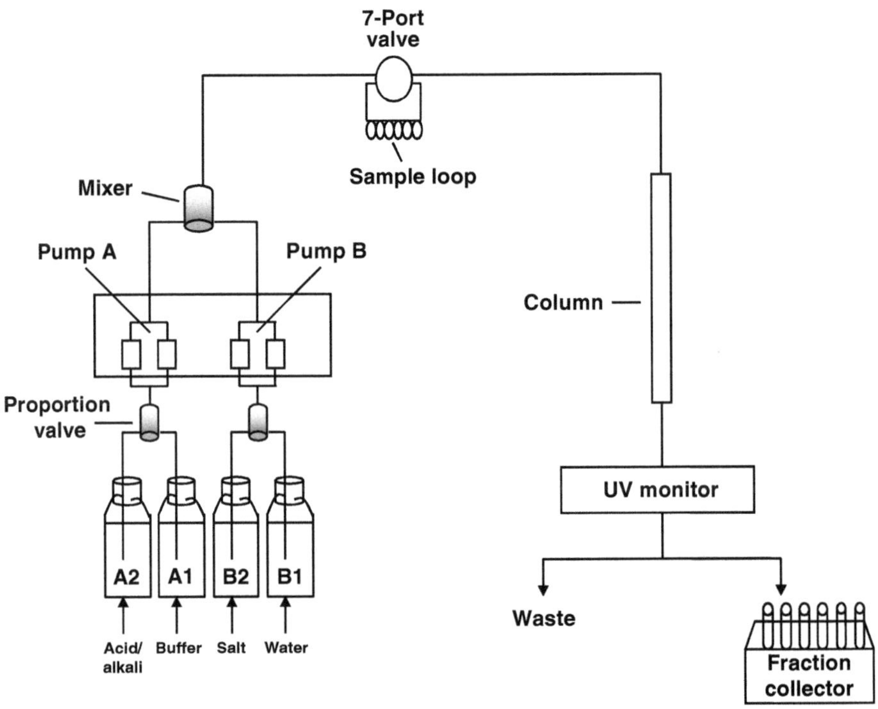

Fig. 1. Schematic outline of ÄKTA explorer FPLC system. Accessories and other information are available (12).

2. Materials

1. The FPLC system (ÄKTA FPLC explorer; Amersham Pharmacia Biotech AB, Uppsala, Sweden) is the instrument used here. This typically consists of a program controller, up to four P-900 pumps (two each for buffers A and B, see Fig. 1), a mixer (M-925), a prefilter, a seven-port M-7 valve, an autosampler or loading loops, a column, a UV-900 UV monitor, a flow restrictor, and a fraction collector. The system contains a compact separation unit including modules and components, and a personal computer running UNICORN™ software version 3.2 or higher to control the separation unit. A fraction collector (Frac-900 or Frac-950) is also included. Flow rates of up to 20 mL/min and pressures up to 5 MPa are possible. In its standard configuration, the sample is applied by using a sample loop. Superloops and a sample pump are available as optional components for the application of larger sample volumes. After loading the sample into the loop, gradients are mixed using two separate buffers connected to the A and B pump modules of P-920. The output flow from the pump is routed to Mixer M-925. The mixer is supplied with a 0.6 mL mixer chamber. It can be used at all flow rates up to 20 mL/min (12). During pump priming, the column is replaced by plastic tubing.

   All reagents for buffer preparation (see Notes 1 and 2) are of Analar grade. Prepare buffers in HPLC or milli-Q-grade water, and filter and degas them immediately before use.
2. Buffer A: 10 mM Tris–HCl, pH 7.0. Filter (0.22 μm filter) and degas.
3. Buffer B: 10 mM Tris–HCl, pH 7.0, 1 M NaCl. Filter (0.22 μm filter) and degas.
4. Buffer C: 50 mM Tris–HCl, pH 7.0, 100 mM KCl.

3. Methods

3.1. Sample Preparation

Generally, FPLC chromatography is performed on material that has already been subjected to some preliminary chromatography steps. For anion exchange, the sample must be desalted in buffer A (see Note 3). This may be accomplished by passing it through a column of Sephadex G-25 (45 × 3 cm in diameter, total bed volume 318 mL) (GE Healthcare, Little Chalfont, Bucks, UK):

1. Define Sephadex G-25 resin by aspiration before packing as a slurry in water into a glass column with a no. 1 glass sinter.

2. Equilibrate the column in three- to four-column volumes of buffer A. Equilibration is complete when the pH and conductivity of the eluate are the same as buffer A at the same temperature. Load the sample (50 mL) under gravity flow. Pass buffer A through the column. Monitor the protein by measuring A_{280}, and collect as a single large peak.
3. Measure the conductivity with a conductivity meter. The conductivity and pH of desalted protein should be the same as that of buffer A at the same temperature.

Sephadex G-25 gel filtration is performed at 4°C in a cold room or cabinet (see Note 4). The sample is centrifuged for 3–5 min before application to the column.

The first step in FPLC separation is the binding of soluble proteins onto the chromatography medium. After that, unbound proteins are washed out with urea-containing eluent buffer A. The separation of the proteins is carried out by gradient elution. This involves a gradual increase in counter-ion concentration (i.e., chloride) by increasing the percentage of eluent buffer B. Within this elution procedure, the proteins are eluted according to their effective net charge in the buffer solution. The elution time of the various proteins depends on their relative charge differences. The less negatively charged proteins are eluted at rather low salt concentrations, while the highly charged proteins require much higher salt concentrations. They are then desalted after chromatographic separation of the fractions (7).

3.2. FPLC Modes

3.2.1. Ion-Exchange FPLC

Operate the FPLC system at room temperature, taking care to hold samples on ice and return eluted protein fractions to ice as soon as possible (see Note 4):

1. With no column in the system (this may be replaced with some plastic tubing), prime P-900 pumps A and B with filtered (0.22 μm filter) and degassed buffers A and B, respectively (see Note 5).
2. Set the pressure limits on both P-500 pumps well below the maximum for the column in use (see Notes 6 and 7).
3. Equilibrate the Mono Q column (1 mL volume) with five volumes of buffer A and ten volumes of buffer B followed by five volumes of buffer A (see Notes 8 and 9).
4. Wash the loading loop with buffer A.
5. Load the sample (0.5–10 mL) (approximately 1 mg/mL), and wash with buffer A. Collect the wash-through and assay for the protein of interest.
6. If protein has not bound, immediately replace the Mono Q column with the Mono S column and repeat the procedure described in steps 3–5 (generally at pH 7.0, proteins that do not

bind to one resin will bind to the other). If protein still does not bind, follow the procedures given in Subheading 3.2.2.

7. The Mono Q (or Mono S for cation exchange) column is developed with a gradually increasing concentration of buffer B. This usually involves a linear gradient (e.g., 0–100%) of buffer B. The increase in concentration of chloride counter-ion (sodium counter-ion for Mono S) allows selective exchange for individual proteins at different points in the gradient. Elution of protein is determined by monitoring A_{280} while the protein of interest is assayed.
8. Regenerate the column by washing with ten volumes of buffer B, followed by five volumes of buffer A. The column is now ready to receive another sample (see Note 10).

3.2.2. Scouting FPLC Methods

The aforementioned procedure describes a basic FPLC ion-exchange chromatography experiment (see Fig. 2 for a typical separation). If the protein of interest has not previously been purified using FPLC, a new method of purification may be needed. Usually, 25–500 μL loadings are used to scout a new procedure. The general approach used is as follows:

1. By varying 0% B at different time-points, create gradients of varying degrees of shallowness (down to and including "isocratic," i.e., constant % B). Knowing the % B at which the protein of interest elutes, separation from other proteins can be improved by varying the gradient program around this value (13).
2. Having identified a column to which the protein of interest binds, carry out chromatography at different pH values using various buffer systems (see Table 1; see Notes 1 and 2).

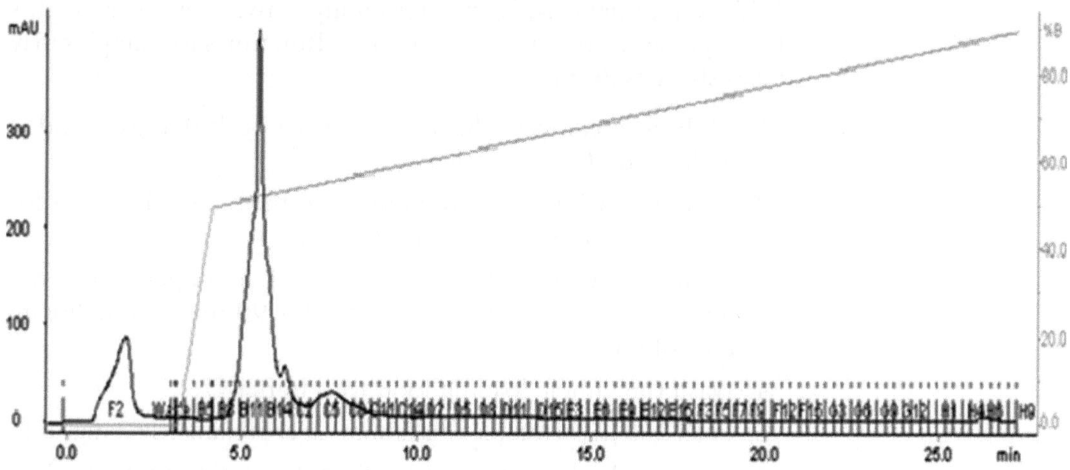

Fig. 2. Cation exchange of casein peptides on Mono Q anion exchange.

Table 1
Some buffers suitable for ion-exchange FPLC

pH range	Buffer A	Column
3.8–4.3	50 mM Na formate–formic acid	Mono S
4.5–5.2	20 mM Na acetate–acetic acid	Mono S
5.5–6.0	20 mM Histidine–HCl	Mono Q
6.0–7.6	20 mM Na phosphate	Mono S/Mono Q
7.6–7.8	50 mM HEPES	Mono S
7.0–8.0	10 mM Tris–HCl	Mono Q
8.0–9.0	50 mM Tris–HCl	Mono Q
9.0–9.5	20 mM Ethanolamine–HCl	Mono Q

By carrying out rechromatography of a single sample at a number of different pH values on a column, this can often separate two proteins that may elute close to each other at a single pH.

3. Having identified a suitable buffer and gradient program for optimum separation of the protein of interest, the 500 μL sample loading loop may be replaced with the superloop. This allows the loading of up to 10 mL sample volumes (20 mg protein) on a column. Fractions may be collected for further analysis.
4. Final adjustments to the gradient may be required for the effect of scaling up the loading.

3.3. Gel Filtration FPLC

Gel filtration is often used as a final "polishing" step to remove minor contaminants from partially purified protein (see Note 11). It is also especially useful in determining native molecular weights for oligomeric proteins (see Note 12). Run times for samples may be as short as 30 min:

1. With no column in the system, prime both pumps A and B with buffer C (see Note 5).
2. Set pressure limits on both pumps to the same value (that for the column in use; see Notes 6 and 7).
3. Connect the Superdex 200 10/300GL and operate the system at a flow rate of 0.5 mL/min for 90 min to equilibrate the column.
4. Wash the sample loading loop with buffer C.
5. Load 200 μL of sample (maximum) at approximately 2 mg/mL.
6. Monitor A_{280} as before and collect the eluting protein. Note the elution volumes at which the protein peaks appear.

3.4. Automated AKTA FPLC

1. Fill the sample loop either partially or completely with sample. Do the former when high recovery is required and the sample volume loaded should be at a maximum of 50% of the loop volume. In the latter case, for analytical reproducibility, a sample volume five times the volume of the sample loop should be used.
2. Use two separate buffers connected to the A and B pump P-920 modules for mixing gradients. Route the output flow from the pump to Mixer M-925.
3. AKTA FPLC has been used for the preliminary purification of bioactive peptides obtained through microbial or enzymatic hydrolysis. In a typical procedure, bioactive solutions are first centrifuged, the supernatants obtained are then ultrafiltered using a cellulose 3 kDa cutoff membrane (e.g., Amicon Ultra, and Millipore) by centrifugation at 4,000×*g* for 30 min, and the permeates thus obtained are used for peptide separation chromatography. Typically, 1 mL of permeate is applied to a quaternary ammonium anion-exchange column with 15 μm particle size and polystyrene/divinylbenzene matrix (RESOURCE Q; GE Healthcare) connected to the ÄKTA FPLC system. Peptides are eluted at 5 mL/min with a gradient of buffer B in buffer A from 0 to 90% B in 20 min. The column is equilibrated for 20 min with buffer A before each chromatographic run. The eluate is monitored at 280 nm and successive 250 μL fractions collected. Peak fractions are then combined together, desalted using a HiTrap desalting column (GE Healthcare) and dried under vacuum.

4. Notes

1. The purpose of using a fairly neutral pH to begin with is to allow the assessment of both anion- and cation-exchange columns at the same pH value. It is much quicker to swap columns than to reprime the entire system with new buffers at different pH values.
2. Although it is possible to carry out ion-exchange FPLC across a wide pH range, extremes of pH should be avoided if possible unless it is known that the protein of interest is stable at these pH values.
3. If the protein of interest is stable, then desalting may also be achieved by dialysis against three changes of 1 L buffer A (minimum of 6 h/change). Centrifuge samples (10,000×*g*, 30 min) after dialysis to remove any precipitated material. If the protein is particularly unstable then the rapid desalting

HR 10/10 column may be included in the system before the ion-exchange column shown in Fig. 1. This allows for desalting of samples in as little as 4 min.

4. Operation of the FPLC system in a cold room at 4°C may be essential for the purification of some particularly labile proteins. However, permanent location in the cold may lead to deterioration of the system. A convenient approach to this problem is to mount the system on a trolley with the components electrically connected to an extension lead on the trolley. This may be wheeled into the cold room, the extension lead connected to a power point, and the system operated once it has cooled to the temperature of the room. After use, the system may be easily wheeled out of the cold room again.
5. Prime pumps with water to remove 20% ethanol in which the system is stored. This avoids any risk of NaCl precipitation on contact of buffer B with 20% ethanol.
6. Flow rates of 1 mL/min are routinely achievable with newly purchased ion-exchange columns, but even with rigorous cleaning, back-pressure in the system soon rises appreciably after a number of uses. It is often better to accept a comparatively slow flow rate of 0.5 mL/min and modest back-pressures, since this seems to give better chromatography and to extend column life-times.
7. Using decreased flow rates usually improves the resolution of gel-filtration chromatography.
8. It is sometimes desirable to operate the system (gel filtration or ion exchange) in the presence of urea. It is not advisable to use high concentrations (e.g., 6–8 M) of urea though, because urea may easily precipitate, clogging valves, and leading to excessive wear on seals and gaskets. Urea is rarely used at concentrations higher than 6 M, and special care has to be taken if the system is being operated at 4°C. It is essential when chromatography in the presence of urea is complete, that the entire system is washed extensively with water before storing it in 20% ethanol, otherwise the urea will precipitate on contact with the ethanol solution. Also, it is important not to allow air (e.g., air bubbles) to come into contact with buffers containing urea, since this will lead to crystal formation, and once formed such crystals are very difficult to remove.
9. It is sometimes advisable to rechromatograph peaks of interest. This is achieved by rapidly desalting the peak into buffer A and reapplying it to the column. This often produces a considerable improvement in purification, since contaminants "move" to one side of the chromatogram away from the peak of interest.

10. When FPLC is complete, wash the system with water. Fill the pumps with 20% filtered, degassed ethanol, and wash the loading loop with 20% ethanol.
11. Although gel filtration is often used toward the end of a purification procedure, it may be advantageous to use it at an early stage in the particular case of immunoglobulin purification from serum (14).
12. Calibration of gel filtration columns should always be performed whenever a different chromatography buffer is used. A quick way to achieve this is to load standard proteins with widely separated molecular masses (e.g., cytochrome c and ovalbumin) in pairs (i.e., as a single sample).

References

1. Liu H., Finch, J.W., Lavallee M.J., Collamati, R.A., Benevides C.C., and Gebler, J.C. (2007) Effects of column length, particle size and flow rate on peak capacity of nano-scale liquid chromatography for peptide separations. *J. Chromatogr. A* **1147**, 30–36.
2. Sheehan D. (2009) Chromatography in Physical Biochemistry: Principles and Applications. Wiley-Blackwell: Chichester, UK, second edition, 11–51.
3. Cabrera K. (2004) Applications of silica-based monolithic HPLC columns. *J. Sep. Sci.* **27**, 843–852.
4. Richey J. (1982) FPLC: a comprehensive separation technique for biopolymers. *Am. Lab.* **14**, 104–129.
5. Tangvarasittichai S., Tangvarasittichai O., and Jermnim N. (2009) Comparison of fast protein liquid chromatography (FPLC) with HPLC, electrophoresis & microcolumn chromatography techniques for the diagnosis of β-thalassaemia. *Indian J. Med. Res.* **129**, 242–248.
6. Glanc P., Galbas M., Dullin P., Szalata M., and Slomski R. (2003) The application of FPLC chromatography (Mono Q) for the purification of wheat RNA polymerase II and III. *Cell. Mol. Biol. Lett.* **8**, 825–830.
7. Plank J., Andres P.R., Krause I., and Winter C. (2008) Gram scale separation of casein proteins from whole casein on a Source 30Q anion-exchange resin column utilizing fast protein liquid chromatography (FPLC). *Protein Expr. Purif.* **60**, 176–181.
8. BoeszeBattaglia K., Kong F.S., Lamba O.P., Stefano F.P., and Williams D.S. (1997) Purification and light-dependent phosphorylation of a candidate fusion protein, the photoreceptor cell peripherin/rds. *Biochemistry* **36**, 6835–6846.
9. Riaz M., Perveen R., Javed M.R., Nadeem H., and Rashid M.H. (2007) Kinetic and thermodynamic properties of novel glucoamylase from *Humicola* sp. *Enzyme Microb. Technol.* **41**, 558–564.
10. Miyazaki O., Fukamachi I., Mori A., Hashimoto H., Kawashiri M.A., Nohara A., Noguchi T., Inazu A., Yamagishi M., Mabuchi H., and Kobayashi J. (2009) Formation of pre beta 1-HDL during lipolysis of triglyceride-rich lipoprotein. *Biochem. Biophys. Res. Commun.* **379**, 55–59.
11. Moon P.G., Hwang H.H., Boo Y.C., Kwon J., Cho J.Y., and Baek M.C. (2008) Identification of rat urinary glycoproteome captured by three lectins using gel and LC-based proteomics. *Electrophoresis* **29**, 4324–4331.
12. ÄKTA FPLC system manual (2000) Amersham Pharmacia Biotech AB, Uppsala, Sweden.
13. Fitzpatrick, P.J., and Sheehan, D. (1993) Separation of multiple forms of glutathione S-transferase from the blue mussel, *Mytilus edulis*. *Xenobiotica* **23**, 851–862.
14. Havarstein, L.S., Aasjord, P.M., Ness, S., and Endresen, G. (1988) Purification and partial characterisation of an IgM-like serum immunoglobulin from Atlantic salmon (*Salmo salar*). *Dev. Comp. Immunol.* **12**, 773–785.

Chapter 26

Shotgun Proteomics: A Qualitative Approach Applying Isoelectric Focusing on Immobilized pH Gradient and LC-MS/MS

Laurent Geiser, Ali R. Vaezzadeh, Jacques M.P. Deshusses, and Denis F. Hochstrasser

Abstract

Shotgun proteomics is a rapid and near universal strategy to identify proteins in complex protein mixtures. After protein digestion, the resulting peptide mixture is submitted to two orthogonal techniques: peptides are first separated according to their isoelectric point (p*I*) by isoelectric focusing (IEF) on immobilized pH gradient (IPG); after peptide extraction, they are then separated in the second dimension according to their hydrophobic properties by reverse phase liquid chromatography (RPLC). Finally, they are detected by tandem mass spectrometry (MS/MS) and proteins are matched by means of bioinformatics software.

Key words: Protein identification, Peptides, Isoelectric focusing, Liquid chromatography, Tandem mass spectrometry

1. Introduction

"Shotgun proteomics" refers to the field of bulk proteome digestion followed by multidimensional separation (1–3). Among the advantages of shotgun proteomics over more traditional gel-based proteomics methods, the following can be mentioned: simplified sample handling, more exhaustive digestion due to the ease of applying chaotropic agents, avoidance of sample loss in the gel matrix, and increased speed and throughput due to automated data acquisition.

As for the widely used gel-based proteomics, this approach is a near universal one: a wide variety of samples can be analyzed using shotgun proteomics including biofluids (e.g., blood, urine),

Dermot Walls and Sinéad T. Loughran (eds.), *Protein Chromatography: Methods and Protocols*, Methods in Molecular Biology, vol. 681, DOI 10.1007/978-1-60761-913-0_26, © Springer Science+Business Media, LLC 2011

tissues, cells, organelles, and protein complexes. Because of the great dynamic range of protein abundance and the wide range of protein properties present in a protein population, it is advantageous to work with purified organelles or other enriched subproteomes to reduce the complexity of the system.

In shotgun proteomics, a protein population is prepared from a biological source. The integrated workflow is outlined in Fig. 1: (1) enzymatic or chemical digestion to convert proteins into peptides and to generate a complex mixture of peptides: trypsin is typically the enzyme of choice because it is a stable, cost-effective, and relatively specific protease; (2) the peptides generated are subsequently separated using an isoelectric focusing technique; (3) before being analyzed by reverse phase liquid chromatography (RPLC) coupled to tandem mass spectrometry (MS/MS); (4) finally, the resulting peptide sequence data generated from MS/MS spectra are compared through a battery of sophisticated mathematical algorithms against protein databases to identify proteins present in the original population.

Steps (1) and (2), i.e., protein digestion and peptide separation in the first dimension by isoelectric focusing, are carried out according to our protocols, with an IPGphor instrument (GE

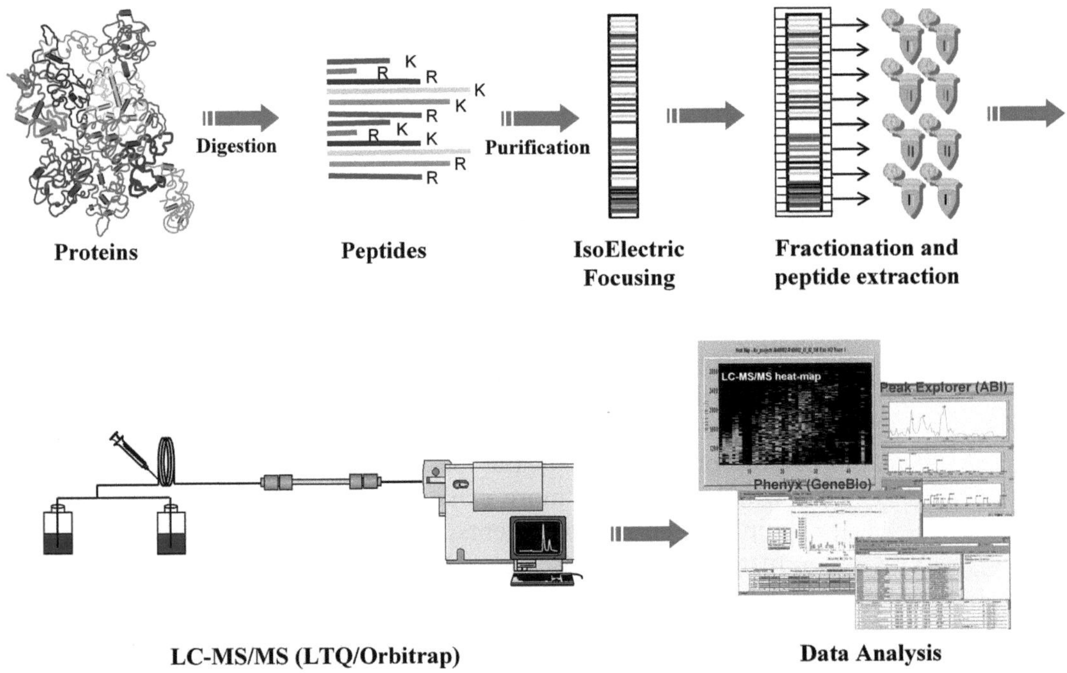

Fig. 1. Shotgun IPG-IEF workflow. Samples are first digested and then separated on an IPG strip by Isoelectric Focusing. The strip is then cut in several fractions and peptides are extracted. The second dimension of separation consists of LC from which the peptides are eluted and sprayed directly into LTQ Oribtrap for analysis. The data obtained are then used for protein identification and visualization.

Healthcare) (see Note 1). On the other hand, steps (3) and (4), i.e., LC-MS/MS analysis and database searching, rely exclusively on the instrumentation and identification software at the disposal of the individual laboratory (see Note 2). For instance, LC-MS/MS analysis can be performed online or off-line, peptides can be ionized either by electrospray ionization (ESI) or by matrix-assisted laser desorption/ionization (MALDI) in conjunction with various tandem mass analyzers (e.g., quadripole, time-of-flight), and database searching can be performed using many bioinformatics tools such as Phenyx software (from GeneBio) and/or Mascot (from Matrix Sciences). Previously mentioned techniques have been tested, but only protocols for one instrumental configuration and identification software are presented in this chapter.

1.1. LC-MS

Currently, most liquid chromatography (LC) separations in proteomics are performed in reverse phase high performance LC mode, because of its compatibility with MS. Proteins or peptides are separated by LC according to their hydrophobic properties and can be introduced directly into the mass spectrometer for identification and analysis. The mobile phase in RPLC normally is a mixture of high-purity water with a miscible organic solvent such as acetonitrile (ACN). Acid (formic, acetic, or trifluoroacetic) is added to the mobile phase to render proteins and peptides positively charged, denatured, and to reduce unwanted ionic interactions with the stationary phase.

1.2. Immobilized pH Gradient-Isoelectric Focusing

Given the potential complexity of peptide mixtures, sample fractionation prior to LC-MS/MS should be highly beneficial. The concept of using IEF, where molecules are separated on the basis of their p*I*, has been widely applied to the separation of protein mixtures not only in the first dimension of separation in 2-dimensional electrophoresis (2-DE) experiments, but also for preparative purifications of proteins in both liquid (4) and gel media (5). More recently, IEF platforms similar to those used for protein mixtures have been used for the analysis of tryptic peptides (6). Peptides are thus separated by their p*I* in the first dimension and by their retention time in LC for the second dimension.

Shotgun IPG-IEF (Immobilized pH gradient-isoelectric focusing) has been applied to a large range of biological materials (e.g., (7, 8)). In order to generate a fractionation scheme resulting in an almost even distribution of peptides per fraction with an IPGphor instrument, our group developed a software tool named "pICarver" (see Note 3). It was shown that this tool increased the throughput of the approach by reducing the number of fractions and merging the peptide-poor regions without any loss of information (9) (Fig. 2).

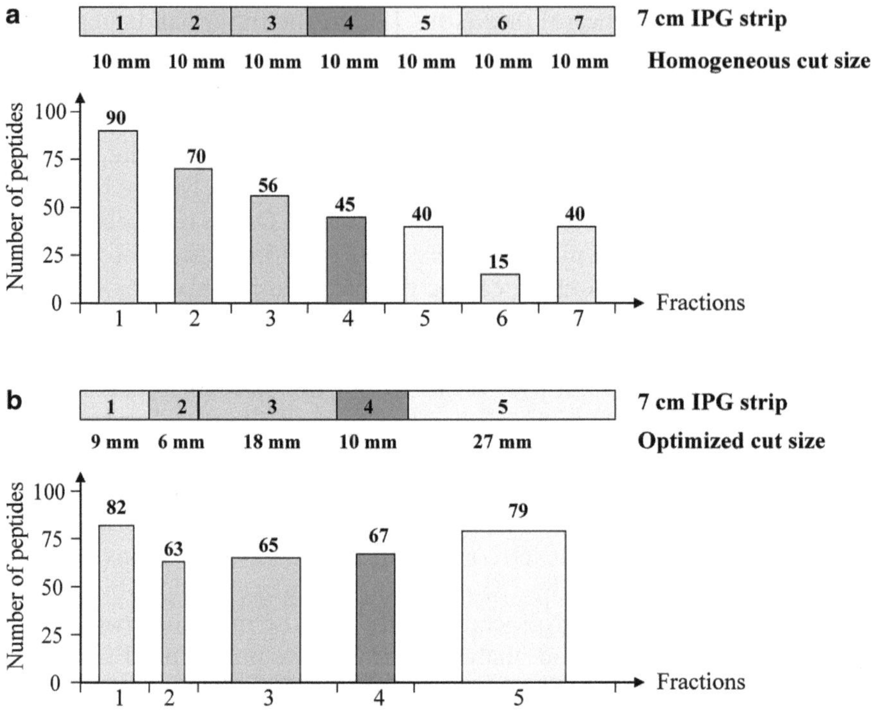

Fig. 2. Example of IPG strips with different cuts. (**a**) Strip cut in seven fractions of same length (10 mm), with completely inhomogeneous peptide distribution (from 15 to 90 peptides per fraction). (**b**) Strip cut according to the pICarver software, into fewer fractions (five), with a more homogeneous peptide distribution (from 63 to 82 peptides per fraction).

2. Materials

All chemicals should be of high-purity grade; in particular, high -purity water (H_2O) and ACN compatible with MS have to be used for the preparation of all buffers and solvents.

2.1. Sample Preparation and Digestion

1. Buffer solution for digestion (used in 2–5 below): 100 mM (i.e., 7.9 mg/mL) ammonium bicarbonate (AB) in H_2O (see Note 4).
2. Denaturation solution: 0.05% (w/v) sodium dodecyl sulfate (SDS) in buffer solution (see Note 5).
3. Reducing reagent solution: 100 mM (i.e., 15.4 mg/mL) 1,4-dithioerythritol (DTE) in buffer solution (see Note 6).
4. Alkylating agent solution: 100 mM (i.e., 18.5 mg/mL) iodo-acetamide (IAM) in buffer solution (see Note 7).
5. Digestion solution: 100 μg/mL trypsin (Promega, Madison, WI) in buffer solution. Cool the buffer solution at around 4°C (e.g., in the fridge, on ice) before adding it to the trypsin (see Note 8).

2.2. Peptide Purification

1. Washing solution: 95:5 H_2O:ACN (v/v) with 0.1% formic acid.
2. Eluting solution: 40:60 H_2O:ACN (v/v) with 0.1% formic acid.
3. Oasis HLB 1 cc/10 mg solid phase extraction cartridge (Waters, Milford, MA), or an equivalent C_{18} phase support for peptide purification and concentration.
4. SpeedVac system (Thermo Scientific) or an equivalent system for solvent evaporation by vacuum.

2.3. IPG-IEF

1. Ettan IPGphor II system (GE Healthcare, Piscataway, NJ).
2. IPG dry strips (GE Healthcare, Piscataway, NJ).
3. Ampholines solution (GE Healthcare, Piscataway, NJ).
4. IEF buffer: 4 M urea (2.4 g) and 0.2% (v/v) of the appropriate pH range Ampholines (20 μL) in 10 mL H_2O (see Note 9).

2.4. Peptide Extraction

1. High boiling point petroleum ether (Fluka, Buchs, Switzerland).
2. Trifluoroacetic acid (TFA) solution: 0.1% TFA (v/v) in H_2O.
3. The software "pICarver" (see Note 3).
4. Eppendorf polypropylene tubes.

2.5. LC-MS and Database Searching

1. HPLC buffer A: 95:5 H_2O:ACN (v/v) with 0.1% formic acid.
2. HPLC buffer B: 5:95 H_2O:ACN (v/v) with 0.1% formic acid.
3. LTQ-Orbitrap mass spectrometer (Thermo Electron, Waltham, MA).
4. TriVersa Nanomate (Advion Biosciences, Norwich, UK).
5. Agilent 1100 nano HPLC system (Agilent Technologies, Waldbronn, Germany).
6. Trapping microcolumn ZORBAX 300SB C_{18} (5 mm × 300 μm ID, 5 μm) (Agilent, Palo Alto, CA).
7. Reversed-phase nanocolumn ZORBAX 300SB C_{18} column (75 μm ID × 15 cm, 3.5 μm) (Agilent, Palo Alto, CA).
8. Phenyx software (Genebio, Switzerland).

3. Methods

3.1. Sample Preparation and Digestion

1. If the samples contain a high amount of salts or other detergents, or if the proteins of interest are in low concentration, one or more preliminary steps are advised: e.g., protein precipitation by an organic solvent, extraction on a column, and/or protein depletion (see Note 10).

2. Dilute around 100 μg of proteins in 200 μL of the denaturation solution (0.05% SDS) (see Note 11).
3. Add 40 μL of reducing reagent solution (100 mM DTE) and vortex.
4. Incubate for 1 h at 37°C.
5. Add 90 μL of alkylating agent solution (100 mM IAM) and vortex.
6. IAM is light sensitive, incubate for 30 min in the dark. For instance, wrap samples in aluminum foil during incubation.
7. Add 20 μL of digestion solution (100 μg/mL trypsin) (see Note 12).
8. Incubate at 37°C overnight (around 15 h).

3.2. Peptide Purification

1. Add 800 μL of washing solution to the sample. Verify the pH to be around 2–3. If not, add 1–5 μL concentrated formic acid to adjust the pH.
2. Equilibrate the Oasis HLB extraction cartridge 1 cc/10 mg with 1 mL of eluting solution.
3. Condition the column with 1 mL of washing solution.
4. Pass the total amount of sample (around 1.15 mL) through the column.
5. Wash with 1 mL of washing solution.
6. Elute the peptides with 700 μL of eluting solution.
7. Evaporate to dryness with a SpeedVac system.
8. Add 50 μL of water. Vortex, spin down, and reevaporate to dryness. This step removes the remaining formic acid, which can interfere with IEF.

3.3. IPG-IEF

1. Dilute the total amount of purified peptides in the IEF buffer. The volume depends on the length of the IPG strip (see Note 13).
2. Spread the sample in one of the wells of the reswelling tray.
3. Rehydrate the IPG strip for at least 6 h. The strips must be covered with paraffin oil to prevent evaporation (see Note 14).
4. Run the IEF according to the manufacturer's protocol for the size and the pH zone of the IPG strip (see Note 15).

3.4. Peptide Extraction and Purification

1. After isoelectric focusing, wash the IPG strips three times for 10 s in three distinct high boiling point petroleum ether baths at room temperature in order to remove the paraffin oil.
2. Manually cut the IPG strip with a scalpel (see Note 16). To avoid contamination, wash the scalpel in 20% isopropanol after each cut.

3. Place the gel pieces in Eppendorf polypropylene tubes containing the TFA solution. A minimum volume of 100 μL is recommended for a 0.5 cm long fraction.
4. Incubate for 30 min without vortexing or sonicating (see Note 17).
5. Repeat steps 3 and 4 one more time and merge extracts together.
6. Follow the same procedure for peptide purification as described in Subheading 3.2, steps 2–7 (i.e., peptide purification through Oasis HLB extraction cartridge).
7. Dilute each sample in 25 μL HPLC buffer A, and store the samples at −20°C.

3.5. LC-MS/MS and Database Searching

1. Load 5 μL of peptides onto a trapping microcolumn in H_2O:ACN 97:3 (v/v) + 0.1% formic acid at a flow rate of 10 μL/min.
2. After 5 min of trapping, start the gradient to elute the peptides from the precolumn and to separate them on the reverse phase column at a flow rate of 300 nL/min (see Note 18).
3. For ESI-MS/MS, perform data acquisition in the range 350–1500 m/z (see Note 19).
4. Convert MS/MS data to a format compatible with your database identification software (see Note 20).
5. Merge peak lists of all the fractions of the same strip together, and proceed to protein identification with your database identification software: select trypsin as the enzyme, oxidized methionine as the variable modification, and carbamidomethylation of cysteine as the fixed modification (see Note 21). Isoelectric point values can be used to validate the peptides, which should be identified in their expected p*I* window.

4. Notes

1. In Chapter 27, an alternative instrument for "Off-Gel electrophoresis" is described. The benefits and drawbacks of each instrument are also briefly discussed.
2. Previous experience with MS/MS data acquisition and protein identification software is assumed throughout this chapter. Note that similar MS/MS equipment and bioinformatics software can be used with shotgun proteomics as for any other protein identification technique.
3. The software "pICarver" is freely available at http://expasy.org/tools/picarver/.

4. AB is used as a pH buffer to keep the pH value at around 8 (optimal pH for protein digestion with trypsin); triethyl ammonium bicarbonate (TEAB) could be used as an alternative here. Denaturation, reducing reagent, alkylating agent, and digestion solutions would then all be prepared by diluting the appropriate compound in this buffer solution.
5. SDS is used as denaturant (to unfold proteins) and detergent (to favor protein solubilization); it could be replaced by urea at a concentration of up to 2 M.
6. Different reducing reagents can be used: e.g., 1,4-dithiothreitol (DTT) or tris(2-carboxyethyl)phosphine (TCEP).
7. Other alkylating agents can be used, such as methyl methane thiosulfonate (MMTS) or acrylamide.
8. Trypsin autodigestion should be prevented to avoid additional peptides (tryptic peptides); thus, trypsin solution should be prepared at low temperature, where this enzyme is mostly inactive. This solution must be used immediately after its preparation.
9. Peptides naturally play the role of ampholines. However, a low concentration of ampholines covers the pH gaps in the IPG strip's pH gradient and ensures a better conductivity along the strip. Therefore, a much lower quantity of ampholines is used compared to 2D-Gel Electrophoresis. Additionally, 0.1% Bromophenol Blue (10 μL) can be used in the IEF buffer to trace the quality of focusing.
10. High amounts of salts or detergents may interfere with the digestion procedure. Highly abundant proteins (such as albumin in blood plasma) should also be removed if they are of no interest in the experiment. Discarding superfluous and excessive materials from the sample at this step is essential in shotgun proteomics and depends on the experimental objectives and on the sample matrix.
11. If the proteins are already in solution, adjust the volume and concentration of the denaturation solution: e.g., dilute 100 μL of a 1 mg/mL protein solution in a 100 μL denaturation solution of 0.1% SDS, 200 mM AB.
12. In the sample, the trypsin to protein ratio is around 1:50. A ratio between 1:20 and 1:100 is optimal to ensure efficient protein digestion as well as negligible trypsin autodigestion.
13. For 7 cm long strips, dilute the peptides in 100 μL IEF buffer; for 13 and 18 cm, in 200 μL, and for 24 cm, in 250 μL.
14. Two other sample application methods are also widely used: Paper Bridge and Cup Loading. Paper Bridge loading is used for the application of high sample volumes. The IPG is rehydrated with the IEF buffer. The sample is applied on the

paper bridge between the one of the electrodes and the IPG strip before focusing. Cup loading is very similar to Paper Bridge loading and is usually used for sample volumes of up to 150 μL. The sample is applied into a loading cup on the anodic or cathodic end of the strip before focusing.

15. Focusing protocols for proteins and peptides are quite similar. However, Krijgsveld et al. found that focusing of peptides occurs even at low voltages (10). The use of internal peptide standards is advised as a measure of focusing efficiency (11).
16. Since the theoretical and observed distribution of peptides on the IPG strip is not homogenous (Fig. 2a), the fraction size can be adjusted to obtain an almost even number of peptides per fraction using the "pICarver" software, resulting in an increase in the resolution or diminution of the number of fractions (Fig. 2b) (9).
17. The use of organic solvents or vigorous vortexing or sonicating results in the extraction of nonpolymerized acrylamide and of other unwanted chemicals from the IPG strip, which can interfere with the later LC step.
18. Gradient conditions should be optimized, according to the LC equipment at one's disposal, to elute the peptides through the LC column. The next steps are necessary, but specified times may need some adjustments: for the first few minutes, ACN is kept to a minimum to ensure the retention of all peptides (e.g., by keeping solvent B at 0% over the first 5 min); then, the ACN concentration is increased linearly to elute peptides (e.g., by increasing solvent B from 0 to 50% over the next 45 min); the ACN concentration is set to a high value to ensure elution of all peptides and clean the column of any impurities in the sample (e.g., by increasing solvent B from 50 to 90% over the next 5 min and keeping solvent B at 90% in the column for 10 min); finally, the solvent composition is returned to its original conditions and the column is equilibrated before the next injection (e.g., by decreasing solvent B from 90 to 0% over the next 5 min and then keeping B at 0% in the column for 15 min). Total time between each sample injection is set at 90 min (including trapping time) in this example, but may be shorter or longer depending on the LC equipment, depending mainly on the instrument's dwell volume (for more information about this issue, see e.g., (12)).
19. For ESI-MS/MS, we used a 400 nozzle ESI Chip (Advion Biosciences) with a voltage of 1.65 kV; the mass spectrometer capillary transfer temperature was set at 200°C; the four most intense precursor ions detected in the full MS survey performed in the Orbitrap (resolution 60,000 at m/z 400) were selected (according to Xcalibur 2.0 software from Thermo Electron)

and fragmented; only precursors with a charge higher than one were selected for CID (collision-induced dissociation) fragmentation.

20. From raw files, MS/MS spectra were exported as dta (text format) files using the extract_msn.exe script of BioWorks 3.2 software (Thermo Electron). Dta files were then merged by a Perl script (available at http://www.matrixscience.com/help/instruments_xcalibur.html) into a Mascot Generic File (mgf) for database searching.

21. Phenyx (GeneBio, Switzerland) was used as database searching engine. With this software, two search rounds were selected. In the second round deamidation was also selected as variable modification. In the first round, one missed cleavage with normal cleavage mode was selected, whereas in the second round three missed cleavages with half-cleaved mode were selected. "Turbo" was only selected in the first round. The minimum peptide length allowed was six amino acids. Parent ion tolerance was 10 ppm. The acceptance criteria were slightly lowered in the second round search (round 1: AC score 8.0, peptide *Z*-score 6.5, *p* value $1E^{-7}$; round 2: AC score 8.0, peptide *Z*-score 6.0, *p* value $1E^{-7}$).

References

1. Link, A.J., Eng, J., Schieltz, D.M., Carmack, E., Mize, G.J., Morris, D.R. et al. (1999) Direct analysis of protein complexes using mass spectrometry. *Nat. Biotechnol.* **17**, 676–682.
2. Washburn, M.P., Wolters, D., Yates, J.R., 3rd. (2001) Large-scale analysis of the yeast proteome by multidimensional protein identification technology. *Nat. Biotechnol.* **19**, 242–247.
3. Wolters, D., Washburn, M.P., Yates, J.R., 3rd. (2001) An automated multidimensional protein identification technology for shotgun proteomics. *Anal. Chem.* **73**, 5683–5690.
4. Egen, N.B., Bliss, M., Mayersohn, M., Owens, S.M., Arnold, L., Bier, M. (1988) Isolation of monoclonal antibodies to phencyclidine from ascites fluid by preparative isoelectric focusing in the Rotofor. *Anal. Biochem.* **172**, 488–494.
5. Righetti, P.L., Gianazza, E. (1987) Isoelectric focusing in immobilized pH gradients: theory and never methodology. *Methods Biochem. Anal.* **32**, 215–278.
6. Cargile, B.J., Talley, D.L., Stephenson, J.L., Jr. (2004) Immobilized pH gradients as a first dimension in shotgun proteomics and analysis of the accuracy of pI predictability of peptides. *Electrophoresis* **25**, 936–945.
7. Scherl, A., François, P., Charbonnier, Y., Deshusses, J.M., Koessler, T., Huyghe, A. et al. (2006) Exploring glycopeptides-resistance in Staphylococcus aureus: a combine proteomics and transcriptomics approach for the identification of resistance-related markers. *BMC Genomics* **7**, 296.
8. Chick, J.M., Haynes, P.A., Molloy, M.P., Bjellqvist, B., Baker, M.S., Len, A.C. (2008) Characterization of the rat liver membrane proteome using peptide immobilized pH gradient isoelectric focusing. *J. Proteome Res.* **7**, 1036–1045.
9. Vaezzadeh, A.R., Hernandez, C., Vadas, O., Deshusses, J.M.P., Lescuyer, P., Lisacek, F., Hochstrasser, D.F. (2008) PICarver: a software tool and strategy for peptides isoelectric focusing. *J. Proteome Res.* **7**, 4336–4345.
10. Krijgsveld, J., Gauci, S., Dormeyer, W., Heck, A.J. (2006) In-gel isoelectric focusing of peptides as a tool for improved protein identification. *J. Proteome Res.* **5**, 1721–1730.
11. Vaezzadeh, A.R., Simicevic, J., Chauvet, A., François, P., Zimmermann-Ivol, C.G., Lescuyer, P. et al. (2008) Imaging mass spectrometry using peptide isoelectric focusing. *Rapid Commun. Mass Spectrom.* **22**, 2667–2676.
12. Dolan, J.W. (2006) Dwell volume revisited. *LCGC North Am.* **24**, 458–466.

Chapter 27

Shotgun Proteomics: A Relative Quantitative Approach Using Off-Gel Electrophoresis and LC-MS/MS

Laurent Geiser, Loïc Dayon, Ali R. Vaezzadeh, and Denis F. Hochstrasser

Abstract

Shotgun proteomics originated as a strategy to identify proteins in complex protein mixtures, but it is also possible to obtain information on relative quantitation with some adjustments to the procedure. After protein digestion, the resulting peptide mixture is labelled with isobaric tags. Then, labelled peptides are submitted to two orthogonal techniques: first, peptides are separated according to their isoelectric point (p*I*) by Off-Gel electrophoresis (OGE), a relatively new isoelectric focusing (IEF) technique; after peptide purification, they are then separated in a second dimension according to their hydrophobic properties by reversed-phase liquid chromatography (RPLC). Finally, following detection by mass spectrometry (MS) and sequencing by tandem mass spectrometry (MS/MS), proteins are matched by means of bioinformatics software, and protein ratios are calculated by comparing isobaric tagged reporter fragments to highlight the different expression of one protein in one sample relative to other samples.

Key words: Protein quantification, Peptide labelling, Isobaric tags, Off-gel electrophoresis, Tandem mass spectrometry

1. Introduction

Shotgun proteomics has been explained and its use illustrated for qualitative purposes in the Chapter 26. In the present chapter, a relative quantitative approach is presented, using an alternative peptide fractionation strategy with an Off-Gel electrophoresis (OGE) instrument.

Biomarkers are commonly sought by comparing two sample populations, e.g. healthy versus diseased patients, medicated versus non-medicated patients, and so on. In qualitative proteomics, a protein must be present in one population and absent in the other to be used as a biomarker. This on/off system is, however, of limited use as it frequently occurs that one protein is over- or

Dermot Walls and Sinéad T. Loughran (eds.), *Protein Chromatography: Methods and Protocols*, Methods in Molecular Biology, vol. 681, DOI 10.1007/978-1-60761-913-0_27, © Springer Science+Business Media, LLC 2011

under-expressed in one set of samples. In order to recognize this over- or under-expression, relative quantitative proteomics is necessary. Data acquired on tandem mass spectrometers are very powerful for polypeptide sequence identification, but provide limited quantitative information. For example, peptide ionization and consequently mass spectrometric (MS) signal intensity is dependant on co-eluting substances which may differ from one sample to another, generating matrix effects. Labelling techniques have been introduced to bypass these problems (1–3). While quantitative proteomics with label-free techniques are also widely used, they are not addressed in this chapter (to be found elsewhere, e.g. (2, 3)).

As mentioned in the previous chapter, a wide variety of samples can be analysed including biofluids (e.g. blood, urine), tissues, cells, organelles, and protein complexes, and the workflow of relative quantitative shotgun proteomics is similar to the one described for qualitative shotgun proteomics in that chapter (see Chapter 26, Fig. 1). After an optional sample purification step, the following steps are performed: (1) enzymatic digestion

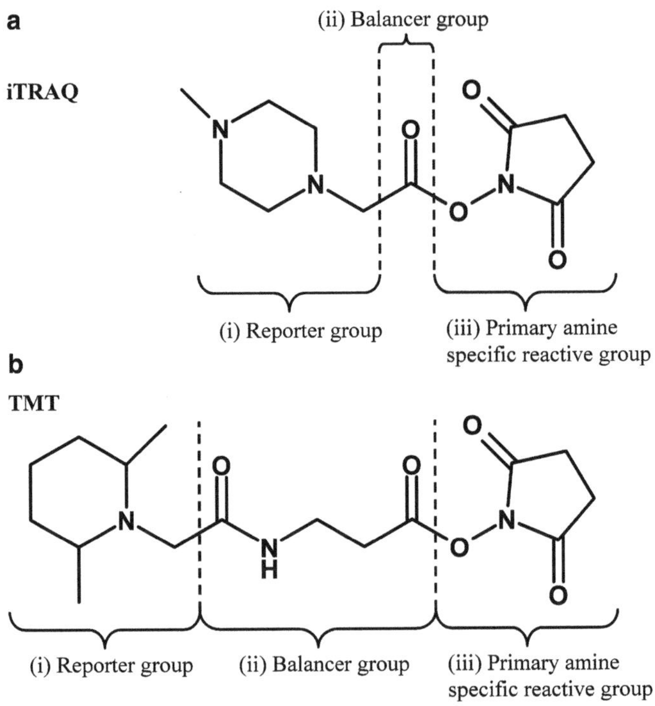

Fig. 1. Commercially available isobaric chemical tags: (**a**) 4-Plex iTRAQ, with masses of reporter groups of 114–117, balancer group of 28–31, for a constant total of 145; note that the 8-plex iTRAQ (not shown) has reporters groups of 113–121 (skip of 120) and balancer groups of 184–192 (skip of 185). (**b**) 6-Plex TMT with masses of reporter groups of 126–131, balancer group of 99–104, for a constant total of 230. Carbon, nitrogen, and/or oxygen isotopes are included in the reporter and/or balancer groups to obtain the required mass.

to convert proteins into peptides by trypsin; (2) labelling of the peptides with isobaric tags; (3) separation of the peptides by isoelectric focusing (IEF); (4) before analysis by reversed-phase liquid chromatography (LC) coupled to tandem mass spectrometry (MS/MS); (5) the resultant peptide sequence data generated from tandem mass spectra are compared with protein databases to identify proteins present in the original population, and (6) the ratios of proteins to each other are inferred from the relative ratios of isobaric tag reporter ions . This analysis sequence differs from the qualitative approach with the two additional steps of labelling the peptides (i.e. step 2) and calculation of the protein ratio (i.e. step 6). Furthermore, OGE is used as the IEF separation technique. In addition, some chemicals may interfere with the peptide labelling step and have to be replaced as highlighted in the following protocol.

Isobaric chemical tags are used in labelling techniques to enable the simultaneous identification and relative quantification of proteins using MS/MS. They are small chemical molecules with three distinct parts: (1) a reporter group, (2) a balancer group, and (3) a primary amine specific reactive group (see Fig. 1). The reporter group is used for relative quantification, the balancer group ensures identical behaviour of peptides in peptide separation methods and first MS dimension, and the reactive group permits covalent attachment to the free amino-termini and epsilon-amino functions of lysine residues of peptides and proteins.

1.1. Isoelectric Focusing by Off-Gel Electrophoresis

Due to the complexity of the peptide mixtures, sample fractionation prior to LC-MS/MS should be beneficial. The concept of using the immobilized pH gradient (IPG) as IEF technique was illustrated with an IPGphor instrument in Chapter 26: as an alternative, an OGE instrument can be used for peptide fractionation.

OGE involves placing the sample in a liquid chamber on top of an IPG gel strip (see Fig. 2) (4, 5). Following the application of an electric field, proteins or peptides are separated on the basis of their isoelectric points (p*I*), migrating until they reach a position where their p*I* corresponds to the pH of the fraction above the IPG gel. This technique has been successfully applied to peptide fractionation for improved protein identification (6). Once the separation is finished, peptides are in a liquid solution and can be recovered very easily for subsequent treatment, such as peptide extraction by solid phase extraction.

1.2. IEF: Comparison of IPGphor and OGE

The IPGphor and OGE instruments each has its advantages. One major advantage of the IPGphor instrument in combination with the "pICarver" software is the possibility of obtaining an almost even distribution of peptides per fraction. Hence, peptide separation is optimized, which may increase the number of peptides detected

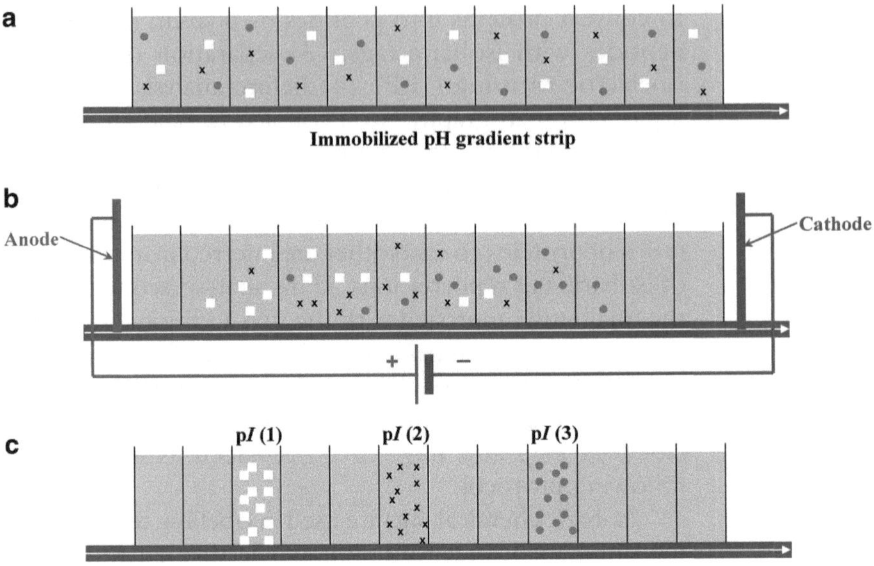

Fig. 2. Principle of OGE instrument, illustrated with three peptides of different isoelectric points: p*I*(1) < p*I*(2) < p*I*(3). (**a**) A sample is placed in the wells frame, on top of the IPG strip. (**b**) Electric field is applied and peptides migrate according to their p*I*. (**c**) Peptides have been separated and can be recovered in liquid solution.

by MS/MS, which is essential for qualitative purposes. On the other hand, sample handling is more convenient with OGE and is thus very appropriate for fast quantitative purposes (see Note 1).

2. Materials

All chemicals should be of high-purity grade; in particular, high-purity water (H_2O) and acetonitrile (ACN) compatible with MS have to be used for the preparation of all buffers and solvents.

Currently, two brands of peptide tags are commercially available and successfully applied for relative quantification: isobaric tags for relative and absolute quantification (iTRAQ, Fig. 1a) from Applied Biosystems (7) and tandem mass tags (TMT, Fig. 1b) from Thermo Scientific (8). They are both intended for the same purpose and we have obtained very similar results with both. At the time of writing, iTRAQ is available in 4-plex and 8-plex, while TMT is available in 0-plex (for method optimization at lower cost), 2-plex, and 6-plex (see Note 2).

2.1. Sample Preparation and Digestion

1. Internal standard solution: 100 μg/mL bovine β-lactoglobulin (LACB) (Sigma-Aldrich, St. Louis, MO) in H_2O.
2. Denaturation solution: 0.1% (w/v) sodium dodecyl sulphate (SDS) and 200 mM (i.e. 32.6 mg/mL) triethyl ammonium bicarbonate (TEAB) in H_2O.

3. Buffer solution for digestion (used in items 3–5 below): 100 mM (i.e. 16.3 mg/mL) TEAB in H_2O (see Note 3).
4. Reducing reagent solution: 100 mM (i.e. 28.6 mg/mL) Tris-(2-carboxyethyl)-phosphine (TCEP) in buffer solution (see Note 4).
5. Alkylating agent solution: 100 mM (i.e. 18.5 mg/mL) iodoacetamide (IAM) in buffer solution (see Note 5).
6. Digestion solution: 100 μg/mL trypsin (Promega, Madison, WI) in buffer solution. Cool the buffer solution at around 4°C (e.g. in the fridge, on ice) before adding it to the trypsin (see Note 6).
7. SpeedVac system (Thermo Scientific) or an equivalent system for solvent evaporation by vacuum.

2.2. Peptide Labelling

1. Isobaric tags: iTRAQ (Applied Biosystems, Foster City, CA) or TMT (Thermo Scientific, Rockford, IL).
2. Hydroxylamine solution: 5% (w/v) of hydroxylamine in H_2O.

2.3. Peptide Purification 1

1. Washing solution: 95:5 H_2O:ACN (v/v) with 0.1% formic acid.
2. Eluting solution: 40:60 H_2O:ACN (v/v) with 0.1% formic acid.
3. Oasis HLB 1 cc/30 mg solid phase extraction cartridge (Waters, Milford, MA), or an equivalent C_{18} phase support for peptide purification and concentration.

2.4. OGE Separation

1. OFFGEL Fractionator (Agilent Technologies, Waldbronn, Germany).
2. Immobiline DryStrips (IPG strips) pH 3–10, 13 cm (Agilent Technologies, Waldbronn, Germany).
3. Ampholyte buffer solution pH 3–10 (Agilent Technologies, Waldbronn, Germany).
4. Glycerol solution 50% (Agilent Technologies, Waldbronn, Germany).
5. Mineral oil (Agilent Technologies, Waldbronn, Germany).

2.5. Peptide Purification 2

1. Washing solution: 95:5 H_2O:ACN (v/v) with 0.1% formic acid.
2. Eluting solution: 40:60 H_2O:ACN (v/v) with 0.1% formic acid.
3. Oasis HLB 1 cc/10 mg solid phase extraction cartridge (Waters, Milford, MA), or an equivalent C_{18} phase support for peptide purification and concentration.

2.6. LC-MS/MS and Database Searching

1. HPLC buffer A: 95:5 H_2O:ACN (v/v) with 0.1% formic acid.
2. HPLC buffer B: 5:95 H_2O:ACN (v/v) with 0.1% formic acid.
3. LTQ-Orbitrap mass spectrometer (Thermo Electron, Waltham, MA) or a similar system.
4. Phenyx software (Genebio, Geneva, Switzerland).

3. Methods

3.1. Sample Preparation and Digestion

1. If the samples contain a high amount of salts or other detergents, or if the proteins of interest are in low concentration, one or more preliminary steps are advised: e.g. protein precipitation by an organic solvent, extraction on a column, and/or protein depletion.
2. For each sample, protein concentration is estimated using the Bradford protein assay (see Note 7; see Chapter 13).
3. Each sample is diluted in H_2O (90 μL of final volume) to have the same concentration of protein in all samples: protein concentration should be approximately 550 μg/mL.
4. To 90 μL of sample (around 50 μg of proteins), add 10 μL of internal standard solution (i.e. 1 μg of LACB) (see Note 8). Proceed in the same way for all samples.
5. Add 100 μL of the denaturation solution (0.1% SDS and 200 mM TEAB).
6. Add 40 μL of reducing reagent solution (100 mM TCEP) and vortex.
7. Incubate for 1 h at room temperature.
8. Add 90 μL of alkylating agent solution (100 mM IAM) and vortex.
9. Incubate for 30 min in the dark (IAM is light sensitive). For instance, wrap samples in aluminium foil during incubation.
10. Add 20 μL of digestion solution (100 μg/mL trypsin).
11. Incubate at 37°C overnight (around 15 h).
12. Evaporate to dryness with a SpeedVac system.
13. Dilute each sample in 30 μL of buffer solution (100 mM TEAB).

3.2. Peptide Labelling

1. Each sample is labelled with a tag with a different reporter group. Proceed according to the manufacturer's protocol. Usually, the procedure consists of diluting isobaric chemical tags in 70 μL of ethanol, mix with the sample (vortex), and incubate it at room temperature for 1 h.

2. Block the reaction by adding 15 μL of hydroxylamine solution to each tube. Vortex and incubate for 15 min (see Note 9).
3. Pool all samples in one new tube (see Note 10).
4. Evaporate to dryness with a SpeedVac system.

3.3. Peptide Purification 1

1. Add 1 mL of washing solution to the sample. Verify the pH to be around 2–3. If not, add 1–5 μL concentrated formic acid to adjust the pH.
2. Equilibrate the Oasis HLB extraction cartridge 1 cc/30 mg (see Note 11) with 1 mL of eluting solution.
3. Condition the column with 1 mL of washing solution.
4. Pass the total amount of sample through the column.
5. Wash with 1 mL of washing solution.
6. Elute the peptides with 700 μL of eluting solution.
7. Evaporate to dryness with a SpeedVac system.

3.4. OGE Separation

1. Follow the OGE protocol, preparing 10 mL of OGE stock solution containing 5% (v/v) of glycerol and 1% of ampholytes (w:v) in water.
2. Add 400 μL of H_2O and 1,600 μL of OGE stock solution to the sample and vortex.
3. Place a 13 cm long strip in one tray of the OGE instrument, and assemble the 12 wells frame on top of it (see Note 12).
4. Hydrate the strip according to the manufacturer's instructions and wait for 30 min.
5. Place 150 μL of sample in each well (Fig. 2a).
6. Assemble the tray, mineral oil, and electrodes according to the manufacturer's protocol.
7. Start the separation procedure (Fig. 2b).
8. When separation has finished (see Note 13), recover each of the 12 fractions in individual vials.

3.5. Peptide Purification 2

1. Add 800 μL of washing solution to each of the 12 fractions and verify the pH to be around 2–3. If not, add 1–5 μL concentrated formic acid to adjust the pH.
2. Purify the peptides on an Oasis HLB extraction cartridge 1 cc/10 mg, following the procedure for peptide purification described in Subheading 3.3, steps 2–7.
3. Add HPLC buffer A to the sample, and store the sample at −20°C. The volume of buffer A should be 300 μL for experiments with a 6-plex isobaric chemical tags, and adjusted accordingly for other tags (see Note 14).

3.6. LC-MS/MS and Database Searching

1. For LC-MS/MS analyses, follow similar conditions to those described in Chapter 26, Subheading 3.5.
2. Merge peak lists of all the fractions together, and proceed to protein identification with your database identification software (see Note 15): select trypsin as enzyme, oxidized methionine and the isobaric tag (i.e. TMT or iTRAQ) as variable modifications on the N-terminus and as fixed modifications on lysine, and carbamidomethylation of cysteine as a fixed modification.
3. Estimate the ratio of tagged peptides to untagged peptides (see Note 16).
4. Repeat the second step, but with isobaric tag (i.e. TMT or iTRAQ) as fixed modification on both N-terminus and lysine. All other conditions are the same.

3.7. Quantitative Data Analysis

1. Tagged peptides are used to quantify protein ratios. After fragmentation, they produce mass reporters according to the tagging procedure (e.g. reporters of 114–117 in MS/MS data for 4-plex iTRAQ). The MS/MS intensities of these reporters are used for data treatments. In order to discuss further treatments, numerical examples are given in Tables 1 and 2 (see Note 17). For the sake of clarity, this descriptive example only considers two mass reporters, but similar data treatments are applicable with additional mass reporters (see Note 18).
2. Use the peptides of the internal standard (LACB) to determine a correction factor, as illustrated in Table 1 (see Note 19).
3. Select the proteins of interest for relative quantification (see Note 20).
4. Calculate the ratio for each protein by:
   (a) Identifying all peptides corresponding to this protein, and report them in a table (see Note 17);
   (b) Applying the correction factor when necessary, as illustrated for protein no. 1 in Table 1.
   (c) Calculating the ratio of all these peptides (see Note 21);
   (d) Optionally eliminating one or two potential ratio outliers (see Note 22);
   (e) Calculating the mean peptide's ratio and its confidence interval (see Note 23).
5. Determine which proteins are under- or over-expressed in one sample relative to another (see Note 24).
6. Conclusions can be drawn from the experiments; additional experiments could be carried out to help decide whether a

Table 1
Example for protein no. 1: determination of its ratio in two samples using the internal standard

LACB				Protein no. 1				
	Reporter at MS/MS: 126	Reporter at MS/MS: 127	Reporter ratio 126/127		Reporter at MS/MS: 126	Reporter at MS/MS: 127	Corrected reporter: 127	Reporter ratio 126/corr. 127
Peptide no. 1	520	490	1.06	Peptide no. 1	220	215	226[b]	0.97
Peptide no. 2	450	435	1.03	Peptide no. 2	160	150	158[b]	1.01
Peptide no. 3	440	435	1.01	Peptide no. 3	135	135	142[b]	0.95
–	–	–	–	Peptide no. 4	100	90	95[b]	1.06
Peptide no. 10	110	100	1.10	Peptide no. 5	80	80	84[b]	0.95
–	–	–	–	Peptide no. 6	75	70	74[b]	1.02
Peptide no. *N*	–	–	–	Peptide no. 7	40	45	47[b]	0.85
Correction factor:			1.05[a]	Mean ± confidence interval ($N=7$, $\alpha=0.05$)				0.97 ± 0.05[c]

[a]Correction factor is calculated from the mean of ratio 126/127 for the ten most abundant peptides
[b]Calculated by multiplying the correction factor by the intensity of MS/MS reporter 127 (e.g. 226 = 1.05 × 215)
[c]Data demonstrate that protein no. 1 is neither over- nor under-expressed in one sample, the value of 1.0 being included in the confidence interval

Table 2
Example for protein no. 2 and no. 3: determination of their ratio in two samples after correction by the internal standard (as in Table 1)

Protein no. 2	Reporter at MS/MS: 126	Corrected reporter: 127	Reporter ratio 126/corr. 127	Protein no. 3	Reporter at MS/MS: 126	Corrected reporter: 127	Reporter ratio 126/corr. 127
Peptide no. 1	410	115	3.57	Peptide no. 1	360	220	1.64
Peptide no. 2	255	75	3.40	Peptide no. 2	150	–	–
Peptide no. 3	195	50	3.90	Peptide no. 3	–	115	–
Peptide no. 4	148	40	3.70	Peptide no. 4	110	105	1.05
Peptide no. 5	95	–	–	Peptide no. 5	–	85	–
Peptide no. 6	60	–	–	Peptide no. 6	100	55	1.82
Peptide no. 7	45	–	–	Peptide no. 7	50	–	–
Mean ± confidence interval ($N=4$, $\alpha=0.05$)			3.64 ± 0.21[a]	Mean ± confidence interval ($N=3$, $\alpha=0.05$)			1.50 ± 0.46[b]

[a]Data demonstrate that protein no. 2 is over-expressed in one sample (i.e. the one tagged with reporter 126), the value of 1.0 being significantly outside the confidence interval

[b]Data are not conclusive: protein no. 3 may be slightly over-expressed in one sample (i.e. the one tagged with reporter 126). Additional experiments could be conducted to draw firm conclusions

protein in the grey area (such as protein no. 3) is differentially expressed or not (see Note 25).

4. Notes

1. Both IPGphor and OGE are fully compatible with either qualitative or quantitative applications. It is thus fully justifiable to perform all experiments, whether qualitative or quantitative, on one single instrument for economical or practical reasons. If so, OGE would be particularly convenient due to its handling convenience and could be used according to Note 12 for qualitative purposes.
2. Selection of either iTRAQ or TMT should be done according to price consideration and/or considering the number of samples that need to be compared. For instance, if one sample must be compared to another, using a 2-plex TMT is probably the cheapest option, while an iTRAQ 8-plex is necessary if eight samples must be compared within one analysis. In addition, a further interest of using 6-plex or 8-plex tags is reported in Note 25, i.e. to enhance the data precision.
3. Primary amines would react with iTRAQ and TMT reagents, competing with peptide derivatization. Hence, TEAB must be used in lieu of ammonium bicarbonate.
4. Different reducing reagents such as 1,4-dithiothreitol (DTT) or 1,4-dithioerythritol (DTE) are not advised if isobaric tags must be used afterwards.
5. Methyl methane thiosulfonate (MMTS) is fully compatible with isobaric tags and can be used instead of IAM.
6. Trypsin autodigestion should be prevented to avoid additional peptides (tryptic peptides); thus, trypsin solution should be prepared at low temperature, where this enzyme is mostly inactive. This solution must be used immediately after its preparation.
7. The Bradford protein assay is a colorimetric and spectroscopic method used to measure the total concentration of proteins in a sample. It gives a good estimation of this concentration: measuring the protein quantity in each sample is essential, the protocol being compatible with protein quantities up to 100 μg (higher protein amounts would produce an incomplete reaction with the reactive isobaric tag quantities used).
8. It is possible to perform relative quantification without addition of an internal standard, but our experience suggests

improved results by using it. Any protein not present in the sample could be used as internal standard.

9. Isobaric chemical tags are in excess in solution. Hydroxylamine blocks these tags.
10. For an optimal recovery of all labelled peptides, wash each sample tube twice with 30 μL of water (vortex each time) and add them to the new sample tube.
11. As peptides from all samples have been merged in one sample, the peptide quantity at this step may be rather important. It is thus advisable to use an extraction support compatible with large peptide quantities.
12. For increased peptide separation, a 24 cm long IPG strip could be used. Sample volumes should be adjusted, and a 24 wells frame be used. As a result, 24 fractions will be obtained instead of 12, and twice as much sample will be purified and injected into the LC-MS/MS system. For relative quantifications purposes, fractionating peptides into 12 fractions is often sufficient. On the other hand, 24 wells frame is strongly advised for qualitative experiments if OGE is preferred in place of IPGphor. It was shown elsewhere that going from 12 to 24 wells frame increased the number of proteins identified by about 20% (9).
13. If samples are properly purified, the separation procedure should take less than 12 h. Performing it overnight is optimal. During analysis with electric current limit set at 50 μA, the voltage increases; once peptides are focused (Fig. 2c), the voltage reaches a plateau and the electric current may decrease. So, voltage and electric current can be used as indicator of peptide separations.
14. In order to have approximately the same peptide concentration, volume of buffer A should be adapted to the quantity of peptide in the sample. The sample must be concentrated enough for good quantification, but over-concentrated samples must be avoided to prevent column clogging during LC separations. Typically, for 5 μL injections in the LC, we dilute in a volume of $x \times 50$ μL, x being the number of compared samples (i.e. 100 μL for 2-plex, 200 μL for 4-plex, 300 μL for 6-plex, and 400 μL for 8-plex, respectively).
15. Phenyx is used as database searching engine. "Turbo" is selected, minimum peptide length allowed is six amino acids, and parent ion tolerance was set at 10 ppm (this value is dependant on the MS analyzers). Identification parameters are adjusted to have a false discovery rate (FDR) of around 2%.
16. The ratio of tagged peptides/untagged peptides is a good indicator of tagging efficiency. This value should be close to 90%; if the value is inferior to 70%, it indicates problem related with tagging, and protocol should be evaluated and maybe

adapted to increase this value. Low tagging efficiency does not completely prevent relative quantification thanks to the use of an internal standard. However, the less peptides being tagged, the fewer reporter values are available and relative quantification on some low abundant proteins will be impossible.

17. For each protein, peptide intensities (arbitrary units) are reported in a table, ranging from the most intense peptide to the one with lowest intensity, as in Tables 1 and 2.
18. At the time of writing, no unique data treatment procedure for relative quantification is accepted in the scientific community. Data treatment can be performed manually, or with different software: e.g. Mascot (Matrix Science Inc., Boston, MA, USA), Pro Quant (Applied Biosystems, Foster City, CA, USA), or Libra (TPP, Institute for System Biology, Seattle, WA, USA). More information about this topic can be found in the literature (e.g. (10)). The descriptive example given in this chapter applies the Mascot procedure.
19. Each protein produces different number of peptides. Theoretically, one peptide per protein could be sufficient to calculate a protein ratio. However, at least 3 peptide ratios per protein should be used to ensure acceptable quantitative results. Up to 10 peptide ratios per protein are used if such a number of data are available (e.g. LACB in Table 1). Using more peptide ratios is possible, but it should have negligible impact on the estimated protein ratio.
20. All proteins could be selected at this step, or the quantitative analysis could be limited to some specific proteins.
21. Only peptides present in all samples are considered, i.e. if a reporter is not present for a peptide, this peptide is not considered for relative quantification, as illustrated in Table 2.
22. In order to eliminate some ratio values, statistical tests can be performed (e.g. Dixon, Grubbs); by acquiring experience, the operator may decide to reject values independently of statistical tests results.
23. As ratios are averaged, a geometric mean should be calculated. However, results are comparable between geometric and arithmetic means. Thus, we apply herein the later one as a first approximation in order to simplify computation. Confidence intervals are calculated as $t(\alpha; N_{-1}) \times s \times N^{-0.5}$, where N is the number of peptide ratios, $t(\alpha; N_{-1})$ is a constant (available in statistical t-tables or in Excel software) depending on a probability (often set at $\alpha = 0.05$), and s is the standard deviation.
24. Determining at which ratio a protein is considered significantly under- or over-expressed in one sample is open for discussion. Cut-offs could be arbitrarily set at around 0.5 or 2 (i.e. values above 2 indicate a protein over-expression, inferior to 0.5 values indicate the opposite), but in our opinion fixing

a cut-off is the responsibility of the operator and depends on different criteria, such as the data quality and the risk the operator prefers to take. In our examples, protein no. 1 is obviously neither over- nor under-expressed, and protein no. 2 is over-expressed. However, such a definite statement cannot be made for protein no. 3, which stays in the grey area.

25. Repeating the experience or tagging the same sample with different tags is very useful to set a cut-off. For instance, if two samples must be compared and enough samples are at disposal, using 6-plex or 8-plex isobaric chemical tags would allow 3 or 4 tagging of the same sample. As a result, additional data are available to determine if a given protein (such as protein no. 3) is statistically differentially expressed or not.

References

1. Tao, W.A., Aebersold, R. (2003) Advances in quantitative proteomics via stable isotope tagging and mass spectrometry. *Curr. Opin. Biotechnol.* **14**, 110–118.
2. Bantscheff, M., Schirle, M., Sweetman, G., Rick, J., Kuster, B. (2007) Quantitative mass spectrometry in proteomics: a critical review. *Anal. Bioanal. Chem.* **389**, 1017–1031.
3. Mueller, L.N., Brusniak, M.Y., Mani, D.R., Aebersold, R. (2008) An assessment of software solutions for the analysis of mass spectrometry based quantitative proteomics data. *J. Proteome Res.* 7, 51–61.
4. Ros, A., Faupel, M., Mees, H., van Oostrum, J., Ferrigno, R., Reymond, F. et al. (2002) Protein purification by Off-Gel electrophoresis. *Proteomics* **2**, 151–156.
5. Michel, P.E., Reymond, F., Arnaud, I.L., Josserand, J., Girault, H.H., Rossier, J.S. (2003) Protein fractionation in a multicompartment device using Off-Gel isoelectric focusing. *Electrophoresis* **24**, 3–11.
6. Hörth, P., Miller, C.A., Preckel, T., Wenz, C. (2006) Efficient fractionation and improved protein identification by peptide OFFGEL electrophoresis. *Mol. Cell. Prot.* **5**, 1968–1974.
7. Ross, P.L., Huang, Y.N., Marchese, J.N., Williamson, B., Parker, K., Hattan, S. et al. (2004) Multiplexed protein quantitation in *Saccharomyces cerevisiae* using amine-reactive isobaric tagging reagents. *Mol. Cell. Prot.* **3**, 1154–1169.
8. Dayon, L., Hainard, A., Licker, V., Turck, N., Kuhn, K., Hochstrasser, D.F. et al. (2008) Relative quantification of proteins in human cerebrospinal fluids by MS/MS using 6-plex isobaric tags. *Anal. Chem.* **80**, 2921–2931.
9. Hubner, N.C., Ren, S., Mann, M. (2008) Peptide separation with immobilized pI strips is an attractive alternative to in-gel protein digestion for proteome analysis. *Proteomics* **8**, 4862–4872.
10. Lacerda, C.M., Xin, L., Rogers, I., Reardon, K.F. (2008) Analysis of iTRAQ data using Mascot and Peaks quantification algorithms. *Brief Funct. Genomic Proteomic.* 7, 119–126.

Chapter 28

Clinical Proteomics: Liquid Chromatography–Mass Spectrometry Purification Systems

Michael Henry and Paula Meleady

Abstract

Liquid chromatography/mass spectrometry (LC/MS) has become a routine powerful technology in clinical proteomic studies in protein identification, protein characterisation, and the discovery of biomarkers. In this chapter, we describe simple protocols for protein digestion and the analysis of digested peptides produced both by one-dimensional reversed-phase liquid chromatography and two-dimensional liquid chromatography (i.e. strong cation exchange chromatography followed by reversed-phase chromatography), coupled with tandem mass spectrometry.

Key words: Clinical proteomics, Protein identification, Protein digestion, One-dimensional liquid chromatography, Two-dimensional liquid chromatography, Tandem mass spectrometry

1. Introduction

Liquid chromatography–mass spectrometry (LC/MS) is an analytical technique that combines the physical separation capabilities of liquid chromatography with the mass analysis capabilities of mass spectrometry and is a fundamental tool in clinical proteomics (1). Monitoring the protein expression pattern in cells from clinical samples by proteomic technologies offers opportunities to discover potentially new biomarkers for the early detection and diagnosis of diseases including cancer (2). There are several proteomic technologies used for protein expression profiling of clinical samples, including 2-Dimensional Polyacrylamide Gel Electrophoresis (2D-PAGE), Surface Enhanced Laser Desorption/Ionisation Time of Flight (SELDI-ToF) MS, Protein Arrays, and quantitative LC/MS approaches using stable isotope labelling technologies such as Isotope Coded

Dermot Walls and Sinéad T. Loughran (eds.), *Protein Chromatography: Methods and Protocols*, Methods in Molecular Biology, vol. 681, DOI 10.1007/978-1-60761-913-0_28, © Springer Science+Business Media, LLC 2011

Affinity Tags (ICAT), Isobaric Tagging for Relative and Absolute Protein Quantitation (iTRAQ™), and Stable Isotope-Labelled Amino Acids in Culture (SILAC). Quantitative label-free LC/MS-based approaches are also becoming increasingly popular for biomarker discovery. These proteomic tools have been used for the differential analysis of various types of clinical biological samples including tissue (3), serum (4), plasma (5), urine (6), and saliva (7, 8) to better understand the molecular basis of the pathogenesis of disease, and also in the validation and characterisation of disease-associated proteins.

There have been many advances in the use of LC/MS in the proteomic analysis of clinical samples, but many challenges still remain including the dynamic range of proteins present in biological samples such as serum where potential biomarkers of interest may be present at very low concentrations, thus requiring the depletion of large serum proteins, and also the sensitivity and reproducibility of the LC and MS instrumentation (9, 10).

In this chapter we describe two LC-based approaches for analysing proteins from clinical samples. The first approach involves the separation of purified proteins from 1D (one dimensional) or 2D PAGE gels or very simple mixtures of proteins. The second approach involves the analysis of complex protein mixtures by 2D LC/MS, often referred to as "shotgun" proteomics. This online LC/MS method is also often termed multidimensional protein identification technology (MudPIT) (11, 12).

2. Materials

2.1. Sample Preparation from 1D/2D PAGE Gels (In-Gel Digestion)

1. Gel stains: Coomassie Brilliant Blue (CBB) R250, silver stain, fluorescent stains such as Sypro Ruby, Deep Purple (see Note 1).
2. Solution containing 50% methanol and ammonium bicarbonate (40 mM). Prepare fresh on the day of use.
3. Potassium ferricyanide solution (30 mM). Prepare fresh on the day of use.
4. Sodium thiosulphate solution (100 mM). Prepare fresh on the day of use.
5. Ammonium bicarbonate solution (100 mM). Prepare fresh on the day of use (see Note 2).
6. Solution containing ammonium bicarbonate (10 mM) in 10% (v/v) acetonitrile. Prepare fresh on the day of use.
7. Solution containing DTT (10 mM) in ammonium bicarbonate (100 mM). Make shortly before use.
8. Solution containing iodoacetamide (55 mM) in ammonium bicarbonate (100 mM). Make shortly before use.

9. Trypsin solution: Prepare 12.5 ng/μL trypsin in 10 mM ammonium bicarbonate containing 10% (v/v) acetonitrile. Re-suspend the Trypsin Gold (sequencing grade, Promega) at 1 μg/mL in 50 mM acetic acid (100 μg vial re-suspended in 0.1 mL of 50 mM acetic acid buffer). Remove the amount of this 80× trypsin stock needed and re-freeze the unused portion in 10 μL aliquots and store at −20°C. Make the 1× trypsin shortly before use by diluting in 10 mM NH_4HCO_3 containing 10% (v/v) acetonitrile to a concentration of 12.5 ng/μL (1:80 (v/v)).
10. Ammonium bicarbonate solution (20 mM). Prepare fresh on the day of use.
11. Solution containing 50% acetonitrile and 0.1% trifluoroacetic acid. Use a fume hood to prepare this solution.
12. 0.1% Trifluoroacetic acid (TFA). Use a fume hood to prepare this solution.
13. Sonicating water bath.
14. Vacuum centrifuge.

2.2. One-Dimensional Reverse Phase Chromatography for Simple Protein Mixtures

1. Solvent A: 2% acetonitrile in LC/MS grade water containing 0.1% formic acid (Prepare 1 L). Use a fume hood to prepare this solution.
2. Solvent B: 2% LC/MS grade water in acetonitrile containing 0.1% formic acid (Prepare 1 L). Use a fume hood to prepare this solution.
3. Sample loading solution (i.e. trap column mobile phase: 0.1% TFA). Add 1 mL of TFA to 1 L of water. Use a fume hood to prepare this solution.
4. Nano LC System: Ultimate 3000 (LC Packings/Dionex).
5. Mass Spectrometer: LTQ Orbitrap XL (Thermo Fisher Scientific).
6. Column: PepMap C18 capillary column (300 μm × 15 cm, 3 μm particles) (LC Packings/Dionex).
7. Trap column: PepMap C18 trap cartridge (300 μm × 5 mm) (LC Packings/Dionex).
8. Column oven.

2.3. Sample Preparation from an In-Solution Digestion of a Complex Protein Sample

1. Urea solution: 7 M urea, 50 mM Tris–HCL (pH 8.2), 2 mM DTT. Prepare the 50 mM Tris–HCl first and pH the buffer to 8.2 with dilute HCl. Add the urea and DTT to this buffer.
2. Ammonium bicarbonate (100 mM). Prepare fresh on the day of use.
3. Solution of 200 mM DTT in 100 mM ammonium bicarbonate. Make shortly before use.

4. Solution of 1 M iodoacetamide in 100 mM ammonium bicarbonate. Make shortly before use.
5. Trypsin solution: see Subheading 2.1.

2.4. Two-dimensional LC/MS for Complex Protein Samples

1. Solvent A, Solvent B, sample loading solution (see Subheading 2.2).
2. Solution of ammonium formate (500 mM) containing 0.1% TFA.
3. Solution of ammonium formate (100 mM) containing 0.1% TFA.
4. Solution of ammonium formate (50 mM) containing 0.1% TFA.
5. Solution of ammonium formate (25 mM) containing 0.1% TFA.
6. Solution of ammonium formate (10 mM) containing 0.1% TFA.
7. Mass Spectrometer: LTQ-XL (Thermo Fisher Scientific).
8. 2D LC System: MDLC (GE Healthcare).
9. SCX Column: BioBasic ™ SCX, 2.1 × 250 mm (Thermo Fisher Scientific).
10. Trap column: 300SC C18 trap column, 300 μm i.d. × 5 mm Zorbax™ (Agilent).
11. Analytical column: 300SC C18 analytical column, 75 μm i.d. × 150 mm, 3 μm (Agilent).

3. Methods

In order to separate peptides by LC and identify the proteins by mass spectrometry from a 1D or 2D PAGE gel, the sample must be prepared using in-gel digestion. Enzymatic digestion of the protein(s) results in the production of smaller peptides (e.g. tryptic peptides), which are separated and analysed by 1D LC/MS. This 1D LC/MS approach is also suitable for the analysis of simple protein mixtures (e.g. immunoprecipitated proteins). The digestion of proteins into smaller peptide fragments helps to overcome the solubility and handling problems associated with proteins and creates peptide fragments which are easily ionised by mass spectrometry. Although the masses of the tryptic peptides can be used in a direct search against a genome or protein database for protein identification, for example using MALDI TOF MS, high confidence protein identification using amino acid sequence information is obtained using tandem mass spectrometry data.

Two-dimensional peptide separations are based on the utilisation of two independent physical properties to separate their mixtures into individual components. Fractionation of complex peptide mixtures prior to MS analysis using a multidimensional approach that combines strong cation exchange and reversed-phase LC is an effective gel-free approach to analyse complex mixtures of proteins (e.g. whole cell lysates). This method generally starts with the tryptic digestion of proteins in solution (i.e. gel-free) that generates a complex mixture of peptides which can then be analysed by online 1D or 2D LC/MS. In the 2D LC/MS method described here, 1D separation involves retaining the peptides under acidic conditions on a strong cation exchange (SCX) stationary phase. Bound peptides are eluted using a stepwise increase in the ionic strength from the SCX column to the 2D column containing C18 reverse phase resin. The reversed-phased column is re-equilibrated, and the process is repeated with the salt concentration increasing on the SCX column at each salt step. After pre-fractionation of complex protein mixtures, the collected fractions can be further separated and analysed by LC/MS.

3.1. Sample Preparation from 1D/2D PAGE Gels (In-Gel Digestion)

1. Proteins separated by either 1D or 2D gels for protein identification are normally visualised by staining with dyes such as Coomassie Brilliant Blue (CBB) R250 and silver stain (glutaraldehyde-free), or fluorescence-based methods such as Sypro Ruby, Deep Purple, etc. Excise the protein(s) of interest from the gel (see Note 3), and transfer the gel piece to a 96 well plate or a PCR tube (see Note 4).
2. De-stain the gel pieces by washing several times in a solution of 50% methanol in 40 mM ammonium bicarbonate for CBB stained gels or a 1:1 mixture of 30 mM potassium ferricyanide:100 mM sodium thiosulphate for silver-stained gels. De-staining is not necessary for fluorescent-stained gel pieces.
3. Protein reduction and alkylation is only necessary for 1D gels as proteins from 2D gels will have been reduced and alkylated following the Iso-Electric Focusing (IEF) step (see Note 5). Add neat acetonitrile to each gel piece and incubate until the gel pieces shrink and become opaque. Remove the liquid and add 50 μL of 10 mM DTT in 100 mM ammonium bicarbonate solution and incubate for 30 min at 56°C. Cool the samples to room temperature and remove the liquid. Add 50 μL of 55 mM iodoacetamide in 100 mM ammonium bicarbonate solution and incubate at room temperature for 20 min in the dark. Remove the liquid and add neat acetonitrile to shrink the pieces again and then remove the liquid.
4. Add neat acetonitrile to the de-stained gel piece to shrink and remove the liquid. Add enough trypsin solution to saturate the gel piece (see Note 6). Incubate at 4°C for 1 h. Add 20 μL

of 20 mM ammonium bicarbonate solution to cover the gel pieces and keep them wet during the overnight digestion. Incubate the gel pieces overnight at 37°C (see Note 7).

5. To extract the digested peptides from the gel pieces, add 20 μL of 50% acetonitrile/0.1% TFA solution to each piece and incubate for 15 min at room temperature on a shaker. Sonicate the gel plugs for 1 min in a sonicating water bath and then transfer the extracted peptides to a clean tube. Repeat this step twice. Concentrate the extracted peptides using a speed vacuum centrifuge.
6. Add 20 μL of 0.1% TFA to the concentrated peptides. Samples can be analysed straight away or frozen at −20°C for future analysis.

3.2. One-Dimensional Reverse Phase Chromatography for Simple Protein Mixtures

1. Prior to analysis, equilibrate the columns in Solvent A for 10 min and set the column temperatures to 25°C using a column oven.
2. Inject 5 μL of digested protein samples using the injection pickup of the LC system onto a PepMap C18 trap cartridge (300 μm × 5 mm) at a flow rate of 25 μL/min for 5 min to de-salt and concentrate the sample.
3. Elute the peptides from the trap column at a flow rate of 350 nL/min acetonitrile/water gradient (2–50% Solvent B in 30 min) onto a PepMap C18 capillary column (300 μm × 15 cm, 3 μm particles) directly into the electrospray tip. Peptides are eluted directly off the column into the LTQ Orbitrap XL mass spectrometer, which is the instrument that we use in our laboratory.
4. Re-equilibrate the columns in Solvent A for 10 min prior to analysis of the next sample.
5. The scan sequence of the MS is based on a data-dependent method. Acquire full scan in the Orbitrap at a resolution of 60,000 and then acquire subsequent MS/MS scans of the five most abundant peaks in the spectrum in the linear ion trap. Use dynamic exclusion to exclude multiple MS/MS of the same peptide. Set the dynamic exclusion to a repeat count of 1, a repeat duration of 30 s, and an exclusion list of 500. The general MS conditions we use are: electrospray voltage of 1.6 kV, ion transfer tube temperature of 200°C, collision gas pressure of 1.3 mTorr, normalised collision energy of 35%, and an ion selection threshold of 500 counts for MS2. An activation q-value of 0.25 and an activation time of 30 ms for MS2 acquisitions are also used.
6. Identify the peptides using the tandem mass spectrum (MS-MS) data generated by mass spectrometry. Search each tandem mass spectrum against a database using database-searching

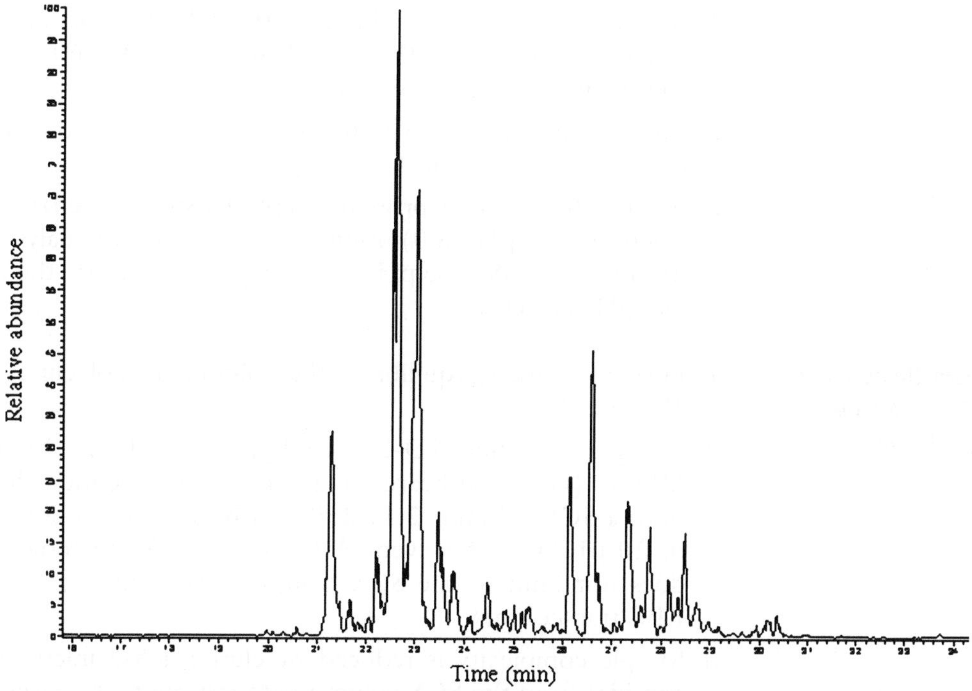

Fig. 1. Mass spectrometry trace showing the total ion intensity from all the mass spectra recorded during a 1D reverse phase liquid chromatography–mass spectrometry run over 35 min, shown as a function of elution time. The sample shown is a serum protein digest from a 2D gel spot.

software packages (see Note 8). Peptide identifications are reported in terms or XCorrelation scores and probability scores, as in the case for the SEQUEST algorithm (13) that we use in our analysis (Thermo Fisher Scientific). If several statistically significant peptides are identified from the same protein, then generally this protein identification is accepted.

7. Figure 1 shows the peptides obtained from a typical 1D LC/MS run from a 2D gel spot of a serum protein sample from a cancer patient.

3.3. Sample Preparation from an In-Solution Digestion of a Complex Protein Sample

1. Dissolve 1–10 mg of the target protein sample in the urea solution. The minimum reaction volume is 25 μL and the maximum volume is 1 mL. Heat at 60°C for 60 min to denature the proteins (see Note 9).
2. After denaturation, allow the reaction to cool and add the appropriate amount of 50 mM ammonium bicarbonate solution to reduce the urea concentration to below 1 M. Bring the final volume to 100 μL with 50 mM ammonium bicarbonate solution.
3. Reduce the sample by adding 5 μL of 200 mM DTT (final concentration, 10 mM DTT) and boil for 10 min.

4. Alkylate the sample by adding 4 μL of 1 M iodoacetamide (final concentration, 20 mM iodoacetamide) for 60 min at room temperature in the dark.
5. Add trypsin solution to a final protease:protein ratio of 1:50 (w/w). Incubate at 37°C overnight.
6. Acidify the digested protein sample to stop the enzymatic reaction (to a pH < 3.0) using 0.1% TFA prior to analysis by 2D-LC/MS-MS. Use pH indicator paper to ensure the correct pH is reached.

3.4. Two-Dimensional LC/MS for Complex Protein Samples

1. Prior to analysis, equilibrate the columns in Solvent A for 10 min.
2. Using the online 2D-LC configuration of the Ettan™ MDLC (GE Healthcare), load the sample peptide digest onto a SCX column (2.1 × 250 mm BioBasic) at a flow rate of 20 μL/min of Solvent A for 10 min. All material that does not bind to the SCX column passes through to the RPC trap column.
3. Sample complexity is reduced by eluting small fractions of peptides from the SCX column with salt plugs at concentrations of 0, 10, 25, 50, 100, and 500 mM ammonium formate containing 0.1% TFA. Elute the peptides off the SCX column by injection with 100 μL applications of 10, 25, 50, 100, and 500 mM ammonium formate containing 0.1% TFA with a flow rate of 20 μL/min and a collection time of 10 min per salt fraction.
4. These SCX fractions containing peptides are then subjected to a reverse phase separation. Load the peptides onto a trap column (300SC C18 trap column, 300 μm i.d. × 5 mm Zorbax™) that was first washed for 10 min with Solvent A. Elute the peptides onto the reverse phase column (300SC C18 analytical column, 75 μm i.d. × 150 mm, 3 μm) by using a linear 60 min gradient to 60% Solvent B and then hold for 10 min using 100% Solvent B before returning to 100% Solvent A for 10 min for column re-equilibration.
5. Elute the peptides into the mass spectrometer. We use an LTQ-XL linear ion trap for this type of analysis. The MS method consists of a cycle combining a full MS scan in profile mode followed with complementary MS-MS for the selected precursor ion using collision-induced dissociation (CID). For CID, set a 35% collision energy and a dynamic exclusion duration of 30 s.
6. Acquire a full scan in the linear ion trap in profile mode with data-dependent CID scans for each of the top three selected ions. Acquire MS/MS scans in consecutive instrument scans in the LTQ ion trap of the three most abundant peaks in the spectrum.

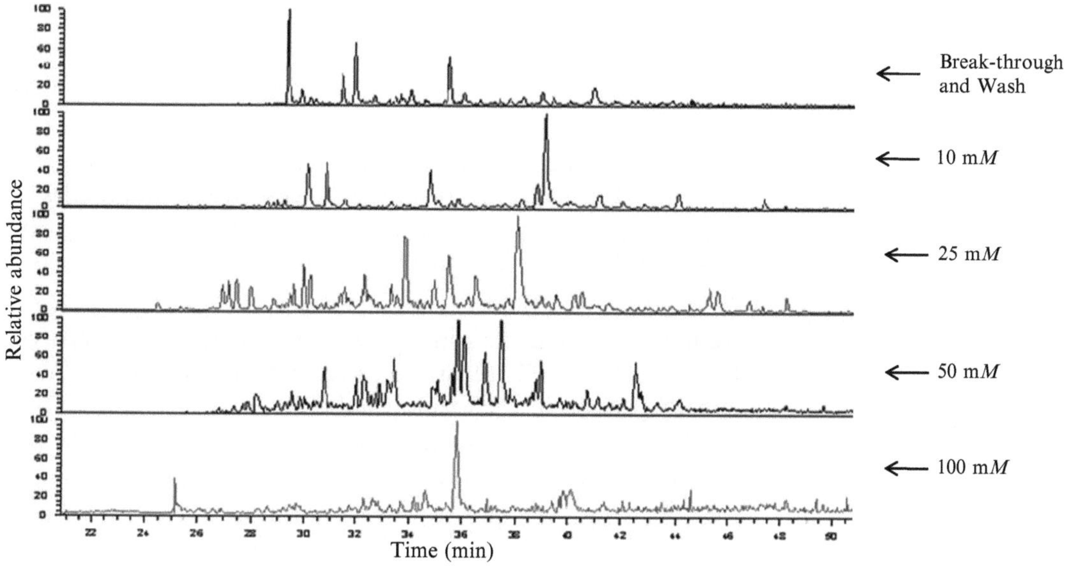

Fig. 2. Mass spectrometry traces overlaid from a 2D-LC run. The total ion intensity from all the mass spectra recorded during a 2D-liquid chromatography–mass spectrometry run over 60 min from the break-through and wash fraction, and then from the 10, 25, 50, and 100 mM salt washes, shown as a function of elution time.

Use dynamic exclusion to exclude multiple MS–MS of the same peptide. Set dynamic exclusion to a repeat count of 1, a repeat duration of 30 s, and an exclusion list of 500. The general MS conditions that we use are: electrospray voltage of 1.7 kV, ion transfer tube temperature of 200°C, collision gas pressure of 1.3 mTorr, normalised collision energy of 35%, and an ion selection threshold of 500 counts for MS2. An activation *q*-value of 0.25 and an activation time of 30 ms for MS2 acquisitions are also set.

7. Figure 2 shows the peptides obtained from different salt fractions of a typical 2D LC/MS experimental run from a complex whole cell lysate of a human cancer cell line.

4. Notes

1. There are many commercially available stains for visualising proteins on 1D or 2D gels. For mass spectrometry analysis, the silver stain method should not use glutaraldehyde.
2. All solvents must be LC/MS grade. All chemicals must be of the highest purity. Water must also be LC/MS grade.
3. When working with SDS–PAGE gels, gloves must be worn at all times to minimise keratin and dust contamination which can affect your LC/MS results. When excising protein from

gels, use sterile scalpel blades and this work should be done in a laminar flow cabinet to minimise the possibility of any hair, dust, or skin flakes contaminating the sample with unwanted keratins.

4. If you are using 96 well plates to carry out protein digests prior to LC, ensure that you use polypropylene-grade plastic and not regular polystyrene plastic plates. The use of polypropylene tubes and tips helps to minimise protein loss by adsorption.
5. Reduction and alkylation are essential for samples prepared by 1D SDS–PAGE separation even when reducing agent is already present in the loading buffer.
6. The most common enzyme used for peptide generation for LC/MS analysis is trypsin which has been modified by reductive alkylation to suppress trypsin autolysis. We use modified porcine trypsin (Promega) in our laboratory.
7. There are enzymes other than trypsin that can be used for protein digestion which have different cleavage specificities, such as lysyl endopeptidase (Lys-C) and chymotrypsin.
8. We use the SEQUEST (Thermo Fisher Scientific) search engine for protein identification. There are other similar search engines including MASCOT (Matrix Science) and X!Tandem.
9. In general, lower sample volumes and higher protein concentrations result in better digestions. Starting protein material should be free of any detergents prior to digestion.

References

1. Abersold, R., and Mann, M. (2003) Mass spectrometry-based proteomics. *Nature* **422**, 198–207.
2. Hanash, S. (2003) Disease proteomics. *Nature* **422**, 226–232.
3. Bouchal, P., Roumeliotis, T., Hrstka, R., Nenutil, R., Vojtesek, B., and Garbis, S.D. (2009) Biomarker discovery in low-grade breast cancer using isobaric stable tags and two-dimensional liquid chromatography-tandem mass spectrometry (iTRAQ-2DLC/MS/MS) based quantitative proteomic analysis. *J. Proteome Res.* **8**, 362–373.
4. Dowling, P., O'Driscoll, L., Meleady, P., Henry, M., Roy, S., Ballot, J., Moriarty, M., Crown, J., and Clynes, M. (2007) Two-dimensional difference gel electrophoresis of the lung squamous cell carcinoma versus normal sera demonstrates consistent alterations in the levels of 10 specific proteins. *Electrophoresis* **28**, 4302–4310.
5. Pernemalm, M., De Petris, L., Eriksson, H., Brandén, E., Koyi, H., Kanter, L., Lewensohn, R., and Lehtiö, J. (2009) Use of narrow-range peptide IEF to improve detection of lung adenocarcinoma markers in plasma and pleural effusion. *Proteomics* **9**, 3414–3424.
6. Quintana, L.F., Campistol, J.M., Alcolea, M.P., Bañon-Maneus, E., Sol-Gonzalez, A., and Cutillas, P.R. (2009) Application of label-free quantitative peptidomics for the identification of urinary biomarkers of kidney chronic allograft dysfunction. *Mol. Cell. Proteomics* **8**, 1658–1673.
7. Dowling, P., Wormald, R., Meleady, P., Henry, M., Curran, A., and Clynes, M. (2008) Analysis of the saliva proteome from patients with head and neck squamous cell carcinoma reveals differences in abundance levels of proteins associated with tumour progression and metastasis. *J. Proteomics* **71**, 168–175.

8. Rao, P.V., Reddy, A.P., Lu, X., Dasari, S., Krishnaprasad, A., Biggs, E., Roberts, C.T., and Nagalla, S.R. (2009) Proteomic identification of salivary biomarkers of type-2 diabetes. *J. Proteome Res* **8**, 239–245.
9. Qian, W.-J., Jacobs, J.M., Liu, T., Camp, D.G. 2nd., and Smith R.D. (2006) Advances and challenges in liquid-chromatography-mass spectrometry-based proteomics profiling for clinical applications. *Mol. Cell. Proteomics* **5**, 1727–1744.
10. Mauri, P., and Scigelova, M. (2009) Multidimensional protein identification technology for clinical proteomic analysis. *Clin. Chem. Lab. Med.* **47**, 636–646.
11. Washburn, M.P., Wolters, D., and Yates, J.R. 3rd (2001) Large-scale analysis of the yeast proteome by multidimensional protein identification technology. *Nat. Biotechnol.* **19**, 242–247.
12. Florens, L., and Washburn, M.P. (2006) Proteomic analysis by multidimensional protein identification technology. *Methods Mol. Biol.* **328**, 159–175.
13. MacCoss, M.J., Wu, C.C., and Yates, J.R. 3rd. (2002) Probability-based validation of protein identifications using a modified SEQUEST algorithm. *Anal. Chem.* **74**, 5593–5599.

Chapter 29

Strategies for the Purification of Membrane Proteins

Sinead Marian Smith

Abstract

Although membrane proteins account for 20–30% of the coding regions of all sequenced genomes and play crucial roles in many fundamental cell processes, there are relatively few membranes prateins with known 3D structure. This is likely due to technical challenges associated with membrane protein extraction, solubilisation, and purification. Membrane proteins are classified based on the level of interaction with membrane lipid bilayers, with peripheral membrane proteins associating non-covalently with the membrane, and integral membrane proteins associating more strongly by means of hydrophobic interactions. Generally speaking, peripheral membrane proteins can be purified by milder techniques than integral membrane proteins, whose extraction requires phospholipid bilayer disruption by detergents. Here, important criteria for strategies of membrane protein purification are addressed, with a focus on the initial stages of membrane protein solublilisation, where problems are most frequently encountered. Protocols are outlined for the successful extraction of peripheral membrane proteins, solubilisation of integral membrane proteins, and detergent removal which is important not only for retaining native protein stability and biological functions, but also for the efficiency of later purification techniques.

Key words: Peripheral membrane protein, Integral membrane protein, Detergent, Protein purification, Protein solubilisation

1. Introduction

Membrane proteins are associated with the membrane of a cell or particular organelle and are generally more problematic to purify than water-soluble proteins. Membrane proteins represent approximately 20–30% of the open-reading frames of an organism's genome (1) and play crucial roles in basic cell functions including signal transduction, energy production, nutrient uptake, and cell–cell communication. Additionally, many drugs and drug candidates target membrane proteins (2, 3). However, less than 2% of the listed 3D structures in the protein data bank (4) are

Dermot Walls and Sinéad T. Loughran (eds.), *Protein Chromatography: Methods and Protocols*, Methods in Molecular Biology, vol. 681, DOI 10.1007/978-1-60761-913-0_29, © Springer Science+Business Media, LLC 2011

membrane proteins, a fact that is likely due to the technical challenges associated with membrane protein solubilisation and purification in sufficient quantities for crystallisation (5).

Membrane proteins are classified into peripheral and integral membrane proteins, which are associated to varying degrees with the phospholipid bilayer (6, 7). Peripheral or extrinsic membrane proteins interact with the membrane surface non-covalently by means of electrostatic and hydrogen bonds. Peripheral membrane proteins can be recruited to the membrane during signalling events or are constitutively localised to the membrane. Integral or intrinsic membrane proteins are more strongly associated with the membrane and interact with hydrophobic moieties of the phospholipid bilayer. They contain one or more characteristic runs of apolar amino acids that span the lipid bilayer (6). Integral membrane proteins are further classified into Type I, which are positioned so that their COOH-terminus is embedded in the cytosol or Type II, which are positioned with the NH_2-terminus in the cytosol. Although there is no single protocol for the purification of membrane proteins, it is the initial stages of membrane protein solublilisation where problems are most frequently encountered. This chapter addresses important criteria for membrane protein extraction and solubilisation. The Subheading 3 describes protocols for the successful extraction of peripheral membrane proteins, solubilisation of integral membrane proteins, and detergent removal which is important not only for retaining native protein stability and biological functions, but also for the efficiency of later purification techniques.

1.1. Considerations for Membrane Protein Purification

The analysis of membrane proteins represents a significant technical challenge in the field of proteomics and there are several reasons why the purification of membrane proteins is more difficult than that of water-soluble proteins. Firstly, endogenous expression of membrane proteins is quite low and usually quite large quantities of protein are required for structural investigations. Additionally, integral membrane proteins are extremely hydrophobic and often require high concentrations of detergents for solubilisation. Membrane proteins have the tendency to form aggregates, even in the presence of detergents, resulting in the reduction of efficiency of subsequent separation techniques (8). The choice of detergent may also affect the efficiency of downstream protein purification procedures. For example, ion-exchange chromatography (see Chapter 12) should not be carried out in the presence of charged detergents, and hydrophobic interaction chromatography (see Chapter 24) can be problematic in the presence of all detergents (8). In such cases detergents can be removed (see Subheading 1.4). Once solubilised, the membrane proteins are often more susceptible to degradation by proteases. Thus, addition of protease inhibitors such as ethylenediamine tetraacetic acid (EDTA), which inactivates metalloproteases, or phenylmethyl sulfonyl fluoride (PMSF), which inhibits serine proteases, needs to be considered.

It is also worthwhile considering the availability of efficient functional assays to detect the protein of interest at different stages during the purification process, for example, measurement of enzymatic activity and immuno- or ligand-binding assays. Given the unique properties of individual proteins, it is usually necessary to determine appropriate assays on a case-by-case basis (8). There is no single protocol for obtaining membrane protein purification; more likely a series of methods are needed, depending on the particular needs of the investigator.

1.2. Peripheral Membrane Protein Extraction

Peripheral membrane proteins can be dissociated using relatively mild techniques that break the electrostatic or hydrogen bonds between the peripheral proteins and the membrane, without total membrane disruption. Common dissociating reagents for the extraction of peripheral membrane proteins are listed in Table 1. Extractions using buffers containing high salts are useful as they decrease electrostatic interactions between proteins and charged lipids (6). Chaotropic ions disrupt hydrophobic bonds present in the membrane surface and promote the transfer of hydrophobic groups from non-polar environment to the aqueous phase (6). Usually, extraction procedures employing high ionic strength NaCl and KCl, alkaline or acidic buffers, and metal chelators result in a relatively distinct separation between solubilised peripheral proteins and membrane-associated integral membrane proteins (7). High pH causes the fractionation of peripheral membrane proteins from integral membrane proteins by disrupting sealed membrane structures without denaturing the lipid

Table 1
Treatments for the extraction of peripheral membrane proteins

Treatment type	Example
Acidic buffers	pH 3.0–5.0
Alkaline buffers	pH 8.0–12.0 (e.g. 100 mM Na_2CO_3, pH 11.3, see Subheading 3.1)
Chaotropic ions	I^-, ClO_4^-, SCN^-
Denaturing agents	8 M urea or 6 M guanidine hydrochloride
Metal chelators	10 mM EDTA or EGTA
Salt solutions/high ionic strength	1 M NaCl or KCl

bilayer or extracting integral membrane proteins (9). The high pH method for extraction of peripheral membrane proteins is described in Subheading 3.1 of this chapter. It is worthwhile determining the effect of the buffer on any enzymatic activity the protein of interest may have, and potential interactions the buffer may have with any column matrix that will be used at later stages in the purification process. Additionally, buffer cost may need to be considered if large-scale preparations are to be carried out.

Following extraction (i.e. breaking of electrostatic and H bonds between peripheral protein and the membrane) in the chosen buffer for 10–30 min, the remaining membrane bilayer and its associated integral proteins are separated by centrifugation (30–60 min, 100,000 × *g*) and the released peripheral membrane proteins are recovered in the supernatant (7, 10).

1.3. Integral Membrane Protein Extraction

In order to solubilise integral membrane proteins, it is necessary to disrupt the lipid bilayer, which may be achieved with organic solvents, but is more commonly accomplished using detergents. Extraction using the organic solvent N-butanol (see Subheading 3.2) uses a biphasic system for solubilising proteins from membranes into dilute aqueous buffers. The low solubility of N-butanol in water, combined with its lipophilicity, minimally denatures proteins (9). Detergents are amphipathic molecules that contain both hydrophobic and hydrophilic moieties and form micelles in water. A micelle is a cluster of detergent molecules in which the hydrophilic head moieties face outward. Detergents solubilise proteins by binding to the hydrophobic parts of the protein on one side and interacting with the aqueous parts on the other side (8). The detergent of choice should sufficiently solubilise the membrane protein without irreversibly denaturing it. Detergents can be ionic, non-ionic, or zwitterionic. A list of commonly used detergents for extraction of integral membrane proteins is shown in Table 2. Selection of a particular detergent depends on the properties of the protein of interest and the given aims of subsequent experiments involving the purified protein. If there is little information in the literature on the purification of similar proteins, or if one is purifying a particular protein for the first time, it is often necessary to screen a number of detergents in order to optimise protein solubilisation. Membrane aliquots should be incubated with various concentrations of commonly used detergents and incubation time, buffer concentration, salt solutions, and temperature conditions necessary for optimal solubilisation should be determined.

When screening potential detergents, it is important to be aware of the unique critical micelle concentration (CMC), which is the concentration of free detergent at which the transition from disperse detergent molecules to a micellar structure occurs (10). Since solubilisation corresponds to the removal of the protein

Table 2
Detergents used for extracting integral membrane proteins

Detergent type	Name	Alternative chemical name	CMC (mM)
Ionic	CTAB	Cetyltrimethylammonium bromide	1.0
	Sodium cholate		~10
	Sodium deoxycholate		~2
Non-ionic	Big Chap	*N*,*N*-bis(3-D-gluconamidopropyl) cholamide	3.4
	$C_{12}E_8$	Octaethylene glycol monododecyl ether	<0.1
	Triton X-100	Nonaethylene glycol octylphenol ether	0.3
Zwitterionic	CHAPS	3-[(3-cholamidopropyl) dimethylammonio]-1-propanesulfonate	3–10
	CHAPSO	3-[(3-cholamidopropyl) dimethylammonio]-2-hydroxypropane-1-sulfonate	4–8
	LDAO	Dodecyldimethylamine oxide	~1

CMC critical micelle concentration

from the membrane into the detergent micelle, the CMC is the minimal concentration of detergent necessary to form the required micellar structure for protein extraction (10). CMC values, some of which are listed in Table 2, vary between different detergents, but are usually available from the detergent manufacturer.

Additional considerations when choosing detergents include evaluating the effects of a given detergent on the structural and functional properties of the protein of interest. The effects of detergents on the protein stability may be checked during preliminary screens using different detergents. The compatibility of the chosen detergent with subsequent purification steps should also be considered as certain detergents may affect the efficiency of certain chromatographic techniques. For example, charged detergents may cause problems using assays based on charge difference, such as ion-exchange chromatography (see Chapter 12), and lectin chromatography which may be used to affinity purify subsets of glycoproteins is especially sensitive to high concentrations of a variety of detergents (7, 8). It is often necessary to remove or replace detergents to overcome these problems, thus the ease at which excess detergent can be removed from the solubilised protein fraction should be considered (see Subheading 1.4).

When solubilising integral membrane proteins, buffered stock solutions at a physiological pH environment should be prepared containing the membrane preparation, detergent, and protease inhibitors, such as EDTA, EGTA, and/or PMSF (10). Membrane preparations are used at a final protein concentration of 1–5 mg/mL and are solubilised by detergent concentrations of 0.1–5% (v/v) (7, 10). The mixture should be stirred gently for 1 h at room temperature or 4°C, followed by centrifugation for 1 h at 100,000×*g* at 4°C. Generally speaking, retention of a membrane protein in the supernatant following centrifugation for 60 min at 100,000×*g* after solubilisation defines this protein as soluble (7). The pellet may subsequently be washed to remove residual detergent and finally re-suspended in the appropriate buffer (10). Protein recovery and activity should be investigated in both the pellet and supernatant at this stage. The procedure for solubilising membrane proteins using the non-ionic detergent Triton X-100 is outlined in Subheading 3.3.

1.4. Removal of Detergents from Membrane Protein Fractions

The high detergent concentrations that are often required during the initial extraction of integral membrane proteins could potentially affect the stability and subsequent analysis of the isolated membrane proteins; therefore, excess detergent should be removed or exchanged for an alternative detergent prior to subsequent purification procedures. Examples of methods used to remove or exchange detergents are listed in Table 3. The choice of technique depends on the unique properties of the detergent used and the concentration range of the protein fraction.

Successful detergent exchange or removal can be achieved using various chromatographic supports, followed by extensive washing with the desired buffer, containing the new detergent if necessary (6). Alternatively dialysis can be carried out to facilitate detergent exchange or removal. The efficiency of dialysis depends on the CMC and micelle molecular weight, which is determined by the aggregation number of detergent molecules (11). Most detergents with linear alkyl hydrophobic groups (e.g. Triton

Table 3
Commonly used techniques for detergent removal/exchange

Technique	Reagent
Affinity chromatography	Ligand immobilised sepharose
Equilibrium dialysis	Appropriate buffer or water
Gel permeation chromatography	Sephadex G-25 (GE Healthcare)
Hydrophobic interaction chromatography	Bio-Beads SM-2 (Bio-Rad)
Ion-exchange chromatography	Dowex 1-X2 (Sigma-Aldrich)
Precipitation	Acetone
Ultrafiltration	High molecular weight cut-off membrane

X-100) have a high micelle molecular weight value and do not pass through dialysis membranes (6). Detergents with a low micelle molecular weight and high CMC (e.g. bile acids and their derivatives) can be removed by dialysis (6). A protocol for dialysis is described in Subheading 3.4 of this chapter. Detergent removal by means of chromatographic supports (see Subheading 3.5) is relatively work-intensive, but is a more rapid procedure than dialysis so can be advantageous in cases where protein stability is an issue.

Following initial extraction of membrane proteins, solubilisation using detergent, and detergent removal or exchange, membrane proteins can then be purified to homogeneity using a variety of protein purification techniques, depending on the particular needs of the investigator and the given properties and abundance of the protein of interest. Because there is no single procedure to characterise membrane proteins, the key importance of membrane purification lies with the initial extraction and solubilisation steps, in order to generate a high yield of pure protein in its native biologically active state. The Subheading 3 of this chapter outlines examples of techniques used during the important initial stages of membrane protein purification.

2. Materials

2.1. Fractionation of Peripheral and Integral Membrane Proteins Using High pH

1. High pH buffer: 100 mM Na_2CO_3, pH 11.3.
2. Dounce homogeniser, e.g. Potter-Elvehjem PTFE pestle and glass tube (Sigma-Aldrich).
3. Ultracentrifuge, e.g. Thermo Scientific Sorvall WX.

2.2. Extraction of Membrane Proteins Using Butanol

1. N-butanol.
2. Cooled bench top centrifuge, e.g. Eppendorf centrifuge 5417R.

2.3. Extraction of Membrane Proteins Using Triton X-100

1. TE buffer: 10 mM Tris–HCl, 2 mM EDTA.
2. 2% Triton X-100 in phosphate buffered saline (PBS) (see Note 1).
3. Ultracentrifuge, e.g. Thermo Scientific Sorvall WX.

2.4. Removal of Non-ionic Detergents by Detergent-Adsorption Chromatography

1. Columns with a bed volume of approx. 5 mL (e.g. Econo-column, Bio-Rad).
2. Commercially available detergent absorption matrix (e.g. Bio-Beads SM-2, Bio-Rad, see Note 2).
3. Blocking buffer: 0.1% (w/v) bovine serum albumin in 50 mM Tris–HCl, pH 7.4, 0.15 M NaCl (see Note 3).
4. Washing buffer: 50 mM Tris–HCl, pH 7.4, 0.15 M NaCl.

2.5. Removal of Detergent with Low Micelle Size and High CMC by Dialysis

1. Dialysis tubing with a molecular weight cut-off of approx. 10,000 Da.
2. Wash buffer: 100 mM $NaHCO_3$, 50 mM EDTA.
3. Dialysis buffer: 20 mM Tris–HCl, pH 7.4, 0.15 M NaCl.

3. Methods

3.1. Fractionation of Peripheral and Integral Membrane Proteins Using High pH

1. Re-suspend the membrane fraction (see Note 4) at a concentration of <2 mg/mL in high pH buffer (see Notes 5 and 6).
2. Homogenise the suspension in a dounce homogeniser using six to eight strokes.
3. Maintain at 4°C for 30 min. Mix by vortexing three times during this period.
4. Pellet the membrane fraction by centrifugation for 60 min at 100,000 × *g* at 4°C and transfer the supernatant, which contains the peripheral membrane proteins, into a fresh tube and assay for protein (see Note 7).

3.2. Extraction of Membrane Proteins Using Butanol

1. Add an equal volume of N-butanol to the membrane fraction (see Note 4) and maintain at 4°C.
2. Centrifuge at 500 × *g* at 4°C for 10 min to separate the mixture into an upper phase containing butanol and membrane lipids and a lower aqueous phase containing solubilised proteins. Lipid rich material is localised to the interface.
3. Separate the upper and lower aqueous phases into separate tubes.
4. Dialyze the aqueous phase against a large volume of water or suitable buffer.
5. Assay the dialysed aqueous phase for protein (see Note 8).

3.3. Extraction of Membrane Proteins Using Triton X-100

1. Re-suspend cells in TE buffer at a concentration of 1×10^7 cells/mL.
2. Centrifuge the cells at 40,000 × *g* for 10 min. Remove the supernatant and add fresh TE.
3. Repeat this step and re-suspend the cells in approximately 1 mL of TE.
4. Add cells drop-wise to the 2% Triton X-100 while stirring (see Note 9).
5. Allow to solubilise for 30 min at 4°C.
6. Centrifuge at 100,000 × *g* for 30 min at 4°C.
7. Transfer the supernatant to a fresh tube and assay for protein (see Note 10).

3.4. Removal of Non-ionic Detergents by Detergent-Adsorption Chromatography

1. Before starting, ensure that the protein fraction containing the non-ionic detergent (e.g. Triton X-100) has a concentration of >1 mg/mL (see Note 11) and that the molecular weight of the protein to be recovered is large enough to avoid entrapment in the pores of the affinity matrix.
2. Apply distilled water to the column matrix, followed by blocking buffer. Next, apply washing buffer to the column and repeat wash step.
3. Transfer the protein fraction to the column matrix (see Note 12).
4. Collect 0.5–1 mL fractions and assay for protein.

3.5. Removal of Detergent with Low Micelle Size and High CMC by Dialysis

1. Prepare the dialysis tubing by boiling a section in wash buffer for 10 min (see Note 13). Then boil the dialysis tubing in distilled water for 10 min, followed by washing thoroughly in distilled water.
2. Transfer the solubilised membrane protein fraction into the dialysis tubing (see Note 14) which is securely closed at one end by either tying a double-knot in the tubing or securing it with a plastic clamp (see Note 15).
3. Remove air bubbles and seal the dialysis tubing, allowing for a volume increase during dialysis. Check the integrity of the seal to ensure no leakage occurs.
4. Place the tubing in a beaker containing a large external volume (approx. 5 L) of the appropriate buffer. Dialyze with gentle stirring at 4°C. Change the external buffer regularly.
5. When the dialysis is finished, remove the dialysis tubing and wash the outside. Hold the tubing and carefully remove the upper clamp. Using a Pasteur pipette, transfer the protein fraction to a new tube (see Note 16).

4. Notes

1. Make a stock solution of 20% Triton X-100 by weighing 2 g Triton X-100 and adding PBS up to 10 mL and stirring gently until fully dissolved. Store the stock solution at 4°C.
2. Bio-Beads are macro-porous polystyrene beads and have a high surface area that adsorbs organics with a molecular weight of <2,000 from aqueous solution. They may be used to remove Triton X-100 from protein fractions. Due to the presence of linear alkyl hydrophobic groups, Triton X-100 has a high micelle molecular weight value and does not pass through dialysis membranes. Detergents with a low micelle molecular weight and high CMC (e.g. bile acids and their derivatives) can be removed by dialysis (see Subheading 3.5).

3. Bovine serum albumin is used as a bulk carrier protein to saturate non-specific protein binding sites and minimise protein loss during this procedure.
4. The starting material depends on the source from which the membrane proteins are being purified. Membrane proteins can be successfully isolated from plant and animal tissues or cell cultures, bacteria, yeast, and fungi. Animal tissues can be broken easily with a mixer or blender. Due to the presence of robust cell walls, unicellular organisms like yeast or bacteria and plant cells are more difficult to disrupt. Different techniques for breaking down cell walls include glass bead milling, grinding mills, homogenization, ultrasonication, osmotic shock, repeat freeze thawing, and enzymatic lysis (8). If possible, the protein should be prepared from sources where it is in high abundance, as a certain amount of protein may be lost during the purification process. The starting material can be enriched if the target protein is known to be associated with the plasma membrane, mitochondria, or nucleus. During initial steps of membrane protein isolation, cytosolic proteins can be removed to obtain an enriched preparation of membranes containing the protein of interest. Soluble cytoplasmic proteins are extracted by cell disruption in a neutral pH, isotonic, and detergent-free buffer (7), followed by differential centrifugation or purification using sucrose gradient centrifugation.
5. The pH of the working buffer should be tested following addition of any protease inhibitors, as addition of such components may alter the final pH of the buffer.
6. It is worthwhile determining the effect of the high pH buffer on any enzymatic activity the protein of interest may have, and considering potential interactions the buffer may have with any column matrix that will be used at later stages in the purification process.
7. A suitable protein concentration assay should be considered. Options include measuring ultra-violet absorbance at 280 nm, or using one of several commercially available dye-binding assays, such as the Bradford assay, the bicinchonic acid (BCA) assay, or other assays (see Chapter 13).
8. It is worthwhile to keep the butanol phase for protein assays as it may contain extremely hydrophobic proteins that are difficult to solubilise.
9. The effect of the Triton X-100 solubilisation procedure on the structural and functional properties of the protein of interest should be evaluated during preliminary screening experiments. In order to maintain catalytic activity, the membrane protein should be dissolved under optimal conditions for stability at a detergent/protein ratio that is not much above the

minimal detergent/protein ratio required for solubilisation (8). Additionally, proteins are more susceptible to protease attack following solubilisation with detergents, so protease inhibitors are necessary to prevent protein degradation. Premixed cocktails of commonly used protease inhibitors are now available commercially from a variety of companies including Roche, Sigma-Aldrich, and Pierce. It is recommended to carry out purification procedures at 4°C in order to minimise proteolysis (see Chapter 4). Additionally, the effects of Triton X-100 on subsequent purification techniques should be evaluated. Replenish protein stabilising additives or protease inhibitors if they are removed or inactivated at any stage in the experiment, for example EDTA is removed by hydroxyapatite chromatography (8). If possible, minimise any purification steps that add new detergents or alter the original detergent/lipid ratio.

10. Due to the presence of aromatic groups, Triton X-100 has substantial UV absorbance at 280 nm, thus an alternative protein concentration assay should be carried out. For the same reason, Triton X-100 is not suitable for subsequent purification steps involving column chromatography with UV monitoring of the fractions. As an alternative, bile salts and their derivatives including CHAPS and CHAPSO can be used for solubilisation.
11. A high concentration is necessary to allow for any loss of protein during the procedure.
12. Use washing buffer to dissolve the protein fraction for optimum detergent binding.
13. As dialysis tubing is susceptible to cellulolytic micro-organisms, gloves should be worn when handling the tubing.
14. A small funnel may be used to aid transfer of the protein fraction into the dialysis tubing.
15. Prior to transferring the protein fraction into the dialysis tubing, the integrity of the membrane and clamp/knot can be tested by applying water or buffer and checking the tubing for leaks.
16. Avoid losing dialyzed samples by carefully opening the tubing over a larger glass beaker to collect any accidental spillage.

References

1. Wallin, E., and von Heijne, G. (1998) Genome-wide analysis of integral membrane proteins from eubacterial, archaean, and eukaryotic organisms. *Protein Sci.* 7, 1029–1038.
2. Mohanty, A. K., and Wiener, M. C. (2004) Membrane protein expression and production: effects of polyhistidine tag length and position. *Protein Expr. Purif.* **33**, 311–325.
3. Gordon, E., Horsefield, R., Swarts, H. G., de Pont, J. J., Neutze, R., and Snijder, A. (2008) Effective high-throughput overproduction of membrane proteins in *Escherichia coli*. *Protein Expr. Purif.* **62**, 1–8.

4. Berman, H. M., Battistuz, T., Bhat, T. N., Bluhm, W. F., Bourne, P. E., Burkhardt, K., Feng, Z., Gilliland, G. L., Iype, L., Jain, S., Fagan, P., Marvin, J., Padilla, D., Ravichandran, V., Schneider, B., Thanki, N., Weissig, H., Westbrook, J. D., and Zardecki, C. (2002) The Protein Data Bank. *Acta Crystallogr. D Biol. Crystallogr.* **58**, 899–907.
5. Doerr, A. (2009) Membrane protein structures. *Nat. Methods* **6**, 1.
6. Ahmed, H. (2005) *Principles and Reactions of Protein Extraction, Purification and Characterization.* CRC, Boca Raton, FL.
7. Ohlendieck, K. (1996) Extraction of membrane proteins. In *Protein Purification Protocols* (Doonan, S., Ed.), Humana, Totowa, NJ.
8. Von Jagow, G., Link, T., and Schager, H. (1994) Purification strategies for membrane proteins. In *A Practical Guide to Membrane Protein Purification* (Von Jagow, G., and Schager, H., Eds.), Academic, San Diego, CA.
9. Rosenberg, I. M. (2005) *Protein Analysis and Purification: Benchtop Techniques.* 2nd. ed., Springer, Boston, MA.
10. Schimerlik, M. I. (2001) Overview of membrane protein solubilization. *Curr. Protoc. Neurosci.* Chapter 5, Unit 5.9.
11. Ohlendieck, K. (1996) Removal of detergent from protein fractions. In *Protein Purification Protocols* (Doonan, S., Ed.), Humana, Totowa, NJ.

Chapter 30

A Multi-Step Chromatographic Strategy to Purify Three Fungal Endo-β-Glucanases

Tracey McCarthy and Maria G. Tuohy

Abstract

Fungi and fungal enzymes have traditionally occupied a central role in biotechnology. Understanding the biochemical properties of the variety of enzymes produced by these eukaryotes has been an area of research interest for decades and again more recently due to global interest in greener bio-production technologies. Purification of an individual enzyme allows its unique biochemical and functional properties to be determined, can provide key information as to the role of individual biocatalysts within a complex enzyme system, and can inform both protein engineering and enzyme production strategies in the development of novel green technologies based on fungal biocatalysts. Many enzymes of current biotechnological interest are secreted by fungi into the extracellular culture medium. These crude enzyme mixtures are typically complex, multi-component, and generally also contain other non-enzymatic proteins and secondary metabolites. In this chapter, we describe a multi-step chromatographic strategy required to isolate three new endo-β-glucanases (denoted EG V, EG VI, and EG VII) with activity against cereal mixed-linkage β-glucans from the thermophilic fungus *Talaromyces emersonii*. This work also illustrates the challenges frequently involved in isolating individual extracellular fungal proteins in general.

Key words: Fungal enzymes, (1-3)(1-4)-β-D-glucanase, Thermostable, *Talaromyces emersonii*, Glycoprotein, Cereal β-glucans

1. Introduction

Microorganisms such as fungi elaborate very efficient enzyme systems to convert the vast amounts of carbohydrates present in the biosphere. However, recent fungal genome sequencing data have revealed that fungal genomes contain multiple genes encoding an individual type of enzyme activity. For example, the genome of the mesophilic filamentous fungus *Aspergillus oryzae* contains >20 exoglycosidases (multiple β-xylosidases and β-glucosidases)

Dermot Walls and Sinéad T. Loughran (eds.), *Protein Chromatography: Methods and Protocols*, Methods in Molecular Biology, vol. 681, DOI 10.1007/978-1-60761-913-0_30,

belonging to a single glycosyl hydrolase family (GH3; (1)). The genome of this fungus is also known to contain 69 exopeptidase and 65 endopeptidase-coding genes, yet relatively few (18) have been biochemically characterised to-date (1).

Filamentous fungi are known producers of enzymes that depolymerise β-glucan polysaccharides. The β-glucan family of polysaccharides represents a large and heterogeneous group of polysaccharides found in abundance in the cell walls of plants, lichens, fungi, yeast, and algae. Cellulose, a homopolymer of up to 15,000 β-1,4-linked glucose units (2), is the most abundant renewable carbon source in the biosphere, while fungal and yeast cell wall glucans are typically comprised of polymers of β-1,3-linked glucose chains with occasional β-1,6-branchpoints (3). Mixed-linkage or (1-3)(1-4)-β-D-glucans are found in abundance mainly in the endosperm tissue of members of the Gramineae, including cereals of commerce such as barley, oats, rice, sorghum, and wheat (3, 4). Depolymerization of cereal β-glucan polysaccharides is brought about by the action of a number of different glycosyl hydrolases (5), which may or may not act cooperatively. These enzymes can be sub-divided into four main classes: (a) true lichenases (EC 3.2.1.73; (1-3)(1-4)-β-D-glucan 4-glucanohydrolase) identified to-date almost exclusively in species of *Bacillus* and plants, although two examples of "lichenases" from fungal sources have recently been reported (6, 7); (b) 1,3(4)-β-D-glucanases (EC 3.2.1.6) active on (1-3)(1-4)-β-D-glucans and (1-3)-β-D-glucans; (c) 1,3-β-D-glucanases (EC 3.2.1.39; (1-3)-β-D-glucan 3-glucanohydrolase), which have low or no activity on (1-3)(1-4)-β-D-glucans; and (d) 1,4-β-D-glucanases (EC 3.2.1.4; (1-4)-β-D-glucan 4-glucanohydrolase, or "cellulase"). Only enzymes belonging to mechanistic class EC 3.2.1.73 are specific for 1,4-linkages adjacent to 1,3-linkages in cellotriosyl and cellotetraosyl repeat units present in the main water-soluble (1-3)(1-4)-β-D-glucan isolated from barley (8, 9). The majority of "lichenases" purified to-date are from bacterial and plant sources (9–13). Enzymes in classes (a), (b), and (c) above have been grouped into glycosyl hydrolase family 16, while plant (1-3)(1-4)-β-D-glucanases belong to glycosyl hydrolase family 17 (6). β-Glucanase enzymes have interesting applications in the production of bioenergy and biorefinery feedstocks, as well as in the food, feed, beverage, functional foods, and nutraceutics industries.

Enzyme multiplicity is a reported characteristic feature of several microbial sources of cereal β-glucan-depolymerizing enzymes, including two fungal sources of 1,3(4)-β-glucanases, *Cochliobolus carbonum* and *Phaffia rhodozyma* (14, 15). Purification of individual isoenzymes of a given type of enzyme activity is essential to determine the unique biochemical and functional properties of each enzyme, to obtain a molecular fingerprint by peptide sequencing techniques and to gain an insight into the role of individual enzymes within a complex enzyme system, through analysis of potential synergy or anti-synergy between isolated enzymes. Furthermore, isolation of individual enzymes is important to understand whether the enzymes (or isoenzymes) are in fact isoforms of a single gene product, generated as the result of differential splicing and/or post-translational modification or are derived from the expression of individual genes. In addition, enzymes from fungal species isolated from unique ecological niches, e.g. thermophilic environments, may be exploited as model systems to investigate biochemical and structural factors that underpin protein stability.

This study reports on the isolation of three distinct endoglucanases with noteworthy thermal stabilities and activity against cereal β-glucans from the thermophilic fungus *Talaromyces emersonii*. *T. emersonii* produces an array of β-glucan hydrolases, and multiple β-glucanases are often present in extracellular enzyme systems produced by this fungus. In this study, we describe a strategy used to isolate three β-glucanases with activity against mixed-linkage cereal β-glucans, from *T. emersonii*. The approach, which involves sequential gel filtration, ion-exchange, hydrophobic interaction and lectin affinity chromatographies, and chromatofocusing (also known as pseudo anion ion-exchange chromatography), yields highly purified preparations of the three enzymes.

The three enzymes, denoted EG V, EG VI, and EG VII, are low molecular weight glycoproteins with acidic isoelectric point values (p*I*) that exhibit maximum activity on (1-3)(1-4)-β-D-glucans from barley and lichenan, and differential preferences for 1,4-β-D-glucans and 1,3-β-D-glucans. The study enabled detailed biochemical characterization of EG V–VII and suggested that EG V belongs to EC 3.2.1.4, while the other two enzymes (EG VI and EG VII) are potentially 1,3(4)-β-glucanases (EC 3.2.1.6). The results also clearly confirmed that EG V–VI are distinctly different from four 35-kDa endocellulase isoforms (EC 3.2.1.4; EG I–IV) (16) and a novel 40.7-kDa "lichenase" type component isolated recently from *T. emersonii* (7).

2. Materials

Unless otherwise stated, all research chemicals were purchased from Sigma-Aldrich (Dublin, Ireland), and all solvents were from Reidel de Haën AG (Hannover, Germany).

2.1. Production of Crude Fungal Enzyme Preparation

1. Solka floc (ball-milled spruce cellulose) was purchased from Brown & Co. (Maine, USA). Include at 2% (w/v) in nutrient medium prior to sterilization for 30 min at 108°C.

2. Yeast extract (YE) and Sabouraud dextrose agar (SDA) medium were from Oxoid Ltd. (Basinstoke, Hants, UK). Use 5 g of YE powder in 1 L 5× nutrient medium. Prepare SDA for routine fungal cultivation on agar plates as instructed on the product label (manufacturer's instructions).
3. Chemicals for the preparation of nutrient (culture) media were ammonium sulphate [$(NH_4)_2SO_4$; use 75 g per L of 5× nutrient medium], boric acid [H_3BO_3; use 0.0625 g per L of 5× nutrient medium], calcium chloride dihydrate [$CaCl_2 \cdot 2H_2O$; use 2.5 g per L of 5× nutrient medium], cobalt chloride hexahydrate [$CoCl_3 \cdot 6H_2O$; use 0.0625 g per L of 5× nutrient medium], corn steep liquor [use 25 mL per 1 L 5× nutrient medium], ferrous sulphate heptahydrate [$FeSO_4 \cdot 7H_2O$; use 0.3125 g per L of 5× nutrient medium], magnesium sulphate heptahydrate [$MgSO_4 \cdot 7H_2O$; use 2.5 g per L of 5× nutrient medium], potassium dihydrogen phosphate [anhydrous KH_2PO_4; use 25 g per L of 5× nutrient medium], potassium iodide [KI; use 0.0625 g per L of 5× nutrient medium], sodium sulphate [Na_2SO_4; use 2.5 g per L of 5× nutrient medium], zinc sulphate heptahydrate [$ZnSO_4 \cdot 7H_2O$; use 0.0625 g per L of 5× nutrient medium].
4. Chemicals for preparation of nutrient (culture) medium obtained from British Drug House (BDH) were manganese sulphate tetrahydrate [$MnSO_4 \cdot 4H_2O$; use 0.0625 g per L of 5× nutrient medium], sodium molybdate [Na_2MoO_4; use 0.0625 g per L of 5× nutrient medium].
5. Sodium hydroxide [NaOH] pellets: Prepare a 1 M NaOH solution to adjust the pH of the nutrient medium prior to sterilization.
6. New Brunswick Innova orbital shaker (New Brunswick Scientific).

2.2. Measurement of Enzyme Activity

1. Barley-β-glucan (BBG; ~25 cst) and Pachyman (Megazyme International): Prepare buffered 1% (w/v) solutions of β-glucans as follows: Heat ~50 mL of 100 mM sodium acetate [NaOAc; 100 mM $CH_3COONa–CH_3COOH$] buffer, pH 5.0 to 60°C in 250 mL Erlenmeyer flasks for each β-glucan (on a hotplate). Add the β-glucan powder gradually with constant stirring until dissolved (see Notes 1 and 2). Remove the solution from the heat, and add buffer to bring the final volume to 100 mL. Stir and allow to cool for a further 30 min.
2. High M_r (~100 cst) and mixed degree of polymerisation (DP) β-glucan preparations were gifts from Kerry Biosciences (formerly Quest International Ireland Ltd.): Prepare 1% (w/v) solutions of these β-glucans in NaOAc buffer, pH 5.0 as described in Subheading 2.2 (step 1) (see Notes 1 and 2).

3. Carboxymethylcellulose (CMC; viscosity of 2% solution at 25°C is 10–20 cps; degree of substitution, 0.65–0.85; DP 200), Laminaran (*Laminaria digitata*) and Lichenan (*Usnea barbata*): Prepare 1% (w/v) solutions of these β-glucans in NaOAc buffer, pH 5.0 as described in Subheading 2.2 (step 1) (see Notes 1 and 2).
4. Pectin (citrus), polygalacturonic acid, and xylan (oat spelts and birchwood): Prepare 1% (w/v) solutions of each polysaccharide in 100 mM NaOAc buffer, pH 5.0 (see Note 3).
5. Avicel was sourced from Merck Ltd: Prepare a 1% (w/v) suspension in 100 mM NaOAc buffer, pH 5.0 approximately 1 h before use. Mix continuously, especially when it is for use in enzyme assays (see Note 2).
6. Rhodymenan (1,3;1,4-β-D-xylan) is from laboratory stocks and was isolated by Dr. M. Tuohy from the red alga *Palmaria palmata*, as described previously (17). Prepare a 1% (w/v) buffered solution of this linear polysaccharide for enzyme assay in 100 mM NaOAc buffer, pH 5.0. Store aliquots of this substrate at −20°C until required [typically 30-40 mL in clearly labelled and dated 50 mL polypropylene centrifuge tubes (Sarstedt Ltd.)].
7. D-(+)-Glucose, D-(+)-galacturonic acid, and D-(+)-xylose: Prepare 1 mg/mL stocks in 100 mM NaOAc buffer, pH 5.0 for standard curves to determine enzyme activity with β-glucans (glucose), pectin and polygalacturonic acid (galacturonic acid), and xylans (xylose).
8. 4-Nitrophenyl (4NP)-β-D-glucopyranoside, 4NP-β-D-galactopyranoside, and 4NP-β-D-xylopyranoside: Prepare 1 mM buffered 4NP-β-D-glycoside solutions in 100 mM NaOAc buffer, pH 5.0 (see Note 3).
9. 4-Nitrophenol standard (10 mM), from Sigma-Aldrich: Prepare suitable dilutions to give a standard range from 0 to 0.2 mM for a standard curve to determine enzyme activity with 4-nitrophenyl-β-D-glycosides.
10. Cellooligosaccharides: Prepare 100 mM stock solutions of the cellooligosaccharides (cellobiose, cellotriose, cellotetraose, and cellopentaose) in 100 mM NaOAc buffer, pH 5.0.

2.2.1. Preparation of the Dinitrosalicylate Reagent

1. Enzyme activity using the polysaccharides as substrates is determined by quantifying reducing sugars released with alkaline dinitrosalicylic (DNS) acid reagent (17, 20, 23). The chemicals required for the DNS reagent were NaOH pellets, 3.5-dinitrosalicylic acid, and potassium sodium tartrate tetrahydrate ($C_4H_4KNaO_6 \cdot 4H_2O$).

2. Solution A: Alkaline DNS Reagent. Dissolve 16 g of NaOH in 200 mL distilled H_2O (2 M NaOH solution), and heat this solution to 80-90°C. Add 10 g of 3.5-dinitrosalicylic acid powder gradually with constant stirring (having the NaOH solution hot induces rapid dissolution of the DNS powder). *Caution*: wear gloves and a mask when handling DNS powder, and wear gloves when dispensing and handling DNS solution as it is caustic.
3. Solution B: Rochelle Salt. Prepare a solution of potassium sodium tartrate tetrahydrate by dissolving 300 g $C_4H_4KNaO_6{\cdot}4H_2O$ in 500 mL distilled H_2O (warm this solution to 30-40°C).
4. Add solution B to solution A in a 1 L volumetric flask and cool. Adjust the final volume to 1 L with distilled H_2O [the DNS reagent is stable for 3-4 weeks when stored in a dark glass bottle].

2.3. Determination of the Protein Content of Samples

1. Protein content was determined using the Bensadoun and Weinstein modification of the method of Lowry and coworkers ((18) see also Chapter 13). Bovine serum albumin (BSA) fraction V powder is prepared as a 200 µg/mL stock solution in double-deionized H_2O (denoted MilliQ water; Milli-RX 20 system).
2. Sodium deoxycholate monohydrate ($C_{24}H_{39}NaO_4 \cdot H_2O$). Prepare a 0.15% (w/v) sodium deoxycholate solution in MilliQ water for solubilisation of all proteins prior to precipitation with trichloroacetic acid (TCA).
3. TCA is prepared as a 72% (w/v) TCA solution in MilliQ H_2O for precipitation of proteins, prior to analysis of protein content. *Caution*: wear gloves when handling crystalline and solubilised TCA, as TCA is highly corrosive and can cause skin irritations.
4. Chemicals required for the Lowry assay method were obtained from Sigma-Aldrich. In MilliQ water, prepare the following solutions: A: 1% (w/v) $CuSO_4{\cdot}5H_2O$ solution; B: 2% (w/v) Na (+) tartrate solution, and C: 2% (w/v) Na_2CO_3 (anhydrous) in 0.1 M NaOH. Prepare the "Combined Reagent," directly before assay, by mixing 1 mL B plus 1 mL A (in that order) plus 100 mL C. Mix the final solution well.
5. A 1 N Folin-Ciocalteau Reagent was prepared by making a half dilution of the stock reagent supplied by Sigma-Aldrich with MilliQ water (the concentration on the label of the stock solution should be checked as products from different suppliers can vary).

2.4. Enzyme Purification

1. Fractions are collected during chromatography using a Redi-Frac fraction collector (GE Healthcare).

2. Ultrafiltration is conducted using an Amicon DC2 unit equipped with a H1P 10-43 hollow-fibre dialyzer (Millipore, formerly Amicon).
3. Pretreat dialysis tubing (25 mm width; MW 12,400; cellulose) by boiling in 1.0 M Na_2CO_3 for 30-40 min, followed by thorough washing with MilliQ water.

2.4.1. Anion-Exchange Chromatography

1. DE-52 (diethylaminoethyl-cellulose; DEAE-cellulose) anion-exchange matrix from Whatman BioSystems Ltd. is supplied as a pre-swollen material.
2. Prepare the following buffers. Buffer A: 500 mM CH_3COONa–CH_3COOH (NaOAc), pH 5.5 buffer (10× Buffer A) and 50 mM CH_3COONa–CH_3COOH (NaOAc), pH 5.5 buffer (1× Buffer A). Buffer C: 250 mM CH_3COONa–CH_3COOH (NaOAc), pH 4.5 buffer (10× Buffer C), and 25 mM CH_3COONa–CH_3COOH (NaOAc), pH 4.5 buffer (1× Buffer C). Buffer E: 250 mM Na_2HPO_4–NaH_2PO_4, pH 7.0 buffer (10× Buffer E) and 25 mM Na_2HPO_4–NaH_2PO_4, pH 7.0 buffer (1× Buffer E).
3. Prepare a slurry of DE-52 (20% w/v) in 10× Buffer A, C, or E as appropriate and wash with gentle swirling periodically over a 1-h period. Allow the gel to settle, decant the buffer and repeat this wash procedure a further two times.
4. Pour the equilibrated gel slurry into a BioRad perspex column (BioRad Laboratories Ltd.) to yield final matrix dimensions of 2.0×4.5 cm (see Note 4 and 7).

2.4.2. Hydrophobic Interaction Chromatography

1. Phenyl Sepharose CL-4B is available as a gel suspension in 20% (v/v) ethanol from Pharmacia Biotech (now GE Healthcare) (see also Chapter 24).
2. Prepare 50 mM CH_3COONa–CH_3COOH (NaOAc), pH 5.0 buffer (Buffer B).
3. Prepare a 20% (v/v) suspension of Phenyl Sepharose CL-4B in Buffer B and pour into a perspex column to give a final gel column size of 1.5×5.5 cm. Equilibrate with three column volumes of Buffer B.

2.4.3. Gel Filtration Chromatography

1. Prepare 100 mM CH_3COONa–CH_3COOH (NaOAc), pH 5.0 buffer (Buffer D).
2. Re-hydrate BioGel P-60 gel filtration matrix, which is supplied in dry powder format (BioRad Laboratories Ltd.), overnight at room temperature in Buffer D. Pour the re-hydrated gel slurry into a perspex column (GE Healthcare; final matrix dimensions, 1.5×75.0 cm; see Note 7) and equilibrate with at least three column volumes of Buffer D.

3. Sephacryl S-100 HR (HR, high resolution) is available as a gel suspension in 20% (v/v) ethanol (Pharmacia Biotech, now GE Healthcare). Prepare a (20% v/v) suspension of Sephacryl S-100 HR in Buffer D prior to pouring into a perspex column (GE Healthcare; final matrix dimensions, 2.3 × 126.6 cm; see Note 7), and equilibrate with at least three column volumes of Buffer D.
4. Standard protein molecular weight calibration kits for gel filtration are available from GE Healthcare.

2.4.4. Lectin Affinity Chromatography

1. Prepare 100 mM CH_3COONa–CH_3COOH (NaOAc), pH 5.0 containing 1.0 M NaCl, 1.0 mM $CaCl_2$, and 1.0 mM $MgCl_2$ (Buffer F).
2. Concanavalin A is available as a gel suspension in 20% (v/v) ethanol from Pharmacia Biotech (GE Healthcare).
3. Prepare a suspension of Concanavalin A-Sepharose 4B in Buffer F before pouring into a perspex column (GE Healthcare, final matrix dimensions, 2.0 × 3.2 cm; see Note 7), and equilibrate with three column volumes minimum of Buffer F.

2.4.5. Chromatofocusing

1. Prepare 25 mM piperazine-HCl buffer, pH 3.5 (Buffer G).
2. Polybuffer exchanger 94 for chomatofocusing (GE Healthcare).
3. Prepare a 20% (v/v) suspension of Polybuffer exchanger 94 in Buffer G in advance of pouring into a perspex column (BioRad Laboratories Ltd.; final matrix dimensions, 1.5 × 14.5 cm; see Note 7) and equilibrate with at least three column volumes of Buffer G.
4. Polybuffer 74 and Pharmalyte (2.5-5.0) stock solutions for elution of adsorbed proteins during chomatofocusing were obtained from Pharmacia Biotech/GE Healthcare.

2.5. Electrophoresis and Zymogram Staining

Use polyacrylamide gel electrophoresis (PAGE) under denaturing conditions in a discontinuous buffer system (in the presence of sodium dodecyl sulphate (SDS) and 2-mercaptoethanol) to assess protein purity, by the method of Laemmli [(22); see also Chapter 13]. The reagents and materials are as follows.

2.5.1. Sodium Dodecyl Sulphate Polyacrylamide Gel Electrophoresis

1. Acrylamide/bisacrylamide. *Caution*: wear gloves at all times when all manipulating acrylamide solutions, as it is a neurotoxin in its unpolymerized liquid form.
2. SDS (also known as lauryl sulphate). *Caution*: wear gloves and a mask when weighing and handling SDS powder as it is a respiratory irritant. Prepare a 10% (w/v) solution of SDS by adding 10 g SDS powder to ~80 mL MilliQ water, allow to dissolve and bring to a final volume of 100 mL. Note: the stock SDS solution may crystallize when stored at 4°C, but if warmed gently or brought to room temperature it will re-dissolve.

3. Trizma base (Tris) was obtained from Sigma-Aldrich and Hydrochloric acid (HCl) from BDH Chemicals Ltd. Two Tris–HCl buffer solutions are required. To prepare 1.5 M Tris–HCl pH 8.8 buffer solution, add 36.30 *g* Trizma base to 100 mL MilliQ water. Allow to dissolve, adjust the pH to 8.8 with 1 M HCl and bring to a final volume of 200 mL with MilliQ water. To prepare a 0.5 M Tris–HCl pH 6.8 buffer solution, dissolve 6 g Trizma base in approximately 60 mL MilliQ water, bring to pH 6.8 with 1 M HCl and adjust to a final volume of 100 mL with MilliQ water. Note: both Tris–HCl buffers should be stored at 4°C for 1-2 weeks only. Check the pH each time before use.
4. Prepare 5× electrode running buffer, pH 8.3 by adding 15 g/L Trizma base, 72 g/L Glycine, and 5 g/L SDS. The final pH of the solution when all components are dissolved in double-distilled deionised H_2O should be 8.3.
5. 10% (w/v) Ammonium persulphate (APS) is prepared by dissolving 100 mg APS [$(NH_4)_2S_2O_8$] in 1.0 mL of MilliQ water in a 1.5 mL microcentrifuge tube. Prepare this solution immediately before use.
6. Sample loading buffer (10×) is prepared by combining the following components: 3.2 mL 10% (w/v) SDS, 0.8 mL 2-mercaptoethanol, 2.0 mL 0.5-M Tris–HCl buffer pH 6.8, 1.6 mL Glycerol, made up to a final volume of 8.0 mL with H_2O. This solution is stable for 4-6 weeks at 4°C, and for months if kept at −20°C.
7. Additional reagents required are isoamyl alcohol (3-methyl-1-butanol) and TEMED (*N*, *N*, *N*′, *N*′-tetramethylene-ethylene-diamine), the catalyst for polymerization.
8. "Low molecular weight" and "High molecular weight" protein electrophoresis calibration kits (GE Healthcare).
9. A Mini-Protean II electrophoresis system, a Power Pac 1000 power supply and a BioRad GelDoc 2000 system were used in this study [BioRad Laboratories Ltd.].

2.5.2. Isoelectric Focusing

1. This IEF Procedure assumes use of the Pharmacia Flat Bed FBE 3000 system (GE Healthcare).
2. IEF is conducted on Ultrathin (1 mm thick) pre-made Ampholine PAG plates (GE Healthcare), over the pH range from 3.5 to 9.5, as per the manufacturer's instructions (Booklet No. 18-1016-67).
3. Reagents for IEF: anode solution, 1 M Phosphoric acid (H_3PO_4) and cathode solution, 1 M NaOH, both from GE Healthcare. Alternatively, solutions of 1 M H_3PO_4 and 1 M NaOH can be prepared in the laboratory using MilliQ water and ultrapure stocks of H_3PO_4 and NaOH. Gloves should be

worn at all times and care should be taken as both solutions are corrosive (acid) or caustic (alkali).

4. Insulating fluid: light mineral oil (Sigma-Aldrich).
5. Standard protein calibration kit over the pH range 3.5–9.5 for IEF (GE Healthcare). The standard protein mixture contains amyloglucosidase (pI 3.5); soybean trypsin inhibitor (pI 4.55); β-lactoglobulin (pI 5.2); bovine carbonic anhydrase B (pI 5.85); human carbonic anhydrase B (pI 6.55); horse myoglobin acidic band (pI 6.85); horse myoglobin basic band (pI 7.35); lentil lectin acidic band (pI 8.15); lentil lectin middle band (pI 8.45); lentil lectin basic band (pI 8.65); trypsinogen (pI 9.3).

2.5.3. Visualisation of Proteins in SDS-PAGE and IEF Gels Using Coomassie Blue R-250

Wear gloves at all times when handling gels to prevent contamination by thumbprints.

1. Staining solution: Coomassie Brilliant Blue R-250, methanol, and acetic acid. Prepare a 0.2% (w/v) solution of Coomassie Brilliant Blue R-250 in methanol/acetic acid/MilliQ water (40:10:50) (see Note 6). Filter before use through large Whatman No.1 filter paper disks.
2. De-staining solution: CH_3OH/CH_3COOH/MilliQ water (40:10:50) solution.

2.5.4. Zymography

1. Wash solution 1: isopropanol (propan-2-ol). Prepare 100 mL 25% (v/v) isopropanol (propan-2-ol) in assay buffer [50 mM NaOAc ($CH_3COONa–CH_3COOH$) buffer, pH 5.0].
2. Wash solution 2: 50 mM NaOAc buffer, pH 5.0.
3. Agar overlay: 1% (w/v) Difco Noble Agar (BD Biosciences) is prepared by adding 1 g of agar powder to 80 mL of assay buffer (100 mM NaOAc buffer, pH 5.0) and heating the suspension to 100°C to dissolve the agar. Once the agar is completely dissolved, remove the solution from the heat and allow to cool (~70°C). Add a 20 mL aliquot of 1.0% (w/v) buffered substrate to the agar solution, bring the volume to 100 mL with buffer. Mix this solution by gentle swirling before transferring to GelBond, an agarose support medium (FMC Bio-products).
4. Staining solution: Congo red. Prepare a 0.2% (w/v) Congo red solution in MilliQ water.
5. De-staining solution: 1 M sodium chloride (NaCl). Prepare 1 L of a 1 M NaCl solution in MilliQ water.
6. Contrast solution: 1 M hydrochloric acid (1 M HCl). Prepare ~500 mL of 1 M HCl.

3. Methods

3.1. Microbial Culture Conditions for Crude Fungal Enzyme Production

1. Inoculate SDA plates with individual plugs of mycelial mat from laboratory glycerol stocks (maintained at -20°C until required). Incubate plates at 45°C for 3-4 days until the agar surface is covered by at least 2-3 cm diameter of mycelial mat (fungus).
2. Prepare "seed" liquid cultures by inoculating 100-800 mL sterilized mineral salts/inducing medium containing 2% (w/v) solka floc (as added carbon source), pH 4.5, in 250 mL-2 L Erlenmeyer flasks with 3-4 ~1 cm^2 pieces of mycelial mat. Place inoculated flasks in a New Brunswick Innova orbital shaker set at 45°C and 250 rpm (19).
3. Large-scale cultures of *T. emersonii* to obtain sufficient enzyme for purification studies are produced in a 10 L New Brunswick bioreactor on sterilized mineral salts/inducing medium (19) with 2% (w/v) solka floc as carbon source.
4. Harvest liquid cultures after growth for a specified time period (i.e. 132 h) by centrifugation at 2,568 ×*g* for 45 min in a Sorvall superspeed centrifuge equipped with a GSA rotor.
5. Filter the supernatant (liquid fraction) through several layers of sterile fine-grade muslin to remove fungal mycelia. This culture filtrate is used for enzyme purification studies. Typical yields of each β-glucan-degrading activity per litre at 132 h are β-1,3;1,4-glucanase 54,806 IU (barley β-1,3;1,4-glucan as assay substrate; BBGase), CMCase 26,521 IU (β-1,4-glucanase using carboxymethyl cellulose; CMC) laminarinase 6,201 IU and pachymanase 13,545 IU, while the corresponding extracellular protein levels typically reach 573 mg.

3.2. Measurement of Enzyme Activity

3.2.1. Polysaccharide Hydrolase Activity

1. For convenience all enzyme assays with polysaccharide substrates, unless otherwise stated, are carried out at 50°C for 15 min in 100 mM sodium acetate (NaOAc) buffer pH 5.0.
2. Transfer 400 μL of substrate to individual 2.0 mL microcentrifuge tubes (Sarstedt) for "test" and "substrate blanks" in triplicate. Add 200 μL buffer to the "substrate blank" tubes (to replace enzyme). The final substrate concentration is 0.66% (w/v).
3. Prepare "enzyme blanks" by transferring 400 μL of buffer (no substrate) to individual 2.0 mL microcentrifuge tubes (triplicates).
4. Prepare "reagent blanks" by dispensing 600 μL of buffer to individual 2.0 mL microcentrifuge tubes (triplicates).
5. Pre-incubate these tubes at 50°C for 5 min.

6. Pre-incubate a stock of suitably diluted enzyme (enzyme dilutions are prepared using the assay buffer) separately for ~4 min.
7. Add 200 μL of the suitably diluted enzyme to the "test" and "enzyme blanks" only.
8. To terminate the reaction and measure reducing sugars released, add 900 μL of DNS reagent to each tube and mix well using a vortex whirlimixer.
9. Boil the tubes for 10 min in a waterbath. Cool the tubes by sitting in cold tap water (in a sink).
10. When cooled, mix the contents of the individual tubes and transfer 250 μL from each tube to individual wells in a high optical quality, low protein-adsorbing 96-well microtitre plate (Sarstedt).
11. Determine the absorbance values of the supernatants at 540 nm (A_{540}) or 550 nm (A_{550}) using a microplate reader equipped with a suitable filter, e.g. Victor™ Multi-label microtitre plate reader (PerkinElmer Ltd).
12. Prepare a range of standard sugar concentrations (0-1 mg/mL), each tube with 600 μL working volume, by varying the volumes of 1 mg/mL sugar stocks and assay buffer. Add 900 μL of DNS reagent to each tube, mix well, boil, cool, and read the absorbance values as described above. Subtract the absorbance for the "0 mg/mL" concentration from all of the other absorbance values and plot absorbance (*y*-axis) vs. concentration (mg/mL *x*-axis).
13. To convert absorbance values to International Units of enzyme activity (IU/mL), subtract the absorbance of the "reagent blank" from all of the other absorbance values. Then subtract the residual "enzyme blank" and "substrate blank" values from the "tests" to give the final corrected absorbance values.
14. Determine the concentration of sugar (mg/mL) corresponding to corrected "test" absorbance values using the equation for the line ($y = mx + c$) obtained for the relevant sugar standard curve. Calculate IU/mL (see Note 5) enzyme using the following equation:
    a. IU/mL = mg/mL sugar "test" × 1/15 × 1/0.2 × 1/0.18 (for hexoses) × enzyme dilution factor
15. The hydrolysis of BBG, CM-cellulose (CMC), pachyman, Avicel, laminaran, lichenan, galactomannan (locust bean gum), and polygalacturonic acid are measured using the method given above.
16. As reaction mixtures containing polygalacturonic acid become turbid on addition of DNS, centrifuge these mixtures at

20,800×*g* for 5 min in an Eppendorf 541C Bench Microcentrifuge (Eppendorf UK Ltd.). Then determine the absorbance values of the supernatants.

17. Hydrolysis of oat spelts and birchwood xylans, and rhodymenan is also quantified by measuring reducing sugars released in a different assay format, as described previously (17, 20).
18. For the assays with the xylans, add 50 μL of suitably diluted pre-incubated enzyme to 450 μL of pre-incubated 1% (w/v) xylan for "tests." Prepare at least three replicates per enzyme sample to be analyzed.
19. Prepare controls as follows: (1) reagent blank: 500 μL of assay buffer, (2) substrate blank: 50 μL of assay buffer plus 450 μL of pre-incubated 1% (w/v) xylan, and (3) enzyme blank: 50 μL of suitably diluted pre-incubated enzyme plus 450 μL of assay buffer. Prepare at least three replicates per control sample.
20. Incubate "test" and blanks for 10 min. Add 750 μL of DNS reagent to each tube, mix the contents of each tube well and boil for 10 min.
21. Cool, mix the contents of the individual tubes and transfer 250 μL from each tube to individual wells in a high optical quality, low protein-adsorbing 96-well microtitre plate and measure the absorbance values at 540 nm (A_{540}) or 550 nm (A_{550}) as described above. Subtract blank absorbance values from the test absorbance values as described previously.
22. Prepare a xylose standard curve over the range 0-1 mg/mL within a working volume of 500 μL. Add 750 μL of DNS reagent to each tube, mix the contents of each tube well, boil for 10 min and cool. Mix the contents of the individual tubes, transfer 250 μL from each to individual wells measure the absorbance values at 540 nm (A_{540}) or 550 nm (A_{550}).
23. Determine the concentration of sugar (mg/mL) corresponding to corrected "test" absorbance values using the equation for the line ($y = mx + c$) obtained for the relevant sugar standard curve. Calculate IU/mL (see Note 5) enzyme using the following equation:

    IU/mL = mg/mL sugar "test" × 1/10 × 1/0.05 × 1/0.15 (for pentoses) × enzyme dilution factor

3.2.2. Exoglycosidase Activity

1. Measure exoglycosidase activity by monitoring the increase in absorbance at 410 nm (A_{410}) due to the release of the 4-nitrophenolate anion after incubating suitably diluted enzyme with 1 mM 4-NP-glycosides in a 110-μL reaction volume. Use 100 mM sodium acetate buffer, pH 5.0 as assay buffer.

2. Transfer 5 μL suitably diluted enzyme to clean wells in a microtitre plate (high optical quality, low protein-adsorbing quality); transfer 5 μL assay buffer to wells designated as "substrate blanks," 100 μL to "enzyme blank" wells and 110 μL to "reagent blank" wells. Add 100 μL 1 mM 4-NP-glycoside to 'tests' and 'substrate blanks. Cover the microplate (use a microplate cover or an strip of adhesive film or aluminium foil) and incubate at 50°C for 30 min.
3. Add a 100 μL volume of 1.0 M Na_2CO_3 to stop the reaction (changes the pH) and develop the characteristic yellow colour of the 4-nitrophenolate anion (alkaline pH).
4. Measure the absorbance at $\lambda = 405$ nm (A_{405}, a reference filter at $\lambda = 550$ nm was also used) due to free 4-nitrophenolate anion using a Victor™ Multi-label microtitre plate reader, equipped with a 405-nm filter.
5. Quantify the concentration of 4-NP released by reference to a standard curve of 4-NP (0.02-0.20 μmol/mL).
6. Subtract the reagent blank from all other absorbance values (blanks and tests). Then subtract the remaining substrate and enzyme blank absorbance values from the tests.
7. Determine the concentration of 4-NP (μmol/mL) corresponding to corrected "test" absorbance values using the equation for the line ($y = mx + c$) obtained for the relevant sugar standard curve. Calculate IU/mL (see Note 5) enzyme using the following equation:

   IU/mL = μmol/mL 4-NP "test" × 1/30 × 1/0.005 × enzyme dilution factor

3.3. Estimation of Protein Content

3.3.1. Method 1: Bensadoun and Weinstein Method

Protein in crude extracts and fractionated samples is determined by the Bensadoun and Weinstein modification of the method of Lowry and coworkers (18). Protein standard (20-200 μg/mL BSA) was treated in the same manner as samples.

1. Transfer 1.0 mL of suitably diluted protein sample (or standard) to clean 1.5 mL Polypropylene microcentrifuge tubes.
2. Add 0.1 mL of 0.15% Na deoxycholate and mix well.
3. Add 0.1 mL of 72% TCA and mix well again (the final volume in each microcentrifuge tube should be 1.2 mL).
4. Allow microcentrifuge tubes to stand on ice for 15 min.
5. Centrifuge at 18,845 × *g*, for 15 min, using a microfuge.
6. Decant the supernatant solutions very carefully and drain the pellets.
7. Resuspend each pellet in 1.0 mL of Reagent C (alkaline reagent required to resuspend acid-precipitated pellets). A dilution of the protein sample can be made at this step if required, prior to analysis.

8. Allow to stand for 10-15 min, then mix well but very gently to minimize protein loss by shearing, etc. (excessive vortexing can result in the latter).
9. Analyze resuspended protein solutions by the Lowry method (see below).
10. Aliquot 0.1 mL of each resuspended standard/sample solution into clean 1.5 mL microcentrifuge tubes (assay in triplicate at least).
11. Add 1.0 mL of freshly prepared Combined Reagent to each tube. Additions should be made in sequence, at fixed time intervals, over a 10-min period.
12. Ten minutes after adding Combined Reagent to the first microcentrifuge tube, add 0.1 mL of 1 M Folin-Ciocalteau reagent. The sequence and times of addition should be identical to those for the Combined Reagent.
13. Measure the absorbance of each solution at $\lambda_{max} = 680$ nm in a spectrophotometer, against a MilliQ water blank (or a plate reader). Absorbance values can be measured 10 min after addition of the Folin-Ciocalteau reagent to the first tube.
14. Determine the protein content of each sample (taking any sample dilutions into consideration) by reference to the BSA standards.

3.3.2. Method 2: Ultraviolet Absorption Analysis

1. For convenience during chromatographic separation, measure the absorbance at 280 nm (A_{280}) of fractions collected during chromatography, which provides data for protein elution profiles.
2. Estimate molar extinction coefficients (mol L/cm) after measuring the UV absorption spectra of purified enzyme samples, in MilliQ water, at 280 nm (21).

3.4. Enzyme Purification Techniques

A strategy involving sequential gel filtration, ion-exchange, hydrophobic interaction, and lectin affinity chromatographies and chromatofocusing is required to obtain highly purified preparations of the novel β-glucanase isoenzymes, EG V, EG VI, and EG VII. Unless otherwise stated, conduct all chromatography steps at 4°C. Record all volumes throughout enzyme purification.

The steps required to obtain homogeneous preparations of EG V, EG VI, and EG VII are outlined in Fig. 1.

1. The initial step involves concentration of the crude culture filtrate by two sequential approaches. The first approach consists of ultrafiltration in an Amicon DC2 unit equipped with a H1P 10-43 hollow-fibre dialyzer at a pressure of <10 psi. (or alternative ultrafiltration system – avoid using cellulose-based membranes, which can be degraded by cellulases in the

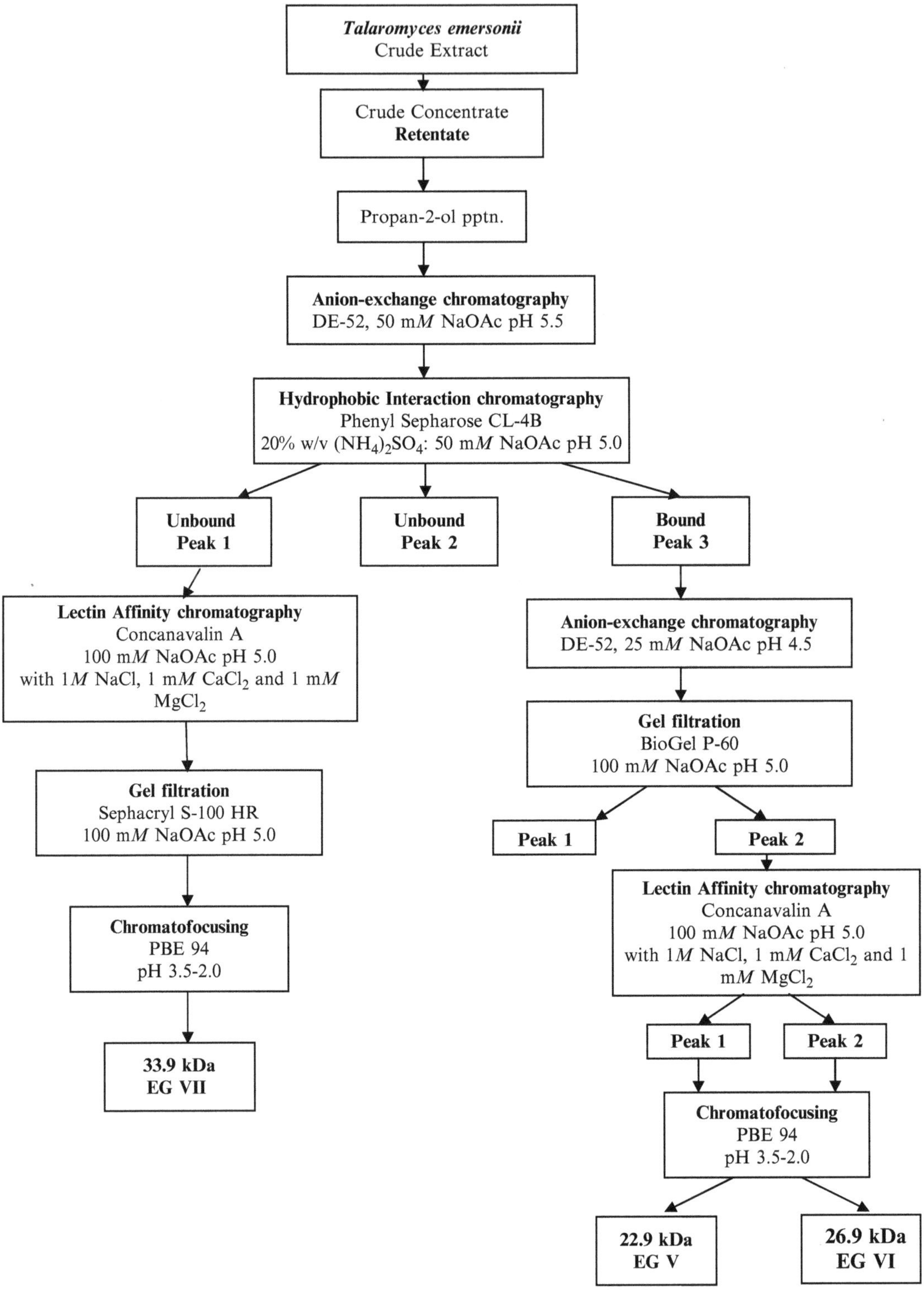

Fig. 1. Schematic overview of the steps required for the purification of EG V, EG VI, and EG VII.

enzyme preparation). Aim to obtain a 60-fold concentrate and minimal loss of β-glucanase activity (BBGase, CMCase, and laminarinase).

2. A variety of methods have been investigated for the second concentration step, but solvent precipitation at 4°C with two volumes of ice-cold (−70°C) propan-2-ol maximizes enzyme recovery, yet facilitates the removal of interfering substances.
3. Centrifuge the precipitate at 15,322 ×*g* for 30 min (SS-34 rotor in a Sorvall superspeed centrifuge) to recover β-glucanase activity in the pellet fraction.
4. Evaporate residual solvent under a stream of air, resuspend the pellet (minimum volume of Buffer D) and clarify by centrifugation at 7,818 ×*g* for 30 min.

3.5. Purification of Isoenzymes EG V and EG VI

3.5.1. Anion-Exchange Chromatography

1. Dialyze the final concentrate against 100 volumes of Buffer A, and apply to the column of DEAE-cellulose (DE-52) anion-exchange matrix, pre-equilibrated with Buffer A.
2. Elute protein at a flow-rate of 25 mL/h; collect 3.0 mL fractions and assay for enzyme activity and protein content (a typical profile is shown in Fig. 2a). In this profile one peak of BBGase activity (~40%) elutes in the application and wash (fractions 40–170) and contains xylanase, β-glucosidase and one peak of CMCase activity.
3. Apply a 500 mL linear salt gradient (0.0-1.0 M NaCl) in Buffer A to elute adsorbed protein. This will yield a second peak of BBGase activity (fractions 260–470) between 0.05 and 0.50 M NaCl (this peak co-elutes with a second peak of CMCase activity).

3.5.2. Hydrophobic Interaction Chromatography

1. Dialyze the second peak from AEC against Buffer B and apply to a Phenyl Sepharose CL-4B, which has been pre-equilibrated with Buffer B containing 20% (w/v) ammonium sulphate [$(NH_4)_2SO_4$].
2. Bring the sample to 20% (w/v) saturation with $(NH_4)_2SO_4$.
3. Elute protein at a flow-rate of 25 mL/h using $(NH_4)_2SO_4$-saturated Buffer B. This will yield three peaks of BBGase activity, one in the application (fractions 40–60), a second peak in the wash (fractions 75–100), and a third larger peak between fractions 159–225 in the simultaneous decreasing $(NH_4)_2SO_4$ (20%→0%) and increasing ethylene glycol (0%→20%) gradient.
4. Pool each of the three peaks, dialyze independently against 100 volumes of MilliQ water, and store at 4°C until required.

3.5.3. Second Anion-Exchange Chromatography

1. Dialyze Peak 3 from hydrophobic interaction chromatography (HIC) against Buffer C and purify further using a second

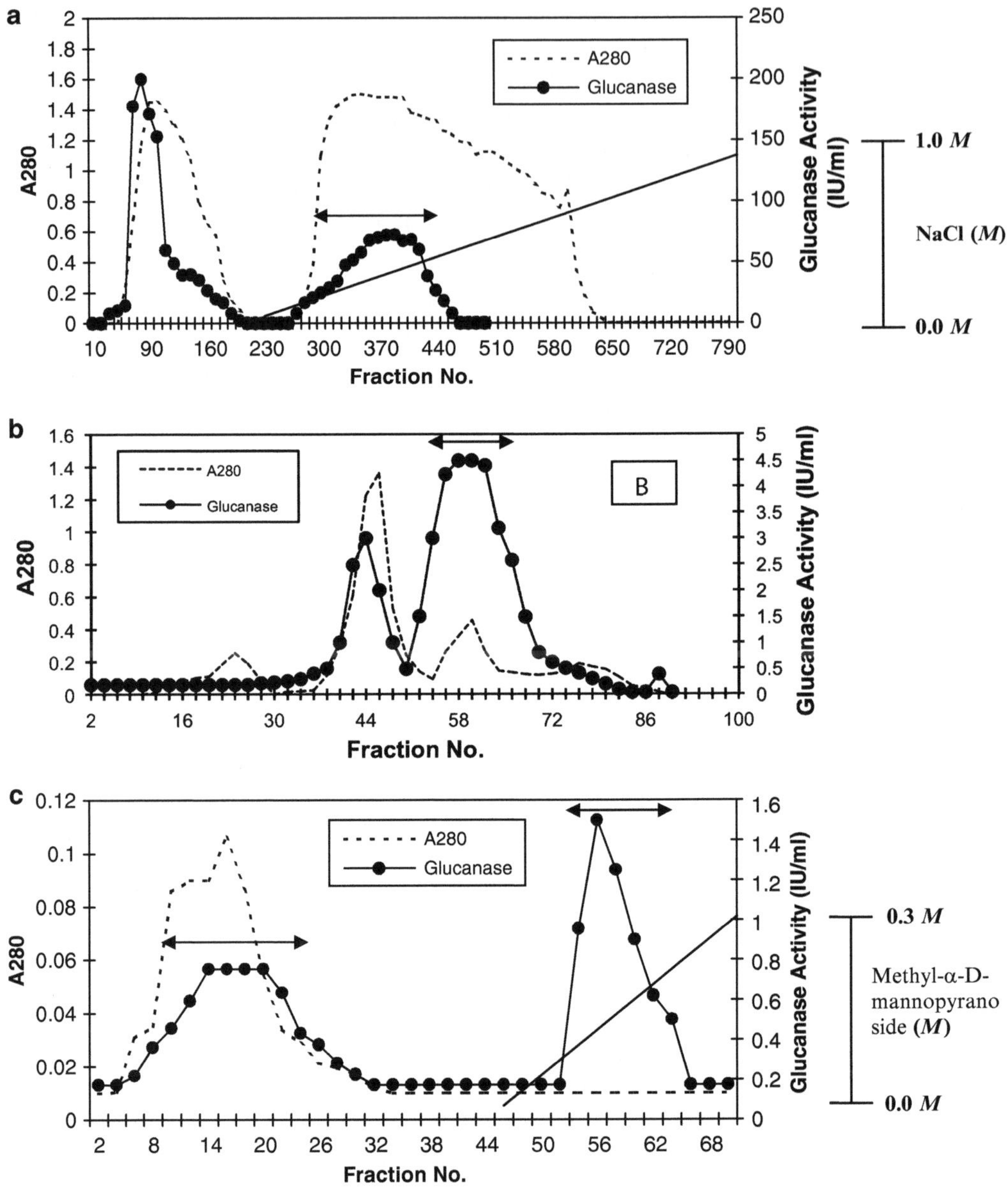

Fig. 2. Anion-exchange (**a**), BioGel P-60 (**b**), and lectin affinity chromatography (**c**) elution profiles for EG V and EG VI during enzyme purification. Peaks pooled for subsequent separation steps are indicated with a *black arrow.*

anion-exchange step on DEAE-cellulose (DE-52) column, pre-equilibrated with Buffer C. This will give two peaks of BBGase, one in the application (fractions 100–200) and a second larger peak in the buffered 0.0-1.0 M NaCl gradient (fractions 300–375).

2. Pool fractions corresponding to the second peak and dialyze against MilliQ water, before dialyzing against 100 volumes of Buffer E.

3. Apply the dialyzed enzyme sample to a DE-52 matrix pre-equilibrated at pH 7.0 with Buffer E. Elute BBGase using a 100 mL linear gradient of 0.0-0.3 M NaCl in Buffer E. AEC at pH 7.0 will yield two protein peaks, the second of which contains BBGase activity. Pool this BBGase peak (fractions 28–56 strictly), dialyze against 100 volumes of MilliQ water and lyophilize (or freeze-dry).

3.5.4. GFC on BioGel P-60

1. Resuspend the lyophilized sample in 2.0 mL Buffer D and fractionate further by gel filtration chromatography on a column of BioGel P-60, pre-equilibrated with Buffer D.
2. Elute protein with the same buffer (collect 1.0 mL fractions) at a flow-rate of 25 mL/h; assay for enzyme activity and protein content as before. Gel filtration chromatography (GFC) will give two peaks of BBGase activity, the first peak (peak 1; higher molecular mass) elutes in fractions 40–50. A second, lower molecular mass peak (peak 2) elutes in fractions 55–70 (see Fig. 2b).

3.5.5. Lectin Affinity Chromatography

1. Pool peak 2 from GFC (higher specific activity) and dialyze overnight against Buffer F.
2. Apply the dialyzed sample to a column of Concanavalin A-Sepharose 4B, pre-equilibrated with Buffer F.
3. Irrigate the column with 54.0 mL of Buffer F, and then a 40.0 mL linear gradient (0.0-0.3 M) of methyl-α-D-mannopyranoside (in Buffer F). Use a flow-rate and fraction size of 17.0 mL/h and 1.0 mL, respectively. BBGase activity will elute in two peaks (see Fig. 2c), a broad, unbound peak (peak 1; fractions 5–30), and a sharp peak that elutes at a methyl-α-D-mannopyranoside concentration of 0.15 M (peak 2; fractions 63–76).

3.5.6. Chromatofocusing

1. Pool both peaks from lectin affinity chromatography (LAC) separately, dialyze overnight at 4°C against MilliQ water, and lyophilize.
2. Apply a 5.0 mL aliquot of pharmalyte (pH 2.5–5.0), diluted 45-fold with MilliQ water and adjusted to pH 2.5 with HCl, to the chromatofocusing column of PBE 94, that was pre-equilibrated with Buffer G prior to sample application to generate a stable, linear, and decreasing pH gradient.
3. Resuspend each lyophilised BBGase peak in 1.0 mL Buffer G and apply to the column in independent runs. Use a 25 mL/h flow-rate throughout.
4. Apply additional diluted pharmalyte to elute adsorbed protein.
5. BBGase activity present in peak 1 from LAC elutes in a single peak in fractions 110–125 (in the most acidic range of the gradient)

and yields homogeneous EG V. Peak 2 from LAC gives a single BBGase active peak in fractions 45–55, near the beginning of the gradient, and represents homogeneous EG VI.

6. Remove pharmalytes from EG V and EG VI samples by bringing each separate fraction pool to 20% (w/v) saturation with $(NH_4)_2SO_4$.
7. Apply each peak to a column (1.5 × 5.5 cm) of Phenyl Sepharose CL-4B, pre-equilibrated with 20% (w/v) $(NH_4)_2SO_4$. Elute adsorbed EG V and EG VI using MilliQ water, lyophilize, and assess purity by electrophoresis [sodium dodecyl sulphate polyacrylamide gel electrophoresis (SDS-PAGE)].

3.6. Purification of Isoenzyme EG VII

The first four steps in the purification of EG VII are identical to those used for EG V and EG VI.

1. Lyophilize peak 1 (unbound; fractions 40–60) from HIC on Phenyl Sepharose CL-4B (Buffer B) and resuspend in 2.0 mL of Buffer F before application to the LAC column (pre-equilibrated with Buffer F).
2. Elute unbound protein with 46.0 mL of Buffer F.
3. Elute adsorbed protein at a flow-rate of 17.0 mL/h in 1.0 mL fractions using a 25 mL linear gradient (0.0-0.3 M) of buffered methyl-α-D-mannopyranoside to obtain a sharp peak of BBGase activity at a methyl-α-D-mannopyranoside concentration of 0.1 M (fractions 54–67).
4. Dialyze this BBGase pool against 100 volumes of MilliQ water, lyophilize, resuspend in 2.0 mL of Buffer D and apply to a column of Sephacryl S-100 HR (GFC), pre-equilibrated with Buffer D.
5. Elute protein from the GFC column at a flow-rate of 25.0 mL/h (in 1.0 mL fractions) and assay for protein content and BBGase activity. GFC will give a single symmetrical BBGase peak between fractions 210 and 260.
6. Pool fractions 210–260 from this GFC, dialyze against MilliQ water at 4°C, and lyophilize.
7. Resuspend in 0.5 mL of Buffer G and apply to the chromatofocusing column as described previously. Fractions 96–104 (end of the gradient) contain a single peak of BBGase.
8. Pool the corresponding peak fractions and remove residual pharmalytes as before assessing purity by electrophoresis (SDS-PAGE).

3.6.1. Preparation of a Purification Table

1. Measure the BBGase activity and protein content of pooled samples prepared during enzyme purification.

2. Use total (sample) volumes pooled at each step during purification to calculate total protein and total BBGase in each sample.
3. Determine specific activity by dividing enzyme activity (IU/mL) by protein content (mg/mL).
4. Table 1 shows a summary of the purification of EG V, EG VI, and EG VII and includes purification parameters such as yield (%).

Table 1
Summary of the purification of EG V, EG VI, and EG VII from *T. emersonii*

Purification step	Total activity (IU)	Total protein (mg)	Specific activity (IU/mg)	Purification factor (*x*-fold)	Yield (%)
Crude extract	54,806.2	573.1	95.6	1.00	100.00
Ultrafiltration (Amicon DC2)	50,613.4	452.2	111.9	1.17	92.30
Propan-2-ol pptn.	50,016.9	364.8	137.4	1.44	91.30
Anion-exchange (DE-52, pH 5.5)	25,064.7	240.1	104.4	1.10	45.70
Phenyl Sepharose CL-4B					
Peak 1	3,862.6	24.9	154.9	1.62	7.10
Peak 2	6,763.5	30.8	219.3	2.29	12.30
Peak 3	14,411.1	42.1	342.3	3.58	26.30
EG VII (Peak 1)					
Concanavalin A	3,120.2	11.3	275.9	2.89	5.70
Sephacryl S-100 HR	670.1	3.2	212.5	2.22	1.20
Chromatofocusing PBE 94, pH 3.5-2.0	400.4	0.53	756.9	7.92	0.71
EG V, EG VI (Peak 3)					
Anion-exchange (DE-52, pH 4.5)	4,662.3	10.40	447.4	4.68	8.50
Anion-exchange (DE-52, pH 7.0)	3,339.2	5.77	579.1	6.06	6.10
BioGel P-60	3,043.8	3.60	855.7	8.95	5.50
Concanavalin A					
Peak 1	1,115.6	1.60	679.1	7.10	2.10
Peak 2	863.5	1.20	738.7	7.73	1.60
Chromatofocusing PBE 94, pH 3.5–2.0					
EG V	415.3	0.54	764.8	7.99	0.76
EG VI	168.7	0.18	947.6	9.09	0.31

Yields and final specific activity values are similar for EG V and EG VII, while the yield of EG VI is lower but this enzyme has a higher final specific activity. While the final yields of the three enzymes (cumulative value of 1.8%) is very low, these values may in part be due to the removal of other BBG-degrading activities during purification some of which may be potentially synergistic with EG V, EG VI, and EG VII.

3.7. Electrophoresis (SDS-PAGE) and Isoelectric Focusing

1. Prepare a 12.0% discontinuous SDS-PAGE separating gel by mixing 20 mL 30% (w/v) acrylamide, 12.5 mL 1.5 M Tris–HCl, pH 8.8, 0.5 mL 10% (w/v) SDS, 0.25 mL 10% (w/v) APS, and 16.725 mL H_2O (final volume is 49.975 mL).
2. Add a 25-μL aliquot of TEMED (*N,N,N′,N′*-tetramethylethylenediamine), the catalyst for polymerisation, directly in advance of pouring the gel (BioRad Mini-Protean casting system). Pour the gel to within ~1.0-1.5 cm of the top of the glass plate.
3. Prepare a 4.0% stacking gel by mixing 5.0 mL Tris–HCl, pH 6.8 plus 2.6 mL 30% acrylamide–bisacrylamide plus 12.2 mL double-distilled deionised H_2O plus 0.2 mL 10% SDS plus 0.1 mL 10% APS. Add 10 μL TEMED, mix gently, and once the separating gel has polymerized, transfer the stacking gel mix into the gel chamber using a pipette and insert the gel comb.
4. Treat samples by boiling an aliquot of sample with sample loading buffer (the final sample loading buffer concentration is 1×) for 5 min. Stand the samples on ice until ready to apply to the gel.
5. Load samples and size markers, and electrophorese (typically at a constant current of 30 mA per gel) until the dye front has reached the bottom of the gel.
6. Stain the gel with Coomassie Brilliant blue R250 for 30 min with constant gentle agitation.
7. After staining, immerse the gel in destain with constant agitation. Refresh the destain four or five times at 1-h intervals until all of the background staining has been removed from the gel. Protein bands should now be visible.
8. Calculate the relative molecular mass (M_r) of each purified enzyme by comparing the relative mobilities of standard (M_r range of standard proteins: 14.4–94.0 kDa) and sample proteins.
9. Determine the pI values for EG V–VII by isoelectric focusing (IEF) on LKB Pharmacia PAG plates (pH range 3.5–9.5), using ampholytes (pH 3.0–10.0) as supplied by GE Healthcare (and procedures recommended in the corresponding GE Healthcare information booklet 1804). Apply a cocktail of standard proteins of known pI (pH 3.5–9.3; GE Healthcare)

to separate application wicks on the same gel. The pI values of EG V–VII from IEF are <pH 3.5.

10. To obtain a more accurate pI value for EG V, EG VI, and EG VII, use chromatofocusing on PBE 94, pre-equilibrated with 0.025 M piperazine-HCl, pH 3.5, to determine accurate pI values for EG V–VII. Elute adsorbed protein using a 1 in 45 dilution of polyampholyte at pH 2.0.
11. By monitoring pH over the elution profile, the pH corresponding to each β-glucanase peak can be determined (pI value). Differences of 0.01 pH units can be detected, which yields accurate pI values.
12. The location of β-glucanase activity in PAGE and IEF gels may be determined using the zymogram overlay technique (Subheading 2.5.4) (7) with 0.2% (w/v) BBG as substrate.

EG V, EG VI, and EG VII give single protein bands, on staining of SDS-PAGE gels, each coincident with bands of BBGase activity following zymogram analysis (see Fig. 3). Silver staining can be used to verify apparent purity. The relative molecular mass (M_r) values (kDa), determined by SDS-PAGE, are 22.9 ± 0.1 for EG V, 26.9 ± 0.2 for EG VI, and 33.8 ± 0.2 for EG VII; M_r values determined by gel filtration on Sephacryl S-100 HR and BioGel P-60 are in very close agreement, or identical, to those obtained by SDS-PAGE. All three enzymes were single subunit glycoproteins following Schiff staining of SDS-PAGE gels (results not shown).

IEF should confirm the homogeneity of each enzyme preparation. However, as protein and enzyme activity bands are located at positions in the pH gradient (3.5-9.5) corresponding to pI < 3.5, it is best to use chromatofocusing on PBE 94, with a decreasing pH gradient from 3.5 to 2.0, to determine individual

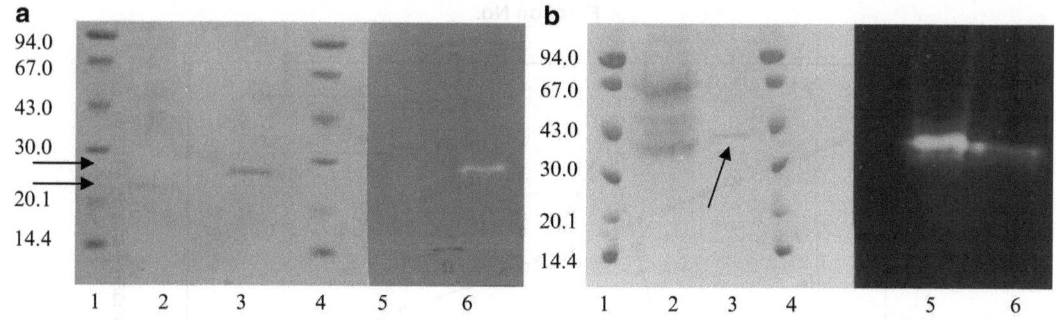

Fig. 3. Electrophoretic separation and zymogram analysis of purified EG V and EG VI (**a**) and EG VII (**b**). In (**a**), *lanes 1* and *4* are Coomassie blue-stained standard proteins, *lanes 2* and *3* are EG V and EG VI, respectively, while *lanes 5* and *6* show enzyme-active bands corresponding to EG V and EG VI obtained by zymogram analysis. In (**b**), *lanes 1* and *4* show Coomassie blue-stained standard proteins, *lane 2* indicates the number of protein bands present in the post-Concanavalin A EG VII-enriched pool and *lane 3* shows the purified EG VII. *Lanes 5* and *6* in the zymogram correspond to *lanes 2* and *3* in the Coomassie blue-stained gel, respectively. Images of PAGE and IEF gels were obtained using a BioRad Fluor-S™ MultiImager.

pI values for EG V, EG VI, and EG VII (see Fig. 4). The pI values of the three enzymes are EG V, 2.45; EG VI, 3.00; EG VII, 2.85. Calculated molar extinction coefficient (ε_{280}) values (mol L/cm) are 1.04×10^{-5} for EG V, 8.80×10^{-6} for EG VI and 7.05×10^{-6} for

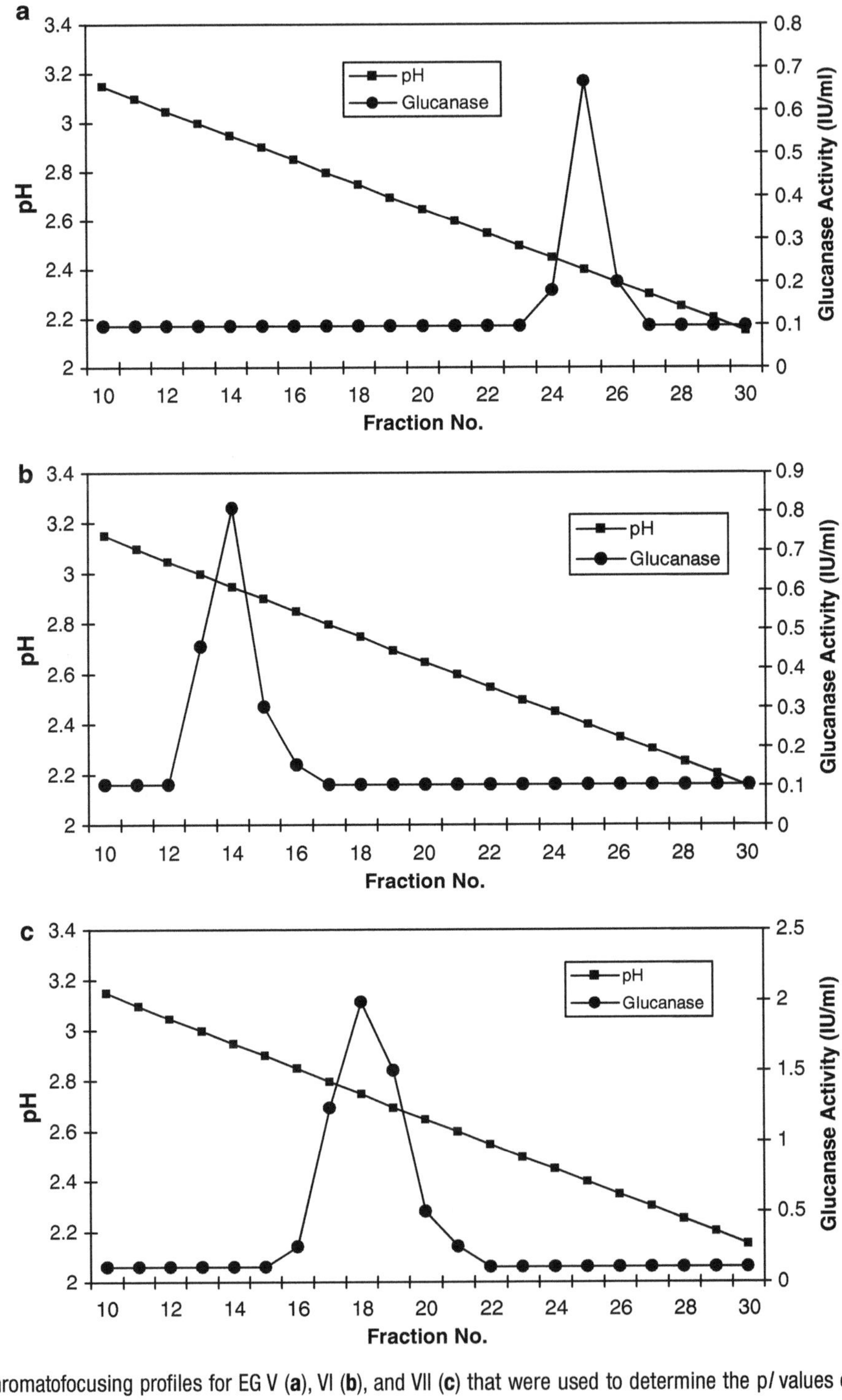

Fig. 4. Chromatofocusing profiles for EG V (**a**), VI (**b**), and VII (**c**) that were used to determine the p*I* values of the three β-glucanases.

EG VII, which may reflect differences in aromatic amino acid and overall protein composition between the three enzymes (21).

3.8. Substrate Specificity on Various Polysaccharides and Synthetic Glycosides

1. Measure the activity of EG V, EG VI, and EG VII against a wide variety of carbohydrates (BBG, lichenan, CMC, other polysaccharides and synthetic glycosides) at 50°C and pH 5.0 (as per Subheading 3.2.1 and 3.2.2).
2. Express enzyme activity as a percentage relative to that observed with BBG (100%).

A representative overview of the type of substrate specificity data that has been obtained previously for EG V, EG VI, and EG VII is shown in Table 2. Crude extracts of *T. emersonii*, grown under the conditions described herein, catalyze the hydrolysis of a variety of polysaccharides including cellulose, CMC, BBG, laminaran, lichenan, xylan, pectin, and a spectrum of synthetic

Table 2
Substrate specificities of purified EG V, EG VI, and EG VII from *T. emersonii*

Substrate	Linkage	% Relative activity[a] 40.7 kDa "lichenase"[b]	EG V	EG VI	EG VII
Barley β-glucan	β-(1,3)(1,4)	100.0	100.0	100.0	100.0
Mixed DP BBG	β-(1,3)(1,4)	159.0	165.0	230.2	107.6
High DP BBG	β-(1,3)(1,4)	118.0	249.9	342.8	144.1
CMC	β-(1,4)	0.5[c]	74.9	58.4	57.7
Lichenan	β-(1,3)(1,4)	118.9	204.3	177.7	166.9
Laminaran	β-(1,3)	0.0	0.0	21.3	18.1
Xylan	β-(1,4)	0.0	1.8	0.7	1.5
Rhodymenan (xylan)	β-(1,3)(1,4)	0.0	35.4	9.3	0.0
Pectin	α-(1,4)	0.0	0.0	0.0	0.0
Mannan	β-(1,4)	0.0	0.0	0.0	0.0
Avicel	β-(1,4)	0.0	0.0	0.0	0.0
Filter paper	β-(1,4)	0.0	0.0	0.0	0.0
4NP-β-D-Glucopyranoside	β-(1,4)	0.0	0.0	0.0	0.0
4NP-β-D-Galactopyranoside	β-(1,4)	0.0	0.0	0.0	0.0
4NP-β-D-Xylopyranoside	β-(1,4)	0.0	0.0	0.0	0.0

[a]% Activity relative to activity against BBG, which is taken as 100.0%
[b]Purified *T. emersonii* lichenase
[c]After 24 h at 50°C

glycoside derivatives. However, as Table 2 shows, the three purified enzymes are not active against 4-NP derivatives of β-D-glucopyranoside, β-D-galactopyranoside, β-D-xylopyranoside, or α-D-galactopyranoside. EG V, EG VI, and EG VII do not display any activity against filter paper, Avicel, locust bean gum galactomannan, and do not catalyse the oxidation of cellobiose, even on extended incubation with substrate. All three enzymes exhibit maximum activity against the mixed-linkage β-glucans, BBG, and lichenan, with markedly more activity against lichenan (Table 2). Trace activity exhibited by all three enzymes with xylan on extended incubation periods may be explained by the fact that the oat spelts xylan preparation used contains minor, contaminating amounts of β-glucan.

4. Notes

1. The β-glucan powder should be added gradually with constant stirring to ensure that lumps of polysaccharide do not form. The polysaccharide solution should not be boiled or heated for lengthy periods of time as this may result in a certain level of depolymerisation or fragmentation of the substrate.
2. Lichenan does not yield a homogeneous solution and the buffered substrate solution will be a suspension and contain a significant amount of insoluble, particulate material. Aliquots should be transferred to assay tubes from a continuously stirring suspension of buffered lichenan (i.e. from a beaker containing the substrate on a stir plate located beside the assay waterbath or incubation chamber).
3. Solutions of pectin should be stored at −20°C until required or prepared freshly on the day of use as buffered solutions of pectin undergo slow, spontaneous depolymerisation; therefore, storage for longer than 1–2 days at 4°C is not recommended. Birchwood xylan will form a homogeneous solution; however, oats spelts xylan will form a suspension with some remaining insoluble material. Therefore, a similar approach should be taken as described for lichenan (see Note 2) when using this substrate in enzyme assays.
4. Buffered solutions of 4-nitrophenylglycosides (e.g. 4-nitrophenyl-β-D-glucopyranoside) are sensitive to light and should be stored in labelled and dated dark bottles or in aluminium foil-covered 15 mL or 50 mL polypropylene centrifuge tubes.
5. In all cases, enzyme activity is expressed as IU/mL, i.e. micromoles of reducing sugar equivalents or 4-nitrophenol released per minute of reaction time per millilitre of enzyme solution.

6. The CH_3OH/CH_3COOH combination in the staining solution also acts as a protein "fixing" agent. However, if gel thickness is <0.5 mm, then gels should be fixed before staining by incubation in a 20% (w/v) TCA solution for 30 min. Gels are then rinsed for 15 min in de-staining solution before staining with Coomassie Blue.
7. It is possible to pour stand alone columns of individual chromatography matrices or to buy prepared columns for gravity feed or low pressure chromatography (LPC) systems, or to pack columns for fast protein liquid chromatography (FPLC). Care must be taken when pouring columns in any of the formats above to eliminate air pockets which would interfere with subsequent fractionation. Addition of ~15% of the total column volume of buffer into the empty Perspex column casing prior to pouring the matrix will prevent air being trapped at the base of the column. It is also helpful to degas buffers prior to pouring columns. It is important to use a fully hydrated, homogeneous (no lumps), equilibrated slurry of the required chromatography matrix and to pour as much as possible of this matrix in one step. Once the matrix has fully settled (i.e. after pouring is completed) additional matrix should not be added to the top of the column bed, as this will create zones in the matrix that will affect protein separation and peak resolution. Additional matrix should only be added to the top of the column bed as the freshly poured column matrix is settling, and should be done by carefully removing the upper buffer layer and pouring additional slurry into the column using a glass rod to direct the flow of the gel down along the wall of the Perspex column casing.

Acknowledgements

This work was funded by an Enterprise Ireland Applied Research Award to M.G.T. T.McC. is grateful for a junior teaching fellowship from NUI, Galway and a postgraduate scholarship from Enterprise Ireland.

References

1. Kobayashi, T., Abe, K., Asai, K., Gomi, K., Juvvadi, P.R., Kato, M., Kitamoto, K., Takeuchi, M., and Machida, M. (2007) Genomics of *Aspergillus oryzae*. *Biosci. Biotechnol. Biochem.* **71**, 646–670.
2. Coughlan, M.P. (1992) Enzymatic hydrolysis of cellulose: an overview. *Bioresour. Technol.* **39**, 107–115.
3. Stone, B.A. and Clarke, A.E. (1992) Chemistry and Biology of 1,3-β-Glucans, La Trobe University Press, Bundoora, Australia.
4. McCleary, B.V. and Matheson, N.K. (1986) D-Glucans based on a (1-3)-β-backbone and on (1-3)-β-chains. *Adv. Carbohydr. Chem. Biochem.* **44**, 266–273.

5. Planas, A. (2000) Bacterial 1,3-1,4-β-glucanases: structure, function and protein engineering. *Biochim. Biophys. Acta* **1543,** 361–382.
6. Chen, H., Li, X.-L., and Ljungdahl, L.G. (1997) Sequencing of a 1,3-1,4-β-D-glucanase (lichenase) from the anaerobic fungus Orpinomyces strain PC-2: properties of the enzyme expressed in *Escherichia coli* and evidence that the gene has a bacterial origin. *J. Bacteriol.* **179,** 6028–6034.
7. Murray, P.G., Grassick, A., Laffey, C.D., Cuffe, M.M., Higgins, T., Savage, A.V., Planas, A., and Tuohy, M.G. (2001) Isolation and characterization of a thermostable endo-β-glucanase active on 1,3-1,4-β-D-glucans from the aerobic fungus *T. emersonii* CBS 814.70. *Enzyme Microb. Technol.* **29,** 90–98.
8. Lloberas, J., Querol, E., and Bernues, J. (1988) Purification and characterization of endo-β-1,3-1,4-D-glucanase activity from *Bacillus licheniformis. Appl. Microbiol. Biotechnol.* **29,** 32–38.
9. Akiyama, T., Kaku, H., and Shibuya, N. (1996) Purification and characterization of a (1→3,1→4)-β-glucanase from rice *Oryza sativa* L. *Biosci. Biotechnol. Biochem.* **60,** 2078–2080.
10. Borriss, R., Noack, D., and Geuther, R. (1982) β-1,3-1,4-Glucanase in spore-forming microorganisms. VI Genetic instability of β-glucanase production in a high producing strain of *Bacillus amyloliquefaciens* grown in a chemostat. *Z. Allg. Mikrobiol.* **22,** 293–298.
11. Schimming, S., Schwartz, W.H., and Staudenbauer, W.L. (1991) Properties of a thermoactive β-1,3-1,4-glucanase (Lichenase) from *Clostridium thermocellum* expressed in *Escherichia coli. Biochim. Biophys. Acta,* **177,** 447–452.
12. Heinemann, U., Aÿ, J., Gaiser, O., Müller, J.J., and Ponuswamy, M.N. (1996) Enzymology and folding of natural and engineered bacterial beta-glucanases studied by X-ray crystallography. *J. Biol. Chem.* **377,** 447–454.
13. Hrmova, M., and Fincher, G.B. (1998) Barley β-D-glucan exohydrolases. Substrate specificity and kinetic properties. *Carbohydr. Res.* **305,** 209–221.
14. Görlach, J.M., Van Der Knaap, E., and Walton, J.D. (1998) Cloning and targeted disruption of *MLG1*, a gene encoding two of three extracellular mixed-linked glucanases of *Cochliobolus carbonum. Appl. Environ. Microbiol.* **64,** 385–391.
15. Bang, M.L., Villadsen, I., and Sandal, T. (1999) Cloning and characterization of an endo-β-1,3(4)glucanase and an aspartic protease from *Phaffia rhodozyma* CBS 6938. *Appl. Microbiol. Biotechnol.* **51,** 269–272.
16. Coughlan, M.P., and Moloney, A.P. (1988) Isolation of 1,4-β-D-glucan 4-glucanohydrolases of *Talaromyces emersonii. Methods Enzymol.* **160,** 363–368.
17. Tuohy, M.G. and Coughlan, M.P. (1992) Production of thermostable xylan-degrading enzymes by *Talaromyces emersonii. Bioresour. Technol.* **39,** 131–137.
18. Bensadoun, A., and Weinstein, D. (1976) Assay of proteins in the presence of interfering materials. *Anal. Biochem.* **70,** 241–250.
19. Moloney, A.P., Hackett, T.J., Considine, P.J., and Coughlan, M.P. (1983) Isolation of mutants of *Talaromyces emersonii* CBS 814.70 with enhanced cellulase activity. *Enzyme Microb. Technol.* **5,** 260–264.
20. Tuohy, M.G., Puls, J., Claeyssens, M., Vrsanska, M., and Coughlan, M.P. (1993) The xylan-degrading enzyme system of *Talaromyces emersonii*: novel enzymes with activity against aryl-β-D-xylosides and unsubstituted xylans. *Biochem. J.* **290,** 515–523.
21. Sakellaris, G., Nikolaropoulos, S., and Evangelopoulos, A.E. (1989) Purification and characterization of an extracellular polygalacturonase from *Lactobacillus planatarum* strain A.A. II. *J. Appl. Bacteriol.* **67,** 77–85.
22. Laemmli, U.K. (1970) Cleavage of structural proteins during the assembly of the head of bateriophage T4. *Nature (London)* **227,** 680–685.
23. Miller, G.L. (1959) Use of dinitrosalicylate acid reagent for the determination of reducing sugar. *Anal. Chem.* **31,** 426–428.

Index

A

Affinity chromatography 7, 38, 97–98, 112, 153, 157, 160, 283, 285–286, 303, 369–381
Affinity tag 97, 152, 153, 157–166, 168, 259, 262, 263, 268–270, 288, 289, 311, 313, 349, 350, 353, 365
Ammonium sulphate precipitation 7, 182, 210, 213, 217, 222, 224, 297, 301–302, 373–375, 377–379
Amylose-agarose 153, 283–285, 287, 290, 292
Anion-exchange 139, 216–221, 223, 286, 287, 299, 305, 306, 440, 441, 443, 445, 505, 515–517
Antibody 7, 35, 73, 87, 112, 152, 216, 230, 263, 297, 313, 369, 383, 403

B

Baculovirus 159, 167, 271, 295–308
Biacore 403–416
Bicinchoninic acid (BCA) 49, 220, 223, 224, 229, 230, 232–233, 240, 242–243, 254, 496
Biological activity 103–106, 122, 125–127, 181, 189, 196, 213, 434, 436
Biosensor analysis 404
Bovine brain 434
Bradford 49, 223, 224, 229–232, 240–241, 254, 466, 471, 496
Butyl Sepharose 431, 432

C

Cation-exchange 216–219, 221, 299, 305, 306, 443, 445, 479
Cereal β-glucans 500, 501
Chaotropic agents 46, 48, 162, 165, 432, 449
Chemical modification 106, 123–125, 127–128, 146
Chicken 42, 88, 93–98, 383–400
Chromatography 7, 25, 35, 61, 74, 87, 112, 137, 152, 187, 210, 215, 260, 282, 298, 311, 337, 360, 369, 414, 419, 431, 439, 451, 463, 475, 488, 499
CID/ETD MS/MS spectra 346
Clinical proteomics 475–484

D

Deae-Sepharose® Fast Flow 220, 223, 224
Detergent 6, 28, 106, 118, 157, 162, 210, 211, 240, 242, 244, 247, 254, 255, 261, 268, 273, 277, 288, 289, 291, 322, 332, 333, 341, 355, 357, 359, 361, 363, 364, 432–434, 436, 437, 456, 484, 488, 490–497
Downstream processing 7, 73–84, 207, 211, 329

E

Electron transfer dissociation (ETD) 337–348
Epitope 88, 89, 91, 93, 165, 323, 386
Excipients 122, 181, 193–195, 199

F

Factor Xa 262, 268, 269, 271, 273, 275, 278, 279, 288, 329, 332, 351, 352, 361
Fast protein liquid chromatography (FPLC) 439–447
Folding 4, 7, 73, 106, 112, 114, 116–119, 121, 123, 124, 128, 129, 151–153, 160, 167, 168, 180, 212, 263, 281, 282, 290, 313, 350
Freeze-drying 181, 182, 185, 189–191, 195, 199, 517
Freezing 180–182, 189, 191–195, 197, 199, 200, 209, 213, 392
Fungal enzymes 123, 501–502, 509
Fusing protein tags 274
Fusion protein 38, 44, 90, 119, 153, 157–161, 164, 165, 260–280, 288, 290, 292, 313, 315, 319–321, 325, 332, 349–366, 372
Fusion tags 152, 159, 161, 329

G

Gel-exclusion 25
Gel-filtration chromatography 25–32
Gel-permeation 25, 216
Glucanase 123, 499
Glutathione S-transferase (GST) 119, 157–158, 160, 167, 259–280
Glycerol gradients 433, 434, 436
Glycoprotein 26, 491, 501, 521

Dermot Walls and Sinéad T. Loughran (eds.), *Protein Chromatography: Methods and Protocols*, Methods in Molecular Biology, vol. 681, DOI 10.1007/978-1-60761-913-0, © Springer Science+Business Media, LLC 2011

H

His-tag 97, 161–163, 168, 308, 312, 313, 315, 318–320, 323, 324, 326–331
Hydrophobic Amino Acids 4, 7, 273

I

Immobilised metal affinity chromatography (IMAC) 39, 47, 157, 158, 160–164, 166, 290, 304, 311, 313, 315, 318–319, 326–329, 331, 337–348
Immobilization 97, 98, 122–123, 164, 165
Immunoaffinity chromatography (IAC)........... 35–52, 165
Inclusion body solubilisation 208
Insect cell expression system................................ 295, 296
Intact proteins 365, 420, 421, 424, 425
Integral membrane protein............................210, 488–494
Ion exchange.......................7, 143, 157, 164, 187, 215–227, 236, 250, 260, 261, 270, 282–284, 286–287, 292, 299, 303, 305–306, 365, 433, 435, 440, 442–446, 488, 491, 501, 513
Ionic strength 7, 28, 146, 182, 217, 221, 222, 227, 238, 289, 291, 305, 306, 338, 407, 408, 415, 433, 435, 479, 489
Isobaric tags.................................. 463–465, 468, 471, 476
Isoelectric focusing (IEF) 7, 49, 139, 141, 142, 144–147, 211, 256, 449–458, 463–464, 479, 507–508, 520–523

K

Kinetics...........................44, 105, 106, 110, 112, 118, 120, 123, 124, 157, 180, 186, 196, 200, 276, 277, 291, 405, 409–415

L

Lab-on-a-chip.. 137, 143
LC-MS/MS..449–458, 461–474
Liquid chromatography (LC).....................9, 12, 215, 273, 338, 347, 360, 419–428, 439–447, 450, 451, 463, 475–484
Lowry 49, 229–232, 240–242, 254, 504, 512, 513
Lyophilisation..31, 179–200, 213

M

Maltose binding protein (MBP).............................97, 109, 110, 153–157, 160, 167, 281–292
Mass spectrometry (MS) 12, 142, 146, 147, 337–348, 419–428, 449–458, 461–484
Matrix 3, 9, 25–31, 35, 41–46, 50, 88, 97, 143, 161, 164, 165, 190, 204, 212, 216–222, 251, 259, 262, 267–269, 278, 303, 304, 311, 313, 314, 330, 339, 341, 345, 369, 431, 432, 435, 445, 449, 451, 456, 462, 473, 484, 490, 493, 495, 496, 505, 506, 515, 517
Microfluidics 137–147, 283
Molecular mass estimation 31, 257
Molecular-sieve 25
Monoclonal 38, 44, 48, 51, 52, 73, 79, 83, 88–90, 93–95, 246, 300, 301, 317, 321, 372, 374–380, 405–409, 416

N

Ni-NTA affinity chromatography 296, 298, 304
Non-polar...120, 323, 489

O

Octyl Sepharose.. 433–435
Off-Gel Electrophoresis (OGE) 455, 461–474
One-dimensional liquid chromatography.......478, 479, 481
Operating conditions.. 26–28, 345

P

Peptide labelling 463, 465–467
Peptides 4, 29, 38, 61, 91, 142, 161, 180, 218, 338, 350, 445, 450, 461, 478, 500
Peripheral membrane protein 488–490, 494
pGEX 157, 259, 261, 270–273
pH 7, 26, 42, 62, 83, 94, 105, 139, 161, 179, 204, 216, 232, 261, 283, 297, 313, 338, 353, 373, 384, 405, 432, 441, 446, 463, 477, 489, 502
Phage display.......................................38, 87–98, 113, 384
Phenyl Sepharose 433–435, 505, 515, 518
Phosphopeptide purification 340, 343–344
Phosphoproteomics 337
Polyclonal 37, 41, 88, 89, 94, 246, 372–378
Post-translational modification 5, 7, 73, 158, 167, 271, 282, 295, 313, 337, 338, 501
Prolyl oligopeptidase 215–227, 434
Protease 6, 28, 61, 113, 153, 184, 210, 259, 282, 302, 329, 338, 349, 450, 482, 488
Protein A....................................38, 40, 42, 44, 48, 79, 82, 370, 372, 374–376
Protein aggregation and soluble aggregates 281
Proteinase inhibitor buffer....................................... 67, 205
Protein digestion 450, 456, 481, 484
Protein engineering 6, 37, 111–118
Protein expression......................... 63, 74, 75, 96, 141, 152, 158, 159, 166, 167, 208, 245, 261, 263–268, 271, 273–275, 278, 282, 283, 295, 296, 299, 312–314, 321, 325, 326, 331, 332, 475
Protein expression and purification 74, 75, 208, 261, 263–267, 273, 282, 283, 295, 296, 299, 312–314, 321, 325, 326, 331, 332
Protein G....................38, 40, 112, 370–372, 375, 378–379

Protein identification 9, 10, 12, 296, 372, 419, 450, 455, 463, 468, 476, 478, 479, 481, 484
Protein L 38, 40, 41, 371, 372, 375, 380
Protein–protein interactions 9, 46, 212, 403–416
Protein purification 3–22, 68, 69, 73–84, 97, 152, 158, 159, 162–164, 184, 207, 218, 260, 266, 269, 278, 283–287, 296, 304, 306, 311, 314, 315, 326, 329, 361, 440, 488–489, 493, 506
Protein quantification.....................................229–257, 471
Protein refolding........................... 161, 162, 165, 205–206, 208, 213, 326, 329
Proteins 3, 26, 36, 61, 73, 87, 103, 137, 151, 179, 203, 216, 229, 259, 281, 295, 311, 337, 349, 370, 383, 403, 419, 431, 439, 450, 461, 475, 487, 501
Protein solubility 7, 152, 158, 185, 212, 263, 318, 324–326
Protein solubilization..................... 205, 208, 212, 303, 307, 456, 488, 490, 493
Protein stability 61, 103–131, 180, 186, 203, 488, 491, 493, 497, 501
Proteolysis 29, 61–70, 104, 113, 129, 147, 164, 166, 180, 181, 183–185, 187, 210, 265, 267, 270, 271, 273, 274, 278, 291, 321, 332, 333, 350, 497
Proteolytic cleavage 5, 166, 272, 329, 349
Proteome 9, 88, 166, 338, 449
Purification.. 3, 28, 34, 62, 73, 94, 112, 151, 179, 207, 215, 231, 259, 281, 295, 311, 338, 357, 369, 384, 434, 440, 453, 462, 475, 487, 500
Purification techniques.................... 35, 160, 204, 216, 282, 488, 493, 497, 513–515

Q

Quantitative immuno-PCR (qIPCR) 230, 231, 235–238, 248–253

R

Recombinant antibodies............. 38, 88, 89, 91–92, 98, 370

S

Scale-up.. 29, 73–84, 216, 273, 352
Separations 4, 27, 36, 69, 74, 112, 138, 152, 194, 212, 216, 244, 266, 338, 362, 369, 420, 431, 439, 449, 463, 475, 488, 513
Shotgun proteomics.................. 12, 449–458, 461–474, 476
Single-chain fragment variable (scFv) 37, 91–93, 95–98, 372, 380, 383–400, 405, 406, 409–416
Site-specific proteolysis 349, 351–353
Size-exclusion................................. 25, 27, 29, 30, 282, 365
Size-exclusion reaction chromatography (SERC) 31
Solubility 7, 95, 120, 141, 151–168, 185, 194, 203, 210–212, 261, 263, 274, 281, 282, 288, 290, 296, 297, 299–301, 305, 318, 319, 324–326, 333, 351, 478, 490
Stabilising additives..185–189, 497
Storage of pure proteins... 179–200
Surface plasmon resonance (SPR)...................... 46, 49, 94, 97, 403, 404

T

Tag.......................... 151–168, 259, 262, 263, 268–270, 282, 296, 311, 313, 329, 349–366, 405–407, 409, 410, 463, 466, 468, 471
Talaromyces emersonii 501
Tandem mass spectrometry..................... 12, 337–348, 450, 462, 463, 478, 480
TEV protease 69, 288, 289, 353, 366
Thermal stability measurement 103
Thermophile.................................. 104, 106, 108, 109, 118
Thermostable 109–111, 114, 121
Thrombin 262, 268, 269, 271, 273, 275, 278, 279, 288, 329, 332, 351–352, 361
Trichloroacetic acid precipitation 205, 207, 211, 300, 504
Two-Dimensional Liquid Chromatography........ 9–12, 483

U

Ultra-performance liquid chromatography (UPLC) 419–428

V

Viral capsid proteins.. 295

MIX
Papier aus verantwortungsvollen Quellen
Paper from responsible sources
FSC® C105338

If you have any concerns about our products,
you can contact us on
ProductSafety@springernature.com

In case Publisher is established outside the EU,
the EU authorized representative is:
Springer Nature Customer Service Center GmbH
Europaplatz 3, 69115 Heidelberg, Germany

Printed by Libri Plureos GmbH
in Hamburg, Germany